This book presents the theory of the electrodynamic phenomena which occur in the magnetosphere of a pulsar. It also provides a clear picture of the formation and evolution of neutron stars. The authors address the basic physical processes of electron–positron plasma production, the generation of electric fields and currents, and the emission of radio waves and gamma rays. The book also reviews the current observational data, and devotes a complete chapter to a detailed comparison of these data with accepted theory and with some recent theoretical predictions. Tables containing the values of the physical parameters of all observed radio pulsars are also provided. The book is illustrated throughout and is fully referenced.

Graduate students and researchers in astrophysics and plasma physics, working in the field of radio pulsars will find this book to be of great value.

PHYSICS OF THE PULSAR MAGNETOSPHERE

PHYSICS OF THE PULSAR MAGNETOSPHERE

V. S. BESKIN, A. V. GUREVICH & Ya. N. ISTOMIN
P. M. Lebedev Physical Institute, Moscow

Translated from the Russian by M. V. Tsaplina

CAMBRIDGE
UNIVERSITY PRESS

CAMBRIDGE UNIVERSITY PRESS
Cambridge, New York, Melbourne, Madrid, Cape Town, Singapore, São Paulo

Cambridge University Press
The Edinburgh Building, Cambridge CB2 2RU, UK

Published in the United States of America by Cambridge University Press, New York

www.cambridge.org
Information on this title: www.cambridge.org/9780521417464

First published 1993
This digitally printed first paperback version 2006

A catalogue record for this publication is available from the British Library

Library of Congress Cataloguing in Publication data

Beskin, V. S.
Physics of the pulsar magnetosphere / V. S. Beskin, A. V. Gurevich &
Ya. N. Istomin ; translated from Russian by M. V. Tsaplina.
p. cm.
Includes bibliographical references and index.
ISBN 0-521-41746-5
1. Pulsars. 2. Magnetosphere. 3. Electrodynamics.
4. Astrophysics. I. Gurevich, A. V. (Aleksandr Viktorovich), 1930–
II. Istomin, Ya. N. III. Title.
QB843.P8B47 1993
523.8′874—dc20 92-18417 CIP

ISBN-13 978-0-521-41746-4 hardback
ISBN-10 0-521-41746-5 hardback

ISBN-13 978-0-521-03253-7 paperback
ISBN-10 0-521-03253-9 paperback

PHYSICS OF THE PULSAR MAGNETOSPHERE

V. S. BESKIN, A. V. GUREVICH & Ya. N. ISTOMIN

P. M. Lebedev Physical Institute, Moscow

Translated from the Russian by M. V. Tsaplina

CAMBRIDGE
UNIVERSITY PRESS

CAMBRIDGE UNIVERSITY PRESS
Cambridge, New York, Melbourne, Madrid, Cape Town, Singapore, São Paulo

Cambridge University Press
The Edinburgh Building, Cambridge CB2 2RU, UK

Published in the United States of America by Cambridge University Press, New York

www.cambridge.org
Information on this title: www.cambridge.org/9780521417464

First published 1993
This digitally printed first paperback version 2006

A catalogue record for this publication is available from the British Library

Library of Congress Cataloguing in Publication data

Beskin, V. S.
Physics of the pulsar magnetosphere / V. S. Beskin, A. V. Gurevich &
Ya. N. Istomin ; translated from Russian by M. V. Tsaplina.
p. cm.
Includes bibliographical references and index.
ISBN 0-521-41746-5
1. Pulsars. 2. Magnetosphere. 3. Electrodynamics.
4. Astrophysics. I. Gurevich, A. V. (Aleksandr Viktorovich), 1930–
II. Istomin, Ya. N. III. Title.
QB843.P8B47 1993
523.8′874—dc20 92-18417 CIP

ISBN-13 978-0-521-41746-4 hardback
ISBN-10 0-521-41746-5 hardback

ISBN-13 978-0-521-03253-7 paperback
ISBN-10 0-521-03253-9 paperback

Contents

Introduction

Pulsars, or more precisely radio pulsars – sources of pulsed cosmic radio emission – were discovered in 1967 and almost immediately identified with rotating neutron stars. Such stars must originate from catastrophic gravitational contraction (collapse) of ordinary stars that have exhausted the stores of their nuclear fuel. In neutron stars, the gravitational forces are brought to equilibrium not by plasma pressure, as in ordinary stars, and not by pressure of degenerate electrons, as in white dwarfs, but by pressure of strongly compressed neutron matter. They are, so to say, huge clusters of nuclear matter, and although their mass is of the order of the solar mass, their radius is only about 10 km.

An experimental discovery of pulsars, i.e. the neutron stars predicted by Baade and Zwicky as far back as 1934, is rightly regarded as one of the greatest discoveries in astrophysics. For this discovery, Hewish was awarded the Nobel Prize in 1974. There is a very large amount of literature devoted to radio pulsars – in a little more than two decades after their discovery, about 5000 papers have been published. The basic methods of study and the results of observations of radio pulsars are summarized in the excellent monographs by Manchester & Taylor, and Smith, both titled *Pulsars* and published in 1977, and in *Pulsar Astronomy* by Lyne & Smith published in 1990.

Radio pulsars are now under study in practically all of the biggest observatories of the world. In particular, the list of sources discovered is increasingly large. By 1992 more than 520 radio pulsars were already known. The total number of radio pulsars in our Galaxy is of the order of 100 000. The number of extinct pulsars, i.e. neutron stars not radiating in the radiofrequency band, is three orders of magnitude larger.

The most remarkable property of radio pulsars is the very high stability of frequency of succession of radiation pulses. The time interval between pulses arriving at the Earth is called the pulsar period P. The periods of all known pulsars range between 0.00156 and 4.31 s, the majority of them lying between 0.3 and 1.5 s. The period P for each pulsar is known up to six or more digits. Such a strict periodicity of pulse succession was a decisive factor in the construction of the pulsar model presented in Fig. 0.1. According to this model, the pulsar period P is equal to the period of the neutron star's rotation. Only the rotation of exceedingly compact stars can account for both a very small pulsar period and the high stability of pulse succession.

It is of importance that the periods of all pulsars increase gradually. The rate of period variation $\dot{P}$ is stable and for most pulsars is between 10^{-14} and 10^{-16}. An increase in the period implies deceleration of the neutron star's rotation. The energy released in this process always exceeds the pulsar radio emission energy.

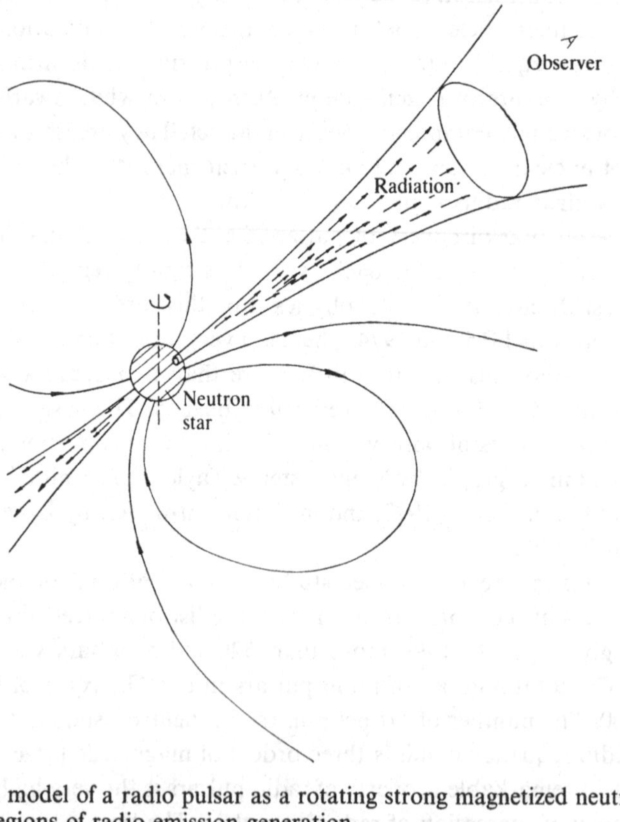

Fig. 0.1. A model of a radio pulsar as a rotating strong magnetized neutron star, showing regions of radio emission generation.

We see that the basic observational data allow us to establish both the nature of pulsars and the energy source for their activity. It is natural to assume also that the high coherence and directivity of observed pulsar radio emission is due to the presence of a magnetic field (see Fig. 0.1) and plasma in the vicinity of the neutron star, i.e. the pulsar magnetosphere in which this emission is generated.

At the same time, many questions naturally arise. What is the reason for a neutron star's deceleration? How is it realized? What is the plasma in the pulsar magnetosphere and where is it generated? In what way is part of the star's deceleration energy converted into radiation energy? What is the mechanism of coherent generation and highly directed radio emission?

Theoreticians have made great efforts to give an exhaustive answer to these questions. Recently, the situation has been clarified. It has become possible to formulate a consistent theory of the physical processes in the magnetosphere of a neutron star. From a unique standpoint, this theory explains the deceleration of neutron star rotation, the release of rotational energy in active regions, the generation of plasma and directed radio emission. This monograph is devoted to the presentation of that theory.

The physical ground for the whole complex of processes proceeding in the pulsar magnetosphere proves to be very simple and natural: all this results from fast rotation of a strong magnetized conducting body, i.e. a neutron star. The surrounding medium and the peculiarities of the surface of the body are of little importance. This means that the theory developed is physically closed and contains, in fact, no model assumptions.

The first three chapters of the book are introductory. The first chapter presents a detailed review of basic observational data, the second chapter presents a modern conception of the structure and properties of neutron stars. The main physical processes in the pulsar magnetosphere are discussed in the third chapter at a qualitative level. A detailed quantitative theory is presented in Chapters 4–6. In Chapter 7, theoretical results are compared with observations. The predictions of the theory are shown to be in good agreement with the basic observational characteristics of pulsars.

It should be noted that the comparison refers only to stationary characteristics of radio emission averaged over many pulses. Observations show, besides, the presence of a complicated space–time structure of radio signals caused by various non-stationary processes. The origin of non-stationary processes should be sought in the instabilities developed in the pulsar magnetosphere. They are briefly discussed in Chapter 7.

Now we would like to say a few words about our list of references. We have tried to point out all the most important observational papers. As far as theoretical papers are concerned, we included only those to which reference was necessary for a consistent presentation of modern theory. At the same time, some other ideas are considered in the literature, but for the reason mentioned above we could not include them in the main list. It seemed reasonable to present a separate list, which we have called Additional Literature and which reflects the whole variety of approaches to the problems of pulsar theory.

In conclusion we wish to express our gratitude to Professor V. L. Ginzburg for his support and interest in our work. We are grateful to V. S. Imshennik, D. A. Kirzhnits, M. V. Popov and K. A. Postnov who looked through several chapters of the manuscript and made some constructive comments. We are also grateful to members of the group of pulsar studies of the Radioastronomic Station of the Lebedev Physical Institute of the Russian Academy of Sciences for the possibility of permanent discussions with them of the whole range of questions connected with pulsars. We are indebted to I. V. Bessmertnaya, N. V. Orlova and N. M. Demina for their help in preparation of the book.

List of Notation

A	Observability function
A_r	Moment of inertia of the neutron star
$\mathbf{A}$	Vector potential of the magnetic field
$\mathbf{A}_T$	Vector potential of the current magnetic field
A_p	Plasma parameter
a	Plasma-curvature parameter
a_0	Bohr radius
a_B	Atomic radius in the magnetic field
$\mathbf{a}$	Acceleration
$\dot{\mathbf{a}}$	Derivative of the acceleration
a_m	Multiplicity coefficient for the secondary plasma generation
$a_f, a_\rho, a_\beta, a_\psi, a'_\psi$	Constants
$\bar{a}$	Power coefficients
$\mathbf{B}$	Magnetic field
B_0	Magnetic field strength on the star's surface
B_c	Magnetic field strength at the light cylinder
$B_\hbar$	Quantum magnetic field
b_0	$B_0/B_\hbar$
$\mathbf{B}_T$	Magnetic field of the current
$\mathbf{b}$	Unit vector along the magnetic field
B_b, B_k	Characteristic values of the magnetic field
b_θ	Some function of the angle θ
b, b_1	Functions
$b(K)$	The multiplication function for photons
C, C_1	Coefficients
c	Speed of light
$c_\perp, c_\parallel$	Constants
c_1, c_2, c_3	Constants

D	The death function of photons
$\mathscr{D}$	Electric induction of the electric field
D_c	Debye radius
DM	Dispersion measure
d	Distance to the pulsar
d_{br}	Characteristic value of the distance
D_ε	Function
$\bar{d}$	Power
$D(\lambda)$, $D_0(\lambda)$, $D_1(\lambda)$	Functions
$\mathscr{D}_{\alpha\beta}$	Quadrupole moment
$\mathbf{E}$	Electric field
$\mathbf{E}_c$	Corotation electric field
$E_{\parallel}$	Electric field component along the magnetic field line
$\tilde{E}_\nu$	Mean energy in the pulse
E_x	Function
$\mathscr{E}$	Energy
$\mathscr{E}_\gamma$	Energy of the gamma-quanta
$\mathscr{E}_{\text{cur}}$	Energy of the curvature photon
e	Electron charge
$\mathbf{e}_\phi$	Unit vector in the azimuthal direction
F	Distribution function of particles
$F_{\parallel}$	Longitudinal function of particles
$\mathbf{F}_A$	Ampère force
$F_{\text{ac}}, F_{\text{cc}}$	Autocorrelation and cross-correlation functions
f	Magnetic surface; frequency
f^*	Separatrix
$f^\pm$	Perturbation of the distribution function
f_r	Magnetic surface on which radiation occurs
$F_{l,m,z}$	Frequencies of the irregularities
$F_d(\chi)$	Function
$\mathscr{F}$	Ai + i Gi Airy function
G	Gravitational constant
$G(f)$	Profile of the current
G_0	Function of the magnetic field
g_0	Ray-optic parameter
$g(f)$	Function of current
$g_{1,2}$	Functions
H	Height of the double layer
h	Height over the star's surface
$\mathscr{H}$	Scalar function of the magnetic field
I_ν	Radio emission intensity

$I_{\nu 0}^{(j)}$	Radio emission intensity on the *j*-mode
I_ν	Radio emission intensity near the star's surface
I_r	Power radiation by the particle
$I_{\min}$	Minimum intensity observed by the antenna
dI	Differential intensity
I	Full electric current
I_c	Critical value of the current
I_1	Electric current
I_p	Invariant
$i_\parallel$	Longitudinal current
i_0	Dimensionless electric current
$\mathscr{J}, \mathscr{J}_1$	Functions
$J_{S,R,f,T}$	Surface currents
$\mathscr{J}_{5,6}$	Coefficients
j_e	Current density
$j_\parallel$	Longitudinal current density
j_c	Critical current density
$\mathbf{K}$	Torque momentum
$K_{m,0,1}, K^\pm$	Multiplication coefficients
$\mathbf{k}$	Wave vector
$k_{c,B0}$	Characteristic energy of the photons
k	Boltzmann constant
k_N	Coefficient
L_r	Radio luminosity
L_X	X-ray luminosity
L_γ	Gamma-ray luminosity
L_{ed}	Eddington luminosity
L_{RL}	Luminosity at the light cylinder
$\mathbf{L}$	$\mathbf{r} - \mathbf{r}'$
L	Coefficient
$\mathscr{L}$	Characteristic length of the inhomogeneity
l	Coordinate along the magnetic line
l	Free path of the photon
$\mathbf{l}$	Binormal
M_{ch}	Chandrasekhar mass
M	Neutron star mass
M_c	Mass of the star's companion
$M_\odot$	Sun mass
m_e	Electron mass
m_n	Neutron mass
$m(r)$	Mass distribution

m	Modulation index
$\mathbf{M}$	Magnetic moment
N	Distribution function of pulsars
N^{obs}	Distribution function of observed pulsars
N_{ph}	Distribution function of the photons
N_f	Total number of pulsars
N_{int}	Distribution function of the interpulse pulsars
n	Refractive index
n_e	Electron density
$n^{\pm}$	Positron and electron densities
n_c	Corotation density
n_B	Density of the equipartition
n_b	Braking index
$\mathbf{n}$	Unit vector of normal
P_{eq}	Equilibrium period
P	Period
$P(r)$	Pressure
$P_{2,3}$	Periods of drifts
P_μ	Period of the microstructure
P_b	Orbital period
$P(\mathbf{k})$	Probability
$\dot{P}$	Pulsar deceleration
p	Parameter
Q	Pulsar parameter
$Q_{N,F,S'}$	Birth operators of the photon
$\mathbf{q}$	Dimensionless momentum
q	Source of the particles
q_f	Parameter
R	Star radius
R_L	Radius of the light cylinder
R_0	Radius of the polar cap
R_s	Radius of the hot spot
$\mathbf{R}^*$	Function of the particle trajectory
$\mathbf{r}$	Vector from the star's centre
$\mathbf{S}$	Pointing vector
S_ν	Flux energy
$S_{c,f,0,d}$	Surfaces
S_S	Particle scattering operator
s	Length
$s_{5,6}$	Coefficients

$\bar{s}$	Power degree
$d\mathbf{s}$	Element of the surface
T	Temperature
t	Time
$t_{1,2}$	Transverse modes
U_ν	Spectral energy
U	Source of the pulsars
u	Power degree
V	Volume
$\mathbf{v}$	Particle velocity
$v_{\parallel,\perp}$	Particle velocity components
v_g	Group velocity
v_t	Pulsar velocity
$\bar{v}$	Power degree
v_{dr}	Drift velocity
W	Probability
$W_{\mathrm{tot,em},H,T,P}$	Powers
W_r	Radio window width
X	$\exp\{i\omega t \cdots\}$
x	Dimensionless function
x_1	Coordinate
$x^{\pm}$	Cyclotron resonance parameter
$Y(z)$	Function
y	Dimensionless function
z	Distance from the galactic plane
z_1	Coordinate
z_G	Gravitational shift
α	Dimensionless parameter
$\alpha(\mathbf{r})$	Angle between **B** and **M**
α_1	Parameter
α_T	Transformation coefficient
$\bar{\alpha}$	Power degree
β	Angle between the line observation and $\mathbf{\Omega}$
$\boldsymbol{\beta}_R$	$(1/c)[\mathbf{\Omega r}]$, Corotation parameter
β_0	Dimensionless potential
$\beta_{\mathrm{ph,p}}$	Angles of the photon and particle with respect to the magnetic field
$\bar{\beta}$	Power degree

Γ	Adiabatic index
Γ_g	Gas constant
Γ_L	Increment
Γ_p	Dimensionless Lorentz factor
$\Gamma(z)$	Gamma-function
γ	Lorentz factor
γ_c	$\langle\gamma^{-3}\rangle^{-1/3}$
Δ	Width of boundary layer
$\Delta(\eta)$	Width of the primary beam
δ	Angle between the separatrices
δ_r	Part of the radiated energy
$\bar{\delta}$	Power degree
ε	$1-\rho_\perp$
$\varepsilon_{1,2}, \varepsilon_d$	Parameters
$\varepsilon_{\alpha\beta}$	Dielectric permittivity tensor
$\varepsilon_{\alpha\beta}^{H,AH}$	Hermitian (antihermitian) parts of ε
η	Parameter
η_{tr}	Transformation coefficient
η_m	Modulation coefficient
$\eta(\cos\varphi)$	Polar cap boundary
θ	Angle between **k** and **B**
$\theta_{\parallel,\perp}$	Characteristic angles for the unstable modes
$\Delta\theta$	Variation of the angle θ
$\Theta(z)$	Theta function
Θ	Angle between **B** and **Ω**
κ	Ratio of the azimuthal component of the magnetic field to the radial one
$\kappa_{\parallel,\perp}$	Heat conductivities
κ_j	Imaginary part of the wave vector
κ_m	Wave vector of the modulation
Λ	Logarithmic factor
$\Lambda(k)$	Logarithmic factor
λ	Multiplicity
λ_f	Wavelength
$\lambdabar$	Compton wavelength
μ	Plasma parameter
μ_F	Fermi energy
μ_j	Absorption
$\mu_{\parallel,\perp}$	Coefficients

μ_c	Polarization factor
$\bar{\mu}$	Power degree
ν, $\nu_{1,2}$	Frequency
ν_{br}	Brake frequency
$\bar{\nu}$	Power degree
ξ	Transverse wave vector of the photon
ξ	Current potential
ξ	Curvature phase
ρ	Mass density
ρ_n	Nuclear density
ρ_c	Charge corotation density
$\rho_\perp$	Distance from the rotation axis
$\rho'_\perp$	Distance from the magnetic momenta axis
ρ_l	Curvature radius
ρ_τ	Radius of the field line torsion
ρ_{in}	Radius of the inner ring
Σ	Polar cap surface
$\Sigma_{\perp,\wedge}$	Surface conductivity
σ	Stefan–Boltzmann constant
$\sigma_{\parallel,\perp}$	Conductivities
$\sigma_{\mathrm{on,off}}$	Square root dispersions
$\sigma_{p,\mathrm{ph}}$	Signs of the velocities
$\sigma_{\alpha\beta}$	Conductivity tensor
τ_{kin}	Kinetic age of the pulsar
τ_d	Dynamic age of the pulsar
$\tau_{a,b,\mathrm{tot}}$	Lifetime
τ_B	Magnetic field extinction time
τ_μ	Microstructure time
τ_r	Radiation time
τ_l	Time of the free path
τ_θ	Variation time of the angle θ
τ_{in}	Time of the instability
τ_e	Time of escape
τ_j	Optical depth
τ_s	Characteristic time of the motion in the polar cap
τ	Time of resonant interaction
ϕ	Longitude, azimuthal angle
Φ	Full electric field potential
Φ_c	Potential of the corotation
Φ_G	Potential of the gravitational field

Φ_n	Nutation phase
Φ_S	Synchrotron function
Φ_0	Arbitrary function
φ	Dimensionless potential
$\phi_1(\alpha_1, k)$	Function
ϕ_p	Position angle
ϕ_c	Angle of the cone
χ	Angle of the pulsar inclination
$\delta\chi$	Amplitude of the nutations
Ψ	Electric field potential
Ψ_0	Plasma generation potential
Ψ_m	Maximum value of the potential in the polar cap
ψ_p	Phase of the wave
Ω	Star rotation frequency
Ω_S	Solid angle
Ω_n	Frequency of the nutations
Ω_m	Modulation frequency
Ω_p	Plasma frequency
ω	Wave frequency
ω_B	Cyclotron frequency
ω_p	Plasma frequency
$\tilde{\omega}$	Doppler shift frequency
ω_c	Curvature photon frequency
ω^*	Characteristic frequency
ω_{br}	Brake frequency

Constants and Quantities

Electron charge	e	=	4.806×10^{-10} ESU = 1.602×10^{-19} C
Electron mass	m_e	=	9.109×10^{-28} g
Speed of light in vacuum	c	=	2.998×10^{10} cm s^{-1}
Planck constant	$\hbar$	=	1.055×10^{-27} erg s^{-1}
Compton wavelength of electron	$\lambda\!\!\!^{-}$	=	$\hbar/m_e c = 3.862 \times 10^{-11}$ cm
Rest energy of electron	$m_e c^2$	=	8.186×10^{-7} erg = 0.511×10^6 eV
Gravitational constant	G	=	6.672×10^{-8} cm^2 s^{-2} g^{-1}
Bohr radius	a_0	=	$\hbar^2/m_e e^2 = 5.292 \times 10^{-9}$ cm
Fine-structure constant	$\alpha_\hbar$	=	$e^2/\hbar c = 1/137.04$
Quantum magnetic field	$B_\hbar$	=	$m_e^2 c^3/\hbar e = 4.4 \times 10^{13}$ G
Boltzmann constant	k	=	1.381×10^{-16} erg K^{-1}
Stefan–Boltzmann constant	σ	=	5.670×10^{-5} erg s^{-1} cm^{-2} K^{-4}
Solar mass	$M_\odot$	=	1.99×10^{33} g
Sun luminosity	$L_\odot$	=	3.86×10^{33} erg s^{-1}
Parsec (pc)	1 pc	=	3.086×10^{18} cm
Year	1 y	=	3.16×10^7 s
Flux density unit 1 Jansky	1 Jy	=	10^{-23} erg cm^{-2} s^{-1} Hr^{-1}

1

Basic observational characteristics of radio pulsars

1.1 Pulsar distribution in the Galaxy

Distances to pulsars Observed radio pulsars are galactic objects which belong to the disc component of the Galaxy. Exceptions are pulsar 0529−66 and source 0540−693 located in the Large Magellanic Cloud at a distance of 55 kpc (McCulloch *et al.*, 1983; Seward *et al.*, 1984). By 1990, nearly 450 such radio sources had been registered (Lyne & Smith, 1990).

The basic method used to find the distance to pulsars is to determine the dispersion measure (DM) responsible for the pulse delay $\Delta t_{1,2}$ at two distinct frequencies ν_1 and ν_2 (Manchester & Taylor, 1977; Smith, 1977):

$$\Delta t_{1,2} = \frac{1}{2\pi} \frac{e^2}{m_e c} (\nu_1^{-2} - \nu_2^{-2}) \cdot DM \qquad (1.1)$$

$$DM = \int n_e \, dl = \langle n_e \rangle d \qquad (1.2)$$

where n_e is the electron density and d is the distance to the pulsar. Such a delay is due to frequency dependence of the group velocity of transverse waves in plasma. The characteristic DM values are 10–100 pc cm^{-3}, and therefore the delay of signals at, say, frequencies of ~ 100 MHz for $\nu_1 - \nu_2 \sim 1$ MHz can reach several seconds.

The accuracy of finding the distances by formulas (1.1) and (1.2) is, however, not high, and in different catalogues the values may differ up to 100%. This divergence is mainly due to the difficulty in finding the electron density n_e on the line of sight.

There exist alternative ways for determining distances to pulsars:

1. Estimation along the hydrogen absorption line (Ables & Manchester, 1976; Heiles *et al.*, 1983; Clifton *et al.*, 1988).

Table 1.1. *Precise distances to radio pulsars*

Pulsar	d (kpc)	Method	Reference[a]
0329+54	2.6 ± 0.3	HI	1
0355+54	1.5 ± 0.5	HI	1
0529−66	55 ± 5	SNR	2
0531+21	2.0 ± 0.2	SNR	1
0736−40	2.0 ± 0.5	HI	1
0823+26	0.37 ± 0.08	Parallax	3
0833+45	0.5 ± 0.1	SNR	2
0950+08	0.127 ± 0.013	Parallax	3
1154−62	11.5 ± 1.0	HI	1
1451−68	0.45 ± 0.06	Parallax	4
1509−58	4.0 ± 1.0	SNR	2
1641−45	4.9 ± 0.4	HI	1
1642−03	0.16 ± 0.10	HI	1
1859+03	20 ± 2	HI	1
1929+10	0.045 ± 0.015	Parallax	5
2319+66	3.5 ± 0.5	HI	1

[a] References: (1) Manchester & Taylor (1977). (2) Taylor & Stinebring (1986). (3) Gwinn *et al.* (1986). (4) Bailes *et al.* (1990). (5) Backer & Sramek (1982).

2. Direct estimation of annual parallax (unfortunately, this is possible only for the nearest objects) (Backer & Sramek, 1982; Gwinn *et al.*, 1986; Bailes *et al.*, 1990).
3. Estimation of distances to other sources (a supernova remnant, a companion star, etc.) associated with a pulsar of interest. Table 1.1 gives a list of the most precise values of independent measurements.

The distribution of observed pulsars as a function of the distance d is presented in Fig. 1.1. The characteristic distances to pulsars in this figure amount to several kiloparsecs and, therefore, greatly exceed the distance to the nearest stars. At the same time, such distances are less than the diameter of the galactic disc by an order of magnitude, so we in fact can now observe only a small part of all pulsars, namely, the pulsars mainly located in the vicinity of the Sun. In particular, it is now rather difficult to judge the galactic distribution of pulsars, especially near the galactic centre. There are only some indications that their density increases towards the galactic centre (Guseinov & Yusifov, 1984; Lyne *et al.*, 1985). The estimate of the total number of active pulsars, equal to about $(1\text{–}3) \times 10^5$ (Lyne *et al.*, 1985), is therefore not exact either.

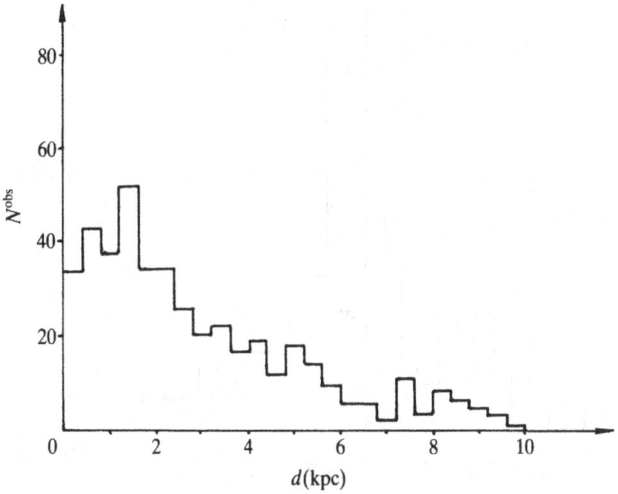

Fig. 1.1. Distribution of observed pulsars as a function of the distance d.

The z-distribution of pulsars One can determine rather well the genuine distribution of observed pulsars as a function of the distance z to the galactic plane (Lyne *et al.*, 1985). It does not differ practically from the observed radio pulsar distribution shown in Fig. 1.2. Fifty percent of all pulsars turn out to concentrate in the range $|z| < 260$ pc and 90% in the range $|z| < 680$ pc. It should be noted that the characteristic half-width of the disc of supernova remnants is about 60 pc.

Proper velocities of pulsars Many of the pulsar characteristics differ from those of usual stars. First, the overwhelming majority of radio pulsars are single objects, whereas the majority of stars are in binary or multiple systems (pulsars belonging to binary systems are described in Section 1.5). Second, proper velocities of pulsars reach 200–400 km s^{-1}, which greatly exceeds characteristic star velocities (Lyne & Smith, 1982; Cordes, 1986). Figure 1.3 shows pulsar distribution as a function of the transverse velocity v_t, i.e. the component of the velocity perpendicular to the line of sight (Cordes, 1986).

Kinematic age For pulsars with a known value of proper velocity, one can introduce the so-called kinematic age

$$\tau_{\text{kin}} = \frac{z}{v_t} \tag{1.3}$$

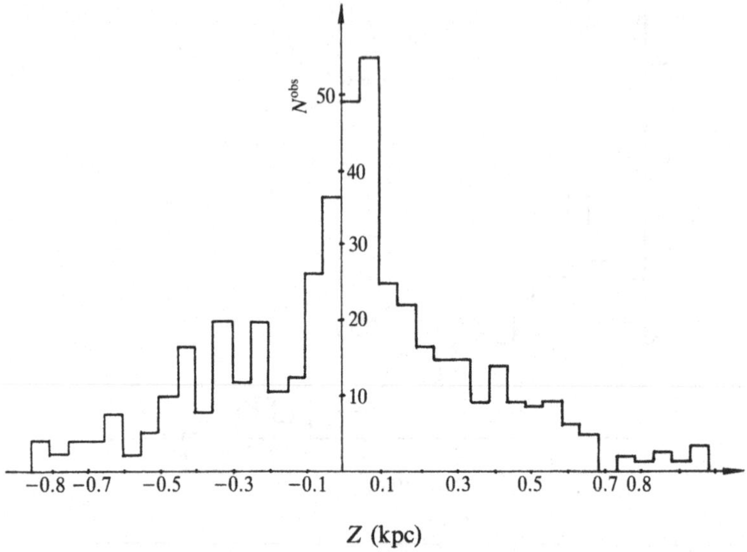

Fig. 1.2. Distribution of observed pulsars depending on the distance to the galactic plane *z*.

whose value customarily equals several million years. The estimate (1.3) thus gives the characteristic active lifetimes of pulsars. As we can see, it is much less than the lifetimes of normal stars. Finally, knowing the overall number of active pulsars and their lifetimes, we can estimate the neutron star birth frequency. According to Lyne *et al.* (1985), it is equal to one birth every 60^{+60}_{-30} years. This value agrees with that of supernova burst frequency, which is thought of as responsible for neutron star formation. This question is considered in more detail in Chapter 2.

Pulsar statistics In concluding this section we emphasize that the observed pulsar sample is non-uniform, since it includes the results of several surveys using different radio antennae that investigated different regions of the sky (Manchester & Taylor, 1977; Damashek *et al.*, 1982; Manchester *et al.*, 1985a; Stokes *et al.*, 1986). The main characteristics of these surveys are presented in Table 1.2. On the other hand, in each survey the true distributions of pulsars $N(P_1, P_2, \ldots, P_N)$, where P_i are parameters characterizing the pulsar, are related to the observed distribution $N^{obs}(P_1, P_2, \ldots, P_N)$ as

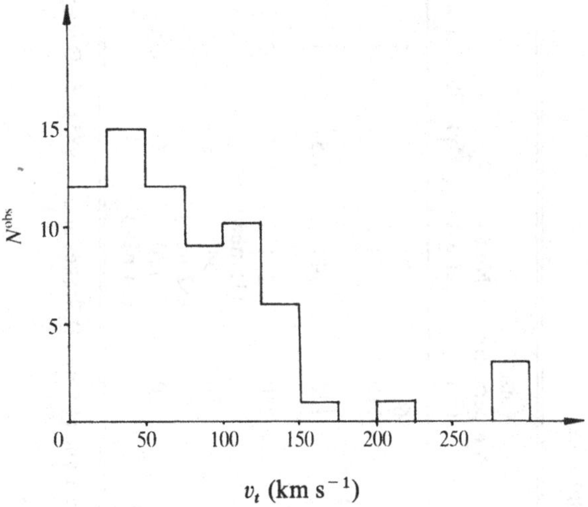

Fig. 1.3. Distribution of pulsars depending on the transverse velocity v_t.

$$N^{obs} = AN \tag{1.4}$$

where $A(P_1, P_2, \ldots, P_N)$ is the observational function, i.e. the probability of the fact that a pulsar with parameters $P_1, \ldots, P_N$ can be registered by a given radiometer (Vivekanand *et al.*, 1982).

Figure 1.4 gives an example of pulsar distributions over distance for two surveys with different threshold sensitivities. One can see that the expected dependence $N \propto d^2$, corresponding to a uniform pulsar distribution in the galactic disc, holds only for the closest sources for which the device is sensitive enough to register all radio pulsars irrespective of their proper luminosity. Given this, the distance d_{br} corresponding to the break of the curve $N(d)$ differs considerably for these two surveys since it depends on the threshold sensitivity.

In the general case, the true pulsar distributions over each of the quantities P_i are to be found from the relation

$$N(P_i) = \int N^{obs}(P_1, \ldots, P_N)/A \cdot dP_1 \cdots dP_{i-1}\, dP_{i+1}\, dP_N \tag{1.5}$$

As is known, the evaluation of the functions N entering the integral relation (1.5) is a rather complicated task. All this leads to uncertainty in statistical quantities discussed in this chapter.

Table 1.2. *Some characteristics of pulsar surveys*

Survey	Frequency (MHz)	Diagram	Area (steradian)	Sky region	Noise temperature (K)	New pulsars	S_{min} (mJy)	Reference[a]
Molonglo I	408	1.5′ × 4°	7	$45° < 1 < 220°$	600	31	80	1
Jodrell Bank I	408	42′	1	$115° < 1 < 250°, \|b\| < 4°$	110	39	15	1
				$-8° < 1 < 115°, \|b\| < 6°$				
Massachusetts Arecibo I	430	10′	0.05	$42° < 1 < 60°, \|b\| < 4°$	130	40	1.5	1
				$182° < 1 < 197°$				
Molonglo II	408	1.4′ × 4.6°	8.4	$-85° < \delta < 20°$	240	224 (115 new)	1–5	2
Jodrell Bank II	1400		0.1	$-4° < 1 < 105°, \|b\| < 1°$	50	54 (32 new)	1	3
Princeton NRAO	390	0.6°	1.8	$15° < 1 < 230°$	30	170 (54 new)	4	4, 5

[a] References: (1) Manchester & Taylor (1977). (2) Manchester *et al.* (1978b). (3) Clifton & Lyne (1986). (4) Stokes *et al.* (1985). (5) Dewey *et al.* (1985).

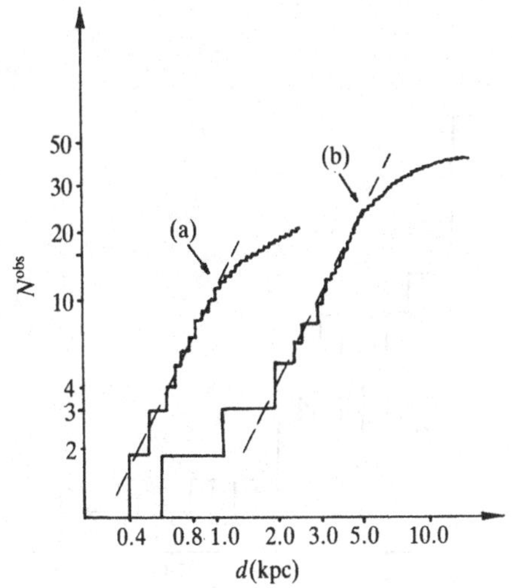

Fig. 1.4. Distribution of pulsars over the distance d for two surveys: (a) Jodrell Bank I, (b) Massachusetts-Arecibo I. The arrows indicate breaking points corresponding to different sensitivities $S_{min}^{JB} = 15$ mJy, $S_{min}^{MA} = 1.5$ mJy.

1.2 Stationary characteristics of pulsar radio emission

1.2.1 Pulsar period

Pulsar radio emission comes to us in the form of separate pulses. The time interval between pulses is called the pulsar period P. This quantity is one of the basic observational characteristics of pulsars.

The periods of all pulsars known at present range within the interval 1.56 ms to 4.31 s, the majority of them lying between 0.3 s and 1.5 s, as shown in Fig. 1.5 (Manchester & Taylor, 1977; Taylor & Stinebring, 1986). It is now definitely established that the deficit of pulsars with small periods (and it is the same for large periods) reflects the real distribution and is not connected with the observational selection effects (Dewey *et al.*, 1985). The properties of pulsars with very small periods $P < 10$ ms, which are apparently a separate class of objects, are discussed in more detail in Section 1.5.

The mean period is constant up to six and more digits. For instance, the period of the fastest millisecond pulsar, 1937+214, is

$$P = 0.001\,557\,806\,448\,8724 \cdots \text{ s}$$

and is now known up to the 13th digit! Such a strict periodicity in pulsar

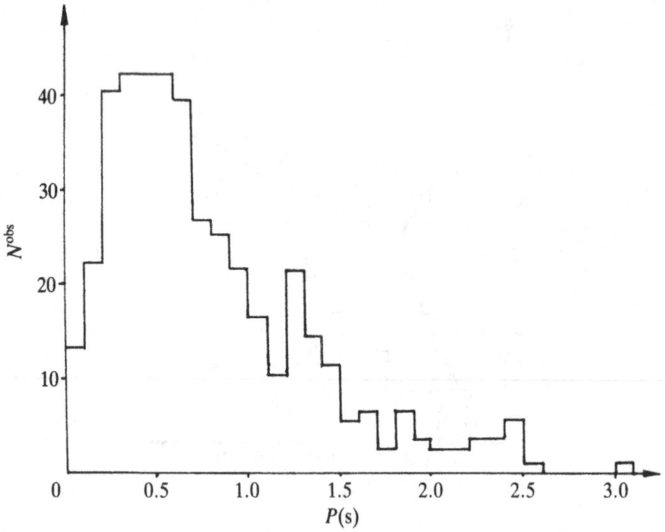

Fig. 1.5. Distribution of observed pulsars as a function of the period *P*.

succession was the decisive factor in constructing the pulsar model (Gold, 1968) that was shown in Fig. 0.1.

Fluctuations of pulsar period are random, but characteristic values of rms deviations do not usually amount to more than several milliseconds (Lyne & Smith, 1990). The statistical analysis shows that this activity is represented by unresolvable glitches of the phase, period *P* and period derivative dP/dt (Downs, 1982). The nature of these glitches is unknown (Cordes & Downs, 1985; Cordes *et al.*, 1988).

These fluctuations are, however, not large and, therefore, as already mentioned, pulsars are exceedingly exact clocks. For example, as shown in Fig. 1.6, the stability of pulsars 0834+06, 1919+21 and 1937+214 on timescales of the order of several years is 10^{-11}–10^{-13}, and for large enough time intervals is compared with the stability of the best modern frequency standards (Backer & Hellings, 1986; Ilyasov *et al.*, 1989). The possibility of introducing a new pulsar time standard making allowance for this remarkable property is now under discussion (Il'in *et al.*, 1984; Rawlay *et al.*, 1988).

1.2.2 Period derivatives

It is exceedingly significant that the period of all pulsars is regularly increasing. The period increase rate dP/dt has already been determined

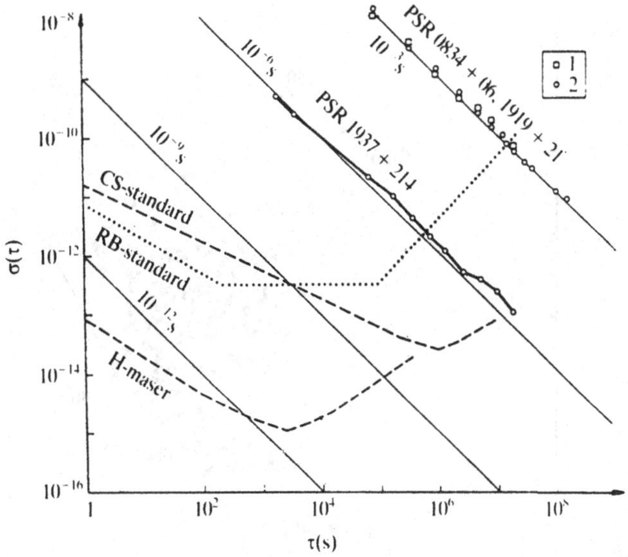

Fig. 1.6. Allen dispersion $\sigma(\tau)$ as a function of the time interval τ for different atomic standards and pulsars. (1) PSR 1919+21; (2) PSR 0834+06. From Ilyasov *et al.* (1989).

for 403 pulsars; the derivative $\dot{P}$ is stable and for the majority of pulsars ranges between 10^{-14} and 10^{-16} (see Table II, Appendix). But the recently discovered pulsar 2127+11, which is part of the globular cluster M15, has a small negative value of the derivative $\dot{P} = -(20 \pm 1) \times 10^{-18}$ (Wolszczan *et al.*, 1989). This $\dot{P}$ value is obviously due to the Doppler shift caused by the neutron star's acceleration in the globular cluster and is not connected with the real pulsar rotation acceleration.

Figure 1.7 shows pulsar distribution on the P–$\dot{P}$ diagram. Circles indicate pulsars in binary systems (for more details see Section 1.5) and crosses indicate pulsars showing considerable irregularities in their radiation (see Section 1.3.3). As shown in the figure, these irregularities are mainly concentrated near the so-called death line. This diagram is analysed in Chapter 7.

The dynamical age τ_d Knowing the values of P and $\dot{P}$, we can find the dynamical age τ_d of pulsars,

$$\tau_d = \frac{P}{\dot{P}} \tag{1.6}$$

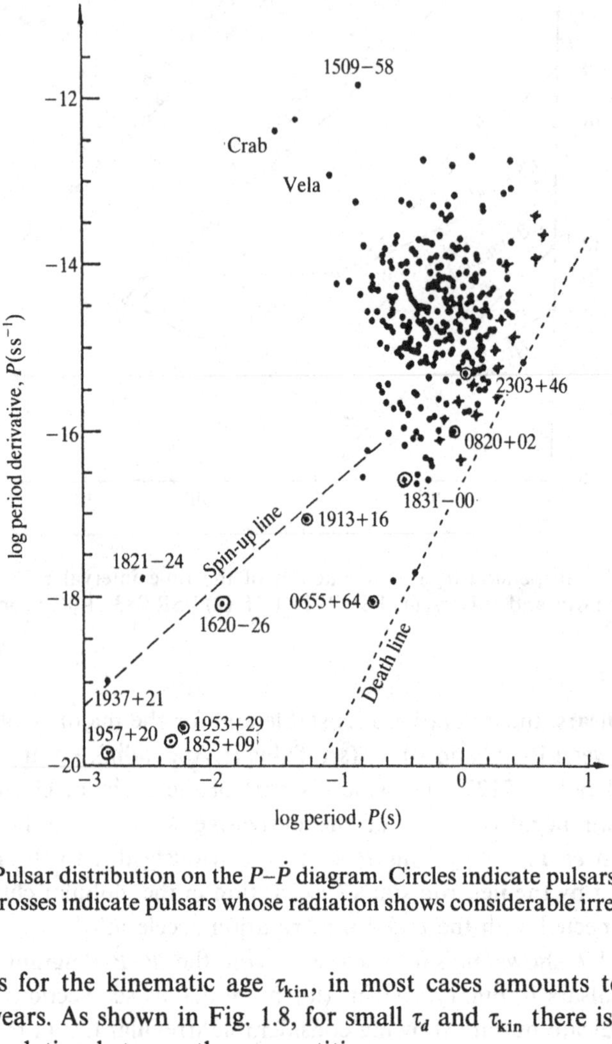

Fig. 1.7. Pulsar distribution on the P–$\dot{P}$ diagram. Circles indicate pulsars in binary systems, crosses indicate pulsars whose radiation shows considerable irregularities.

which, as for the kinematic age τ_{kin}, in most cases amounts to several million years. As shown in Fig. 1.8, for small τ_d and τ_{kin} there is a rather good correlation between these quantities.

Total energy losses W_{tot} Knowing the period P and the period derivative dP/dt, we can find the neutron star's rotation energy losses. They can be estimated by the formula

$$W_{\text{tot}} = -A_r \Omega \frac{d\Omega}{dt} \tag{1.7}$$

where $\Omega = 2\pi/P$ is the angular velocity of pulsar rotation, A_r is the moment of inertia of the pulsar.

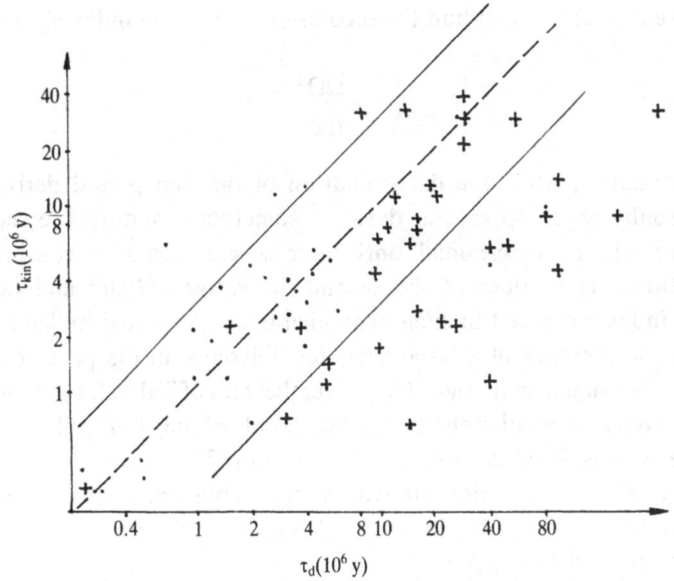

Fig. 1.8. Comparison of the kinematic τ_{kin} and dynamic τ_d ages of radio pulsars. Old pulsars ($Q > 1$, see Section 7.1.2) are marked by crosses.

As shown in Chapter 2, the moment of inertia A_r or real neutron stars amounts to 10^{44}–10^{45} g cm^2. As a result, for the characteristic values $P \sim 1$ s and $dP/dt \sim 10^{-15}$, the total rotation energy losses appear to be of the order of 10^{30}–10^{32} erg s^{-1}, while for fast pulsars, such as 0531+21, 0833−45 and 1509−58, they reach 10^{37}–10^{38} erg s^{-1}. For all pulsars this energy is quite enough to provide the observed radio emission (as well as the emission in the high-frequency spectral region).

1.2.3 Braking index

To discuss higher derivatives of the period, d^2P/dt^2, $d^3P/dt^3, \ldots$, it is convenient to use dimensionless braking indices n_b, $n_b^{(2)}$, etc. The first-order braking index, n_b, is defined as follows

$$n_b = \frac{\ddot{\Omega}\Omega}{(\dot{\Omega})^2} \tag{1.8}$$

It can easily be verified that if $\dot{\Omega} \propto \Omega^n$, then n_b does not depend on time and $n_b \equiv n$. Thus, the braking index n_b provides information on the neutron star evolution law. Accordingly, knowing the third-order period derivative d^3P/dt^3 we can find the second-order braking index $n_b^{(2)}$ by the formula

$$n_b^{(2)} = \frac{\dddot{\Omega}\Omega^2}{(\dot{\Omega})^3} \tag{1.9}$$

Unfortunately, while the determination of the first period derivative dP/dt usually takes up several days, or sometimes months, the second derivative can be determined only after several years of continuous observations. The values of the second derivative d^2P/dt^2 and of the braking index measured in such experiments are presented in Table 1.3, showing the presence of a large irregular difference in the parameter n_b and even the change in its sign. Therefore, the data of Table 1.3 are usually assumed to reflect regular change in the periods of only four pulsars. This question is considered in more detail in Section 7.7.

Finally, the third period derivative was obtained only for pulsar 0531+21 (Lyne *et al.*, 1988b). It is $\dddot{P} = (-2 \pm 6) \times 10^{-29}\ \mathrm{s}^{-2}$, which corresponds to $n_b^{(2)} = 10 \pm 1$.

1.2.4 Observed frequency range

Pulsar radio emission is registered in a wide frequency range of $\nu \gtrsim$ 16 MHz to $\nu \lesssim 25$ GHz, i.e. at practically all frequencies propagating through the Earth's atmosphere and ionosphere. At the same time, as is seen from the example of Fig. 1.19, the majority of pulsars whose spectra have been rather thoroughly analysed exhibit high-frequency ($\nu_{max} \simeq$ 1.4–10 GHz) and low-frequency ($\nu_{min} \simeq 50$–300 MHz) cutoff (Izvekova *et al.*, 1981; Kuzmin *et al.*, 1986; Slee *et al.*, 1986). According to Malov & Malofeev (1981), there exists the following dependence of the frequencies ν_{max} and ν_{min} on the pulsar period P:

$$\nu_{min} \simeq 100P^{-(0.38 \pm 0.09)}\ \ \mathrm{MHz} \tag{1.10}$$

$$\nu_{max} \simeq 3P^{-(0.62 \pm 0.19)}\ \ \mathrm{GHz} \tag{1.11}$$

Their values are given in Table I (Appendix). Figure 1.9 shows the distribution of observed pulsars over the frequencies ν_{min}.

1.2.5 Radio pulse width

The mean profile One of the most important characteristics of radio pulsars is the mean profile shape obtained by summing a sufficiently large number of successive pulses. In most cases, when we sum several hundred

Table 1.3. *Braking index, n_b*

Pulsar	P (s)	$\dot{P}_{-15}$	n_b	Reference[a]
0531+21	0.033	421	2.515 ± 0.001	1
0540−693	0.050	479	2.01 ± 0.02	2
0833−45	0.089	124	42 ± 13	3
1509−58	0.150	1490	2.83 ± 0.03	4
0329+54	0.714	2.05	$(4.81 \pm 0.18) \times 10^3$	5
0611+22	0.335	59.6	3.5×10^2	5
0823+26	0.531	1.72	-1×10^4	5
0950+08	0.253	0.23	$-(5.2 \pm 0.4) \times 10^4$	5
1508+55	0.740	5.03	3.25×10^3	5
1541+09	0.748	0.43	$-(2.5 \pm 0.2) \times 10^5$	5
1604−00	0.422	0.31	$-(1.8 \pm 0.4) \times 10^4$	5
1859+03	0.655	7.49	$(5.50 \pm 0.05) \times 10^3$	5
1900+01	0.729	4.03	$(5.0 \pm 1.2) \times 10^3$	5
1907+00	1.017	5.51	$-(8.7 \pm 0.7) \times 10^3$	5
1907+02	0.990	2.76	$-(9.5 \pm 3.1) \times 10^2$	5
1907+10	0.284	2.64	$-(5.5 \pm 0.2) \times 10^3$	5
1915+13	0.195	7.20	$-(4.2 \pm 0.3) \times 10^1$	5
1929+10	0.226	1.16	$-(2.8 \pm 0.1) \times 10^3$	5
2002+31	2.111	74.6	$(1.20 \pm 0.05) \times 10^2$	5
2020+28	0.343	1.89	$(1.2 \pm 0.1) \times 10^3$	5

[a] References: (1) Groth (1975). (2) Manchester & Peterson (1989). (3) Downs (1981). (4) Manchester *et al.* (1985b). (5) van den Heuvel (1984).

pulses, we obtain a stable profile which does not depend on the storage time (Backer, 1976; Cordes, 1979). Such an averaging is shown in Fig. 1.10.

The radio window width The mean profile of pulsar radio emission usually occupies only a small part of the period. To characterize the profile, an equivalent window width W_r is customarily used, which is defined as the width of a rectangular pulse with the same maximal luminosity I_ν and the same total energy per pulse as in a real signal. One also uses width at 0.5 and 0.1 of the maximal luminosity I_ν. If total period corresponds to 360°, the radio emission pulse duration usually corresponds to values $W_r \simeq 6$–30°. It is only in some rare cases (pulsars 0826−34, 0950+08, 1648−42) that a noticeable radio emission is registered during practically the whole of the period (Biggs *et al.*, 1985; Taylor & Stinebring,

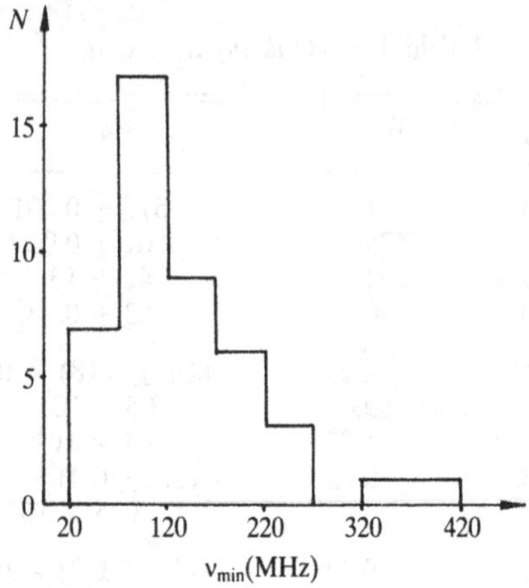

Fig. 1.9. Pulsar distribution over the quantity ν_{min}. From Izvekova *et al.* (1981).

1986). The window width W_r is measured in degrees throughout the book.

The dependence of the window width on the frequency of observation has a rather complicated form. As is seen from Fig. 1.11, the window width W_r gradually decreases on the average with increasing frequency (Rankin, 1983b). But at high frequencies there usually exists a break, such that the rate of variation becomes smaller, and sometimes the pulse width W_r even starts to increase with increasing frequency (Kuzmin *et al.*, 1986; Slee *et al.*, 1987; Izvekova *et al.*, 1989a, 1989b). Also, at some frequencies we observe a sharp decrease in W_r, as if a part of the pulse were absorbed. An example of such a dependence is also shown in Fig. 1.11.

Nonetheless, in a sufficiently wide frequency range the dependence $W_r(\nu)$ is assumed to be exponential:

$$W_r(\nu) \propto \nu^{-\bar{\beta}} \tag{1.12}$$

The values (averaged over many pulsars) of the exponent as a function of frequency are presented in Table 1.4 (Kuzmin *et al.*, 1986). Individual $\bar{\beta}$ values are listed in Table 1 (Appendix).

The dependence of the window width W_r on the period P is also conveniently represented as

$$W_r \propto P^{\bar{\upsilon}} \tag{1.13}$$

Table 1.4. *Frequency dependence*[a] *of W_r on ν*

Frequency interval (GHz)	0.1–0.4	0.4–1.7	1.7–4.6	4.6–10.7
Average of $\bar{\beta}$	0.25 ± 0.04	0.13 ± 0.05	0.16 ± 0.05	0.02 ± 0.08

[a] From Kuzmin *et al.* (1986).

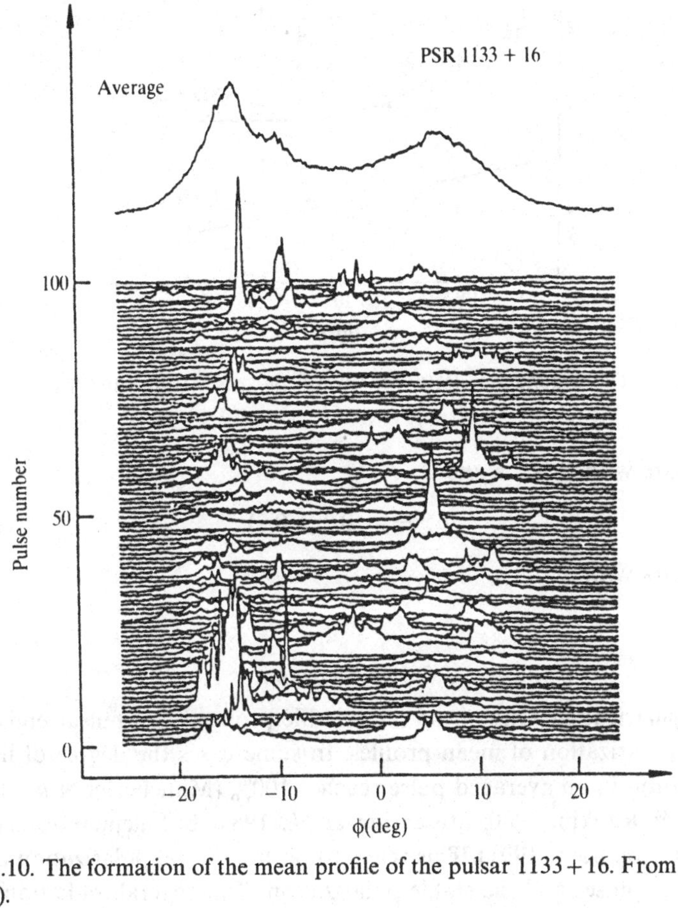

Fig. 1.10. The formation of the mean profile of the pulsar 1133 + 16. From Cordes (1979).

The mean value $\langle \bar{v} \rangle$ is ~ -0.3. Note that the value $\langle \bar{v} \rangle$ differs considerably for fast and slow pulsars. According to Malov & Sulejmanova (1982),

$$\langle \bar{v} \rangle = -(0.43 \pm 0.19) \tag{1.14}$$

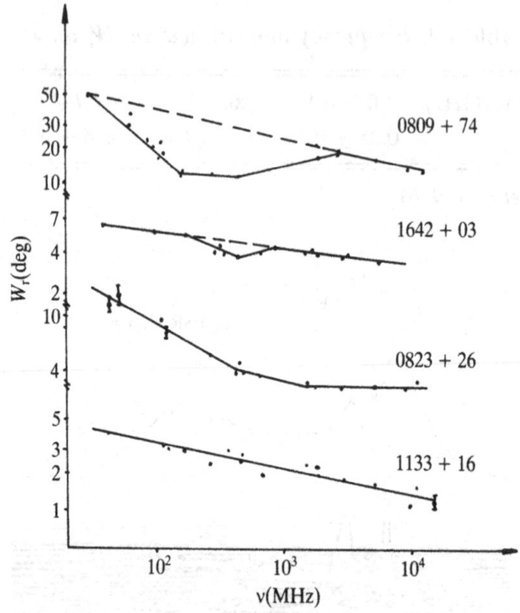

Fig. 1.11. The dependence $W_r(\nu)$ for several pulsars. From Rankin (1983b).

for pulsars with $P > 1$ s and

$$\langle \bar{v} \rangle = -(0.2 \pm 0.2) \qquad (1.15)$$

for pulsars with $P < 1$ s.

1.2.6 Polarization

Linear polarization One of the remarkable properties of pulsar emission is high polarization of mean profiles. In some cases the degree of linear polarization in an averaged pulse reaches 100% (Manchester *et al.*, 1975; Backer & Rankin, 1980; Stinebring *et al.*, 1984a,b; Sulejmanova *et al.*, 1988; Sulejmanova, 1991). This refers to both the total polarization of all individual pulses and the stable polarization of the overall radiation at a given frequency. In contrast to the majority of radio sources, the degree of linear polarization of pulsars usually decreases with increasing frequency (Manchester & Taylor, 1977).

Linear polarization is characterized by the position angle. The position angle dictates orientation of the electric field vector of a wave in the plane orthogonal to the direction of wave propagation, i.e. in the so-called picture plane. The position angle commonly varies rather continuously along the mean profile. The run of the position angle is a stable characteristic of radio emission. It is shown in Fig. 1.12 (see also Fig. 1.15a–d). The total variation of the position angle along the profile does

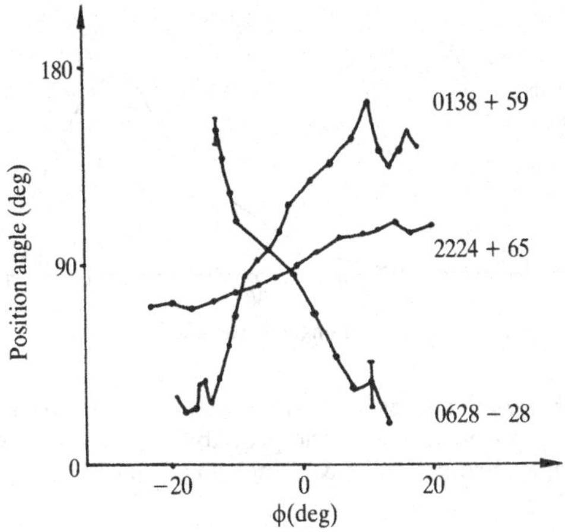

Fig. 1.12. Run of the position angle as a function of the longitudes ϕ at 102.5 MHz. From Sulejmanova, 1989.

not exceed 180° for all pulsars. At the same time, for some pulsars the run of the position angle is not continuous. Moreover, within times of the order of 1 ms, position angles exhibit sharp jumps simultaneously at all frequencies (Stinebring *et al.*, 1984a,b), the radiation intensity remaining approximately the same. Such an irregular run of the position angle is observed, for example, in pulsars 0329 + 54 and 2002 + 34 (Rankin, 1983a).

Two orthogonal modes A sharp variation of the position angle which occurs even in the mean profile suggests that pulsar radio emission consists of two radiation modes with possible fast transitions between them. This conclusion is confirmed by thorough polarization observations.

Figure 1.13 shows the values of the position angle for pulsar 0950 + 08 (Stinebring *et al.*, 1984a). The hatching density is proportional to the probability of observing radiation with a given position angle. One can distinguish two radiation modes with their position angles differing by 90°. In other words, they have orthogonal directions of the electric field vectors, for which reason two such modes are called orthogonal. It is

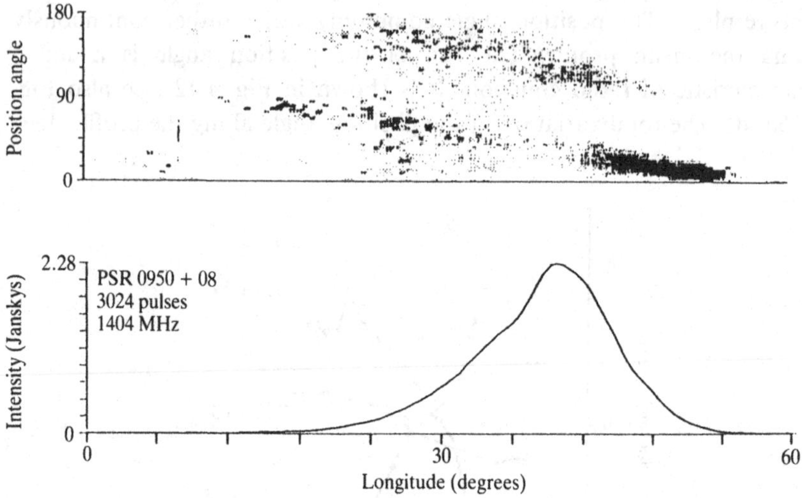

Fig. 1.13. Two orthogonal modes with radio emission of pulsar 0950+08. The density of shading is proportional to the probability of a signal registering with a given position angle. From Taylor & Stinebring (1986).

noteworthy that in each of these modes the position angle varies continuously, the run of the position angle of one of them repeating the run of the other. According to Stinebring *et al.* (1984a,b), two orthogonal modes contain up to 90% of the total pulsar radio emission energy. Only a small fraction corresponds to the randomly polarized component.

Circular polarization Circular polarization is generally not high – it rarely exceeds 20% of the total intensity (Manchester *et al.*, 1975; Rankin, 1983a). There often prevails one direction (either right or left) of circular polarization, but in many pulsars the direction of circular polarization along the profile may reverse once or several times. Examples of polarization reversal are given in Fig. 1.15. It is noteworthy that sign reversals of circular polarization correlate with maxima of the mean profile of a pulsar.

1.2.7 Mean profile shape. Rankin's classification

The shapes of the mean profiles of pulsars are very diverse, and they depend essentially on the frequency of observation. Nevertheless, they have been thoroughly classified with allowance made for both frequency and polarization characteristics of radio emission. This classification, proposed by Rankin (1983a,b, 1986, 1990) is based on the idea that, as shown in Fig. 1.14, the directivity pattern of pulsars consists of two components – the core and the conal. The whole variety of mean profiles of pulsars depends on the relative intensity of these two components as well as on the purely geometric factor determining the minimal distance of the line of sight from the pattern centre. As shown in Fig. 1.14, the mean profile can be divided into five main groups:

S_d Single-humped profile whose emission is determined by the conal component. In their properties they are close to double profiles. At low frequencies the mean profile becomes double. The width of the mean profile is generally rather large, $W_r > 10°$.

The position angle changes little along the mean profile. At high frequencies the polarization is low. The edges of the profile are weakly polarized. On the $P\dot{P}$ diagram such pulsars are located near the 'death line' (see Fig. 1.7).

S_t Single-humped profiles determined by the core component. In their properties they are close to triple ones. At high frequencies, at the edges of such mean profiles, there appear two satellites, so that the mean profile becomes triple. The width of the mean profile is not large, $W_r < 10°$, nor is the variation of the position angle large. A strong circular polarization, exceeding the linear one, is sometimes observed. It is in such pulsars that circular polarization reverses sign. On the $P\dot{P}$ diagram these pulsars lie far from the 'death line', i.e. have relatively low P and high $\dot{P}$ values.

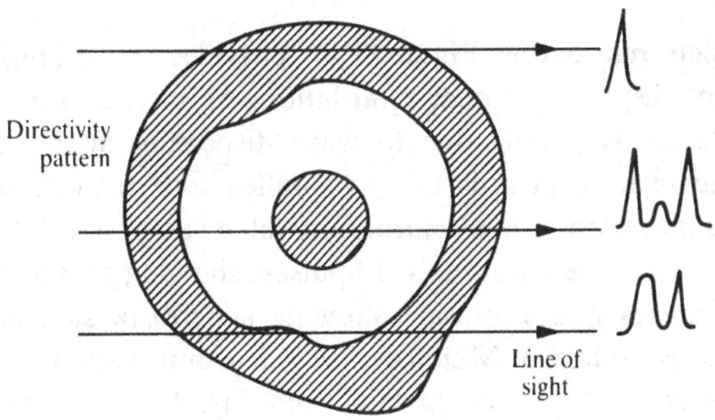

Fig. 1.14. The core and conal components suggested by the Rankin classification. The mean profiles due to rotation of such a directivity pattern are also shown.

T Triple profiles. They are determined by both the core and the conal components. Their main properties are as follows: the edges of external peaks are depolarized; there is a strong change of the position angle in the core component; circular polarization is observed only in the core component.

D Double-humped profiles. They are mainly determined by the conal component. The position angle changes strongly, up to 180°. A detail is sometimes present in the centre with properties close to those of the core component of triple pulses. It is here that we observe not only the greatest change of the position angle, but also the greatest polarization.

M Complex (multicomponent) profiles. Here one can always distinguish components with properties close to those of class S_t and to those of the core components of pulses of class T.

Pulsars of the latter three classes also lie not far from the 'death line' on the $P\dot{P}$ diagram.

Figure 1.15 shows the characteristic mean profiles of all the five classes of pulsars, as well as the values of the linear and circular polarization and the run of the position angle (Rankin, 1983a; Rankin *et al.*, 1989). Figure 1.16 gives examples of the frequency dependence of the shape of the mean profile for pulsars of classes S_t and S_d (Rankin, 1983a; Hankins & Rickett, 1986). The relative intensity of different components is seen to change substantially with frequency. As has already been mentioned, the overall profile with W_r on the whole decreases with frequency by the law of (1.12). The distance between the two components of the mean profile varies in a similar way in pulsars of class D.

Superdispersion retardation Figure 1.17 provides an example of the so-called superdispersion signal retardation. As has already been said, pulse retardation is mainly due to wave dispersion in the interstellar medium. The relation (1.1) is usually fulfilled to a high accuracy. But in some pulsars we observe a retardation value somewhat different from that given by (1.1). There now exist 14 pulsars showing such a retardation at low frequencies and 9 pulsars showing superdispersion advance at high frequencies (Shitov & Malofeev, 1985; Kuzmin, 1986; Kuzmin *et al.*, 1986; Shitov *et al.*, 1988). As is seen from Fig. 1.17, the characteristic value of superdispersion retardation is equal to several tens of milliseconds.

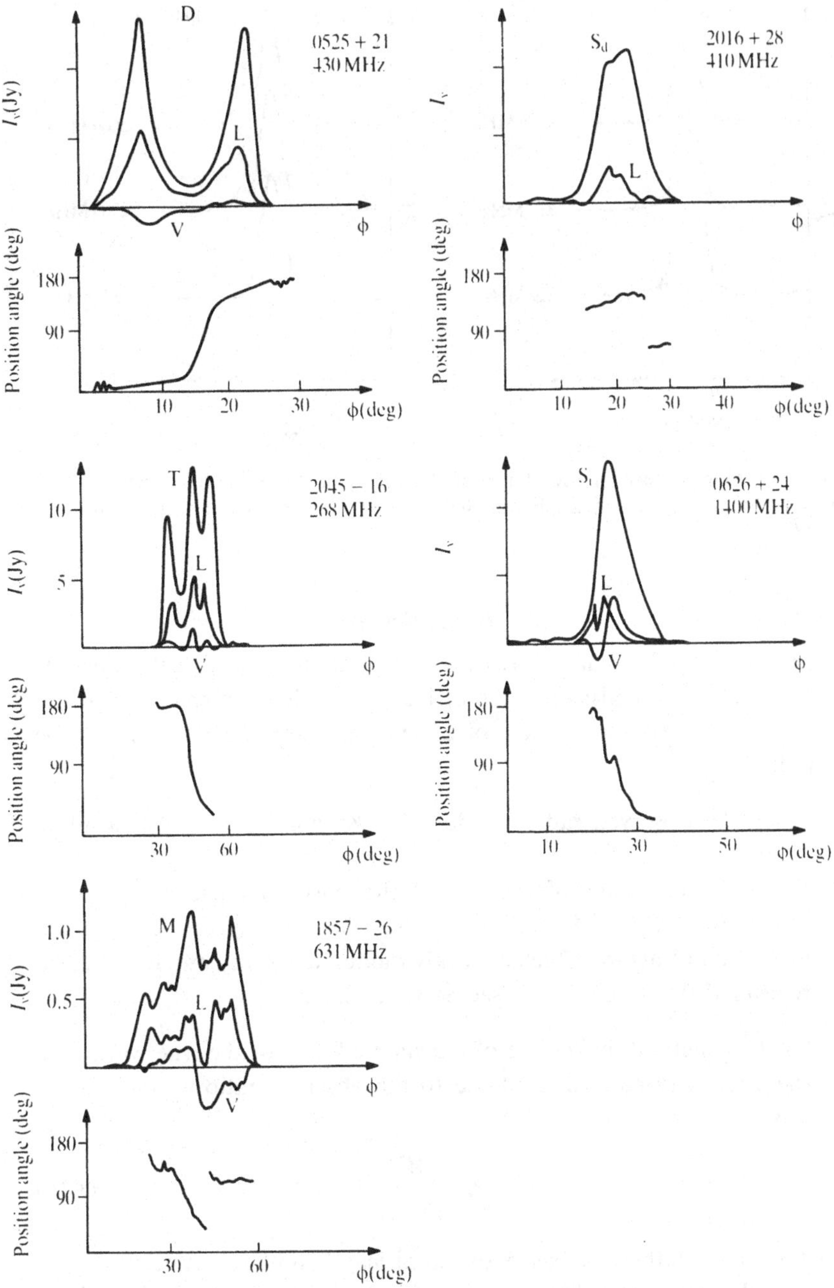

Fig. 1.15. Characteristic mean profiles of all the five classes of pulsars, as well as the values of the linear (L) and circular (V) components and the run of the position angle. From Manchester (1971), McCulloch *et al.* (1978) and Rankin (1983a).

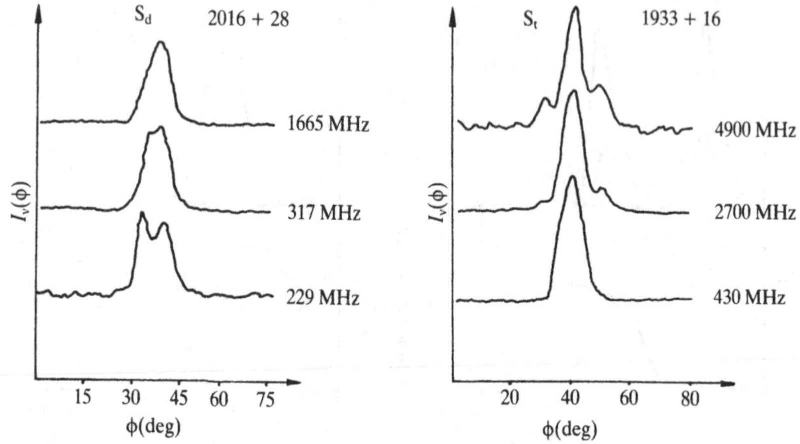

Fig. 1.16. Dependence of the shape of the mean profile on the frequency ν for pulsars of classes S_d and S_t. From Sieber *et al.* (1975) and Hankins & Rickett (1986).

1.2.8 Inclination angle

An important pulsar characteristic determining the geometrical properties of observed radio emission is the inclination angle between the magnetic axis and the axis of rotation of a neutron star. The angle χ can be found:

1. From the observed radio window W_r (Kuzmin *et al.*, 1984; Rankin, 1990).
2. From the character of the run of the position angle ϕ_p (Lyne & Manchester, 1988; Malov, 1990). Also, the angle χ can be estimated for those pulsars for which a weakly modulated X-ray radiation is also registered (for more details see Section 7.2.3).

The first method makes use of the simple fact that the true width W_r^{tr} of the directivity pattern is related to the observed ratio window width simply as

$$W_r = \frac{W_r^{\text{tr}}}{\sin \chi} \tag{1.16}$$

The second method is based on empirical hypothesis concerning the vector character of radio emission (Radhakrishnan & Cocke, 1969), according to which the direction of linear polarization is uniquely determined by the projection of the magnetic field onto the picture plane.

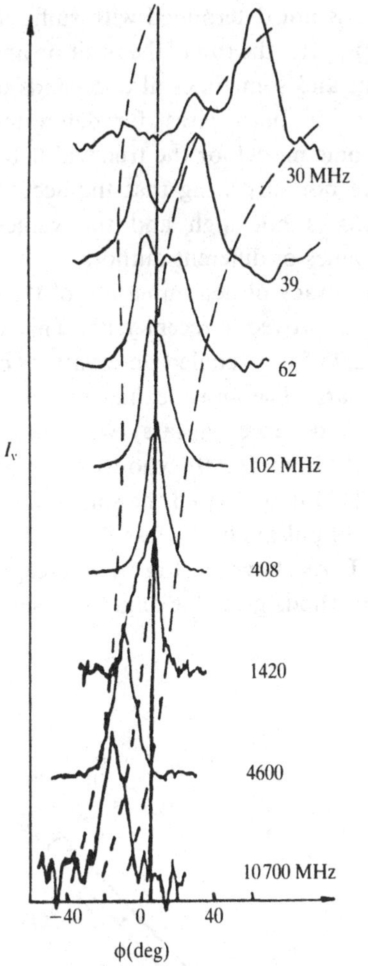

Fig. 1.17. Superdispersion pulse delay. From Kuzmin (1986).

For the dipole magnetic field there exists the simple dependence

$$\tan \phi_p = \frac{\sin \chi \sin \Omega t}{\sin \beta \cos \chi - \cos \beta \sin \chi \cos \Omega t} \tag{1.17}$$

relating the position angle ϕ_p and the time t to the angle χ and the angle β between the axis of rotation and the direction to the observer.

In spite of the fact that the angle β is actually unknown, the observed run of the position angle of some pulsars agrees so closely with equation (1.17) that both the angles β and χ can be found. In the majority of cases the polarization curve is not determined with sufficient accuracy. Moreover, as is seen from Fig. 1.15, the run of the position angle often undergoes noticeable fluctuations and sometimes also depends on the frequency of observation. Therefore, in both cases, for determining χ it is in fact necessary to choose some model for the true width W_r^{tr} of the directivity pattern. It is therefore not surprising that the accuracy with which the angle χ is determined is not high and the values sometimes differ substantially in catalogues of different authors.

Nevertheless, the accuracy of determination of the inclination angle χ has been considerably improved in recent years. This has become possible due to an increase in data from detailed observations of polarizations and mean profiles of pulsars. The most exhaustive catalogues (providing information on 100 and more pulsars) were presented by Lyne & Manchester (1988) and Malov (1990), who used the polarization method, as well as by Rankin (1990), who used the simple model for the width of the directivity pattern of pulsars from class S_t and the central component of pulsars from class T. As shown in Fig. 1.18, except for several pulsars for which one of the methods gives the angle χ close to 90°, there is close

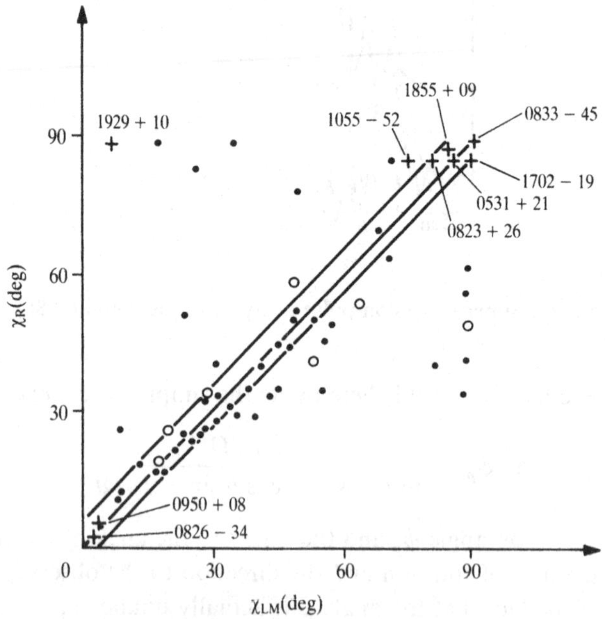

Fig. 1.18. Comparison of inclination angles of axes, χ, determined by Rankin (1990) and Lyne & Manchester (1988). Interpulse pulsars are marked by crosses.

agreement between the results of Lyne & Manchester (1988) and Rankin (1990). The agreement with the results of Malov (1990) is much weaker.

Pulsars for which independent χ measurements carried out by different methods give close results are listed in Table 1.5. We can see that the inclination angle χ of radio pulsars may assume all possible values from 2–4° (when the radiation modulation depends on the structure of the directivity pattern) to 90° (when the radio emission can be observed from two magnetic poles (see Section 1.2.11)).

1.2.9 Radio emission spectrum

The power of pulsar radio emission is characterized by the mean flux density S_ν at a frequency ν,

$$S_\nu = \frac{\tilde{E}_\nu}{P} \tag{1.18}$$

defined as the ratio of the energy density per pulse,

$$\tilde{E}_\nu = \int_0^P I_\nu(t)\,dt = \frac{P}{2\pi}\int_0^{2\pi} I_\nu(\phi)\,d\phi \tag{1.19}$$

to the pulsar period P (Manchester & Taylor, 1977; Smith, 1977). At frequencies of 400 MHz such a flux usually makes up several tens of milli-Janskys, the weakest fluxes (limited to the receiving-device sensitivity) measuring only 1–3 milli-Janskys (Manchester & Taylor, 1977; Manchester *et al.*, 1978b; Damashek *et al.*, 1982; Dewey *et al.*, 1985; Stokes *et al.*, 1985; Clifton & Lyne, 1986).

Figure 1.19 presents typical frequency spectra of a number of pulsars. Such spectra usually have a power-law form in a wide frequency range (Sieber, 1973; Izvekova *et al.*, 1981; Slee *et al.*, 1986):

$$\tilde{E}_\nu \propto \nu^{-\tilde{\alpha}} \tag{1.20}$$

In this figure, the low-frequency cutoffs already discussed in Section 1.2.4, are clearly seen.

According to Malov & Malofeev (1981), the mean value of the spectral index $\tilde{\alpha}$ determined for 43 pulsars in the high-frequency spectral region is

$$\langle\tilde{\alpha}\rangle = 2.0 \pm 0.1 \tag{1.21}$$

Note that at a frequency $\nu = \nu_{\rm br} \sim 1$–3 GHz the spectrum often shows a break and becomes steeper at higher frequencies $\nu > \nu_{\rm br}$. According to

Table 1.5. *Pulsars with known inclination angle* χ[a]

Pulsar	P (s)	$\dot{P}_{-15}$	χ (deg)
0105+65	1.284	13.1	27 ± 1
0136+57	0.272	10.7	50 ± 4
0329+54	0.715	2.0	35 ± 5
0450−18	0.549	5.7	29 ± 2
0531+21	0.033	421.	86 ± 2
0611+22	0.335	59.6	31 ± 3
0736−40	0.375	1.6	17 ± 1
0823+26	0.531	1.7	84 ± 3
0826−34	1.849	1.0	6 ± 2
0835−41	0.752	3.6	52 ± 2
0940−55	0.664	22.7	23 ± 3
0950+08	0.253	0.2	7 ± 2
1039−19	1.386	1.1	32 ± 2
1055−52	0.197	5.8	78 ± 3
1154−62	0.400	3.9	17 ± 2
1237+25	1.382	1.0	50 ± 2
1451−68	0.263	0.1	24 ± 2
1541+09	0.748	0.4	5 ± 1
1642−03	0.388	1.8	69 ± 1
1700−32	1.212	0.7	42 ± 5
1702−19	0.299	4.1	83 ± 5
1737+13	0.803	1.4	41 ± 1
1738−08	2.043	2.3	27 ± 2
1749−28	0.563	8.1	52 ± 5
1821+05	0.753	0.2	30 ± 2
1831−04	0.290	0.2	9 ± 2
1857−26	0.612	0.2	22 ± 1
1859+03	0.655	7.5	37 ± 2
1905+39	1.236	0.5	33 ± 1
1907+03	2.330	4.5	6 ± 1
1911−04	0.826	4.1	67 ± 3
1914+09	0.270	2.5	51 ± 2
1917+00	1.272	7.7	83 ± 3
1919+14	0.618	5.6	24 ± 1
1919+21	1.337	1.3	45 ± 1
1920+21	1.078	8.2	45 ± 2
1929+10	0.226	1.2	7 ± 2
1933+16	0.359	6.0	57 ± 7
1946+35	0.717	7.0	37 ± 5
2003−08	0.581	0.04	15 ± 4
2020+28	0.343	1.9	80 ± 5
2045−16	1.962	11.0	32 ± 3
2111+46	1.015	0.7	10 ± 2
2319+60	2.256	7.0	17 ± 2

[a] From Rankin (1990), Lyne & Manchester (1988).

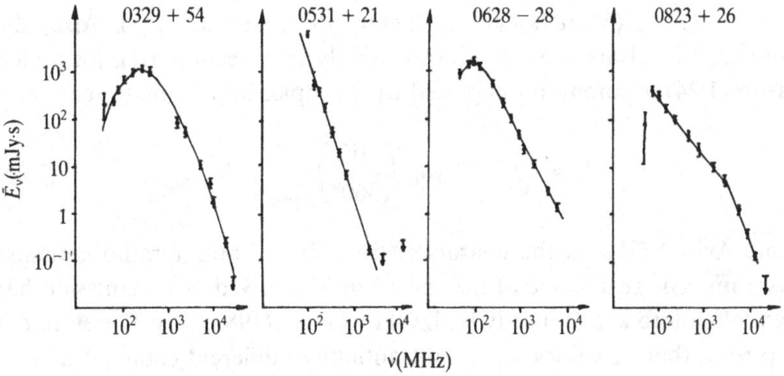

Fig. 1.19. Spectra of radio pulsars. From Izvekova *et al.* (1981).

Kuzmin *et al.* (1986), for 21 pulsars whose spectra had breaks,

$$\langle \bar{\alpha} \rangle = 1.7 \pm 0.3 \tag{1.22}$$

for frequencies $\nu < \nu_{br}$ and

$$\langle \bar{\alpha} \rangle = 3.1 \pm 1.1 \tag{1.23}$$

for frequencies $\nu > \nu_{br}$. The frequency ν_{br} on the average increases with decreasing pulsar period P.

We should emphasize once again that we speak of the mean spectra of pulsars. The point is that the radio emission intensity undergoes stronger fluctuations than the mean-profile shape. It may have substantial variations on scales of the order of days or even months (Manchester & Taylor, 1977), and therefore determination of pulsar spectra is a fairly complicated task (Malofeev, 1989).

1.2.10 Radio emission intensity

Radio luminosity of pulsars The total radio luminosity of a pulsar can be defined by the formula

$$L_r = \pi^3 \frac{d^2}{P} \int \tilde{E}_\nu \left(\frac{W_r}{360°} \right) d\nu \tag{1.24}$$

where d is the distance to the pulsar and W_r is the window width, which is generally frequency dependent. Note that the definition (1.24) involves the model assumption that the directivity pattern is a circular cone with an opening W_r (Manchester & Taylor, 1977; Smith, 1977). Also, the distances to pulsars are not known exactly (see Section 1.1), for which reason (1.24) is commonly replaced by a simpler approximate relation

$$L_r = \pi^3 d^2 S_{408\,\mathrm{MHz}} \left(\frac{W_r}{360°}\right)_{408\,\mathrm{MHz}} \Delta\nu \qquad (1.25)$$

where $\Delta\nu \sim 1$ GHz is the characteristic width of pulsar radio emission spectrum. The real shape of the spectrum of received radio emission has been taken into account only by Izvekova *et al.* (1981). It is therefore not surprising that L_r values differ substantially in different catalogues.

The observed pulsar distribution over the L_r values is presented in Fig. 1.20. The majority of pulsars there have radio luminosities equal to 10^{27}–10^{29} erg s^{-1}, which is four to six orders of magnitude smaller than the total luminosity of the Sun. Therefore, among the observed sources of cosmic radio emission, pulsars can be considered weak ones. At the same time, on the Earth scale their radiation is of course exceedingly large. It exceeds the overall power of all modern radio stations by many orders of magnitude.

The distribution function of pulsars over the L_r value (Fig. 1.21) reconstructed using (1.4) and (1.5) is now reliably defined (Lyne *et al.*, 1985), and one can say with certainty that the spectral power of pulsar radio emission ranges within the limit $3\,\mathrm{mJy\cdot kpc^2} < L_r < 3 \times 10^3\,\mathrm{mJy\cdot kpc^2}$, which corresponds to the total luminosity $10^{27}\,\mathrm{erg\,s^{-1}} < L_r < 10^{30}\,\mathrm{erg\,s^{-1}}$. In particular, the deficit of pulsars with $L_r < 3\,\mathrm{mJy\cdot kpc^2}$

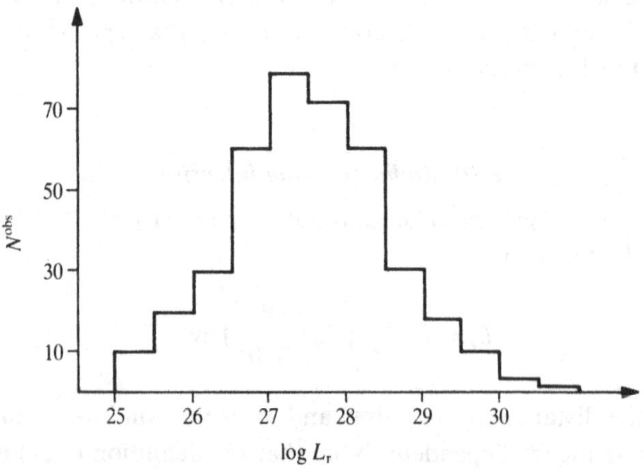

Fig. 1.20. Distribution of observed pulsars depending on total radio luminosity L_r.

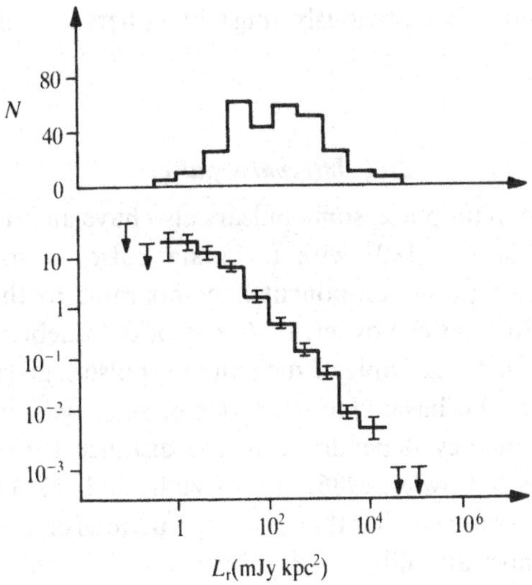

Fig. 1.21. True distribution of pulsars depending on total radio luminosity L_r, established using relations (1.4) and (1.5). From Lyne, Manchester & Taylor (1985).

is also real. Note that this value agrees with the position of the breaks shown in Fig. 1.4.

Coherence of radio emission Making use of the observed radiation power S_ν (1.18), one can estimate spectral density I_ν^0 of radio emission flux near the pulsar surface,

$$I_\nu^0 = \frac{S_\nu d^2}{R^2}$$

where R is the pulsar radius. In radio astronomy it is customary to use not the flux spectral density I_ν^0 but the brightness temperature T_b related to I_ν^0 as

$$I_\nu^0 = \frac{2\pi k T_b}{c^2} \nu^2 \tag{1.26}$$

This relation describes black-body radiation in the Rayleigh–Jeans (low-frequency) region of the spectrum (Lang, 1974). One can readily see that the brightness temperature of radio emission is exceedingly high. It is equal to 10^{23}–10^{25} K and in some pulses even to 10^{30} K. Such temperatures are absolutely unrealistic. It is for this reason that the mechanism of pulsar radio emission obviously must be coherent (Ginzburg *et al.*, 1969).

1.2.11 Interpulse pulsars

In addition to the main pulse, some pulsars also have an interpulse which makes an angle of 120–180° with the main pulse; in some cases the intensity of the interpulse component is comparable to the intensity of the main pulse (Hankins & Fowler, 1986; Taylor & Stinebring, 1986; Lyne & Pritchard, 1987). An example of such an interpulse is given in Fig. 1.22. Table 1.6 presents the basic characteristics of such pulsars. Figure 1.23 illustrates the frequency dependence of the distance between the components (Hankins & Fowler, 1986). As is seen from Table 1.6, the periods of interpulse pulsars are smaller than the mean periods of ordinary pulsars. In particular, almost all millisecond pulsars have interpulses.

The origin of interpulse radiation may be of two kinds. It is most natural

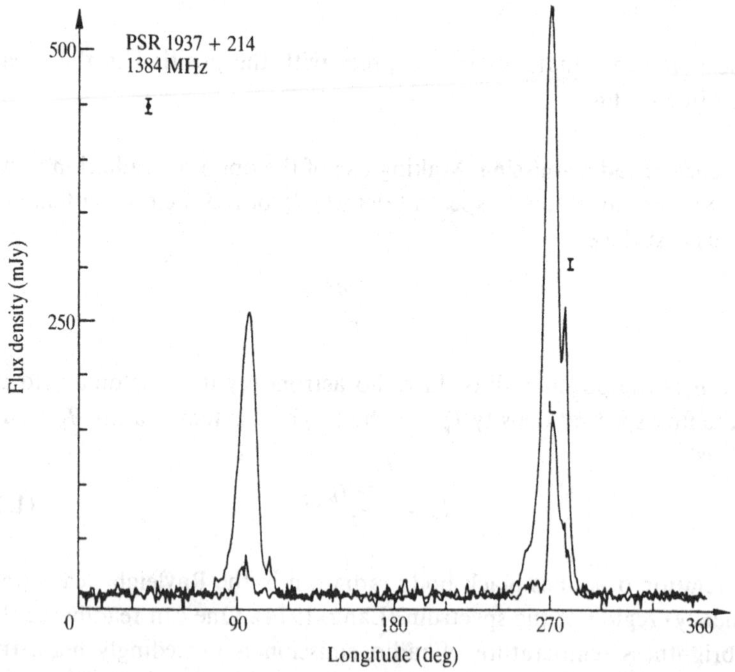

Fig. 1.22. Interpulse radiation of radio pulsar 1937+214. From Taylor & Stinebring (1986).

Table 1.6. *Pulsars with interpulse*

Pulsar	P (s)	$\dot{P}_{-15}$	$\Delta P/P$ (deg)	I_{int}/I_{max}	χ (deg)	Reference[a]
0531+21	0.033	421	216	0.69	86	1
0823+26	0.531	1.7	180	0.005	84	1
0826−34	1.849	1.0	210	0.8	4–8	2
0906−49	0.107	15.2	180	0.26		3
0950+08	0.253	0.2	154	0.012	5–9	1
1055−52	0.197	5.8	154	0.54	78	1
1702−19	0.299	4.1	179	0.20	83	2
1736−29	0.332	7.8	~180	~0.5		4
1821−24	0.003	0.001	120	~0.6		5
1822−09	0.769	52.3	185	0.050		1
1848+04	0.284	0.002	154	0.25		1
1855+09	0.005	0.00002	~180	~0.3		1
1929+10	0.226	1.2	175	0.019	5–9	1
1937+214	0.0016	0.0001	172	0.49	~80	1
1944+17	0.441	0.02	185	0.01		1
1953+29	0.006	0.00003	~180	~0.1		6
1957+20	0.0016	0.00001	~180	~0.4		7

[a] References: (1) Taylor & Stinebring (1986). (2) Biggs *et al.* (1985). (3) D'Amigo *et al.* (1988). (4) Clifton & Lyne (1986). (5) Lyne *et al.* (1987). (6) Boriakoff *et al.* (1983). (7) Fruchter *et al.* (1988).

to associate it with pulsars whose radio emission can be registered from both magnetic poles of the neutron star. Clearly, the inclination angle χ of such pulsars must be close to 90°, the condition of interpulse observability having the form

$$|\chi - \pi/2| < W_r \tag{1.27}$$

In this case the distance between the main pulse and the interpulse must not generally depend on the radiation frequency ν. Second, as shown in Fig. 1.24, interpulse radiation can be imitated in case $\chi \simeq 0$, when the two radiation pulses are due to the double intersection of the conal component of the directivity pattern with the line of sight. In this case the condition

$$\chi < W_r \tag{1.28}$$

must hold. There can exist also a noticeable frequency dependence of the distance between the main pulse and the interpulse connected with the frequency dependence of the directivity pattern with W_r.

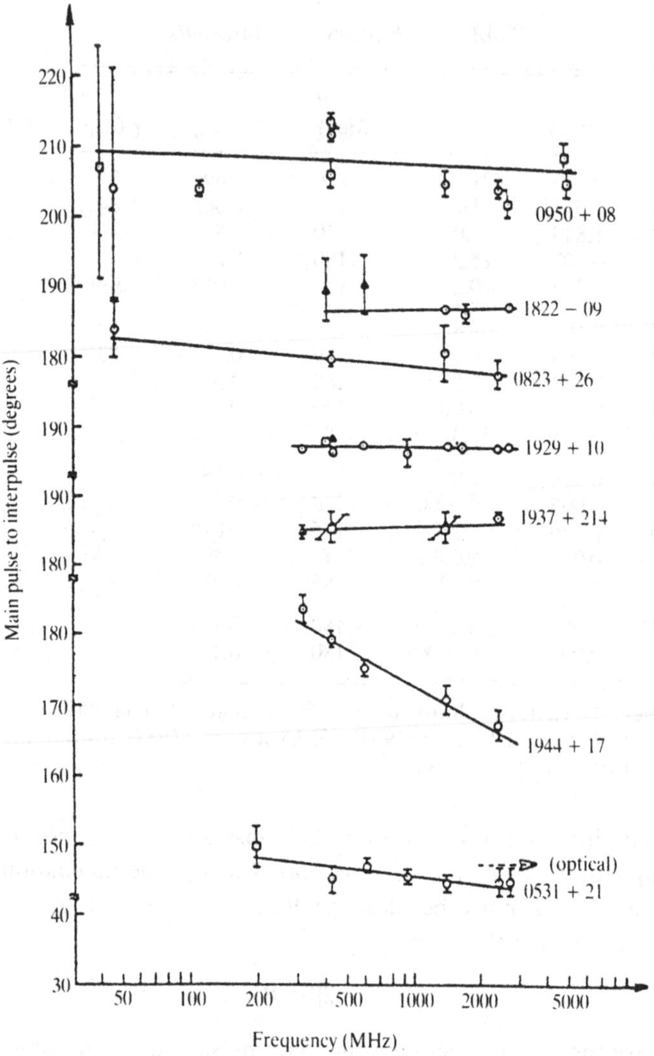

Fig. 1.23. Dependence of the distance between the main pulse and interpulse on the frequency ν. From Hankins & Fowler (1986).

As can be seen from Table 1.6, the simple model presented above is in agreement with observational data. The inclination angles χ of interpulse pulsars actually concentrate near the values $\chi = 0$ and $\chi = 90°$. Pulsars 0531+21, 0823+26 and 1937+214, which have angles $\chi \simeq 90°$, do not

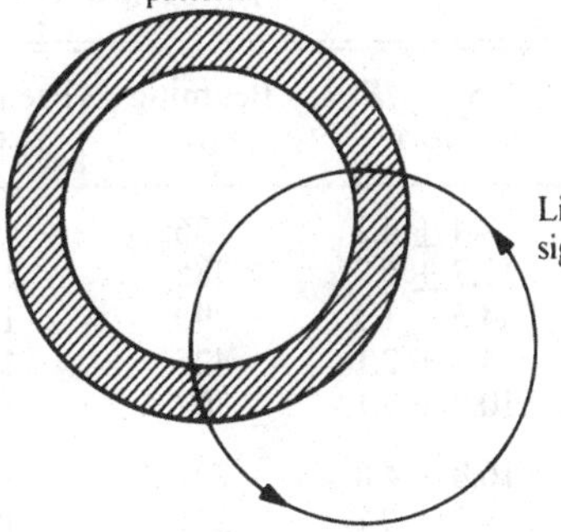

Fig. 1.24. Imitation of interpulse for a pulsar with $\chi \approx 0$. The bursts of radio emission correspond to two intersections of the conical directivity pattern.

show a noticeable dependence of the interpulse distance on the observation frequency, whereas pulsar 0950+08 ($\chi \simeq 10°$) does show this dependence.

Unpulsed radiation In recent years much attention has been paid to the study of unpulsed radiation, i.e. a quasi-constant component occupying a region outside the main pulse (Perry & Lyne, 1985; Smirnova & Shabanova, 1988). Such radiation has already been registered for 16 pulsars. The energy contained in the unpulsed component comprised 2–19% of the main pulse energy. Since unpulsed radiation is stable, it is obviously likely to be generated in the vicinity of a neutron star.

The basic properties of unpulsed radiation are listed in Table 1.7. It should be noted that the relative role of the unpulsed component increases towards low frequencies, so that on decameter waves pulsar radio emission is usually registered for a substantial part of the total period (Bruck & Ustimenko, 1977).

1.3 Non-stationary characteristics of radio emission

1.3.1 Fine structure of individual pulses

Subpulses As mentioned in Section 1.2.5 (see Fig. 1.10), individual pulses of pulsar radio emission are generally quite unlike integral mean profiles. First, the majority of investigated pulses have a pronounced subpulse structure (Rickett *et al.*, 1975; Soglasnov *et al.*, 1981), shown in Fig. 1.25. Subpulses (whose characteristic duration makes up several milliseconds, which corresponds on the angular scale to 2–6°) are obviously elementary

Table 1.7. *Pulsars with unpulsed component*

Pulsar	P (s)	$\dot{P}_{-15}$	S_{unpul}/S (%)	Beaming ratio	Frequency (MHz)	Reference[a]
0031−07	0.943	0.41	8.4 ± 4.2	136	102	1
0138+59	1.223	0.39	12.7 ± 5.7	285	102	1
0329+54	0.714	2.05	4.3 ± 1.3	890	102	1
0809+74	1.292	0.17	4.4 ± 2.1	425	102	1
0823+26	0.531	1.72	10.2 ± 3.1	302	102	1
0943+10	1.098	3.53	16.8 ± 4.8	202	102	1
0950+08	0.253	0.23	5.1 ± 0.7	229	102	1
1508+55	0.740	5.03	6.7 ± 1.4	581	102	1
1541+09	0.748	0.43	15.8 ± 7.9	85	102	1
			16.3 ± 1.3	60	408	2
1604−00	0.422	0.31	10.3 ± 2.9	280	408	2
1642−03	0.388	1.78	36.3 ± 12.8	50	102	1
1749−28	0.563	8.15	33.5 ± 8.8	76	102	1
1929+10	0.226	1.16	4.9 ± 0.5	670	408	2
2016+28	0.558	0.15	2.9 ± 1.4	1000	408	2
2217+47	0.538	2.76	6.0 ± 0.3	419	102	1

[a] References: (1) Smirnova & Shabanova (1988). (2) Perry & Lyne (1985).

cells of radiation. In particular, they correlate well at different frequencies. Given this, the subpulse duration usually depends neither on the observation frequency nor on the position of this subpulse in the mean pulse. On the other hand, the distance between different subpulses increases with decreasing frequency, the same as for the mean profile width. Subpulses are mostly observed in pulsars of class S_d or D as well as in conal components of pulses of pulsars of class T (Rankin, 1983a). It should be noted that subpulses appear incidentally within the mean pulse of radio emission.

Micropulse structure An individual burst of radio emission is composed of individual subpulses, which, in turn, consist of micropulses. Such small-scale structure obviously makes up a substantial part of the radiation, although the modulation depth of the microstructure decreases with increasing frequency (Soglasnov *et al.*, 1981). Figure 1.26 gives an example of micropulse structure at a resolution of 0.8 μs (Hankins & Boriakoff, 1978), which is currently the best. It should be noted that an increase in time resolution results, of course, in a decrease in the receiver's sensitivity. Therefore, the microstructure of only several of the brightest

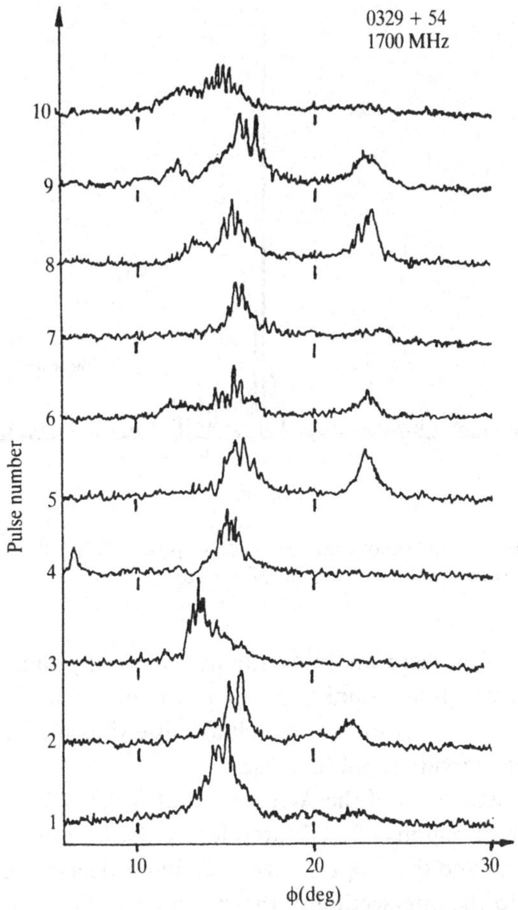

Fig. 1.25. Radio emission subpulses in pulsar 0329+54 at 1700 MHz, time resolution 70 ms.

pulsars, such as 0809+74, 0950+08, 1133+16, has been investigated (Hankins, 1971; Rickett, 1975; Ferguson & Seiradakis, 1978; Hankins & Boriakoff, 1981; Boriakoff, 1983; Smirnova *et al.*, 1986).

To study the fine structure of individual pulses, it is convenient to use the analysis of mean autocorrelation and cross-correlation functions. The *autocorrelation function* (ACF) $F_{ac}(\tau)$ is defined in terms of the product of two intensities I_ν displaced by the time delay τ:

$$F_{ac}(\tau) = \left\langle \int_{\Delta T} I_\nu(t) I_\nu(t+\tau)\, dt \right\rangle \tag{1.29}$$

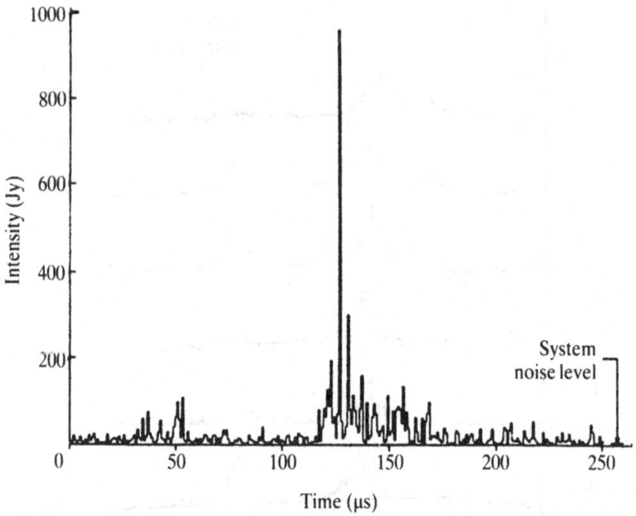

Fig. 1.26. Micropulse structure of radio emission of pulsar 0950+08 with resolution equal to 0.8 μs. From Hankins & Boriakoff (1978).

The angular brackets here imply averaging over many radiation pulses. In other words, when determining an ACF one has to average over the time ΔT within the limits of an individual pulse (integration over dt in (1.29)) and over individual pulsar pulses.

The characteristic form of the ACF of pulsar 1133+16 is presented in Fig. 1.27. Different regions of ACF are identified with the receiver noise, the microstructure and the subpulse structure. In particular, the retardation corresponding to the intersection of the micropulse and subpulse regions determines the timescale of the microstructure.

Similarly, *the cross-correlation function* (CCF) F_{cc} is defined as a product of two different signals (averaged over many pulses), e.g., for two different observation frequencies

$$F_{cc}(\tau) = \left\langle \int_{\Delta T} I_{\nu_1}(t) I_{\nu_2}(t+\tau)\, dt \right\rangle \tag{1.30}$$

or for two different pulses. The CCF for different frequencies makes it possible, for example, to answer the question of the frequency bandwidth for a given scale of the microstructure.

Measurements have shown first of all that the characteristic duration τ_μ of the microstructure is not strictly constant for each pulsar (Hankins, 1971; Cordes, 1976a; Soglasnov *et al.*, 1981; Popov *et al.*, 1987). As an

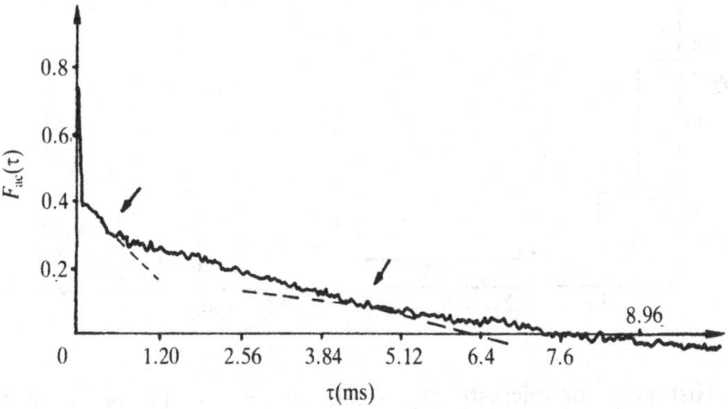

Fig. 1.27. Mean autocorrelation function of pulsar 1133+16. Two breaks (arrows) corresponding to microstructure and substructure details are readily seen (Kardashev *et al.*, 1987).

example, Fig. 1.28 presents a histogram of τ_μ distribution for pulsar 0950+08 (Popov *et al.*, 1987). At the same time we can see that there are two timescales in microstructure. Large-scale bursts with the characteristic duration of 0.2–1 ms (the corresponding angular scale ranges within much smaller limits of $0.27 \pm 0.06°$ (Kuzmin, O. A., 1985)) are wideband ones, i.e. they correlate within a wide frequency band of $\gtrsim 20$ MHz (Rickett *et al.*, 1975; Popov *et al.*, 1987). The characteristic durations of such microstructure details are given in Table 1.8.

The duration of the large-scale microstructure does not depend on frequency. It is only the number of pulses with a given timescale that decreases with decreasing frequency. In contrast to subpulses, the time delay of large-scale microstructure at different frequencies strictly follows the dispersion law (1.1) (Boriakoff, 1983; Popov *et al.*, 1987). This property makes it possible to determine most exactly the dispersion measure (1.2) of such pulsars (Smirnova *et al.*, 1986).

On the other hand, the small-scale structure with $\tau_\mu \lesssim 300$ μs no longer correlates when the frequency difference reaches ~ 1.5 MHz (Popov *et al.*, 1987). This structure is observed in only 20–30% of pulses. It should be noted that, as shown in Fig. 1.28, the distribution over τ_μ for the small-scale microstructure increases as τ_μ decreases to the minimal possible time resolution. This means that even the shortest observed micropulses are the result of summation of radio emission from many shorter individual sources. Note, finally, that periodicity of microstructure with a period $P_\mu \sim 2\tau_\mu$ observed in some cases is actually a short-lived (the *Q*-factor is about 5–10) and non-resonance process with a broad

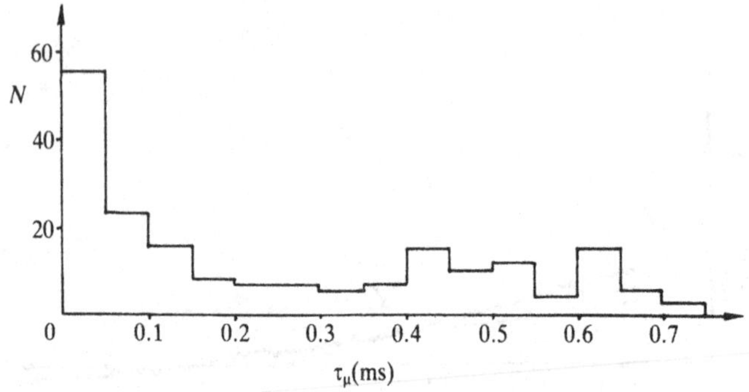

Fig. 1.28. Histogram of microstructure scale distribution for pulsar 0950+08 (Popov *et al.*, 1987).

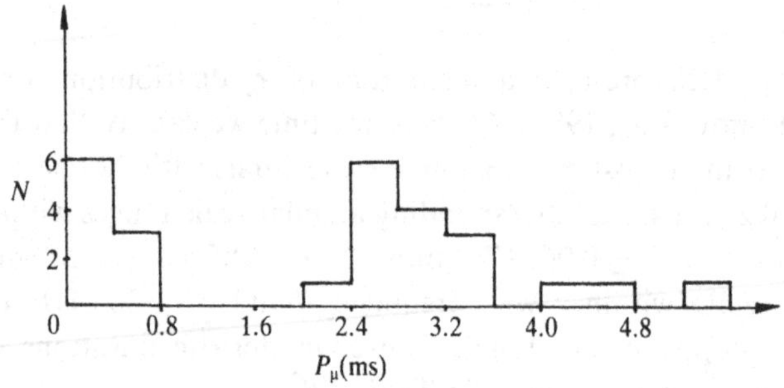

Fig. 1.29. Distribution histogram for microburst repetition periods P_μ for pulsar 0809+74 (Popov *et al.*, 1987).

distribution of periods P_μ (see Fig. 1.29). In particular, the periods P_μ do not correlate at different frequencies for short-scale microstructure (Kuzmin, O. A., 1985).

The amplitude-modulated noise model The general form of the ACF of micropulse radio emission is well understood on the basis of the amplitude-modulated noise model according to which the time structure of the pulse can be described by the expression

$$I_\nu(t) = a(t)n(t) \tag{1.31}$$

where $n(t)$ is the noise filling and $a(t)$ is the envelope (Rickett, 1975; Cordes, 1976b; Hankins & Boriakoff, 1978; Cordes & Hankins, 1979). The presence of the noise filling implies that the micropulses themselves do not directly reflect the coherent mechanism of pulsar radio emission in the

Table 1.8. *Characteristics of long-time microstructure*

Pulsar	P (s)	$\dot{P}_{-15}$	τ_μ (μs)	τ_μ (deg)	Reference[a]
0809+74	1.292	0.2	1300	0.36	1
0834+06	1.274	6.8	1050	0.30	2
0950+08	0.253	0.2	200	0.28	1
			175	0.25	2
1133+16	1.188	3.7	500	0.15	1
			575	0.17	2
1919+21	1.337	1.3	1220	0.33	2
2016+28	0.558	0.1	290	0.19	2

[a] References: (1) Popov *et al.* (1987). (2) Manchester & Taylor (1977).

time interval $\tau \gtrsim \tau_\mu$. The envelope $a(t)$ – in other words, the sequence of large-scale micropulses and their interrelation within a subpulse – in a number of cases (~20%) is described well by the model of determinate chaos with a small correlation dimension $r \lesssim 5$ (Zhuravlev & Popov, 1990). This suggests that the formation of an irregular sequence of large-scale micropulses is in these cases due to a non-linear dynamic system characterized by a relatively small number of independent parameters. However, in microstructure a set of independent unresolved pulses mostly prevails, which must be typical of a system with a well-developed turbulence.

1.3.2 Fluctuations

Intensity fluctuations As has already been mentioned in Section 1.2.9, the total radiation energy in a pulse, $\tilde{E}_\nu$, fluctuates, changing substantially from pulse to pulse. It has been observed that each pulsar has its own histogram of pulse energies $\tilde{E}_\nu$ typical of a particular pulsar. Examples of such distributions are given in Fig. 1.30 (Smith, 1977). According to the histogram, the radiation of pulsar 1642−03 has a narrow distribution near a certain energy value $\langle\tilde{E}_\nu\rangle$. The radiation energy histogram for pulsar 0950+08 is quite different: here the maximal probability falls on a pulse energy close to zero and then we observe a monotonic decrease corresponding to comparatively rare powerful pulses.

Exceedingly powerful pulses are also observed in the pulsar in the Crab Nebula (Manchester & Taylor, 1977). The energy of some pulses registered from this pulsar was 10^3 times higher than the mean value, so that the

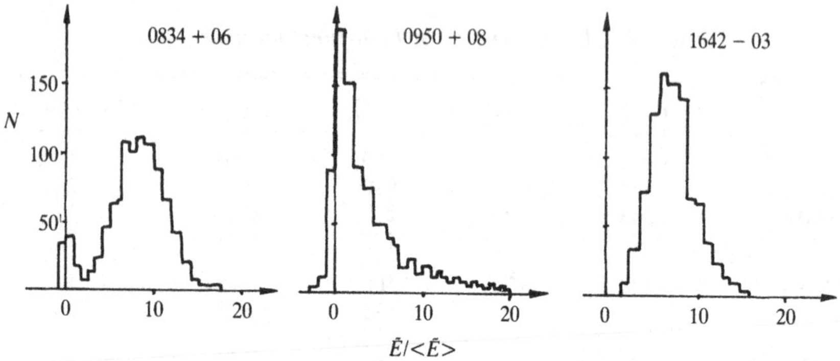

Fig. 1.30. Pulse energy histograms for three pulsars (Smith, 1977).

corresponding brightness temperature reached 10^{31} K. The typical width of such pulses was equal to 100 μs. At those moments the radio emission intensity of the pulsar exceeded that of the whole Crab Nebula.

Nullings The presence of the maximum at zero intensity (see the histograms of Fig. 1.30) shows that radio emission of the pulsars indicated in the graphs dies away for some time. Nullings have proved to be sudden and simultaneous at all frequencies (Ritchings, 1976). Pulsars with multicomponent mean profiles exhibit the intensity fall in all components, then the intensity sharply resumes its value. There have been no successful attempts to register radio emission in a nulling state. Such a state usually lasts from two or three to several tens of periods P and the relative time in a nulling state reaches 10–30%. Pulsars 0826−34 and 1944−17 are unique in this respect since their radio emission is cut off (not registered) for about 70% of the total time (Durdin *et al.*, 1979; Biggs *et al.*, 1985). The occurrence of on and off succeeded in finding the dependence of 'silence' on other parameters of pulsars.

As shown in Fig. 1.7, nulling pulsars group in the region of death lines, i.e. they generally have large enough periods P and low values of the derivative dP/dt. In particular, nullings have never been observed in pulsars of class S_t, but they have been observed in most of the pulsars of classes D, T and S_d (Rankin, 1986). Nullings have now been definitely observed in 24 pulsars.

Modulation index The degree of pulse intensity modulation is represented by the modulation index m defined (Manchester & Taylor, 1977) as

$$m = \frac{(\sigma_{\text{on}}^2 - \sigma_{\text{off}}^2)^{1/2}}{\langle I_\nu \rangle} \tag{1.32}$$

where σ_{on} is the rms pulse intensity deviation from the mean value $\langle I_\nu \rangle$ and σ_{off} is the rms random noise outside the pulse. The observed values of modulation index (after elimination of interstellar scintillation effects) for different pulsars range approximately from 0.5 to 2.5 (Bartel *et al.*, 1980). The modulation index m usually decreases with increasing frequency, but in some pulsars of class S_t the dependence $m(\nu)$ has a break, and at high frequencies the modulation index begins to increase with frequency ν.

For pulsars with a strong modulation, the pulse intensity histograms have approximately exponential form with maximum at zero. For weakly modulated sources, the histograms have a sharp peak near the mean value, so that high-energy radiation is improbable for these sources.

Correlation of fluctuations The mutual correlation of radio emission for different longitudes of the mean pulse are analysed using the CCF,

$$F_{\text{cc}}(\phi_1, \phi_2) = \langle I_\nu(\phi_1) I_\nu(\phi_2) \rangle \tag{1.33}$$

relating the intensities at two distinct longitudes ϕ_1 and ϕ_2. Here again the angular brackets imply averaging over many pulses.

Figure 1.31 presents examples of such a CCF for pulsar 0329+54 (Popov, 1986). Naturally, the correlation is strong at longitudes ϕ_1 close to ϕ_2. The correlation is also strong when the intensities $I_\nu(\phi_1)$ and $I_\nu(\phi_2)$ refer to two different pulses shifted by one period or even more. This means that the active radio-emitting region is generally rather long-lived and in any case is capable of having 'memory' over several periods P. At the same time, no correlation is observed in emission in the parts of the pulse with strongly different ϕ. Moreover, on cross-correlation maps for two successive pulses, when the shift $\Delta\phi$ is maximum, one can clearly see an anticorrelation region reaching 10%. It is also noteworthy that at longitudes of the core component the region of maximal correlation is shifted relative to the maxima of the mean profiles so that subpulse intensity fluctuations at early longitudes of a preceding pulse correlate better with fluctuations at later longitudes of a subsequent pulse.

The scheme of Fig. 1.32 provides an explanation of the basic features of the observed picture of intensity cross-correlation (Popov, 1986). If a discharge region stable for two or three periods appeared along the OA radius, then, owing to the fact that the excitation region does not

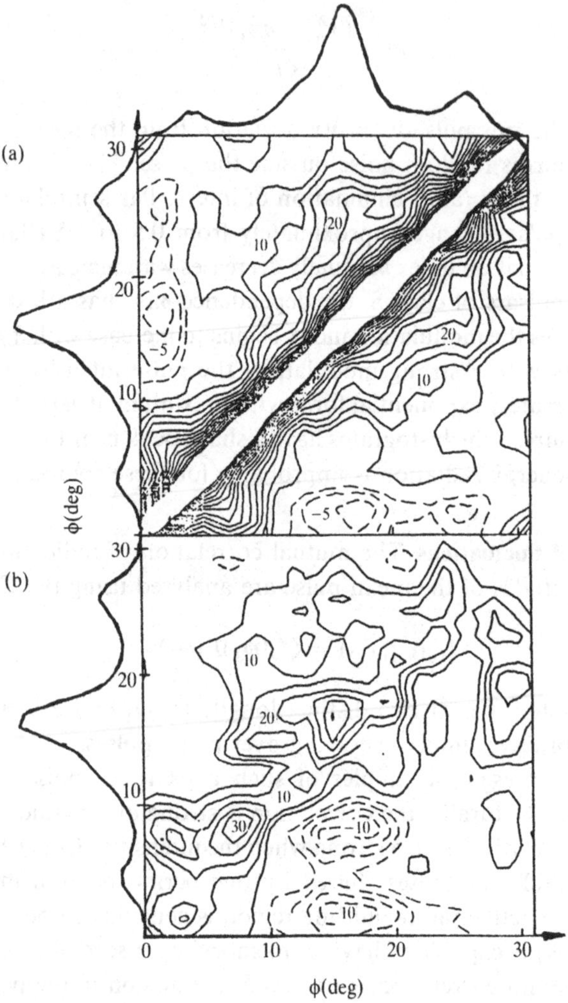

Fig. 1.31. Correlation maps of intensity fluctuations for pulsar 0329+54. Longitudes in degrees are plotted on the axes. Contours show correlation coefficients (F_{cc}) in per cent. (*a*) Identical profiles; (*b*) profiles displaced by a period. From Popov (1986).

propagate through the point O, only intensity fluctuations within the range of longitudes of the first component will be correlated. On the contrary, the OB region will provide correlation between the core and the last components of the mean profile. The physical nature of such a structure of the active region is considered in more detail in the section that follows.

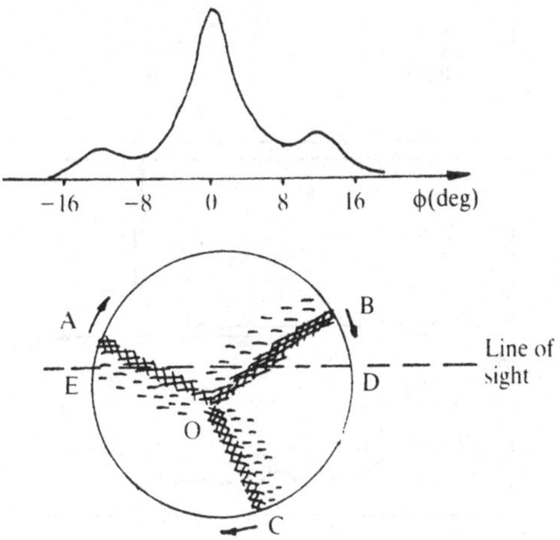

Fig. 1.32. The scheme of explanation of subpulse correlation for pulsar 0329+54. Arrows indicate the rotation direction of radiating regions leading to the observed drift of subpulses. From Popov (1986).

1.3.3 *Quasi-stationary processes*

The fluctuation spectrum To discover quasi-periodic processes in pulsar radio emission it is convenient to make use of the so-called fluctuation spectrum defined as a Fourier-harmonic of the intensity $I_\nu(\phi)$ measured in N pulses ($N \to \infty$) at a given longitude ϕ:

$$I_\phi(f) = \sum_{n=0}^{N} e^{ifn} I_\nu(\phi + 2\pi n) \tag{1.34}$$

Clearly, owing to the stroboscopic effect, the frequency f ranges from 0 to 0.5 cycles per period. Examples of fluctuation spectra are given in Fig. 1.33 (Smith, 1977).

The fluctuation spectra show that narrow spectral lines reflecting a strong periodic modulation are rather frequently met, especially in the case of long-period S_d and D type pulsars, as well as in conal components of T class pulsars. Short-period pulsars are characterized by spectra without details, flat spectra (which reflect random fluctuations) or spectra with a rise in the direction of low frequencies (which reflect a relatively slow aperiodic modulation) (Manchester & Taylor, 1977).

Drift of subpulses The fluctuation spectra discussed in the preceding section show that variations of the intensity I_ν are rather stable in some pulsars. In some cases, such a structure is quasi-periodic and so stable

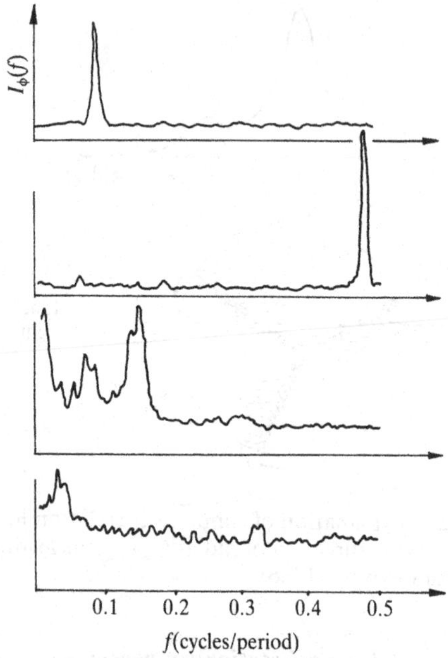

Fig. 1.33. Fluctuation spectra of some pulsars (Smith, 1977).

that it is well pronounced even on individual pulses of radio emission. An example of such activity, called a drift of subpulses, is shown in Fig. 1.34. It is natural to associate the nature of the drift of subpulses with rotation of radiating regions, as is shown in Fig. 1.32. This interpretation is also supported by the displacement of the cross-correlation picture discussed in Section 1.3.2 (see Fig. 1.31).

The regular drift of subpulses is characterized by two periods – the period P_2 (measured in degrees) determining the interval between subpulses and the period P_3 (measured in the units of period P) determining the distance between two drift bands. As can readily be verified, the period P_3 is related to the frequency of the maximum in the fluctuation spectrum f_1 simply as

$$P_3 = \frac{1}{f_1} \tag{1.35}$$

The basic properties of pulsars which have a regular drift are presented in Table 1.9 (Rankin 1986). We can see that a regular drift is observed only in pulsars of class S_d – both drift directions (towards the front and back pulse edges) being registered (Shitov *et al.*, 1980; Wright, 1981;

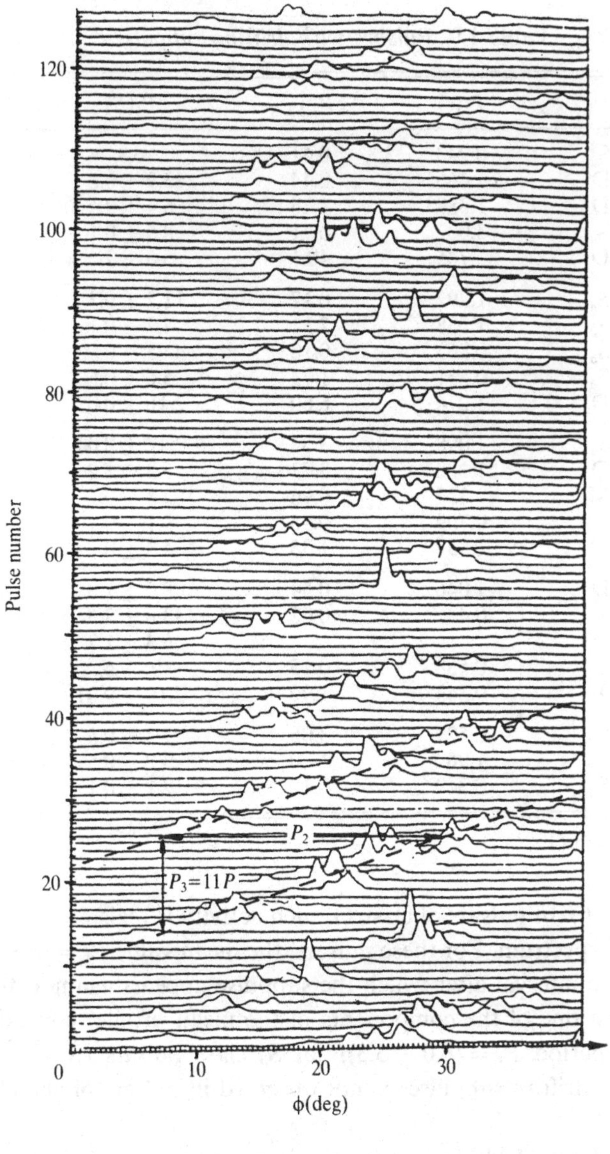

Fig. 1.34. Pulse number vs the interpulse time diagram for 128 successive pulses of pulsar 0809 + 74 with drifting subpulses at 102.5 MHz.

Table 1.9. *Pulsars with drifting subpulses*[a]

Pulsar		P (s)	$\dot{P}_{-15}$	$P_3(P)$	Sense
0031−07	S_d	0.943	0.41	6.9	Neg
0148−06	D	1.465	0.44	13.9 ± 0.8	Neg
0301+19	D	1.388	1.26	6.4 ± 2.	–
0320+39	S_d	3.032	0.71	8.5 ± 0.1	Pos
0525+21	D	3.745	40.1	5.6 ± 2.	–
0809+74	S_d	1.292	0.17	11.2 ± 0.3	Neg
0818−13	S_d	1.238	2.11	4.7 ± 0.1	Neg
0820+02	S_d	0.865	0.10	4.93 ± 0.05	Pos
0823+26	S_d	0.531	1.72	5.6 ± 0.6	–
0834+06	D	1.274	6.80	2.16 ± 0.02	–
0943+10	S_d	1.098	3.53	2.17 ± 0.04	–
1133+16	D	1.188	3.73	5.3 ± 1.2	–
1237+25	M	1.382	0.96	2.8 ± 0.1	–
1540−06	S_d	0.709	0.88	3.1 ± 0.1	Pos
1919+21	M	1.337	1.35	4.3 ± 0.2	–
1942−00	D	1.046	0.54	8.5 ± 1.0	–
1944+17	T	0.441	0.02	13.6 ± 2.5	–
				6.3 ± 1.2	–
2016+21	S_d	0.558	0.15	9. ± 6.	Neg
2020+28	T	0.343	1.90	2.10 ± 0.01	–
2021+51	S_d	0.529	3.05	7.4 ± 0.5	–
2045−16	T	1.962	11.0	3.2 ± 0.5	–
2310+42	S_d	0.349	0.12	2.1 ± 0.1	–

[a] From Rankin (1986).

Rankin, 1986). In pulsars of class D, such activity is registered only in the fluctuation spectrum. For these pulsars, the mean value $P_3 = (6.3 \pm 3.1)P$ (i.e. the time interval after which the subpulse appears again at the same longitude in one of the components of the mean profile) coincides with the mean period $P_3 = (6.0 \pm 3.3)P$ of S_d class pulsars (Rankin, 1986). Finally, the drift of subpulses is not observed in pulsars of class S_t.

Mode switching Mode switching is a sharp variation in the shape of the mean profile of a pulsar (Wright & Fowler, 1981; Bartel *et al.*, 1982; Sulejmanova & Izvekova, 1984; Rankin, 1986). An example of such an event is given in Fig. 1.35. As with nullings, the change in the shape of the profile proceeds simultaneously at all frequencies. The duration of the switching itself is rather difficult to establish since the formation of the mean profile requires much time. There are indications, however, that

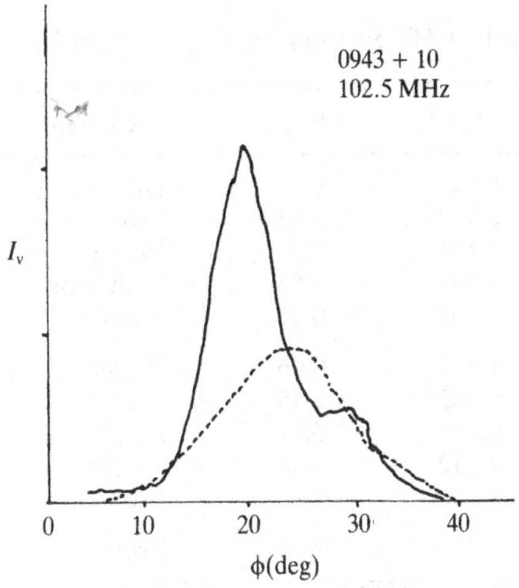

Fig. 1.35. Mode switching in pulsar 0943+10. Mean profiles of radio emission in two different states are shown (Sulejmanova & Izvekova, 1984).

such a switching proceeds literally within a pulsar period (Rankin *et al.*, 1988).

We know 11 pulsars possessing mode switching. i.e. those whose mean profile is unstable (Rankin, 1986). All of them belong to classes S_d, T and M and, therefore, also group together near the death line on the $P\dot{P}$ diagram. The properties of such pulsars are listed in Table 1.10. It should be noted that two modes are customarily distinguished – normal and special – a pulsar being in the normal mode up to 90% of time. It is only in pulsar 0943+10 that the appearance of both these modes is close to equiprobable (Sulejmanova & Izvekova, 1984). Switching pulsars sometimes have several 'special' modes (Deich *et al.*, 1986).

Polarization characteristics, especially values of the position angle, change during switching. Also, the frequency dependence of the profile shape is appreciably stronger in the special mode than in the normal mode (Rankin, 1986; Rankin *et al.*, 1988). If subpulses drift, the drift rates P_2 and P_3 may be different in the two modes. In particular, in pulsar 0943+10 the drift is observed in only one of the two modes (Sulejmanova & Izvekova, 1984). Finally, in pulsar 0950+08 the two modes differ only in the degree of linear polarization of radio emission (Sulejmanova, 1989).

Glitches A number of pulsars exhibit a jump-like decrease of the period P (Boynton *et al.*, 1972; Lohsen, 1975; Downs, 1981; McCulloch *et al.*,

Table 1.10. *Pulsars with mode switching*

Pulsar		P (s)	$\dot{P}_{-15}$	Change	Reference[a]
0031−07	S_d	0.943	0.41	Drift mode	1
0329+54	T	0.714	2.05	Profile	2
0355+54	S_t	0.156	4.39	Profile	3
0943+10	S_d	1.098	3.53	Drift mode	4
0950+08	S_d	0.253	0.23	Polarization	5
1237+25	M	1.382	0.96	Profile	3
1737+13	M	0.803	1.45	Profile	2
1822−09	T	0.769	52.3	Profile	3
1917+00	T	1.272	7.68	Profile	2
1926+18	T	1.220	–	Profile	3
2319+60	T	2.256	7.04	Profile	3

[a] References: (1) Wright & Fowler (1981). (2) Rankin (1986). (3) Bartel *et al.* (1982). (4) Sulejmanova & Izvekova (1984). (5) Sulejmanova (1989).

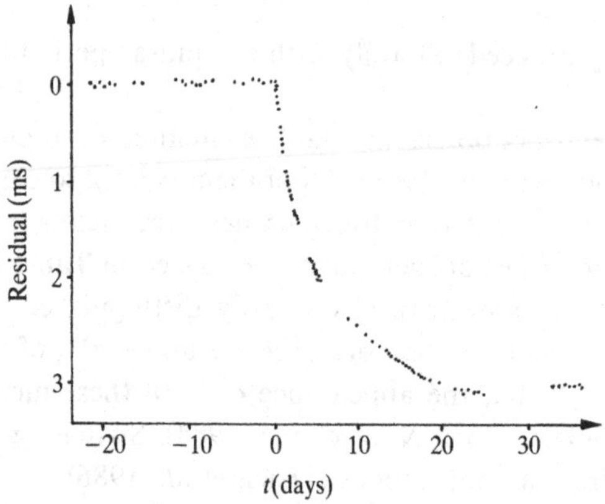

Fig. 1.36. The glitch in pulsar 0531+21 (Crab). From Lyne & Pritchard (1987).

1983; McKenna & Lyne, 1990). This is believed to result from inner reconstructions of neutron stars (Shapiro & Teukolsky, 1983; Alpar *et al.*, 1984a,b). An example of such an event is given in Fig. 1.36. Period discontinuities were accompanied by glitches of the derivative dP/dt, the relaxation process lasting in all the cases from several days to several months. During this relaxation the derivative dP/dt assumed a new stable value. The basic properties of 'glitches' are presented in Table 1.11, and the phenomenon itself is discussed in Chapter 2.

Table 1.11. *Glitch characteristics*

Pulsar	P (s)	$\dot{P}_{-15}$	Year	$\Delta\Omega/\Omega$	$\Delta\dot{\Omega}/\dot{\Omega}$	τ (days)	Q	Reference[a]
0355+54	0.156	4.4	1986	4.4×10^{-6}	0.10 ± 0.04	–	–	1
0525+21	3.745	40.1	1974	10^{-9}	4×10^{-3}	150	–	2
0531+21	0.033	421	1969	$(0.9 \pm 0.4) \times 10^{-8}$	$(1.6 \pm 0.9) \times 10^{-3}$	4.8 ± 2.0	0.92 ± 0.07	3
			1971	$(3.72 \pm 0.08) \times 10^{-8}$	$(2.1 \pm 0.2) \times 10^{-3}$	15.5 ± 1.2	0.96 ± 0.03	4
			1975	3.7×10^{-8}	–	15	0.96	4
			1986	$(0.92 \pm 0.01) \times 10^{-8}$	$(2.5 \pm 0.2) \times 10^{-3}$	5.5 ± 0.5	–	5
0833−45	0.089	124	1969	$(2.33 \pm 0.02) \times 10^{-6}$	$(8.1 \pm 0.2) \times 10^{-3}$	75 ± 20	0.03 ± 0.01	6
			1972	$(2.00 \pm 0.01) \times 10^{-6}$	10^{-2}	60 ± 10	0.03 ± 0.01	6
			1975	1.2×10^{-6}	7.5×10^{-3}	80 ± 20	0.55 ± 0.20	6
			1978	3×10^{-6}	–	55 ± 5	0.02 ± 0.01	6
			1981	10^{-6}	7.2×10^{-3}	233	0.18	7
			1982	2×10^{-6}	1.0×10^{-3}	60	0.04	8
			1985	10^{-6}	6.2×10^{-3}	397	0.17	8
			1988	10^{-6}	8×10^{-3}	–	–	8
1641−45	0.455	20.1	1977	$(1.91 \pm 0.01) \times 10^{-7}$	$(1.6 \pm 0.5) \times 10^{-3}$	31	–	9
1737−30	0.606	466	–	$(4.2 \pm 0.2) \times 10^{-7}$	$(2.8 \pm 0.8) \times 10^{-3}$	–	–	10
			–	$(3.3 \pm 0.5) \times 10^{-8}$	$(1.7 \pm 0.4) \times 10^{-3}$	–	–	10
			–	$(0.7 \pm 0.5) \times 10^{-8}$	$(0.0 \pm 1.2) \times 10^{-2}$	–	–	10
			–	$(3.0 \pm 0.8) \times 10^{-8}$	$(0.0 \pm 0.4) \times 10^{-2}$	–	–	10
			–	$(6.009 \pm 0.006) \times 10^{-7}$	$(2.0 \pm 0.2) \times 10^{-3}$	–	–	10

[a] References: (1) Lyne (1987). (2) Downs (1982). (3) Boynton *et al.* (1972). (4) Lohsen (1975). (5) Lyne & Pritchard (1987). (6) Downs (1981). (7) McCulloch *et al.* (1983). (8) Lyne & Smith (1990). Manchester *et al.* (1978c). (10) McKenna & Lyne (1990).

1.4 Young energetic pulsars. High-frequency radio emission

The age of a neutron star can be estimated with good accuracy in some cases. Such a possibility may, for example, be due to establishment of the age of the supernova remnant with which the radio pulsar is connected. Table 1.12 presents the basic parameters of the five sources for which this connection is beyond doubt. As is known, for pulsar 0531+21 the age is established exactly since the Crab Nebula, in which this pulsar is located, is connected with the historic supernova that exploded in 1054.

We can see that the properties of young pulsars are close to each other. They have small periods P, high deceleration rates dP/dt and high values of radio luminosity. For three sources the dynamical age of the pulsar (1.6) is in agreement with the true age of the supernova remnant. At the same time, for pulsar 1509−58 the age of the nebula and the dynamical age (1.6) are essentially different. Consequently, the dynamical age may generally be a poor indication of the real duration of the active lifetime of a pulsar.

All young pulsars possess another exceedingly important property. In addition to the radiofrequency range, their pulsed emission is also registered in the high-frequency spectrum range. In particular, source 0540−693 has not yet been observed as a radio pulsar (Seward *et al.*, 1984). The emission from this source is received only in the optical and X-ray ranges. However, all the other parameters of this source (period, period derivative, etc.) are close to the parameters of the pulsars tabulated in Table 1.12. The absence of radio emission can easily be explained since the distance to source 0540−693 is 55 kpc (it is located in the Large Magellanic Cloud), which is much more than the mean distance to pulsars. It is interesting that pulsar 1509−58 was also registered for the first time in the X-ray range (Seward & Harnden, 1982), and only after this was it confirmed as a radio pulsar (Manchester *et al.*, 1982).

At the same time, in the high-frequency range young pulsars show substantial differences. Thus, pulsar 0531+21 (Crab) shows pulsed radiation in the optical, X-ray and gamma-ray ranges, up to very high energies of 10^{12}–10^{13} eV (Cocke *et al.*, 1969; Knight, 1982; White *et al.*, 1985; Bhat *et al.*, 1986; Clear *et al.*, 1987). As shown in Fig. 1.37, the high-frequency radiation spectrum is power-law in a wide energy range. Such a spectrum is inherent, for example, in pulsar 0833−45 (Vela), but no pulsed X-ray radiation from this pulsar has been registered in the range 0.1–2 keV (Pravdo *et al.*, 1976; Manchester *et al.*, 1980; Bhat *et al.*, 1987). For other

Table 1.12. *High-energy pulsed radiation from radio pulsars*

Remnant	S_{1000} (Jy)	d (kpc)	Pulsar	P (s)	$\dot{P}_{-15}$	L_r (erg s^{-1})	L_{opt} (erg s^{-1})	L_X (erg s^{-1})	L_γ (erg s^{-1})	L_γ ($>10^{12}$ eV) (erg s^{-1})
Crab	1040	2.0	0531+21	0.033	421	10^{29}	10^{33}	10^{36}	10^{36}	2×10^{34}
Vela	1750	0.5	0833−45	0.089	124	8×10^{28}	10^{28}	–	10^{34}	3×10^{32}
MSH 15-52	60	4.2	1509−58	0.150	1490	3×10^{27}	–	2×10^{34}	–	–
SNR0540-693	–	55.	0540−693	0.050	479	–	10^{34}	2×10^{36}	–	–
CTB 80	120	1.5	1951+32	0.039	5.9	8×10^{29}	–	10^{34}	–	–

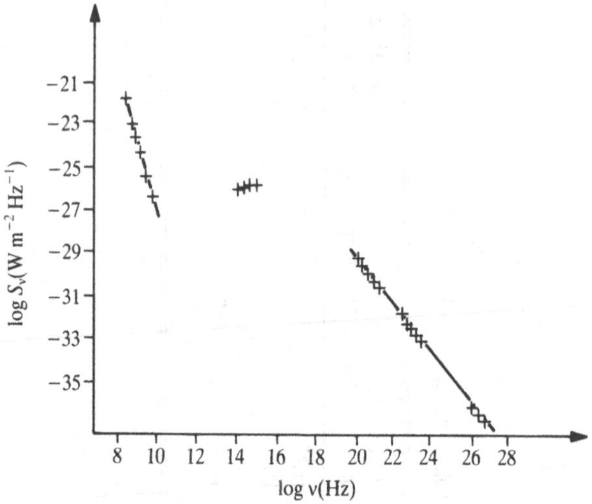

Fig. 1.37. Crab Pulsar radiation spectrum.

sources listed in Table 1.12 no radiation with energies exceeding 4–10 keV has been registered. The radiation powers in all energy ranges are also given in Table 1.12.

We can see that the high-frequency radiation intensity of young pulsars substantially exceeds the radioluminosity of ordinary pulsars and can reach very high values. In particular, gamma-ray luminosity of the pulsar in the Crab Nebula exceeds by thousands of times the total luminosity of the Sun! It should be noted, however, that gamma-ray emission is not highly stable (Clear *et al.*, 1987; Bhat *et al.*, 1986). In contrast, optical and X-ray emissions of the pulsar in the Crab Nebula are exceedingly stable (Cocke *et al.*, 1969; Beskin, G. M., *et al.*, 1983). The mean profiles of high-frequency pulsar emission are shown in Fig. 1.38.

Another specific feature is that, as is seen from Figs. 1.38 and 1.39, the X-ray emission of pulsars 0531+21 (Crab) and 0833−45 (Vela), as for the radio emission, is realized as short bursts, whereas, for instance, the X-ray emission of sources 1509−58 and 0540−693 has the form of weakly modulated signals with a substantial (up to 50%) constant component (Seward & Harnden, 1982; Seward *et al.*, 1984; Kulkarni *et al.*, 1988).

The latter fact is highly important. The point is that neutron stars can also be sources of constant X-ray emission since the bulk thermal emission of their surface (the expected temperature is 10^5–10^6 K) falls just on the X-ray range. Special studies have accordingly been carried out to seek not only a pulsed but also a weakly modulated X-ray emission of neutron stars.

At the present time such an emission is observed from ten radio pulsars (Seward & Wang, 1988). The basic component that might correspond to the thermal radiation of the whole of the neutron star surface was observed only in the pulsars listed in Table 1.12. In the majority of cases it was only the upper boundary of the surface temperature that was determined. For usual radio pulsars it amounts to 10^5–10^6 K (Helfand, 1984; Seward & Wang, 1988).

Observations have been made to seek optical emission of radio pulsars (Manchester *et al.*, 1978a; Lebedev *et al.*, 1983) and emission within the range 10^{-2}–10^{-1} GeV (Graser & Schönfelder, 1983). Such an emission

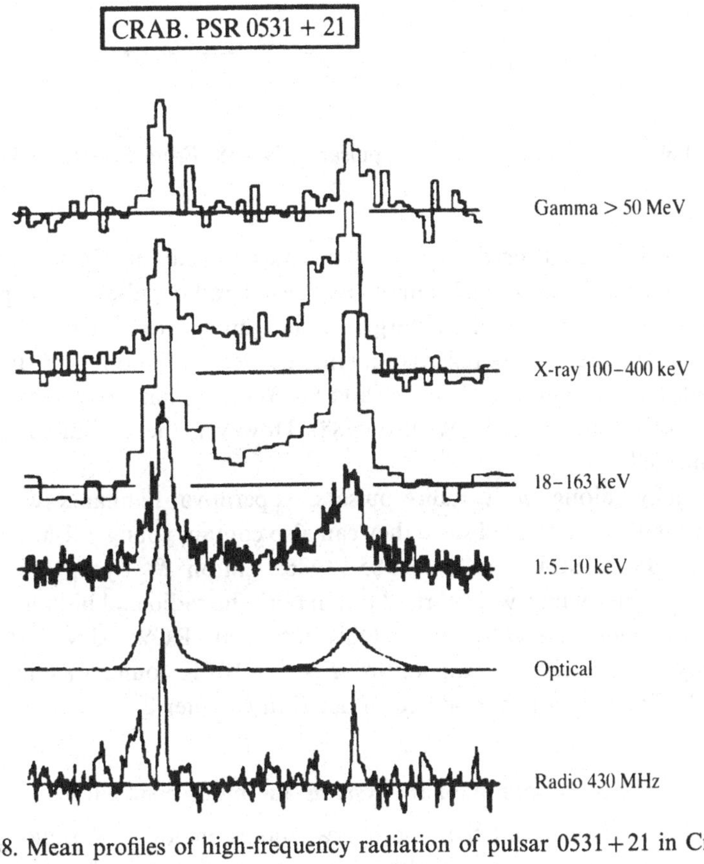

Fig. 1.38. Mean profiles of high-frequency radiation of pulsar 0531+21 in Crab Nebula.

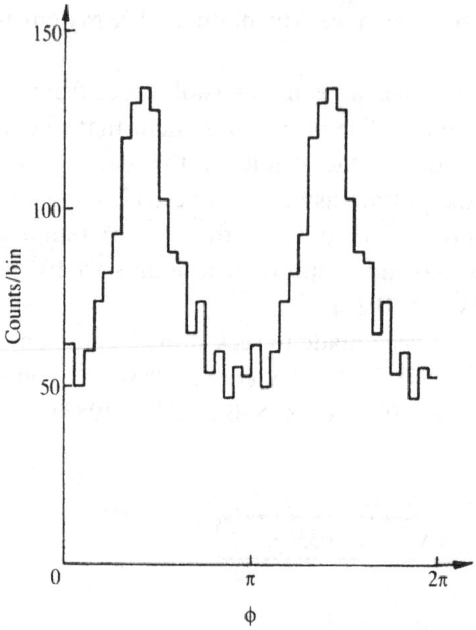

Fig. 1.39. X-ray emission of radio pulsar 1509−58. From Seward & Harnden (1982).

has not been registered, except for the pulsars listed in Table 1.12. This fact suggests that optical emission of usual radio pulsars with periods $P \sim 1$ s does not exceed 10^{27} erg s^{-1}, and gamma-ray emission does not exceed 10^{30} erg s^{-1}. As regards superhigh-energy radiation (10^{12}–10^{13} eV), in addition to pulsars 0531+21 and 0833−45, it has been registered in three other fast pulsars (Weeks, 1988). However, these results are to be confirmed.

Finally, along with radio pulsars, supernova remnants were also thoroughly investigated since they can also contain young pulsars (Dewey *et al.*, 1985; Stokes *et al.*, 1985, 1986; Clifton & Lyne, 1986). The observations, which were carried out in both the radio and high-frequency ranges, yielded a negative result. Only in remnant RCW103 was a thermal X-ray source with a temperature $T \sim 2 \times 10^6$ K found (Tuohy *et al.*, 1983). This result is considered in detail in Chapter 2.

1.5 Pulsars in binary systems and millisecond pulsars

We have already said that the overwhelming majority of radio pulsars are single neutron stars. Only a few pulsars enter into binary systems

(Taylor & Stinebring, 1986; Lyne *et al.*, 1987, 1988a; Fruchter *et al.*, 1988). The basic properties of such sources are presented in Table 1.13.

The binary systems containing pulsars are rather diverse. There exist wide (the orbital period $P_b > 3$ y) and very close ($P_b \simeq 1.4$ h) pairs having circular ($e < 10^{-4}$) and strongly elongated ($e \sim 0.6$) orbits. The masses M_c of companion stars estimated by the value of the mass function

$$f(M_c, M_p, i) = \frac{M_c^3 \sin^3 i}{(M_p + M_c)^2} \tag{1.36}$$

(i is the slope of the plane of the orbit; the value of $f(M_c, M_p, i)$ is found directly from observations (Lang, 1974)) also vary widely.

In some cases, optical radiation from a companion star has been registered. In system 1855+09 the companion is a white dwarf with mass $M_c \sim 0.3M_\odot$ and surface temperature $T \sim 5200$ K (Wright & Loh, 1987), and temperatures of the white dwarfs in systems 0655+64 and 0820+02 are respectively 8×10^3 K and 2×10^4 K (Kulkarni, 1986). On the other hand, no optical radiation from the companion has been discovered in system 1913+16.

Of particular interest is source 1913+16 with a small orbital period $P = 7.75$ h and an exceedingly elongated orbit ($e = 0.62$). The effects of the general theory of relativity appear to be rather substantial in such close binary systems (Taylor & Weisberg, 1982, 1989). For example, the velocity of periastron motion in 1913+16 is equal to 4.2 degrees/year (Taylor & Weisberg, 1982). (Recall that the velocity of perihelion motion of Mercury is only 43 seconds of arc/century). On the other hand, a pulsar in such a system is an exact clock located in a variable gravitational field. A many-year analysis of the fine effects of signal delay in the system of pulsar 1913+16, which are connected with the effects of general relativity, made it possible to estimate accurately the masses of both companions and the inclination angle of the orbit, parameters that cannot be evaluated using only the classical methods of stellar astronomy. The masses of both companions turned out to be equal to approximately 1.4. At the present time, their masses are the most precise estimates of neutron star mass (see Table 2.2).

Furthermore, using the mass values and the orbit parameters thus estimated, we can find the rate with which a binary system loses energy due to radiation of gravitational waves. Such losses, predicted by general relativity, must lead, in particular, to a decrease of the period of orbital motion and to a decrease of the orbit size, so that, e.g. for pulsar 1913+16, the stars will fall onto one another in 200 million years. A decrease of the

Table 1.13. *Millisecond and binary pulsars*

Pulsar	P (ms)	$\dot{P}_{-18}$	$a_1 \sin i$ (light second)	P_b (days)	e	$f(M)$ $(M_\odot)$	$M_c(M_\odot)$	Companion
1937+214	1.558	0.105	–	–	–	–	–	
1957+20	1.607	0.012	0.09	0.38	0.001	0.52×10^{-5}	0.02	WD(5500K)
1821–24	3.054	1.55	–	–	–	–	–	
1855+09	5.362	0.017	9.2	12.3	0.00002	0.0056	0.2–0.4	WD(5200K)
1953+29	6.133	0.03	31.4	117	0.0003	0.0027	0.2–0.4	?
1620–26	11.08	0.82	64	191	0.025	0.008	0.2–0.4	?
1913+16	59.0	8.64	2.3	0.32	0.617	0.13	1.4	NS
0655+64	196	0.677	4.12	1.03	$<10^{-4}$	0.0712	0.7–1.3	WD(8000K)
1820–11	280	?	200.6	357	0.796	0.068	0.6–0.8	NS
1831–00	521	15.0	0.73	1.81	$<10^{-3}$	0.0012	0.06–0.13	?
0820+02	865	105	162.1	1232	0.0119	0.0030	0.2–0.4	WD(10^4K)
2303+46	1066	569	32.7	12.34	0.6584	0.246	1.2–2.5	NS

orbital period has been discovered, the predictions of the theory being in agreement with experiment to an accuracy of about 1% (Backer & Hellings, 1986). This fact is the first experimental confirmation of the existence of gravitational waves.

As is seen from Fig. 1.7, pulsars entering into binary systems make up a separate group of objects. They are characterized by small periods P and small period derivatives dP/dt. In particular this group includes six of the eight known millisecond pulsars. As will be shown in Chapter 2, such properties are due to the evolutional details of pulsars entering into binary systems (Bisnovatyi-Kogan & Komberg, 1974, 1976; Yungelson & Masevich, 1983; van den Heuvel, 1984; Taam & den Heuvel, 1986).

The interest in millisecond pulsars is also explained by the fact that, as shown in Fig. 1.4, the periods of all young pulsars known to us are anomalously small. We could therefore assume that many pulsars must be born with small enough periods. To seek such young pulsars, Jodrell Bank-II and NRAO-II surveys were specially undertaken that could register pulsars with periods down to 1 ms (see Table 1.2). In particular, as has already been mentioned, young supernova remnants were specially investigated. In spite of all these attempts, only a few millisecond pulsars were discovered.

This result is indicative of the fact that the period of a radio pulsar at the moment of its birth is not anomalously small for the majority of pulsars.

1.6 Other sources connected with neutron stars

After radio pulsars, other sources of cosmic radiation were discovered whose activity is connected with the processes proceeding on neutron stars. After the discovery of radio pulsars in 1967 (Hewish *et al.*, 1968), X-ray pulsars were registered in 1971 (Schreier *et al.*, 1972), and sources of X-ray and gamma-ray bursts in 1973–1975 (Klebasadel *et al.*, 1973; Grindlay *et al.*, 1976). Some other peculiar objects (SS 433, Cyg X-3, GEMINGA and others) were discovered as well (Giacconi *et al.*, 1967; Margon *et al.*, 1979; Townes *et al.*, 1983). In what follows, in the analysis of the basic parameters of neutron stars we shall need some properties of their sources and therefore we give here their basic characteristics.

X-ray pulsars These are sources of pulsed X-ray radiation. Nearly 20 such objects are known at the present time (Joss & Rappoport, 1984; Lamb, 1989; Nagase, 1989). Figures 1.40 and 1.41 show the spectrum and the time variation of X-ray radiation. Let us pay attention to the spectral

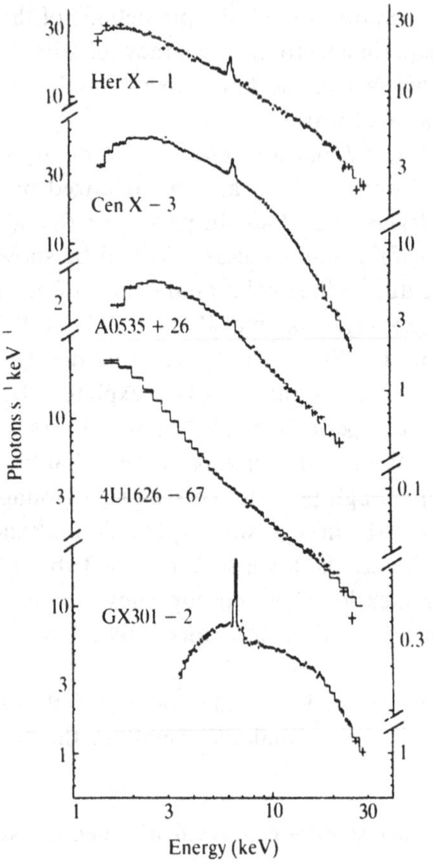

Fig. 1.40. Spectra of X-ray pulsars. From Nagase (1989).

detail near 7 keV observed in the majority of X-ray pulsars and also to the cutoff in the vicinity of 20 keV. Two sources exhibit a spectral singularity near 30–50 keV (Joss & Rappoport, 1984).

The pulse periods range from 69 ms to 835 s, and the characteristic radiation energy is equal to 10^{37}–10^{38} erg s^{-1}. All X-ray pulsars are members of binary systems (the orbital periods range from several hours to several months), their activity being connected with the release of gravitational energy of the matter flowing over from the normal companion star to the compact neutron star. Pulsations of X-ray radiation are caused by non-uniformity in the surface heating of the neutron star and are connected with the strong magnetic field due to which accretion proceeds

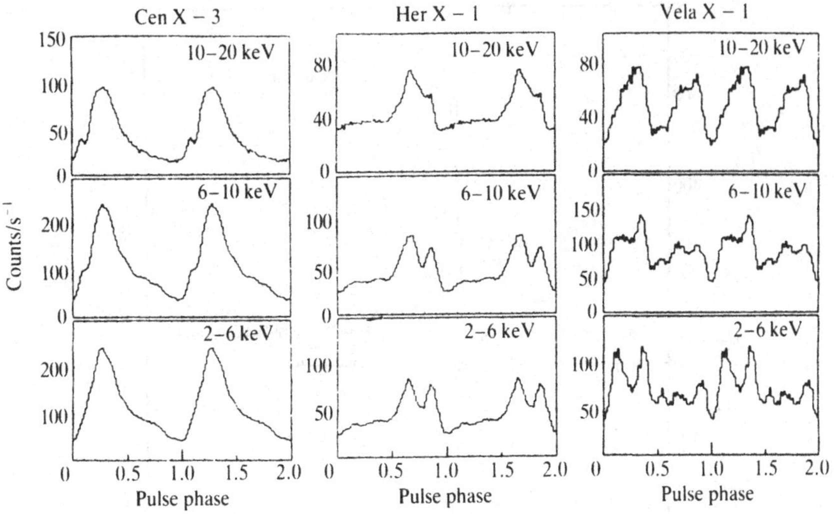

Fig. 1.41. X-ray light-curves at various wavelengths of several X-ray pulsars. From Nagase (1989).

along magnetic field lines towards the region of magnetic poles, where the energy is released (for more details see Lamb (1989)).

The sources of X-ray bursts (*bursters*) are bursting objects with no regular period (Ergma, 1983; Joss & Rappoport, 1984). Their characteristic time profile is shown in Fig. 1.42. The duration of separate bursts is usually several seconds and the time of signal rise is about 1 s. The energy in the maximum reaches 10^{38} erg s^{-1}, and the energy release is 10^{39} erg. The X-ray radiation spectrum (the characteristic energies are equal to 2–20 keV) is close to the thermal one with $T \sim (1\text{–}7) \times 10^7$ K. It should also be noted that in the emission of X-ray bursts there exists a substantial continuous component whose total energy release exceeds that of the pulsar component nearly 100 times. The characteristic time between bursts is 10^4–10^5 s and for each source varies within the range of 20–50%. The power of a burst turns out to be directly proportional to the time that passed after a preceding burst. The thermonuclear model of X-ray bursts is generally accepted; according to this the bursts are caused by combustion of the matter accreting on the surface of a neutron star that possesses a weak magnetic field. The burst-like character of the combustion is connected with matter degeneration, as a result of which the increase of the temperature does not entail pressure increase. Therefore, the energy release from combustion and the corresponding temperature

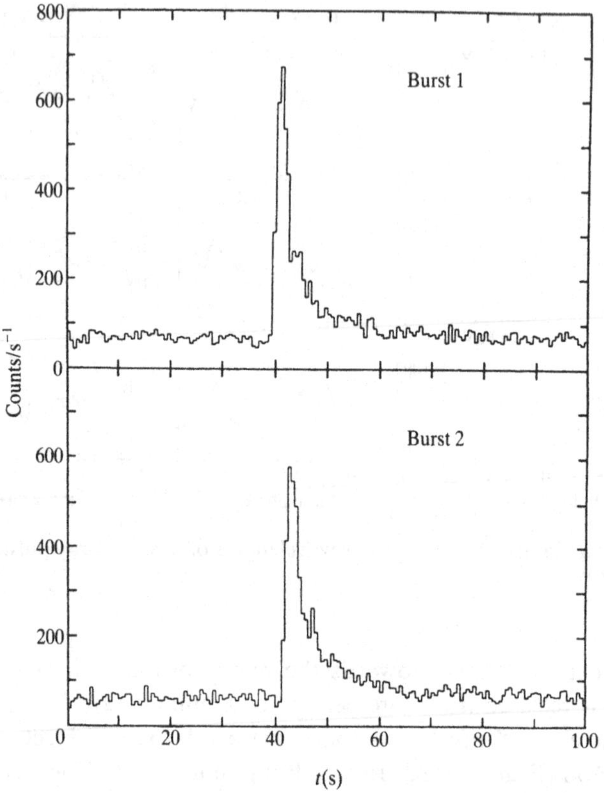

Fig. 1.42. Time histories of X-ray burst 4U 1746–37. From Sztajno *et al.* (1986).

rise do not cause the expansion of the layer in which nuclear reactions go and, accordingly, do not give rise to cooling but, on the contrary, to a further increase in the energy release and temperature. As a result, according to the calculations made by Ergma (1983), on reaching the temperature $T \sim 3 \times 10^8$ K the hydrogen (10^{21} g) accumulated between bursts burns away within several seconds, releasing energy of 10^{18} erg g^{-1}. Owing to the weakness of the magnetic field, the matter is not concentrated near the magnetic poles but falls on the whole of the neutron star surface, so that the radiation is isotropic, as observed. Nearly 50 such sources are known at the present time (for more details see Joss & Rappoport, 1984).

Finally, *gamma-ray bursts* are also sporadic objects (Mazets, 1988). The time dependence and the spectrum of characteristic bursts are presented in Figs. 1.43 and 1.44. We can see that, as in the case of X-ray bursts, the pulse duration is several seconds. It should be also noted that the spectrum shows minima and maxima near 400–450 keV and 10–50 keV.

There is at present no unique point of view concerning the distance of these objects. Therefore the observed fluxes of 10^{-5}–10^{-7} erg cm^{-2} s^{-1} may correspond to energy losses of 10^{37}–10^{38} erg s^{-1} if the mean distance to a source is 100 pc, or 10^{42}–10^{44} erg s^{-1} if the distance is 10 kpc. The question is also quite open of the nature of the activity of gamma-ray bursts. In spite of such uncertainty, the analysis of radiation of gamma-ray bursts (already 400 events of the kind have been registered) allows us, as will be shown in Chapter 2, to draw some conclusions about the properties

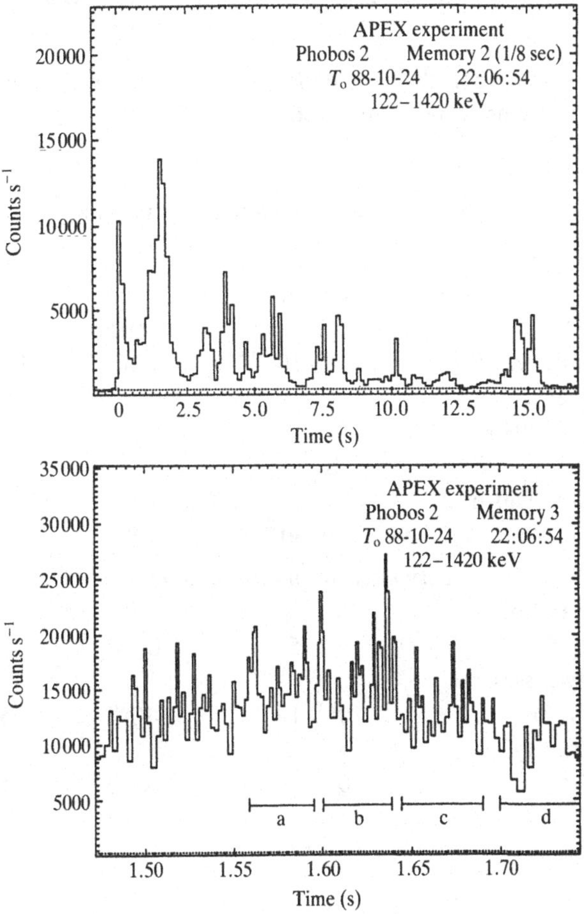

Fig. 1.43. Time history of gamma-ray burst GB881024. From Atteia *et al.* (1991).

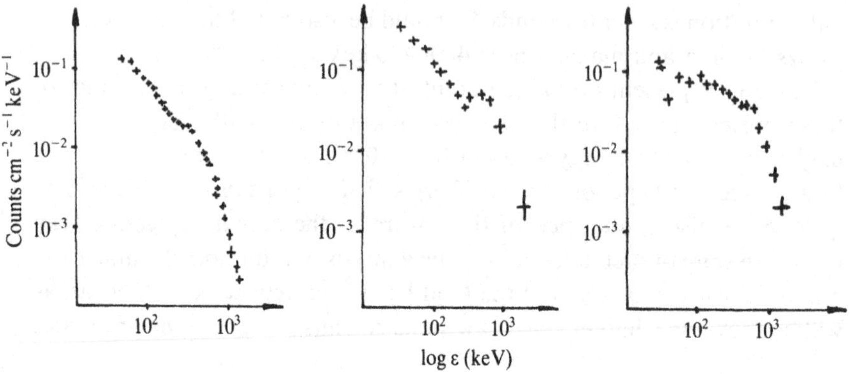

Fig. 1.44. Spectra of several gamma-burst sources. From Mazets (1988).

of neutron stars, since the connection between gamma-ray bursts and neutron stars seems to be established.

The basic characteristics of radio pulsars

Quantity		Min	Max
	Observed		
Period, P (s)	0.3–1.5	0.00156	4.31
Period derivative, dP/dt	10^{-14}–10^{-16}	3×10^{-20}	1.5×10^{-12}
Energy flux at a frequency of 408 MHz, I_ν (mJy)	10–100	1	5000
Radio window width W_r (deg)	6–30°	4°	145–360°
	Estimated from observations		
Distance to radio pulsar, d (kpc)	1–5	0.045	55
Total energy losses, W_{tot} (erg s^{-1})	10^{31}–10^{34}	10^{30}	5×10^{38}
Radio luminosity, L_r (erg s^{-1})	10^{27}–10^{29}	3×10^{25}	3×10^{30}

2
Neutron stars

Since the connection between radio pulsars and neutron stars is reliably established, it seems reasonable to discuss briefly the basic properties of such stars, which will provide insight into the nature of the pulsar activity. We therefore consider in this chapter the formation of neutron stars, their evolution in binary systems, the inner structure, the properties of the surface and other questions. The estimation of the basic characteristics of neutron stars directly from observations is discussed in detail. This range of questions is given a more thorough consideration in monographs by Shapiro & Teukolsky (1983), Imshennik & Nadezhin (1983) and Lipunov (1991).

2.1 The formation and evolution of neutron stars

2.1.1 Origin of neutron stars

Neutron stars are formed under a gravitational collapse of the central regions of normal stars, i.e. under their catastrophic contraction due to the loss of hydrostatic equilibrium. Such processes proceed in the following objects (Shapiro & Teukolsky, 1983; Imshennik & Nadezhin, 1983).

1. Small-mass ($M_{\text{core}} \lesssim 2M_{\odot}$) core of massive ($M \sim 10$–$25M_{\odot}$) stars of the main sequence.
2. White dwarfs with mass exceeding (e.g. due to accretion) the Chandrasekhar limit $M_{\text{Ch}} = 1.44M_{\odot}$.

The hydrostatic equilibrium condition for a non-relativistic star is written in the form

$$\Gamma = \left|\frac{\partial \ln P}{\partial \ln V}\right| > \frac{4}{3} \qquad (2.1)$$

where P is pressure and V is volume. A single star of mass $M \approx (10\text{–}25)M_\odot$ evolves as follows (Imshennik & Nadezhin, 1983). As oxygen and then heavier elements in the central regions of a star burn out, there finally forms an iron core whose mass and temperature are gradually increasing. After about 10^7 years the core mass becomes of the order of $10M_\odot$, the density in the centre reaching $10^9\text{–}10^{10}\,\mathrm{g\,cm^{-3}}$ and the temperature 10^{10} K, which corresponds to the energy of the order of 1 MeV. At this time the electrons responsible for the pressure P become relativistic, affecting the equation of state. Inasmuch as for a relativistic gas $P \propto V^{-4/3}$, we now have $\Gamma = 4/3$, which in itself brings the central region to the boundary of stability. Furthermore, the photodisintegration reaction

$$\gamma + {}^{56}_{26}\mathrm{Fe} \rightleftarrows 13\,{}^{4}_{2}\mathrm{He} + 4\mathrm{n} \rightleftarrows 26\mathrm{p} + 30\mathrm{n}$$

which is the starting point of matter neutronization, becomes energetically advantageous in the central region. The neutronization, which is accompanied by a sharp decrease of the electron density (and, therefore, of the pressure) and is irreversible (since neutrinos leave the star freely) takes on the whole responsibility for the loss of hydrodynamic equilibrium.

As a result, the unstable star core reaches a new stable state characterized by $\rho_{\mathrm{core}} \sim 10^{14}\,\mathrm{g\,cm^{-3}}$, $T \sim 10^{12}$ K and $R \sim 10$ km. This is a neutron star. Its mass is $(1.2\text{–}1.6)M_\odot$. A neutron star formed as a result of a 'quiet' collapse of a white dwarf has analogous parameters (van den Heuvel, 1984). A neutron star may acquire a considerable angular velocity through conservation of angular momentum.

A new stable state can, however, be reached only if the mass of collapsing core is not too large. The main reasons for cessation of gravitational collapse are as follows.

1. Cessation of all mutual transformation processes which require energy.
2. Formation of neutrons which increase elasticity of the medium.
3. Non-transparency of the dense neutron matter to neutrons.

If the mass of the central region of the star exceeds $2M_\odot$, the pressure of the neutron matter cannot compensate the gravitational forces. In this case, the gravitational collapse must result in the formation of a black hole. Figure 2.1 presents the scheme of the main possible ways of neutron star formation. We can see that not every gravitational collapse (nor each explosion of a supernova) is accompanied by neutron star formation.

This last fact is exceedingly important since one of the essential problems of the theory of gravitational collapse is the genetic connection

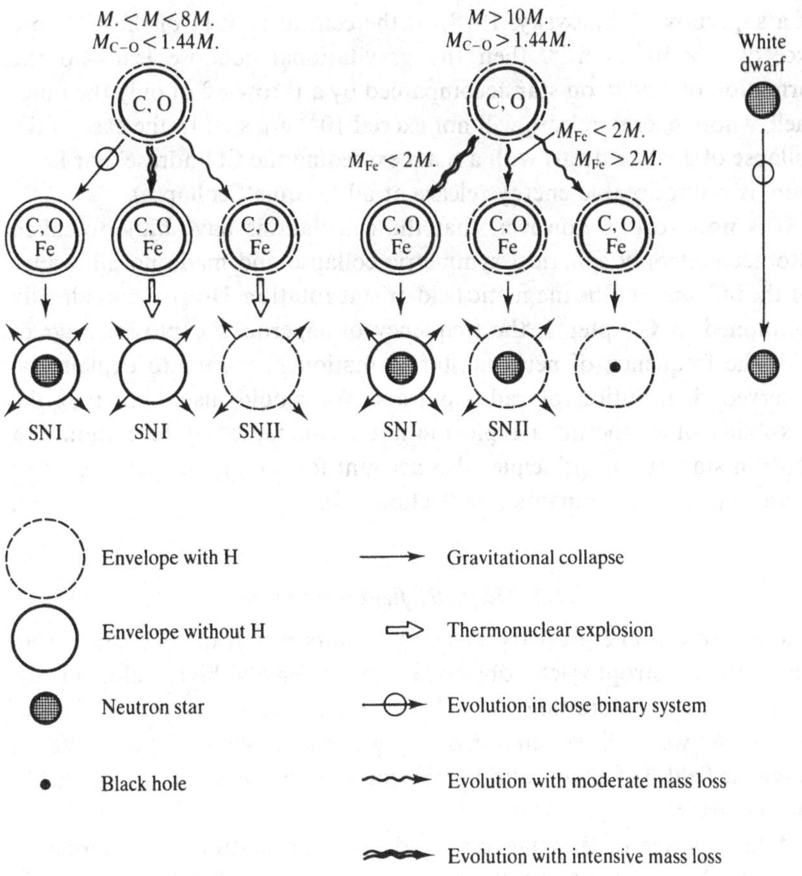

Fig. 2.1. The scheme of possible paths of star evolution and neutron star formation.

between neutron stars and supernova flares. The results of the corresponding analysis based on hydrodynamic calculations for various models of evolution are also shown in Fig. 2.1 (Zeldovich & Novikov, 1971). From the figure we can see that not every gravitational collapse leading to the formation of a neutron star is accompanied by the explosion of a supernova; and conversely, not every supernova is accompanied by the formation of a neutron star. For example, the thermonuclear explosion of a CO-core with core density $\rho_{\text{core}} < 9 \times 10^9$ g cm^{-3} leads to a full expansion of the star matter with the characteristic energy release of up to 10^{51} erg, which fully corresponds to the power released in the explosion of a supernova (Shklovsky, 1978). If the central density of the CO-core exceeds $9 \times 10^9\ \mathrm{g\,cm^{-3}}$, then the gravitational collapse leads to the formation of a neutron star accompanied by a throw-off of only the outer shell, whose energy release will not exceed $10^{49}\ \mathrm{erg\,s^{-1}}$. In the case of the collapse of a white dwarf with a mass exceeding the Chandrasekhar limit, there is no noticeable energy release at all (a 'quiet' collapse).

It is noteworthy, however, that the calculations have hitherto taken into account only spherical symmetric collapse and made no allowance for the influence of the magnetic field or star rotation. However, as already mentioned in Chapter 1, the frequency of supernova explosions agrees with the frequency of neutron star formation necessary to explain the observed distribution of radio pulsars. We would also note that the possibility of a supernova explosion not accompanied by formation of a neutron star can, in principle, also account for the absence of pulsars in young supernova remnants (see Section 1.4).

2.1.2 Magnetic field generation

Another important property of neutron stars which distinguishes them from other astrophysical objects is an extremely high value of the magnetic field, amounting to 10^{11}–10^{13} Gauss for the majority of radio pulsars. As we shall see later, it is owing to the presence of such a strong magnetic field that a fast-rotating neutron star becomes a source of intense radio emission.

A large value of the magnetic field near the neutron star's surface is customarily assumed to be associated with magnetic flux conservation in the course of gravitational collapse (Manchester & Taylor, 1977; Smith, 1977). Indeed, if a neutron star is formed as a result of contraction of a normal star with the strength of the magnetic field at the star surface being $B \sim 10^2$ Gauss and with the mean matter density $\rho_{\mathrm{core}} \sim 1\ \mathrm{g\,cm^{-3}}$, then, since the magnetic field is frozen-in, its strength will increase under contraction as $\bar{\rho}^{2/3}$, so that for the densities $\bar{\rho} \sim 10^{14}\ \mathrm{g\,cm^{-3}}$, the magnetic field can reach a value of about 10^{12} Gauss.

In the literature there is also an indication of the possible field generation mechanism associated with thermomagnetic diffusion (Blandford *et al.*, 1983; Urpin *et al.*, 1986). In this case, a magnetic field is generated by the electron currents induced by the temperature gradient in the outer crust of the neutron star. The specificity of this mechanism is that a substantial magnetic field for the star occurs only some time after its formation. Perhaps this mechanism clarifies the situation with pulsar

1509 – 58 for which the age of the supernova remnant exceeds appreciably the dynamical age of the pulsar itself (see Section 1.4).

The possibility that some mechanisms of magnetic field generation (e.g. those connected with entrainment of protons in superfluid vortices (Sedrakyan & Movsisyan, 1986)) might be realized as well is not excluded. In any case, the existence of strong magnetic fields of 10^{12} Gauss on neutron stars is beyond doubt.

2.1.3 Evolution of binary systems

We now briefly discuss the specifics of the evolution of binary systems containing neutron stars. In spite of some uncertainties, their basic observed properties, discussed in Chapter 1 (see Table 1.13), have been rather convincingly explained.

As is known, most ordinary stars enter into binary systems. The most massive star entering such a system will be the first to stop its evolution on the main sequence, after which it can explode as a supernova and form a neutron star. If such an explosion has not led to a break-up of the binary system (for which purpose at least half of the total system mass should be thrown off), then a pair is formed consisting of a neutron star and a star of the main sequence. Such a pair can also be formed after a 'quiet' collapse of a white dwarf, if its mass exceeds the Chandrasekhar limit. As mentioned in Section 1.6, the existence of such systems is necessary to explain the nature of the activity of many cosmic sources, and there is therefore no reason to doubt their reality. It should be emphasized that, owing to the possibility of an appreciable accretion of the matter flowing over from the companion star to the neutron star, the neutron star is not necessarily a radio pulsar at this stage.

Types of evolution The subsequent evolution of such a system depends on the ratio of masses of the stars entering into the binary system (Bisnovaty-Kogan & Komberg, 1974, 1976; van den Heuvel, 1984). If the mass of the companion exceeds the neutron star's mass and the orbital period is several weeks, then within the time of about 10^3 years after the companion occupied the Roche lobe, the general envelope forms containing the neutron star and the core of the normal massive star. Farther on, within 10^3–10^5 years, the orbital period decreases sharply owing to deceleration of the neutron star, which is moving within the limits of the companion-star envelope. The final orbital period will be several hours. If the mass of the companion does not exceed $\sim 8M_\odot$, the mass of its

CO-core cannot exceed the Chandrasekhar limit, so that after the envelope is scattered a system is formed that contains a radio pulsar and a white dwarf of mass $\sim 1M_{\odot}$ orbiting circularly. Those are just the properties inherent in the system containing pulsar 0655+64.

If the mass of the companion is more than $8M_{\odot}$, then the collapse of its core exceeding the Chandrasekhar limit can result in the formation of a second neutron star. If the binary system remains unbroken, the shape of the orbit will in any case become strongly elongated. Since in a system containing two compact neutron stars there is neither friction due to tidal effects nor ordinary viscous friction, the strongly elongated orbit will retain its shape. In the end, the system will contain two neutron stars of mass $\approx 1M_{\odot}$ with strongly elongated orbits. Such an evolution explains very well the properties of the systems containing pulsars 1913+16 and 2303+46.

Finally, if the mass of the companion is not large ($M_c < 1.2M_{\odot}$, $M_c/M_p < 0.8$) then after the companion has occupied the Roche lobe, the matter starts flowing over to the neutron star. It turns out that in this case, at any instant of time, the companion will fill the Roche lobe, the decrease in its mass being compensated by the increase in the orbit size (van den Heuvel, 1984). The accretion stage lasts 10^6–10^8 years and the accretion rate is equal to $10^{-8}M_{\odot}$/year. As a result, a wide system is formed (with orbital period P_b from several months to several years) which is characterized by small eccentricity and contains a neutron star and a small-mass white dwarf ($M_c \sim 0.2$–$0.4M_{\odot}$). As is seen from Table 1.13, these are precisely the properties inherent in the systems containing radio pulsars 0820+02, 1620−26, 1855+09 and 1953+29.

Millisecond pulsars The high matter flow rate and duration of accretion allow us to explain in the latter case the exceedingly small periods of pulsars 1953+29, 1957+20, 1855+09 and others. The point is that the flow of matter from the companion to the neutron star should be accompanied by angular momentum transfer, which will in turn lead to an increase of the neutron star's rotation rate (Ghosh & Lamb, 1979a,b). The decrease of the rotation period P is actually observed in many X-ray pulsars (Nagase, 1989; Lamb, 1989). Such a decrease may proceed, however, only to a certain minimal period P_{eq} for which the accelerating moment associated with the overflow of matter is balanced by the electromagnetic moment associated with the interaction between the magnetic field of the neutron star and the accretion disc matter. Such an equilibrium period is equal to (Alpar *et al.*, 1982):

$$P_{eq} = 2.4 B_9^{6/7} R_6^{15/7} \left(\frac{M}{M_\odot}\right)^{-5/7} \left(\frac{\dot{M}}{2 \times 10^{-8} M_\odot \text{ y}^{-1}}\right)^{-3/7} \text{ ms} \quad (2.2)$$

where $B_0 = B_9 \times 10^9$ Gauss is the magnetic field on the surface of a magnetic star of radius $10^6 \times R_6$ cm. Many of the observed X-ray sources are possibly just in this stable state.

We can see that for magnetic fields $B_0 \leqslant 10^9$ Gauss the values of the stable period P_{eq} fall in the millisecond range. Since the deceleration velocities observed in millisecond pulsars are anomalously small (see Table 1.13), the values of their magnetic field responsible for deceleration must be just 10^8–10^9 Gauss (see Section 2.1.2), which is much less than the magnetic fields of usual pulsars.

As shown in Fig. 1.7, all radio pulsars entering binary systems do actually have periods P exceeding P_{eq} and, accordingly, they could spin-up owing to matter accretion. This was perhaps just the evolution undergone by pulsars 1821 − 24 and 1937 + 214, which, however, lost their companion stars, for example owing to tidal perturbations or a supernova explosion (van den Heuvel & Taam, 1984). Not everything is clear in this scheme, since there exist alternative models connected, in particular, with the adhesion of stars entering a binary system, which is possible in rather close systems owing to gravitational wave radiation (Taylor & Weisberg, 1982) (see Section 1.5).

Proper velocities In conclusion we say a few words about the nature of high proper velocities of radio pulsars. As has already been mentioned (see Section 1.1), their proper velocities reach 500 km s^{-1}, which exceeds greatly the velocities of ordinary stars. The following mechanisms can lead to such high velocities:

1. A break-up of a binary system owing to an explosion of one of the stars (a supernova) entering this system (Gott, 1970).
2. An asymmetric collapse in the formation of a neutron star (Shklovsky, 1971).

Other acceleration mechanisms have also been proposed (Harrison & Tademaru, 1975; Chugai, 1984).

2.2 Internal structure of a neutron star

2.2.1 The equation of state, the mass and the radius

As mentioned in the preceding section, the modern models of gravitational collapse show that the masses of real neutron stars must range within rather narrow limits from $1.2M_\odot$ to $1.6M_\odot$ (Shklovsky, 1978; Shapiro & Teukolsky, 1983). At the same time, the theory of internal structure gives a much wider mass spectrum in which the neutron star remains stable. In this section we discuss the basic properties of such stationary solutions, and in the section to follow we review the observations allowing us to judge the parameters of real neutron stars.

The majority of models of neutron star structure are based on the solution of the Tallman–Oppenheimer–Volkoff equation (Oppenheimer & Volkoff, 1939):

$$\frac{dP(r)}{dr} = -\frac{G[\rho(r) + P(r)c^{-2}][m(r) + 4\pi r^3 P(r)c^{-2}]}{r^2[1 - 2Gm(r)r^{-1}c^{-2}]}$$
$$\frac{dm(r)}{dr} = 4\pi r^2 \rho(r), \qquad m(0) = 0 \tag{2.3}$$

This is the equation of hydrostatic equilibrium of a spherically symmetric cold star which makes allowance for the effects of the general theory of relativity. The models taking into account the star rotation and the magnetic field have been considered only recently Shapiro *et al.*, 1983; Ray & Datta, 1984; Lindblom, 1986; Friedman *et al.*, 1986).

In equation (2.3) $m(r)$ is the gravitational mass inside a sphere of radius r, $P(r)$ and $\rho(r)$ are the pressure and density corresponding to the radius r. The quantities $P(r)$ and $\rho(r)$ are related by the equation of state, the derivation of which encounters major difficulties. The point is that for densities ρ higher than or of the order of the nuclear density ($\rho_n = 2.4 \times 10^{14}$ g cm^{-3}), there exists a significant uncertainty in the equation of state. It is caused, in particular, by the fact that establishing the properties of matter with $\rho \gtrsim \rho_n$ is a non-trivial task in the many-body problem (Pandharipande *et al.*, 1976; Arnett & Bowers, 1977). Therefore, there exist tens of models ranging from so-called 'soft' equations of state (derived from the models in which for densities of the order of nuclear ones the mean interaction energy corresponds to attraction) to 'stiff' equations (derived for models in which there already exists repulsion for densities below nuclear ones). The basic properties of some of these models are given in Table 2.1 (Shapiro & Teukolsky, 1983; Alcock *et al.*, 1986).

Table 2.1. *Models of the internal structure of neutron stars*

Equation of state	Notation	Interactions	ρ_{core} (g cm^{-3})	Composition	Maximum mass, $M_{max}(M_\odot)$	Moment of inertia, A_r ($M = 1.33M_\odot$)	Red shift, z ($M = 1.33M_\odot$)	$Q_0 = A_{super}/A_r$
Ideal neutron gas	OV	–	$0 < \rho < \infty$	n	0.7			
Reid	R	Soft-core Reid potential	$\rho > 7 \times 10^{14}$	n	1.6	0.9×10^{45}	0.34	0.96
Bethe-Johnson	BJ	Modified Reid potential	$1.7 \times 10^{14} <$ $\rho < 3.2 < 10^{16}$	n, ρ, Λ, Σ^0 Σ^+, Σ^-, Δ^+, Δ^- Δ^0, Δ^{++}	1.9	1.3×10^{45}	0.22	0.77
Tensor interaction	TI	Nucleon attraction due to pion exchange	$\rho > 8.4 \times 10^{14}$	n	2.0	2.1×10^{45}	0.17	0.44
Three-nucleon interaction	TNI	Two- and three-nucleon interactions	$\rho > 1.7 \times 10^{14}$	n	2.0	1.0×10^{45}	0.22	?
Mean field	MF	Attraction due to exchange of scalar particles	$\rho > 4.4 \times 10^{11}$	n	2.7	–	0.18	0.66
'Strange' matter	S	Quark attraction due to gluon exchange	$\rho > 3 \times 10^{14}$	u, d, s	2.0	1.8×10^{45}	0.22	?

Clearly, the stiffer the equation of state, the higher the pressure of a given matter density.

Figures 2.2 and 2.3 show the gravitational mass M as a function of the central density ρ_{core} and radius R of a non-rotating neutron star for different equations of state. We can see that the stiffer equations of state give the higher maximal mass and radius of the star and the lower matter density in the central part. The maximal masses lie within the range

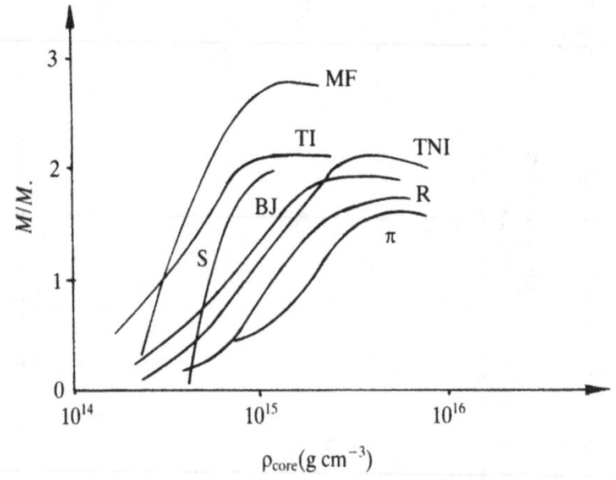

Fig. 2.2. Dependence of the gravitational mass M of a neutron star on the core density ρ. From Baym & Pethick (1979), Alcock *et al.* (1986).

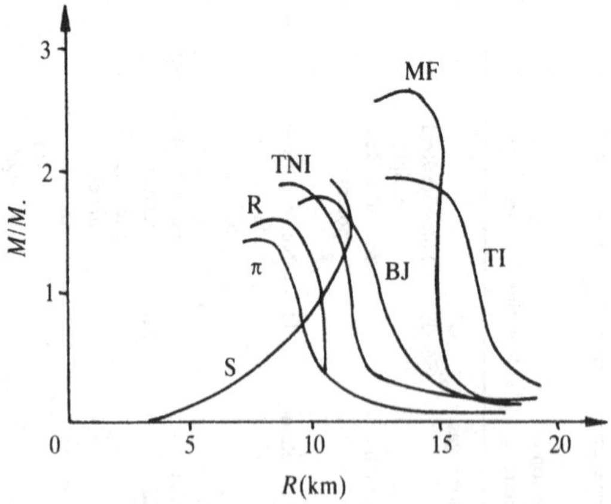

Fig. 2.3. Dependence of the gravitational mass M of a neutron star on the radius R. From Baym & Pethick (1979), Alcock *et al.* (1986).

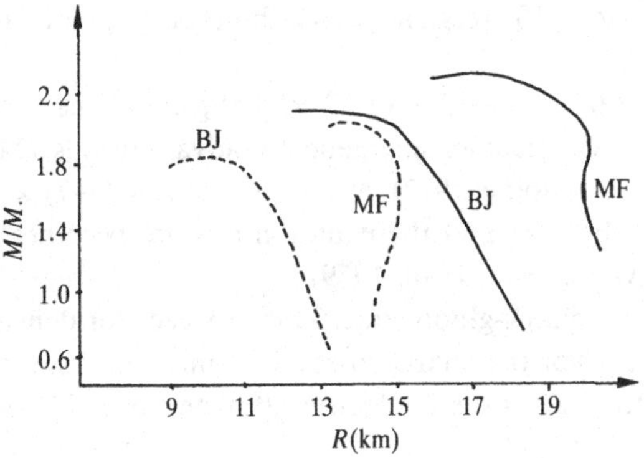

Fig. 2.4. Mass–gravitational radius dependence for an extremely rapid cold neutron star for the stiff (MF) and the softer (BJ) equations of state. The dashed lines show the corresponding dependence for non-rotating stars. From Friedman *et al.* (1986).

$(1.4–2.7)M_\odot$, which corresponds to the radii $R = 7–10$ km. The largest mass of a stable non-rotating cold neutron star, estimated from the stiffest equation of state, is $3.2M_\odot$ (Shapiro & Teukolsky, 1983). Also, in Fig. 2.3 there is a curve corresponding to the so-called strange stars – stars with matter containing equal numbers of u, d and s quarks, whose properties have been intensively discussed in recent years (Witten, 1984; Alcock *et al.*, 1986) (see Table 2.1). We can see that for a mass of $\sim 1M_\odot$ a strange star has the same radius as an ordinary neutron star. Finally, Fig. 2.4 shows the mass as a function of the equatorial radius for a fast-rotating cold neutron star (Friedman *et al.*, 1986). Such stars have larger radii R and lower core densities ρ_{core} than non-rotating stars of the same mass. In particular, the maximum possible mass of a neutron star increases with increasing angular velocity. This is not surprising, since rotation lowers the influence of gravitation. As far as a stable neutron star of minimum possible mass is concerned, it is only $0.1M_\odot$ and has a radius of nearly 150 km.

The basic uncertainties arise in the case of rather massive stars for which, as is seen in Fig. 2.2, the density in the core should exceed substantially the nuclear density. One of the reasons for such uncertainty is that for such densities there may occur various phase transitions, the parameters of these transitions currently being known only within a factor of 2. Such transitions include the following:

1. Crystallization of nuclear liquid. This is the property of stiff equations of state; in the TI model crystallization proceeds for $\rho \sim (7–8) \times$

10^{14} g cm^{-3}, $\mathscr{P} = 10^{38}$ erg cm^{-3} (Kirzhnits & Nepomnyashchii, 1971; Palmer, 1975).
2. Pion condensation. For $\rho \sim (4\text{–}5) \times 10^{14}$ g cm^{-3} the formation of a coherent π-wave becomes energetically advantageous (Migdal, 1972; Sawyer, 1972; Scalapino, 1972). As a result, for $\rho \sim (5\text{–}7) \times 10^{14}$ g cm^{-3} pressure falls by 75% and the equation of state becomes increasingly soft (Grigoryan & Sahakyan, 1979).
3. Transition to a quark-gluon state. This proceeds for densities $\rho \sim 7 \times 10^{14}$ g cm^{-3}, when the characteristic internucleon distances are compared with the size of the nucleon itself (Canuto, 1974, 1975).

We can see that in all cases the transition to a new state proceeds just for densities $\rho > \rho_n$.

2.2.2 *Internal structure*

In spite of the large variety of models, many properties of the structure of a neutron star do not depend on its specific features. In a neutron star, we can distinguish four main regions: the surface, the external crust, the internal crust and the core. The boundary between the external and internal crusts lies in the region of $\rho = 4.3 \times 10^{11}$ g cm^{-3}, when the Fermi energy of neutrons $\mu_F = 25$ MeV, so that the dissociation of nuclei and the formation of free neutrons becomes energetically advantageous. The boundary between the internal crust and the core lies in the region of $\rho_n = 2.4 \times 10^{14}$ g cm^{-3}, when the Fermi energy $\mu_F = 100$ MeV, so that the nuclei dissolve completely in the neutron liquid. Figure 2.5 shows the cross-sections of stars with a mass of $1.4M_\odot$ for the stiff (BJ) and soft (R) equations of state.

The properties of the surface will be considered separately in Section 2.5. We shall only note here that its lower boundary lies in the region of $\rho \leqslant 10^6$ g cm^{-3}, when the influence of the magnetic field and the temperature upon the properties of the matter becomes unimportant.

The external crust (10^6 g cm$^{-3} < \rho < 4.3 \times 10^{11}$ g cm^{-3}) consists of nuclei and electrons, i.e. it is generally ordinary but strongly contracted matter. Stars with a stiffer equation of state have a thicker crust. Owing to a relatively low density, the external crust contains only 10^{-5} of the total star mass. The pressure in the external crust is determined by the electrons, which become relativistic for $\rho > 10^7$ g cm^{-3}. The crystal lattice of the nuclei is in β-equilibrium with the electron gas. For $\rho < 10^7$ g cm^{-3}, the

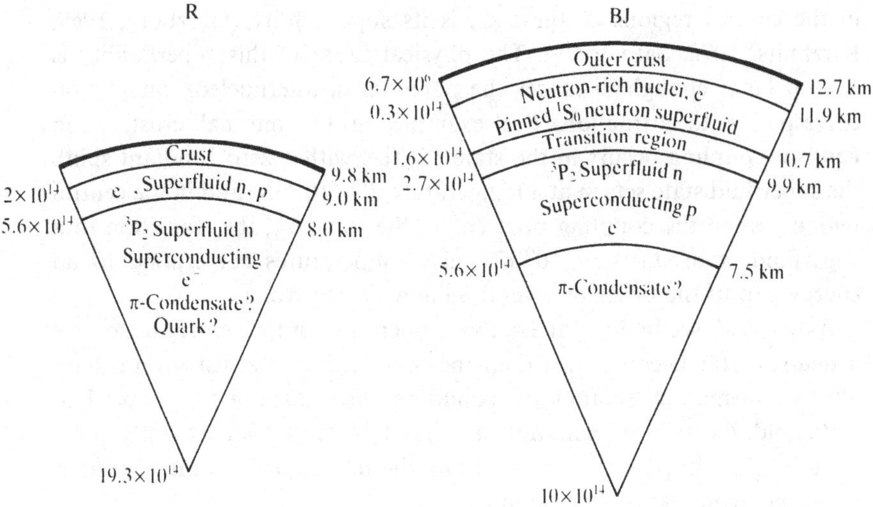

Fig. 2.5. The inner structure of neutron stars with mass $1.4M_\odot$ for the soft (R) and stiff (BJ) equations of state.

ground state is $^{56}_{26}Fe$, while for higher densities nuclei gradually become saturated with neutrons.

The internal crust (4.3×10^{11} g cm$^{-3} < \rho < 2.4 \times 10^{14}$ g cm^{-3}) contains degenerate relativistic electrons and non-relativistic nuclei oversaturated with neutrons, as well as a degenerate non-relativistic neutron liquid as though poured into the crystal lattice. As the density increases, the nuclei become increasingly saturated with neutrons. For densities $\rho < 4 \times 10^{12}$ g cm^{-3} the pressure is determined by the electrons, whereas for $\rho > 4 \times 10^{12}$ g cm^{-3} they are determined by the neutron liquid.

The core ($\rho < 2.4 \times 10^{14}$ g cm^{-3}) does not contain the ion lattice but consists of a three-component liquid (neutrons, protons and electrons) which is in equilibrium under mutual β-transformations. The corresponding number of electrons and protons makes up only about 1% of the number of neutrons. It is only in the star core that the above-mentioned phase transitions are possible. The pressure here is completely determined by neutrons.

Superfluidity and superconductivity A particularly important property of the nucleon liquid (neutrons in the internal crust, neutrons and protons in the central regions of the star) is its superfluidity (Ginzburg, 1969; Kirzhnits, 1972; Sato, 1979). The physical cause of this superfluidity is that at large enough distances the potential of internucleon interaction corresponds to attraction. For example, in the internal crust, when neutron coupling occurs in the state $^1S_0$ (i.e. with a zero resultant spin), the superfluid state sets in at a temperature $T \approx 10^{10}$ K, and in the central regions, when the coupling proceeds in the state $^3P_1$, the transition to a superfluid state starts at 10^9 K. Such temperatures correspond to an energy gap of the order of several million electronvolts.

As we shall see in Section 2.4, the temperature of the central regions of a neutron star becomes less than the indicated values just several days after its formation, so that the conditions of transition to a superfluid state hold. Electrons remain non-superfluid. Note that superfluidity of the proton liquid implies that, in addition, the matter in the central regions of the neutron star is superconducting.

The superproperties lead to some interesting physical effects which become possible in the matter of a neutron star. We shall indicate some of them.

1. In the internal regions of a rotating neutron star there arises a rather dense ($n = 2\Omega m_n/\pi\hbar \sim 10^2$–$10^6$ cm^{-2}) lattice of superfluid filaments, similar to that existing in a 'normal' rotating superfluid liquid. Such filaments will exist in the internal crust where, as has already been said, in addition to the neutron liquid there exists the crystal lattice of atoms with nuclei oversaturated with neutrons (Alpar *et al.*, 1984a,b).
2. A proton liquid may be a type-2 superconductor. In this case, the magnetic field will not be pushed out of the volume filled with superconducting protons, but will, as normal neutrons, be concentrated in vortex filaments (Sedrakyan & Movsisyan, 1986).
3. The heat and electric conduction of the neutron star crust is negligibly small as compared with those of its central regions.

Period discontinuities (glitches) The superfluid properties of the neutron matter can account for the basic characteristics of sharp period jumps observed in some pulsars (see Section 1.3.3 and Table 1.11). Indeed, it is only superfluidity that leads to large relaxation times reaching several days and even months, which is as observed in reality (see Table 1.11).

In the simplest two-component model – a superfluid core with moment of inertia A_S and a solid crust with moment of inertia A_N (Baym & Pethick, 1979) – the period glitch is due to a sharp alteration in the moment of inertia of the crust, induced by its sudden cracking. After the period glitch the angular velocity can be written in the form

$$\Omega(t) = \Omega(t_0) + \Delta\Omega\left\{1 - Q_0\left[1 - \exp\left(-\frac{t - t_0}{\tau}\right)\right]\right\} \tag{2.4}$$

where $Q_0 = A_S/(A_S + A_N)$ and τ is the relaxation time (for more details see Shapiro & Teukolsky, 1983). Table 1.11 presents the values of Q_0 and τ obtained from analysis of observations.

We should note that such a simple model cannot explain all the period glitches observed since, as is seen from Tables 1.11 and 2.1, the theoretical values of Q_0 do not always agree with observations. In addition, the model does not answer the question of the origin of the observed microglitches, which may either increase or decrease the star rotation period (see Section 1.2). There now exists a rather well-developed model in which the interaction between the superfluid core and the normal crust is associated with the interaction between the superfluid filaments and the crystal lattice in the internal crust of a neutron star (Alpar *et al.*, 1984a,b, 1985; Cheng *et al.*, 1988). The model associates the sharp period jumps with the separation of the superfluid filaments from the inner crust nuclei of pulsars.

The effects of the general theory of relativity Concluding this section, we point out the role of the effects of the general relativity in building the models of the neutron star structure. It can readily be verified that the radius of a neutron star is only several times larger than the gravitational radius of a black hole, which for a star of mass $1M_\odot$ equals 3 km. Therefore, the ratio of the gravitational radius to the star radius, $2GM/c^2R$, which enters, in particular, the denominator of eq. (2.3) and characterizes the role of the effects of general relativity, amounts to 20–40% for compact neutron stars. The gravitational field energy correction to the rest mass of a star (we have up to now spoken of the gravitational mass only) must be of the same order. In other words, any calculation of the internal properties of a neutron star accurate to more than 10–20% must make allowance for the general relativistic effects.

Among the other effects of general relativity, we distinguish the following:

1. As is seen from eq. (2.4), relativistic corrections enter both the effective mass and the density, so that the pressure gradient dP/dr will be greater than in the non-relativistic case.

2. All electromagnetic waves emitted from the surface of a neutron star will undergo the gravitational red shift

$$z_G = \frac{\Delta\lambda_f}{\lambda_f} = \frac{GM}{c^2R} \tag{2.5}$$

3. A decrease of the effective temperature of thermal radiation is connected with the gravitational red shift.

2.3 Observed parameters of neutron stars

The data from observations of radio pulsars and other cosmic sources, a brief review of which was made in Section 1.6, allow us to determine a number of neutron star parameters with certainty. In some other cases such 'measurements' are connected with the consistency of a particular model which requires, as its basic element, a compact star of mass $M \sim M_\odot$ and radius $R \sim 10$ km. Such estimates will be called 'astrophysical'.

We consider the basic parameters of neutron stars that can be established from observations..

The mass M

1. The presently most accurate determination of the mass of a neutron star is connected with analysis of the general relativistic effects in the binary system containing radio pulsar 1913+16 (see Section 1.5) (Taylor & Weisberg, 1982, 1989; Backer & Hellings, 1986). Here, $M_{1913+16} = (1.422 \pm 0.003)M_\odot$ (Taylor & Weisberg, 1989). The companion star, which is obviously also a neutron star, has approximately the same mass.

2. Analysis of the radiation of X-ray pulsars and the radial velocities of the companion stars also allows us in some cases to determine the mass of a compact star (Joss & Rappaport, 1984). The results are presented in Table 2.2. We can see that in all the above-mentioned cases the masses of neutron stars agree with the estimate $(1.4 \pm 0.2)M_\odot$ that follows from the theory of gravitational collapse (see Section 2.1).

3. The mass of a compact object can also be estimated by comparing the luminosity of X-ray sources with the so-called Edington luminosity (Lang, 1974),

$$L_{\text{ed}} = 10^{38}\left(\frac{M}{M_\odot}\right) \quad \text{erg s}^{-1} \tag{2.6}$$

which depends on the star mass. The quantity L_{ed} determines the maximal

Table 2.2. *Masses of neutron stars*

Source	Notation	$M/M_\odot$	Reference[a]
Radiopulsar	PSR 1913+16	1.442 ± 0.003	1
Companion		1.386 ± 0.003	1
X-pulsar	Her X-1	0.98 ± 0.12	2
X-pulsar	Vela X-1	1.77 ± 0.21	2
X-pulsar	LMC X-4	1.38 ± 0.5	2
X-pulsar	Cen X-3	$1.06^{+0.56}_{-0.33}$	2
X-pulsar	SMCX-1	$1.06^{+0.33}_{-0.31}$	2
X-pulsar	4U1538-52	$1.87^{+1.33}_{-0.87}$	2

[a] References: (1) Taylor & Weisberg (1989). (2) Nagase (1989).

luminosity for which the radiation pressure does not yet hinder matter accretion. The absence of compact X-ray sources with a radiation power substantially exceeding 10^{38} erg s^{-1} testifies to the fact that their mass does not exceed several solar masses. It should be emphasized, however, that the real calculational formulas should be more complicated, since they should necessarily take into account the effects of general relativity, the uncertainty of the chemical composition (on which the value of L_{ed} depends), the uncertainty of the distance to the source, and so on. Therefore, such an estimate is 'astrophysical'.

The radius R

The determination of the radius of a neutron star is based on the following measurements and astrophysical estimates:

1. Calculation of the spectra of X-ray bursts by the black-body radiation formula

$$L_X = 4\pi\sigma T^4 R^2 \tag{2.7}$$

In all cases the radius of the emitting region proves to be close to the expected neutron star radius and is on the average (8.5 ± 1.5) km (van Paradijs, 1978).

2. In some gamma-ray bursts, the time of signal rise and fall is only 10^{-3}–10^{-4} s. This shows that in the order of magnitude the size of the emitting region is comparable with the size of the neutron star.

3. The upper limit of the neutron star's radius is given by the analysis of stability of fast-rotating radio pulsars (Shapiro *et al.*, 1983; Lindblom, 1986; Friedman *et al.*, 1986). First, the velocity of the star surface cannot

Table 2.3. *Maximum mass and rotation for some equations of state*[a]

Equation of state	P_{min} (ms)	$M_{max}(M_{\odot})$	R_{min} (km)	$\rho_{core}(10^{14}\ \mathrm{g\ cm^{-3}})$
π	0.41	1.74	9.18	44.7
R	0.49	1.94	10.8	32.9
TNI	0.51	2.30	12	25
BJ	0.60	1.94	12.7	27.8
MF	0.83	3.18	17.3	11.1

[a] From Friedman *et al.* (1989).

exceed the velocity of light. This immediately yields the rough estimate

$$R < \frac{cP}{2\pi} \tag{2.8}$$

for example, for the 1.5 ms pulsar 1937+214 we have $R < 50$ km. The results of a more accurate analysis of the stability of real neutron stars are given in Table 2.3 (Friedman *et al.*, 1989) (see also Fig. 2.4). We can see that for millisecond pulsars the limiting radius is 10–20 km.

The magnetic field B_0

The singularities in the spectra of X-ray pulsars and gamma-ray bursts, lying in the range of 10 to 100 keV, can be interpreted as cyclotron lines with energy

$$\varepsilon = \frac{\hbar e B_0}{m_e c} \tag{2.9}$$

In this case, the characteristic magnetic fields of such sources are, in order of magnitude, just 10^{12} Gauss (Trümper *et al.*, 1978; Wheaton *et al.*, 1979; Murakami *et al.*, 1988). As regards radio pulsars, it is only for 0531+21 (Crab) that a singularity was registered in the spectrum for an energy of 77 keV, which corresponds to the magnetic field $B_0 = 7 \times 10^{12}$ Gauss (Strickman *et al.*, 1982). Table 2.4 presents the most reliable estimates of the magnetic fields on the neutron star surface.

The surface temperature T

As mentioned in Section 1.4, continuous X-ray radiation with spectral characteristics close to thermal radiation was registered for several sources. The results of these observations are shown in Table 2.5. In the

Table 2.4. *Magnetic fields of neutron stars*

Source	E_X (keV)	B_{12}	Reference[a]
X-pulsar Her X-1	42	2.3	1
X-pulsar 4U0115+13	20	1.2	2
Pulsar PSR0531+21	77	7	3
Gamma-burst GB880205	20	1.2	4

[a] References: (1) Trümper *et al.* (1978). (2) Wheaton *et al.* (1979). (3) Strickman *et al.* (1982). (4) Murakami *et al.* (1988).

Table 2.5. *Temperatures of neutron star surfaces*

Source	L_X (erg s^{-1})	T (K)	Reference[a]
PSR 0531+21	$<10^{36}$	$<10^6$	1
PSR 0833−45	10^{33}	$(3–9) \times 10^5$	2
RCW 103	6×10^{33}	$(7 \pm 1) \times 10^5$	3
PSR 0540−693	$<2 \times 10^{36}$	$<10^6$	4
PSR 0656+14	10^{32}	$(3–6) \times 10^5$	5
PSR 1055−52	3×10^{32}	$(5–8) \times 10^5$	4
PSR 1509−58	$<2 \times 10^{34}$	$<10^6$	4
PSR 1951+32	6×10^{33}	$(1–2) \times 10^6$	4

[a] References: (1) Helfand *et al.* (1980). (2) Richardson *et al.* (1982). (3) Tuohy *et al.* (1983). (4) Seward & Wang (1988). (5) Cordova *et al.* (1989).

majority of other cases we know only the upper limit of a non-pulsed X-ray radiation, which suggests, however, that within 10^3–10^4 years after the formation of single neutron stars their surface temperature does not exceed 10^5–10^6 K (for more details see Section 2.4). In addition, the table includes pulsars in which pulsed (weakly modulated) X-ray radiation was registered. The nature of this thermal flux is obviously connected with thermal radiation of polar caps heated by energetic particles which are bombarding the surface of a neutron star (Greenstein & Hartke, 1983). The high ($\sim 10^8$ K) surface temperature of accreting neutron stars is also determined by external energy sources and is not connected with the internal structure of the star (see Section 1.6).

Combinations of the parameters R, M, B_0

In many cases observations make it possible to determine not individual parameters of a neutron star but some of their combinations. For example, the estimation of the magnetic field on the basis of the analysis of the deceleration velocity of radio pulsars (see Chapters 3 and 4) implies knowledge of the moment of inertia and the radius of the star. Putting $R = 10$ km and $A_r = 10^{45}$ g cm^2 yields the magnetic fields of radio pulsars ranging within the limits of 10^{11} to 10^{13} Gauss, which agrees with other independent estimates (see Table 2.4). The magnetic field of $B_0 \sim 10^{12}$ Gauss is also obtained from the analysis of the period variation rate for X-ray pulsars (Joss & Rappoport, 1984; Nagase, 1989).

In such estimations, one most often manages to combine the mass and the radius of a neutron star. Of such estimations we shall mention the following.

1. The mean density of a star with rotation period P cannot be less than a certain value $\rho_{\min}$ determined from the star stability condition

$$\rho_{\min} = \frac{\alpha_s}{GP^2} \tag{2.10}$$

where α_s ranges from 30 to 100 depending on the internal structure (Shapiro *et al.*, 1983; Friedman *et al.*, 1986). For millisecond pulsars we have $\rho_{\min} \simeq 10^{14}$ g cm^{-3}. Consequently, their mean density must inevitably be close to the nuclear density.

2. The effects of general relativity must lead to the gravitational red shift of all the lines emitted from the neutron star's surface (see eq. (2.5)). If the emission lines of gamma-ray bursts which lie in the range 400–500 keV are interpreted as the 511 keV line of two-photon annihilation of electron–positron pairs displaced by the red shift, then, according to Liang (1986), such a red shift is on average 0.25 ± 0.10. All the models of internal structure of neutron stars yield precisely this value (see Table 2.1).

3. Assuming the activity of Crab Nebula ($W_{\text{tot}} \sim 5 \times 10^{38}$ erg s^{-1}) and the high-frequency luminosity of pulsar 0531+21 ($W_\gamma \sim 10^{38}$ erg s^{-1}) to be fully determined by the energy lost by the pulsar during deceleration, we obtain the following estimate of the moment of inertia of the neutron star:

$$0.2 \times 10^{45} < A_r < 2.2 \times 10^{45} \quad \text{g cm}^2$$

As is seen from Table 2.1, this value is in close agreement with theory.

4. An astrophysical estimate of the mass and radius of a neutron star can be obtained from the analysis of X-ray burst luminosity. Indeed, if the bursts themselves are due to thermonuclear combustion with an energy release of 10^{18} erg g^{-1}, and the luminosity of the constant component is associated with the gravitational energy of accreting matter

(i.e. $E \sim GM\dot{M}/R$), then the energy release will reach $10^{20}\,\mathrm{erg\,g^{-1}}$ for standard parameters of the neutron star. The gravitational energy source proves here to be 100 times more efficient than the thermonuclear one! As mentioned in Section 1.6, such a situation is precisely what is observed.

In recent years some attempts have been made to estimate more reliably the parameters of neutron stars (Fujimoto & Taam, 1986; Lapidus *et al.*, 1986; Sztajno *et al.*, 1986; Fujimoto *et al.*, 1987). This is mainly connected with the progress in the theory of radiation of some sources, in particular in the theory of X-ray bursts, which made it possible to specify relations (2.6) and (2.7) (London *et al.*, 1984; Kaminker *et al.*, 1989). For example, according to Fujimoto *et al.* (1987), for the source MSB 1636−536,

$$M = (1.45 \pm 0.19)M_{\odot}, \qquad R = 10.2 \pm 1.1\ \mathrm{km}$$

The internal structure parameters

The analysis of observations suggests quite definite conclusions about the internal structure of neutron stars, i.e. about the equation of state of neutron matter.

1. The very existence of compact stars of mass $M \sim M_{\odot}$ contradicts the softest equations of state, for which $M_{max} < M_{\odot}$. In particular, the OV model, i.e. the model of free neutrons, turns out to be unrealistic (see Table 2.1). If it is true that the mass of the X-ray pulsar Vela X-1 exceeds $1.6M_{\odot}$ (Pandharipande *et al.*, 1976), then the R model also contradicts the observations.

2. If neutron liquid were not superfluid, the star would be additionally heated at the expense of internal friction. As a result, the surface temperature would reach 10^7 K which, as has already been said, contradicts the observational data.

3. Important information on the internal structure of a neutron star is yielded by the analysis of sudden jumps (glitches) of the radio pulsar period. So, if we accept the simplest two-component model (see formula (2.4)), we can obtain the following estimate for the time interval between two glitches of the period (Shapiro & Teukolsky, 1983),

$$\Delta t = \frac{2\tilde{A}^2}{\tilde{B}A_r}\frac{\Delta\Omega}{\Omega^2\dot{\Omega}} \tag{2.11}$$

where $\tilde{A}$ characterizes the gravitational energy associated with star deformation, and $\tilde{B}$ is the elastic stress energy. The quantities $\tilde{A}$ and $\tilde{B}$ can be found for each concrete equation of state. It turns out that for pulsar 0531+21 (Crab), formula (2.11) gives $\Delta t = 5$ y for the TI model and $\Delta t \sim 100$ y for the softer R model. Thus, in the framework of the simplest two-component model, stiff equations are preferable. But this model can explain the time between period glitches in pulsar 0833−45 (Vela) for none of the equations of state.

If we accept the period glitch model caused by the interaction between superfluid filaments and the crystal lattice of the internal crust, comparison of theory with observations allows us to find the temperature of the internal regions of a neutron star. According to Alpar *et al.* (1985), this temperature is $(1.5 \pm 1.0) \times 10^7$ K for pulsar 0833−45 and $(4 \pm 2) \times 10^8$ K for pulsar 0531+21. From this we may conclude that their surface temperatures are 1.3×10^5 K and 1.6×10^6 K. As we shall see in the section to follow, such temperatures agree with theoretical predictions concerning neutron star cooling. They also agree with the observed surface temperatures of these pulsars (see Table 2.5).

4. The internal structure can be judged by the character of the frequency spectrum of period fluctuations in X-ray and radio pulsars. Such fluctuations can be caused by variations of the external moment which affect the neutron star's crust (for details, see Cordes & Downs, 1985; Cordes *et al.*, 1988).

2.4 Neutron star cooling

The theory of gravitational collapse discussed in Section 2.1 shows that the temperature of a neutron star at the moment of its formation is extremely high – it reaches 10^{10}–10^{11} K. Such a high initial temperature leads to substantial energy losses and, therefore, to a fast star cooling. We can distinguish here three main stages: neutrino cooling, photon cooling and the isothermal stage (Tsuruta, 1979; Glen & Sutherland, 1980; van Riper & Lamb, 1981; Richardson *et al.*, 1982; Nomoto & Tsuruta, 1986; van Riper, 1988).

Figure 2.6 presents the calculated values of temperature and photon luminosity as functions of time for two models of the internal structure of a neutron star, in one of which the presence of a π-condensate is assumed. The calculations are based on the solution of energy transfer equations which take into account general relativistic effects (for details, see Richardson *et al.*, 1982; van Riper, 1988).

The stage of neutrino cooling is the most efficient, since it is due to volume and not surface losses. The main cooling processes here are the following (Friman & Maxwell, 1979; Maxwell, 1979).

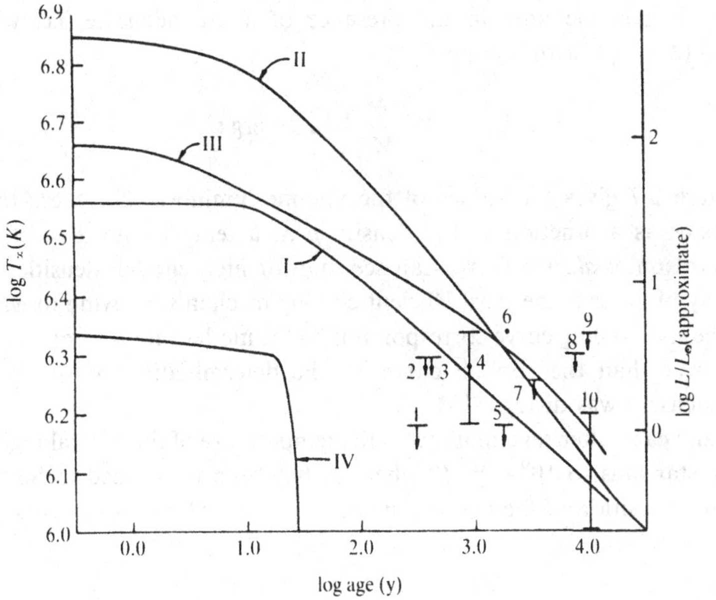

Fig. 2.6. The evolution of the temperature at infinity T_∞ and the photon luminosity L of a cooling neutron star (I = BJ, $0.253M_\odot$; II = BJ, $0.822M_\odot$; III = BJ, $1.54M_\odot$; IV = π, $0.503M_\odot$). Observed temperatures T for Vela pulsar (10) as well as the upper limits for some historic supernovae are also shown. The objects are (1) Cas A, (2) Kepler, (3) Tycho, (4) Crab, (5) RCW86, (6) RCW103, (7) W28, (8) G350.0−1.3, (9) G22.7−0.2. From Richardson *et al.* (1982).

1. The modified URCA processes,

$$\mathrm{n} + \mathrm{n} \rightarrow \mathrm{n} + \mathrm{p} + \mathrm{e}^- + \bar{\nu}_e$$

with total luminosity

$$L_\nu^{\mathrm{mod\,URCA}} = 6 \times 10^{39} \frac{M}{M_\odot} \left(\frac{\rho}{\rho_n}\right)^{1/3} T_9^8 \quad \mathrm{erg\,s^{-1}} \tag{2.12}$$

2. Bremsstrahlung neutrino radiation of electrons on nucleons,

$$\mathrm{e}^- + \mathrm{Z} \rightarrow \mathrm{e}^- + \mathrm{Z} + \bar{\nu}_e + \nu_e$$

with luminosity

$$L_\nu^{\mathrm{brem}} = 5 \times 10^{39} \frac{M}{M_\odot} T_9^6 \quad \mathrm{erg\,s^{-1}} \tag{2.13}$$

3. The β-decay of quasiparticles excited in a pion condensate (naturally,

this is possible only in the presence of a π-condensate, i.e. when $\rho > (2\text{–}3)\rho_n$), with luminosity

$$L_\nu^\pi = 10^{44} \frac{M}{M_\odot} \frac{\rho_n}{\rho} T_9^6 \quad \text{erg s}^{-1} \tag{2.14}$$

Figure 2.7 gives the values of the volume luminosity for these three processes as a function of the density ρ at a temperature $T = 10^{10}$ K (Richardson *et al.*, 1982). We can see that for high enough densities the β-decay of pions is the most efficient cooling mechanism, owing to which fact the star cooling curve corresponding to the model with a π-condensate lies lower than the cooling curve, in the determination of which the π-condensate was disregarded.

In any case, after several minutes the temperature of the central regions of the star falls to 10^9–10^{10} K when, as has been mentioned in Section 2.2, the transition of the neutron matter to a superfluid state occurs.

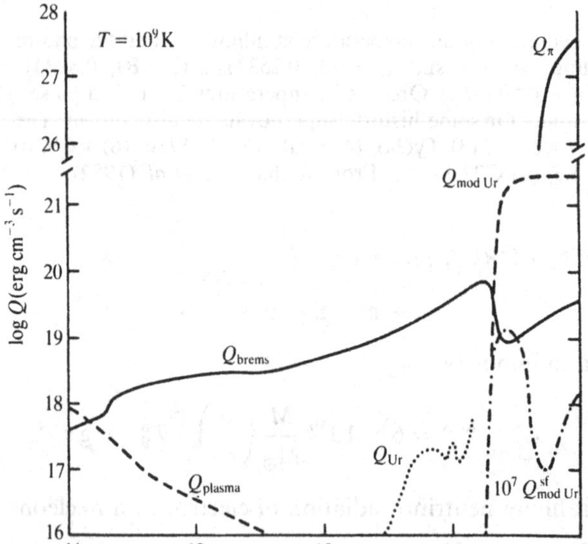

Fig. 2.7. Neutrino energy loss rate per unit volume, Q, for the following neutrino processes: the plasma processes Q_{plasma}, the neutrino bremsstrahlung process $Q_{\text{brems.}}$, the nuclear URCA process Q_{Ur}, the modified URCA process $Q_{\text{mod Ur}}$ for neutrons and protons in the normal state, the modified URCA process $Q^{\text{sf}}_{\text{mod Ur}}$ for neutrons and protons in the superfluid state, and quasi-particle β-decay Q_π. From Richardson *et al.* (1982).

It is of interest that the neutrino losses are so efficient that at a certain stage the temperature of the central regions of the star becomes less than the crust temperature. This is also connected with the fact that the heat conductivity of the normal crust will considerably exceed the heat conductivity of the superfluid core. Only after several months, when the characteristic time of neutrino cooling of the core becomes equal to the characteristic time necessary for energy transfer through the crust, does the second, the photon cooling stage, set in. During this stage, the central regions of the star are heated owing to crust cooling, so the heat flux is mainly directed to the centre of the star. The photon stage of cooling lasts from several months to several years, depending on the model.

After this there comes an isothermal stage of cooling, when the energy flux is everywhere directed from the star centre. Given this, the magnitude of the heat flux itself falls sharply (see Fig. 2.6). It should be emphasized that, owing to relativistic effects, equilibrium is not the state in which the temperature is radius-independent (an isothermal state), but the state in which the quantity $T \exp(\Phi_G/c^2)$ (Φ_G is the gravitational field potential) is constant.

Figure 2.6 also shows the experimental values of the temperature T of sources that exhibited an unpulsed X-ray emission (see Section 1.4), as well as the upper boundaries for a number of young supernova remnants. We can see that in all three cases, when thermal radiation of the surface was registered, there exists close agreement of observations with the model without a π-condensate. These observations do not agree with the model that assumes the presence of a π-condensate.

It should be emphasized that we have considered only the picture of neutron star cooling and neglected possible sources of additional heating. In the first place, these arise under intense matter accretion onto the star surface. In the accreting sources (X-ray pulsars, X-ray bursters and others) the surface temperature is determined by the accretion rate and reaches 10^8 K. In radio pulsars, an analogous heating can be due to fluxes of energetic particles bombarding the neutron star's surface. As shown in Sections 3.2.1 and 5.2, such a particle flux will inevitably arise near the magnetic poles of a pulsar. Variable X-ray radiation due to heating of polar regions is observed in pulsars $1055-52$, $1509-58$, $1951+32$ and source $0540-693$ (see Section 1.4). There may, in principle, exist other sources of neutrino star heating caused by internal friction and by microglitches of period (Shibazaki & Lamb, 1989).

2.5 Neutron star surface

The properties of the surface The properties of the neutron star surface determine the character of its interaction with the magnetosphere and are, therefore, important for the physics of pulsars. The density of the surface layer of a neutron star varies from $\rho \sim 10^3$–10^4 g cm^{-3} on the surface itself to $\rho \sim 10^5$–10^6 g cm^{-3} at a depth of the order of 1–10 cm (Kadomtsev & Kudryavtsev, 1971; Ruderman, 1971). The surface is naturally assumed to consist of iron – the most stable element for such density (Ginzburg & Usov, 1972).

It is essential that the surface is in a solid crystalline state (Ruderman, 1971). Indeed, the melting temperature estimated from the condition $\Gamma_g \simeq 150$ is (Yakovlev & Urpin, 1980)

$$T_M = 3.4 \times 10^7\, Z_{26}^{5/3}(\rho/10^6 \text{ g cm}^{-3})^{1/3}(\Gamma_g/150)^{-1}\mu_e^{-1/3} \quad \text{K} \qquad (2.15)$$

where $\mu_e = Z/A \simeq 0.5$, $\Gamma_g = Z^2e^2/kTa$ is the so-called gas parameter and a is the mean distance between the nuclei. Thus, at temperatures $T < 10^6$ K observed in radio pulsars (see Sections 2.3 and 2.4), the neutron star surface must be solid.

The influence of the magnetic field The properties of the surface are greatly influenced by the magnetic field. Indeed, for densities $\rho \sim 10^4$–10^6 g cm^{-3} the energy $\hbar\omega_B \simeq 10B_{12}$ keV characterizing the effect of the magnetic field is much higher than both the interelectron interaction energy $e^2n_e^{1/3} \sim 10^2\rho_6$ eV and the thermal energy $kT = 10^2T_6$ eV. On the other hand, for densities $\rho \gtrsim 10^4$–10^6 g cm^{-3}, the quantum energy $\hbar\omega_B$ is still significantly lower than the Fermi energy $\mu_F = 10^6\rho_6^{2/3}$ eV. As a result, in the surface regions of a neutron star, not only the lower but also the excited Landau levels with large enough quantum numbers $n \sim \mu_F/\hbar\omega_B \simeq$ 10–100 are occupied. Therefore, although the role of the magnetic field is essential, it does not lead to a complete change in the properties of the surface layers of pulsars. The condition $kT \ll \mu_F$ shows that, as in the external crust, the electrons here are degenerate.

The influence of the magnetic field on the surface structure of a neutron star is determined by the parameter (Ruderman, 1971)

$$\eta = \frac{a_0}{Za_B} \simeq 15B_{12}^{1/2}Z^{-3/2} \qquad (2.16)$$

which is the ratio of the Bohr radius a_0/Z to the size of the electron cloud in the quantizing magnetic field $a_B = (\hbar/m_e\omega_B)^{1/2}$ (Landau & Lifshitz,

1965). The condition $\eta \gg Z^{-3/2}$ corresponds here to the region of strong magnetic fields, when the shell structure of external electrons is determined by the external magnetic field.

Since the condition $\eta \gg Z^{-3/2}$ can be rewritten in the form

$$B_0 > 4.6 \times 10^9 \text{ G} \tag{2.17}$$

it follows that the values $B_0 \sim 10^{12}$ G, which are usual for neutron stars, refer to the region of strong magnetic fields. In this case, individual atoms will be strongly elongated in the direction of the magnetic field, so that their length-to-width ratio can reach 5–10 (Flowers *et al.*, 1977; Flowers & Itoh, 1979). Crystals composed of such atoms generally differ substantially from normal crystals. In particular, there must be a noticeable anisotropy of their properties.

Electric and thermal conductivities Calculation of the coefficients of electric and thermal conductivities for the conditions of the neutron star's surface (Flowers & Itoh, 1976; Raich & Yakovlev, 1982; Yakovlev, 1984) shows their appreciable anisotropy. Figure 2.8 presents the longitudinal,

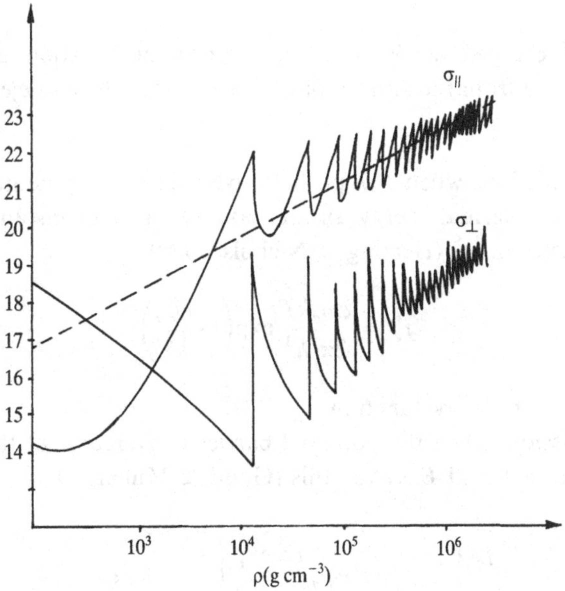

Fig. 2.8. Dependence of the longitudinal $\sigma_\parallel$ and transverse $\sigma_\perp$ conductivities of the surface layer of a neutron star on the density ρ for the magnetic field $B_0 = 10^{12}$ G and the temperature $T = 10^6$ K. From Schaaf (1987).

$\sigma_{\parallel}$ (along the magnetic field), and transverse, $\sigma_{\perp}$, conductivities as functions of the density ρ (Yakovlev, 1984; Schaaf, 1987). According to the Wiedemann–Franz law, the coefficients of thermal conductivity $\kappa_{\parallel}, \kappa_{\perp}$ have similar dependences. The influence of the magnetic field comes down to the appearance of modulation of the quantity $\sigma_{\parallel}$ near the value σ_0, which corresponds to the conductivity in the absence of a magnetic field as well as to a considerable decrease of $\sigma_{\perp}$ as compared with $\sigma_{\parallel}$. The lower the temperature T, the larger the modulation amplitude. This is precisely what we expect because, within the limit $T \to 0$, the dependence $\sigma(\rho)$ assumes a saw-tooth shape typical of a strong magnetic field. Thc role of the temperature is thus reduced to smearing the sharp maxima corresponding to excitation of new Landau levels. The quantity σ_0 itself is (Raich & Yakovlev, 1982)

$$\sigma_0 = 0.6 \times 10^{22}(\rho/10^6 \text{ g cm}^{-3})^{4/3} T_6^{-1} \quad \text{s}^{-1} \tag{2.18}$$

This is two to four orders of magnitude larger than the conductivity of copper. The characteristic value of the coefficient of thermal conductivity is, accordingly,

$$\kappa_0 = 1.7 \times 10^{15}(\rho/10^6 \text{ g cm}^{-3})^{4/3} T_6^{-1} \quad \text{erg s}^{-1} \text{ cm}^{-1} \text{ K}^{-1} \tag{2.19}$$

Emission of charged particles A very important question is charged particle ejection from the surface of a neutron star. Such an ejection may be due to the following:

1. Thermoemission, when the particle overcomes the potential barrier owing to its thermal energy. In this case we have, for instance, for the electron current j_e^{th} (Herring & Nichols, 1949),

$$j_e^{\text{th}} = \frac{e m_e k T}{2\pi^2 \hbar^3} \exp\left(-\frac{e\phi_a}{kT}\right) \tag{2.20}$$

where ϕ_a is the work function.

2. Cold emission, when the potential barrier is overcome at the expense of the external field E. Given this (Good & Müller, 1956),

$$j_e(E) = \frac{e^3}{16\pi^2 \hbar \phi_a} E^2 \exp\left\{-\frac{(18\pi\phi_a^3)^{1/2}}{3\hbar e E}\right\} \tag{2.21}$$

It should be mentioned that for large magnetic fields $B_0 > 10^{12}$ G the expressions for the emission current $j_e^{\text{th}}, j_e(E)$ in the pre-exponential terms could differ from the classical Richardson–Dashman formula (2.20) and Fowler–Nordheim formula (2.21). This is connected with the fact that a strong magnetic field leads to a change in the phase volume of electrons (Beskin, 1982b). But the main exponential factors remain unchanged.

The key question here is to find the work function ϕ_a of particles escaping the pulsar surface, since the emission currents j_e^{th}, $j_e^{\text{th}}(E)$ depend on ϕ_a in a threshold manner. Owing to the role of magnetic field, the value ϕ_a itself can be theoretically determined only on the basis of sufficiently accurate and complicated calculations of the properties of the neutron star's surface. Such calculations have not yet been carried out. The attempts made up to now (Flowers *et al.*, 1977; Jones, 1985a,b, 1986a) yield quite different results and cannot be thought of as reliable (Müller, 1984; Neuhauser *et al.*, 1987).

Parameters of real neutron stars

Mass, M	1.2–$1.6M_\odot$
Radius, R	8–15 km
Moment of inertia, A_r	$\sim 10^{45}$ g cm^2
Magnetic field on the surface, B_0	$\sim 10^{12}$ G
Surface temperature, T	$\lesssim 10^6$ K
Conductivity of the surface regions, σ_0	10^{20}–10^{22} s^{-1}
Coefficient of heat conductivity of surface layers, κ_0	10^{12}–10^{15} erg s^{-1} cm^{-1} K^{-1}

3
Physical processes in the pulsar magnetosphere

In this chapter we consider qualitatively the basic physical processes in the pulsar magnetosphere. A detailed quantitative theory is considered in the chapters that follow.

3.1 Neutron star magnetosphere

Filling with plasma We shall first consider the effect of a strong magnetic field $B_0 \sim 10^{11}$–10^{13} Gauss near a pulsar surface. Since, as has already been said in Section 2.5, the electrical conductivity of the star is large enough, the magnetic field may be assumed to be frozen-in in the neutron star. Therefore, in the internal regions of the star there must hold the condition

$$\mathbf{E}_{\text{in}} + \left[\frac{[\boldsymbol{\Omega}\mathbf{r}]}{c}\mathbf{B}_{\text{in}}\right] = 0 \tag{3.1}$$

where $\mathbf{r}$ is the radius vector from the star centre, $\boldsymbol{\Omega}$ is the angular rotation velocity, c is the velocity of light. Thus, owing to rotation there arises an electric field $\mathbf{E}$ caused by charge redistribution inside the pulsar, which is necessary for fulfilment of the condition (3.1).

In the order of magnitude, the electric field on the star surface is

$$E \sim \frac{\Omega R}{c} B \sim 10^{10}\text{–}10^{12}\ \text{V cm}^{-1} \tag{3.2}$$

where R is the star's radius. It is of importance that the component $E_{\parallel}$ parallel to the magnetic field appears to be of the same order of magnitude. Particles which find themselves in such a strong electric field are accelerated and, moving along the curvilinear magnetic field of the star, emit hard gamma-ray quanta. The latter, absorbed in the magnetic field, generate electron–positron pairs. Thus, there appears the pulsar magnetosphere formed by the electron–positron plasma in the strong magnetic field of the neutron star.

The charge and the corotation current The plasma filling the magnetosphere screens the longitudinal electric field $E_{\|}$ (3.2), i.e. in the magnetosphere,

$$E_{\|} \simeq 0, \qquad \Phi = \Phi(\mathbf{r}_{\perp}) \tag{3.3}$$

Here Φ is the electric field potential, $\mathbf{r}_{\perp}$ is the coordinate orthogonal to the magnetic field lines. Owing to the screening (3.3), the plasma starts rotating along with the star as a solid body. This phenomenon is called corotation (Fig. 3.1). Such a corotation is actually observed in the Earth's and Jupiter's magnetospheres. The corotation, i.e. rotational motion of the plasma across the magnetic field $\mathbf{B}$, is induced by the electric field

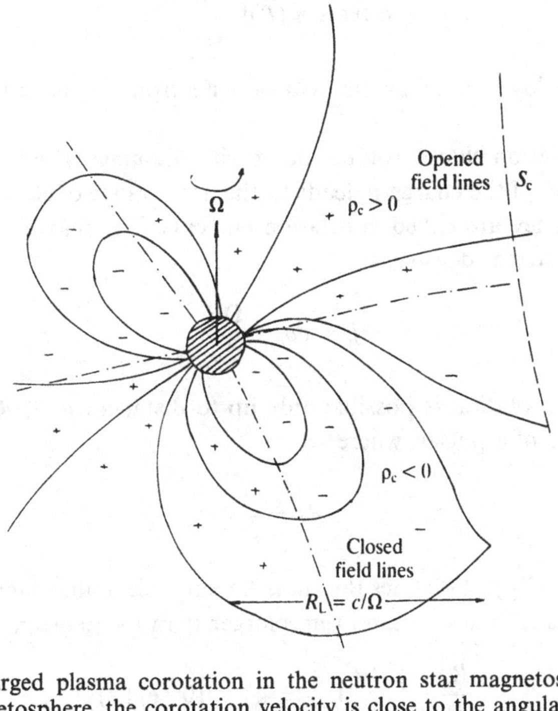

Fig. 3.1. Charged plasma corotation in the neutron star magnetosphere. In a closed magnetosphere, the corotation velocity is close to the angular velocity of pulsar rotation Ω; on open field lines it is generally less than Ω. Corotation currents lead to a magnetic field distortion near the light cylinder $R_L = c/\Omega$, so that the magnetic field differs substantially from the dipole field.

$$\mathbf{E}_c = -\frac{1}{c}[[\mathbf{\Omega r}]\mathbf{B}] \tag{3.4}$$

The field $\mathbf{E}_c(\mathbf{r})$ is generated by polarization of plasma that fills the magnetosphere. The corresponding density of the polarization electric charge

$$\rho_c = \frac{1}{4\pi}\,\mathrm{div}\,\mathbf{E}_c = -\frac{\mathbf{\Omega B}}{2\pi c}, \qquad n_c = \frac{|\rho_c|}{|e|} = \frac{\mathbf{\Omega B}}{2\pi c|e|} \tag{3.5}$$

is called the corotation charge density or the Goldreich–Julian density (Goldreich & Julian, 1969) (here e is the electron charge). In the Earth's magnetosphere, the density $n_c = 10^{-6}$ particles/cm^3 and in the Jupiter magnetosphere $n_c \sim 10^{-5}$ particles/cm^3. In a pulsar magnetosphere the density n_c is much higher: for instance, near the neutron star surface it reaches values of $n_c \sim 10^{11}$–10^{12} particles/cm^3. The corotation density, as well as the magnetic field, decreases rapidly with distance from the star:

$$n_c(r) \simeq n_c(R)\left(\frac{r}{R}\right)^{-3} \tag{3.6}$$

Here and below, the magnetic field of a neutron star is assumed to be dipole.

The corotation charge rotates along with the magnetospheric plasma. This rotation of the charge ρ_c leads to the appearance of electric currents (Fig. 3.1). They are called corotation currents. The maximum value of corotation current density is

$$j_c = c\rho_c \simeq \frac{\Omega B}{2\pi} \tag{3.7}$$

Clearly, corotation is possible only up to distances $\rho_\perp \leqslant R_L$ from the rotation axis of a pulsar, where

$$R_L = \frac{c}{\Omega} \tag{3.8}$$

The quantity R_L just dictates the characteristic size of the magnetosphere. Usually, it is several thousand times larger than the neutron star radius:

$$\frac{R_L}{R} \equiv \left(\frac{\Omega R}{c}\right)^{-1} \simeq 4.8 \times 10^3 (P/1\ \mathrm{s}) \tag{3.9}$$

Magnetic field structure Electric currents deform and perturb the magnetic field of a neutron star. Near the star, the role of such perturbations is not large, whereas at considerable distances from the star, $r \sim R_L$ (3.8), these perturbations become large and even dominant. As shown in Fig. 3.1, magnetic disturbances induced by corotation currents tend to extend the magnetic field lines in the direction orthogonal to the pulsar rotation axis. Owing to the action of these currents, far-field lines extend and, finally, open. Thus, in the magnetosphere there form two essentially different groups of field lines – closed, i.e. those returning to the star surface, and open, i.e. those going to infinity. Such a structure is quite similar to that in the Earth's and planets' magnetospheres.

Open field lines occupy a large region only at considerable distances from the star, $r \sim R_L$ (3.8). Near the star, these lines form a rather narrow conical zone as shown in Fig. 3.2. They emerge from small regions near the magnetic poles of the star. As in the Earth's and planetary magnetospheres, these regions are called *polar caps.* The radius R_0 of the polar cap of a neutron star is not large:

$$R_0 \simeq R\left(\frac{\Omega R}{c}\right)^{1/2}, \qquad R_0 \simeq (1\text{–}3) \times 10^{-2}R \sim 200\ \text{m} \tag{3.10}$$

Plasma motion We shall point out the two basic properties leading to a substantial difference in plasma behaviour on closed and open field lines.

Both ends of closed lines come onto the neutron star. Therefore, on these lines the corotation condition always holds and, accordingly, the electric field potential Φ coincides with the corotation potential Φ_c:

$$\Phi = \Phi_c = \frac{1}{c}\,[\mathbf{\Omega r}]\cdot\mathbf{A} \tag{3.11}$$

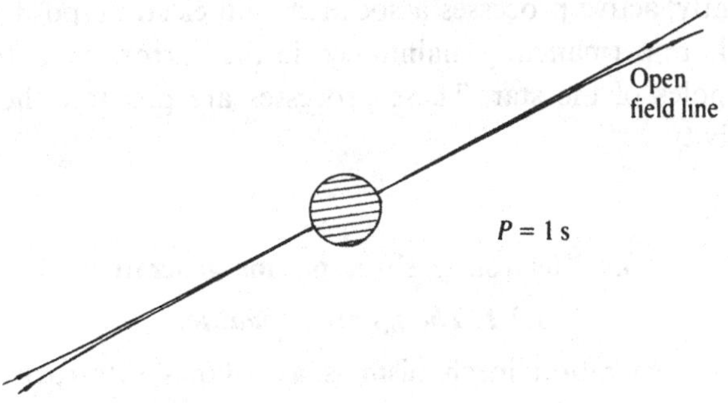

Fig. 3.2. Open field lines near the neutron star surface. The figure shows a real opening of field lines for a pulsar with period $P = 1$ s.

where **A** is the vector potential of the magnetic field **B**. For open field lines, one of their ends comes onto the neutron star, while the other goes to infinity. Hence, on open field lines the electric field potential, which reflects conditions both on the neutron star and far from it, does not necessarily coincide with the corotation potential. It may differ substantially from Φ_c and could, therefore, be represented in the form

$$\Phi(\mathbf{r}_\perp) = \Phi_c(\mathbf{r}_\perp) + \Psi(\mathbf{r}_\perp) \tag{3.12}$$

We have taken into account here that by virtue of (3.3) the electric field potential is always constant along the magnetic field line. The potential $\Psi(\mathbf{r}_\perp)$ reflects the interaction of the magnetic field and currents with plasma; it is the most important characteristic of an open magnetosphere. The presence of a non-zero potential $\Psi(\mathbf{r}_\perp)$ implies that on open field lines the plasma does not rotate along with the star as a solid body. Its rotation velocity is lower; the motion is decelerated. Accordingly, the size of the magnetosphere increases, as shown in Fig. 3.1.

Second, along open field lines plasma may leave the neutron star freely and escape from the magnetosphere. The charge ρ_c (3.5) goes along with plasma, but then the screening condition (3.3) will be violated near the polar caps. In the vacuum region near the star that appears due to escape of plasma there arises a strong potential electric field $E = -\nabla\Psi$. On the scale of the order of the polar cap (3.10), the longitudinal potential difference reaches the value

$$|\Psi_{\max}| \simeq \left(\frac{\Omega R}{c}\right)^{3/2} RB_0 \sim 10^{13}\text{–}10^{15}\ \mathrm{V} \tag{3.13}$$

As has already been said, in such a strong electric field the vacuum proves to be unstable, since an electron–positron plasma is generated in it. Consequently, active processes associated with electron–positron plasma generation are permanently maintained in the narrow conic region near magnetic poles of the star. These processes are precisely the source of pulsar activity.

3.2 Electron–positron plasma generation

3.2.1 The ignition condition

The plasma generation mechanism is as follows (Sturrock, 1971). As mentioned above, owing to the escape of plasma along open field lines, and along with it the corotation charge ρ_c screening the longitudinal electric field, the significant potential difference Ψ (3.13) appears near the polar cap. In this region, the longitudinal electric field

$$\mathbf{E}_{\parallel} = -\nabla_{\parallel}\Psi \tag{3.14}$$

accelerates some particles (say, positrons) in the direction away from the star and other particles (electrons) towards the star, depending on the sign of the charge ρ_c (3.5), as shown in Fig. 3.3. In this process, particles move along a very strong magnetic field.

Since the magnetic field is curvilinear, it follows that on acquiring sufficient energy the particles start emitting high-energy curvature photons, i.e. gamma-ray quanta. These photons are radiated in the direction of particle motion, i.e. along the magnetic field line. But, as shown in Fig. 3.3, owing to curvilinearity of the magnetic field, a photon starts crossing the field lines and gradually reaches the critical electron–positron pair production angle which depends on the photon energy and on the magnitude of the magnetic field. One of the produced particles (positron) goes on moving away from the star while the other particle (electron) becomes accelerated by the electric field in the opposite direction. It also emits curvature photons, which produce pairs near the star's surface (Fig. 3.3). One of the newly born particles (positron) begins acceleration away from the star, and the acceleration and pair production processes are repeated. This is how there arises a chain reaction of multiplication of electrons, positrons and gamma-ray quanta near a neutron star. The multiplication factor is strengthened owing to the fact that particles created at high Landau levels in a magnetic field emit synchrotron radiation, i.e. synchrophotons, which are also capable of pair production.

The double layer The existence of a significant potential difference (3.13) between the star surface and the magnetosphere provides a natural distinction of a layer near the surface, in which there exists a strong electric field (3.14). This layer is similar to the double Langmuir layer near the surface of a body in plasma. It is called a 'double layer' or a 'vacuum gap', since it is precisely in this layer that plasma is absent owing to its escape along the magnetic field lines. In a double layer, particles acquire the high energy necessary for radiation of curvature photons capable of generating electron–positron plasma. Therefore, the main condition for the appearance of vacuum 'breakdown' and for generation of a stationary electron–positron plasma flow are dictated by the double layer.

Curvature radiation is quite similar to ordinary synchrotron radiation, with the only difference that the role of the Larmor radius is played here

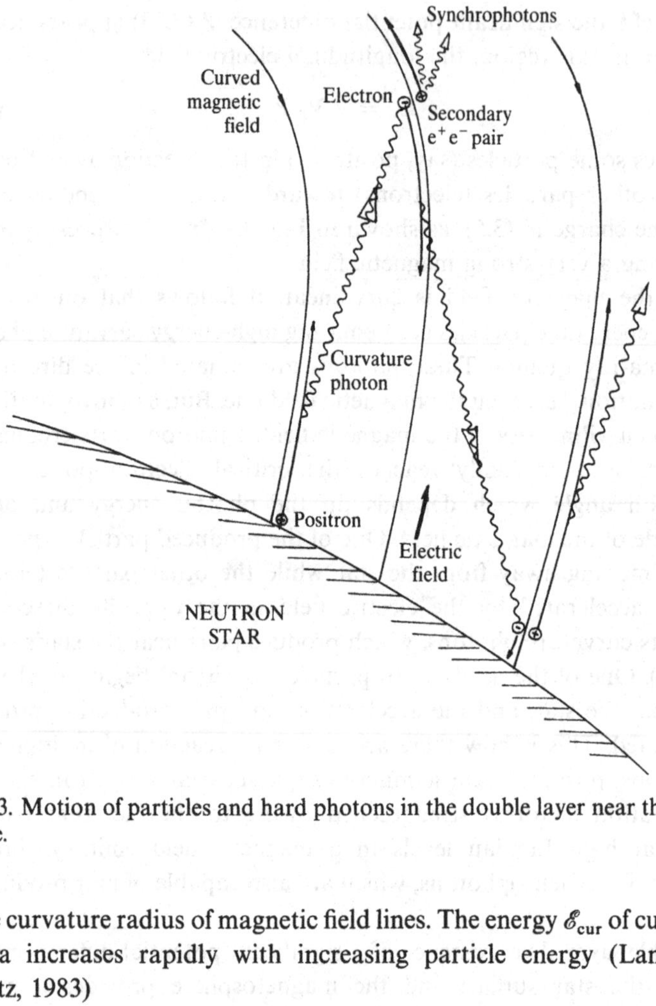

Fig. 3.3. Motion of particles and hard photons in the double layer near the pulsar surface.

by the curvature radius of magnetic field lines. The energy $\mathscr{E}_{cur}$ of curvature quanta increases rapidly with increasing particle energy (Landau & Lifshitz, 1983)

$$\mathscr{E}_{cur} \simeq m_e c^2 \frac{\lambda\!\!\!^{-}}{\rho_l} \gamma^3 \tag{3.15}$$

Here $\lambda\!\!\!^{-} = \hbar/m_e c = 3.9 \times 10^{-11}$ cm is the Compton wavelength, $\gamma = \mathscr{E}_{\parallel}/m_e c^2 = (1 + p_{\parallel}^2/m_e^2 c^2)^{1/2}$ is the Lorentz factor of particle motion along the magnetic field, and ρ_l is the curvature radius of magnetic field lines. Since in the polar region of neutron stars

$$\rho_l \simeq 8 \times 10^7 \text{ cm} \tag{3.16}$$

the photon energy, as follows from (3.15), becomes sufficient for electron–positron pair production, $\mathscr{E}_{cur} \gg m_e c^2$, only when $\gamma \gtrsim 10^6$–10^7, i.e.

$\mathscr{E}_{\|} \sim 10^{12}$–$10^{13}$ eV. This implies that large drops in the field potential Ψ on the double layer near the neutron star surface, $\Psi \sim \Psi_{max}$ (3.13), are actually necessary for vacuum 'breakdown'.

'Breakdown' potential For vacuum 'breakdown', i.e. for stationary plasma generation near a neutron star surface, it is necessary that the following conditions hold. It is necessary that a particle of a definite sign (say, a positron), produced near a neutron star surface $h = 0$, be accelerated in the double layer and emit a high-energy curvature quantum. Before reaching the upper boundary of the double layer $h = H$, the quantum must create a pair (see Fig. 3.3). The electron from this pair must be caught up by the electric field and accelerated towards the body's surface. It is also necessary that this electron emit a curvature quantum, which must, in turn, create a pair before it reaches the body surface $h = 0$. Then the positron from this pair will be accelerated in the direction away from the body, so that the process will repeat. Such a stationary process (which we simply call vacuum 'breakdown') arises if the jump of the potential in the double layer is approximately equal to $\Psi \simeq \Psi_{max}$ (3.13) or, more precisely, if $\Psi = \Psi_0$, where

$$\Psi_0 = c_1 \left(\frac{m_e c^2}{|e|}\right)^{8/7} \left(\frac{\rho_l}{\lambda\!\!\!^-}\right)^{4/7} B_0^{-1/7} c^{-1/7} P^{1/7} \cos^{1/7} \chi \qquad (3.17)$$

Here χ is the inclination angle of the magnetic field $\mathbf{B}_0$ to the rotation axis $\mathbf{\Omega}$ of the star and c_1 is a dimensionless quantity of the order of unity.

From (3.17) it is seen that the breakdown potential Ψ_0 increases with increasing curvature radius ρ_l of the magnetic field line. In particular, near the magnetic axis in the centre of the polar cap $\rho_l \to \infty$, so that $\Psi_0 \to \infty$, and plasma generation is impossible here. The potential Ψ_0 depends rather weakly on the pulsar period and on the magnitude of the magnetic field $\mathbf{B}_0$. It should be noted that positrons or electrons may also appear near a body surface as a result of bombardment of the surface by fast particles. This affects the value of the constant c_1 in (3.17), but under real conditions of pulsars this effect is not large. When $\mathbf{\Omega B}_0 > 0$, i.e. $\chi < \pi/2$, and so near the magnetic poles the signs of ρ_c (3.5) and j_c (3.7) are negative, the thermoelectron emission from the star surface may appear to be more important. It restricts the surface temperature of such pulsars (i.e. pulsars with $\mathbf{\Omega B}_0 > 0$ or $\chi < \pi/2$) by the condition $T \lesssim 10^6$–10^7 K (see Sections 2.4 and 2.5). If this condition is not fulfilled, then the electron current flowing down from the surface may be sufficient to create in the magnetosphere the corotation charge ρ_c, which screens the longitudinal electric field. In this case, the strong electric field necessary for plasma generation may fail to arise.

For the characteristic pulsar parameters

$$\rho_l \simeq 10^7\text{–}10^8 \text{ cm}, \qquad P \simeq 1 \text{ s}, \qquad B_0 \simeq 10^{12} \text{ G} \tag{3.18}$$

the breakdown potential is

$$\Psi_0 \simeq 10^{13}\text{–}10^{14} \text{ V} \tag{3.19}$$

This implies that the characteristic Lorentz factor of particles accelerated in the double layer is

$$\gamma = \frac{e\Psi_0}{m_e c^2} \simeq 10^7\text{–}10^8 \tag{3.20}$$

The ignition conditions The double-layer thickness is

$$H = c_2 \left(\frac{m_e c^2}{|e|}\right)^{4/7} \left(\frac{\rho_l}{\lambda\!\!\!^{-}}\right)^{2/7} B_0^{-4/7} P^{3/7} c^{3/7} \cos^{-3/7}\chi$$
$$c_2 \approx 1 \tag{3.21}$$

so that for the conditions (3.18) the quantity $H = 100$ m. We can see that the double-layer thickness falls with decreasing pulsar period P. Given this, for fast pulsars ($P \ll 1$ s) the double-layer thickness H appears to be much less than the polar cap radius R_0 (3.10). For slow pulsars ($P \gtrsim 1$ s) these dimensions are, in contrast, of the same order of magnitude.

It is of importance that the double layer is a one-dimensional screening layer with an increasing potential only if its size H is less than R_0. The condition

$$H \lesssim R_0 \tag{3.22}$$

therefore determines the limit of potential increase in the layer, i.e. the limit when breakdown and stationary electron–positron plasma generation are possible. We can say, therefore, that the condition (3.22) is the pulsar 'ignition' condition. Making allowance for (3.21) and (3.10), we rewrite this condition in the form

$$PB_{12}^{-8/15} \lesssim c_3 \cos^{2/5}\chi, \qquad c_3 \simeq 1 \tag{3.23}$$

The numerical value of the constant in the right-hand side is given here for the values of the parameters (3.18) typical of neutron stars. We can see that for a magnetic field $B_0 \sim 10^{12}$ G the 'ignition' condition (3.23) corresponds to periods $P \sim 1$ s. It coincides with the pulsar 'death line' shown in Fig. 1.7.

3.2.2 Secondary plasma

Cascade multiplication Outside the double layer, the electron–positron plasma concentration n_c (3.5) increases, so that it can provide screening of the longitudinal electric field $E_{\parallel}$. But the process of electron–positron pair generation does not stop at this stage – plasma concentration goes on increasing up to distances $r \gtrsim R$ to the star surface, when the magnetic field effectively starts decreasing. Given this, pair generation proceeds both due to radiation of curvature photons by a beam of fast particles with $\gamma \sim 10^7$ and due to the new process – radiation of synchrophotons.

Owing to the latter process, secondary plasma generation proceeds in cascades. On passing a distance of the order of the free path l_γ after their birth, curvature photons from a primary beam are absorbed, generating electron–positron pairs which are the first generation of the plasma produced. The electrons and positrons produced find themselves at a non-zero Landau level in the strong magnetic field of the pulsar. Therefore, within a very short time of the order of 10^{-19} s, they pass over to a zero level, emitting synchrophotons whose energy is on average several times (approximately by an order of magnitude) less than the energy of the parent curvature photon. Accordingly, the free path l_γ of synchrophotons of first generation is larger by an order of magnitude than the path of the curvature photons. Pair production by synchrophotons gives plasma of the second generation. In this process, synchrophotons of the second generation are also produced, and their mean energy again decreases approximately by an order of magnitude as compared with the energy of synchrophotons of the first generation, and so on. Usually, plasma generation stops in a second and sometimes in a third generation. Breaking of the cascade process is determined by magnetosphere transparency arising at distances of $r \sim 2$–$3R$ because both the magnetic field B and the curvature radius ρ_l vary significantly with distance from the neutron star.

The distribution function of electrons and positrons Analysis shows that the minimum energy of the particles produced is

$$\gamma_{\min} \simeq \frac{4\rho_l}{R} \sim 300 \qquad (3.24)$$

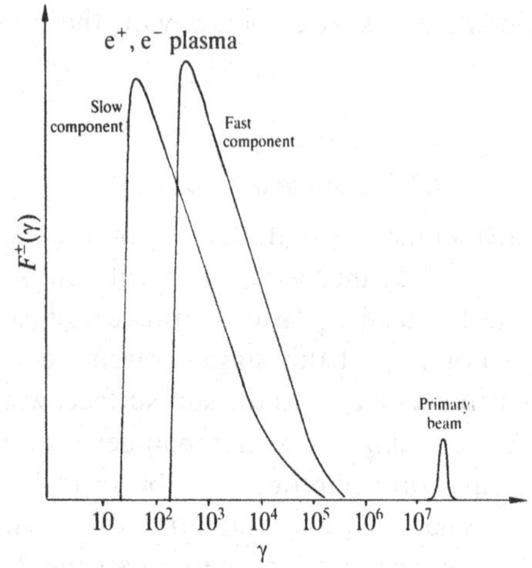

Fig. 3.4. The spectrum of a secondary plasma generated above the double layer and moving along open magnetic field lines.

The quantitative character of the energy spectrum of the plasma generated is shown in Fig. 3.4. We can see that as γ decreases to $\gamma_{\min}$, the distribution function of particles increases exponentially approximately as γ^{-2} and for $\gamma < \gamma_{\min}$ it breaks sharply.

Another essential specific feature is that, as is seen from Fig. 3.4, the distribution functions of electrons and positrons in the magnetospheric plasma do not coincide, for the reason that the magnetospheric plasma screens the longitudinal component of the electric field $E_{\parallel}$ (3.14). To this end it is necessary that the electric charge density in plasma be always close to the corotation charge density ρ_c (3.5). The charge density in the primary beam of particles accelerated in the double layer is generally less than ρ_c. Therefore, in the region of quasi-neutral plasma there automatically occurs a weak longitudinal electric field, with the potential difference

$$\delta\Psi \lesssim \frac{m_e c^2}{|e|}\gamma_{\min} \ll \Psi_{\max} \tag{3.25}$$

which decelerates one of the plasma components and accelerates the other, owing to which the density ρ_c is created and maintained everywhere in the magnetosphere. Those particles whose charge has the same sign as the corotation charge ρ_c are always decelerated. As a result of deceleration, their minimum energy becomes

$$\gamma_{\min} \simeq 30\text{–}100 \tag{3.26}$$

Consequently, their mass appears to be several times less than that of the particles of opposite sign, for which $\gamma_{min} \sim 300$–500 (see Fig. 3.4). In the pulsar magnetosphere, the particles which have the sign of ρ_c therefore play the role of electrons in a normal plasma, while the particles of opposite sign play the role of ions.

The total electron–positron plasma density in the pulsar magnetosphere, which arises in the course of the generation process described, appears to be higher by several orders of magnitude than the corotation density n_c (3.5). It is customarily characterized by the multiplication parameter

$$\lambda \equiv \frac{n_e}{n_c} \sim 10^3\text{–}10^5 \tag{3.27}$$

Note that the plasma energy density $\sim n_e m_e c^2 \gamma_{min}$ in the magnetosphere remains much lower than the energy density of the magnetic field $B^2/8\pi$, so that

$$\mu = \frac{B^2}{8\pi n_e m_e c^2 \gamma_{min}} \sim 10^2\text{–}10^4 \tag{3.28}$$

Gamma-ray emission A directed beam of hard gamma-ray emission is generated together with electron–positron plasma. Its total energy is only a few times less than that of the electron–positron beam energy. The energy spectrum is power-law, and is approximately

$$F(\gamma) \propto \gamma^{-2} \tag{3.29}$$

The spectrum is cut off when

$$\gamma_{max} \sim 10^5 \tag{3.30}$$

The quantity γ_{max} increases with decreasing pulsar period P and with increasing magnetic field B_0. For the pulsar in the Crab Nebula, for example, with $P = 0.033$ s, $B \simeq 10^{13}$ G, $\gamma_{max} \simeq 5 \times 10^6$.

3.3 Radio emission generation

3.3.1 Normal modes

As mentioned above, particles from a primary beam accelerated in the double layer up to high energies of $\gamma \sim 10^7$ radiate curvature photons which generate the electron–positron plasma. The latter has a much lower particle energy of $\gamma \sim 10^2$ (3.24, 3.26) and a higher density than the primary beam density $n_e/n_c \sim 10^3$–10^5 (3.27). Moving in a curvilinear magnetic field, the plasma particles must also produce curvature radiation. As before, it has the characteristic frequency

$$\nu \simeq \frac{c}{\rho_l}\gamma^3 \tag{3.31}$$

and is directed to the narrow cone width

$$\Delta\theta \simeq \frac{1}{\gamma} \tag{3.32}$$

along the direction of particle motion, i.e. along the magnetic field.

However, curvature radiation of primary particles is incoherent – it consists of separate gamma-quanta whose wavelengths are small compared to the mean distance between particles. Here the situation is essentially different: the radiation frequency (3.31) falls in the radio-frequency band $\nu \sim 0.1$–3 GHz. The corresponding wavelengths

$$\lambda_f \sim 10\text{–}300 \text{ cm} \tag{3.33}$$

are larger by many orders of magnitude than the mean distance between plasma particles. In this case, the main role is played by coherence effects. The coherent properties of radiation are determined by collective interaction of waves and particles in the plasma. Radiation in this case is the normal mode of plasma oscillations. The radiation generation is, therefore, the instability of plasma with respect to a given normal mode. Factors in favour of the argument for the coherent character of generation are the high brightness temperature and other properties of the observed radio emission of pulsars (see Section 1.2.10).

Curvature-plasma modes In a relativistic electron–position plasma moving in a homogeneous rectilinear magnetic field, four natural oscillatory modes may exist. Two of them are the usual transverse electromagnetic waves – an ordinary one and an extraordinary one. The other two are magnetoplasma waves. They correspond to drift oscillations propagating along the magnetic field and are analogous to magnetodynamic (Alfven) waves in a usual plasma. Under the conditions of a relativistic plasma flow with the distribution function of electrons and positrons of the type presented in Fig. 3.4, all these modes with wavelengths λ_f (3.33) appear to be stable – oscillations are not excited, radiation is not generated.

The situation changes radically in a curvilinear magnetic field, i.e. in the presence of even a weak curvilinearity

$$\frac{\lambda_f}{\rho_l} \ll 1 \tag{3.34}$$

where λ_f is the radiation wavelength (3.33) and ρ_l is the curvature radius of the magnetic field (3.18). This situation depends greatly on the value of the dimensionless parameter

$$a = 4\pi \frac{\omega_p^2 \rho_l^{4/3} \lambda_f^{2/3}}{\gamma^3 c^2} \tag{3.35}$$

Here $\omega_p \gamma^{-3/2} = (4\pi n_e e^2/m_e \gamma^3)^{1/2}$ is the Langmuir frequency of longitudinal oscillations of the relativistic plasma, and $\rho_l^{2/3} \lambda_f^{1/3}/c$ is the characteristic time determining the influence of the magnetic field curvature upon normal waves. For $a \gg 1$, $a^{1/5} \gg (\lambda_f \gamma^3/\rho_l)^{2/3}$, and as a result of splitting of one of the modes there appear two new oscillatory modes called curvature-plasma waves. For $a \ll 1$, on the other hand, two natural (drift) oscillatory modes disappear. Thus, in a curvilinear magnetic field in the region $a \gg 1$ there may exist six natural modes, whereas in the region $a \ll 1$ there are only two.

Excitation of instability The value of the parameter a depends strongly on the plasma density n_e. In the pulsar magnetosphere, the density n_e, which is proportional to the magnitude of the magnetic field B (see (3.5) and (3.7)), decreases rapidly with increasing distance from the neutron star. The condition $a \gg 1$, $a^{1/5} \gg (\lambda_f \gamma^3/\rho_l)^{2/3}$, therefore, holds only in the internal ($r < r_{max}$) region, i.e. the magnetosphere region closest to the neutron star, where

$$r_{max} \simeq 3.5R \left(\frac{\lambda}{10^4}\right)^{1/4} \left(\frac{B_0}{10^{12}\,\mathrm{G}}\right)^{1/4} \nu_{GHz}^{-1} P^{-1/2} \tag{3.36}$$

Here R is the neutron star radius. In this region there exist six natural oscillatory modes of plasma, two of them (curvature-plasma ones) being unstable.

The polarization of these modes is such that their electric field vector lies in the plane of particle motion and is almost orthogonal to the magnetic field. In a strong homogeneous magnetic field, waves with such a polarization do not interact with plasma particles and, therefore, cannot be amplified. In a curvilinear (inhomogeneous) magnetic field, owing to non-locality of resonance interaction the electric wave field perpendicular to the magnetic field line at a given point will at neighbouring points have a component in the direction of plasma motion, which leads to wave–particle interaction. Since the particle velocity is close to the wave velocity $\omega/k \simeq c$, it follows that practically all the plasma particles appear to be resonant. Wave excitation is therefore connected not only with the curvature, but also with the Čerenkov radiation mechanism.

The joint effect of the Čerenkov and curvature radiation mechanisms leads to a strong instability of curvature-plasma modes. Their time increment Γ_L is positive for the wave vectors **k** lying in a narrow angle cone in the direction of particle motion and is equal to

$$\Gamma_L \simeq \nu^{1/5}\omega_p^{2/5}\left(\frac{c}{\rho_l}\right)^{2/5}\gamma_{\min}^{-3/5} \tag{3.37}$$

where ν is the wave frequency. The amplification cone is here not symmetric about the direction of the magnetic field line: it is extended in the plane of particle motion and has the opening

$$\theta_\parallel = \frac{\omega_p^{4/5}\rho_l^{1/5}}{\nu^{3/5}c^{1/5}\gamma^{4/5}} \tag{3.38}$$

but is contracted in the orthogonal direction:

$$\theta_\perp = \frac{\omega_p^{1/5}c^{1/5}}{\nu^{2/5}\rho_l^{1/5}\gamma^{7/10}} \ll \theta_\parallel \tag{3.39}$$

The region of curvature-plasma wave instability is shown in Fig. 3.5. The amplification increment Γ_L is so large that the total optical depth passed by waves propagating in the instability region appears to be of the order of several hundreds. This does not, of course, mean that such a giant amplification actually takes place because this process is hampered by non-linear processes

3.3.2 Non-linear interaction and transformation of curvature-plasma waves into transverse oscillations

Non-linear stabilization The basic non-linear process is the three-wave interaction of curvature-plasma waves which leads to their scattering and escape from the amplification cone (3.38, 3.39) into the region where they become extinct. Thus, an instability with the increment Γ_L (3.37) sets in. It should be emphasized that the three-wave interaction is absent in an electron–positron plasma if the distribution functions of particles are the

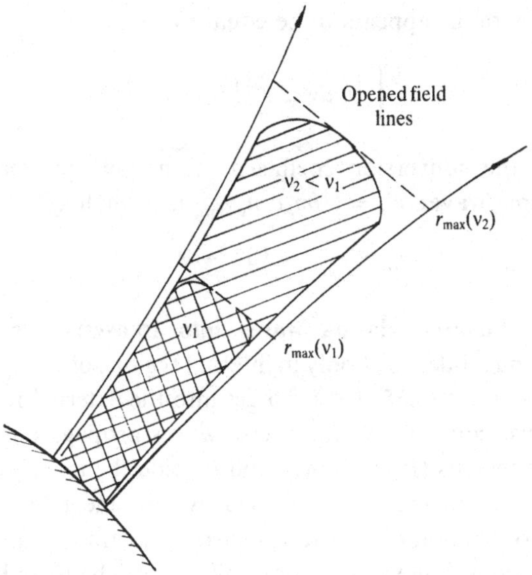

Fig. 3.5. The instability region of curvature-plasma waves. For every ν it spreads from the pulsar surface to the height $r_{max}(\nu)$ (3.36). Higher frequencies are generated closer to the star surface. At a given height, the radiation is therefore broad-band.

same. As mentioned above (see (3.24) and (3.26)), in our case the distribution functions of electrons and positrons are substantially displaced relative to each other (Fig. 3.4) owing to the action of the electric field. Given this, the three-wave interaction becomes significant and the basic electromagnetic properties of plasma are determined for the most part by a less energetic component.

The three-wave interaction determines a stationary spectrum of curvature-plasma waves established within a time of the order of Γ_L^{-1}. The spectral energy density of these waves increases with frequency as

$$U_\nu \propto \nu \tag{3.40}$$

up to the values $\nu_{max}(r)$, where $\nu_{max}(r)$ is determined by the boundary of the instability region $r_{max}(\nu)$ (3.36):

$$\nu_{max}(r) \simeq 3.5\left(\frac{\lambda}{10^4}\right)^{1/4}\left(\frac{B}{10^{12}\ \mathrm{G}}\right)^{1/4} P^{-1/2}\left(\frac{r}{R}\right)^{-1}\ \mathrm{GHz} \tag{3.41}$$

For $\nu > \nu_{max}(r)$, the spectrum is sharply cut off.

As a result, the total energy density of curvature-plasma modes in the entire frequency range appears to be equal to

$$U = \int U_\nu \, d\nu \simeq \lambda^{-1} m_e c^2 n_e \gamma_{min} \tag{3.42}$$

Consequently, the conversion coefficient of moving plasma energy to curvature-plasma waves, $\eta_{tr} = U/m_e c^2 n_e \gamma_{min}$ is a small quantity:

$$\eta_{tr} \simeq \lambda^{-1} \sim 10^{-3}\text{–}10^{-5} \tag{3.43}$$

Conversion of curvature-plasma waves into transverse ones Unstable curvature-plasma modes exist only in internal regions of a magnetosphere, $r < r_{max}$, where $a \gg 1$ (3.35, 3.36). To get into the external region, where $a \ll 1$, they must convert into transverse waves propagating freely both in the external regions ($R < r < R_L$) and outside the magnetosphere.

Rather efficient conversion is provided by non-linear three-wave processes with participation of curvature-plasma and transverse waves. The interaction region is limited to rather small angles θ between **k** and **B**, the non-linear conversion increment being of the same order as the linear one (3.37). The energy is converted both into an ordinary transverse wave ($j = 2$), in which the electric vector lies in the plane **k**, **B**, and into the orthogonal extraordinary wave ($j = 1$). As a result, stationary energy spectra $U_\nu^{(j)}$ are also established for transverse waves in the region of small enough angles θ. For example, for an extraordinary wave,

$$U_\nu^{(1)} \propto \nu \tag{3.44}$$

Such stationary spectra are formed and maintained in the entire instability region (3.36, 3.41).

Radio emission spectrum Since wave propagation in a curved field of a pulsar leads to a continuous increase of the angle θ, transverse waves will gradually leave the interaction cone. This process is stationary, and it generates radio emission going out of the pulsar magnetosphere.

The resultant emission generated in the instability region (3.36, 3.41) and propagating at large distances from the star at a certain angle Θ to the magnetic dipole axis makes it possible to determine the spectral power $I_\nu^{(j)}(\Theta)$ for each of the normal modes j, i.e. the directivity pattern of pulsar radio emission. The radio emission intensity $I_\nu^{(j)}(\Theta)$ is naturally proportional to the stationary energy spectrum (3.44).

The total energy in an observed pulsar pulse is determined as the intensity integral over the region of all directions Θ which come to the observer:

$$\tilde{E}_\nu^{(j)} = \int d\Theta \cdot \Theta \cdot I_\nu^{(j)}(\Theta) \tag{3.45}$$

For example, for an extraordinary wave we have

$$\tilde{E}_\nu^{(1)} \propto \nu^{-2} \tag{3.46}$$

The radio emission spectrum on an extraordinary component is seen to be inversely proportional to the frequency squared. A similar frequency dependence, $\tilde{E}_\nu^{(2)} \propto \nu^{-2.5}$, also takes place for an ordinary wave.

The total radiation power obtained through integration of the spectral density $\tilde{E}_\nu$ over frequencies is conveniently expressed in terms of the total particle energy in the electron–positron plasma flow $W_p = n_e m_e c^2 \gamma_{\min}$:

$$L_r = \alpha_T W_p \tag{3.47}$$

The proportionality coefficient α_T appears here to be equal to (cf. (3.43))

$$\alpha_T \sim \lambda^{-1} \sim 10^{-3}\text{–}10^{-5} \tag{3.48}$$

The radio losses L_r (3.47) are then equal to 10^{26}–10^{29} erg s^{-1}, which is what is observed (see Chapter 1).

Thus, the instability of a relativistic electron–positron plasma flow moving in a curvilinear magnetic field results in an intensive excitation of curvature-plasma modes in the internal region of the magnetosphere ($r \lesssim 3$–$100R$). The conversion in the same region of curvature-plasma waves into transverse modes capable of going freely out of the magnetosphere leads to generation of the observed pulsar radio emission.

3.4 Electric current

Longitudinal current As mentioned above, stationary electron–positron plasma generation near a neutron star appears only when the electric field potential drop Ψ between the magnetosphere Φ and the star surface Φ_c

$$\Psi = \Phi - \Phi_c \tag{3.49}$$

reaches a certain value (3.17) which depends on the pulsar period P, magnetic field B_0 and curvature radius ρ_l of the magnetic field line (Fig. 3.6). The characteristic value Ψ is 10^{13} V (3.19). The potential drop Ψ occurs only in the region of open field lines and has a definite sign which coincides with the sign of the product $\mathbf{\Omega B}$ in the polar cap. Therefore,

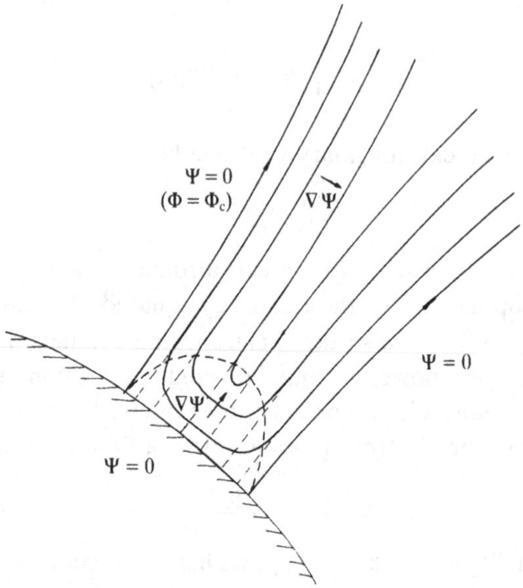

Fig. 3.6. Equipotential surfaces $\Psi = \text{const}$ in the region of open field lines. Over the double-layer boundary (dashed line) the potential Ψ is constant on magnetic field lines. Here the electric field $-\nabla\Psi$ leads only to an additional plasma rotation around the magnetic axis. In the double-layer region, the electric field $\mathbf{E} = -\nabla\Psi$ has a component along the magnetic field, which leads to particle generation and acceleration.

only charges of the same sign as that of the corotation charge ρ_c (3.5) are accelerated in the direction from the star surface and form a primary beam generating the electron–positron plasma. This means that plasma generation is necessarily accompanied by an electric current flowing along magnetic field lines.

The size of the polar cap is small (3.10), and therefore in the entire cap the sign of the product $\mathbf{\Omega B}$ is constant.† Consequently, the longitudinal current above the whole of the polar cap flows in one direction. The magnitude of the longitudinal current $j_\parallel$ does not here exceed the critical current j_c (3.7):

$$j_\parallel \lesssim j_c = \frac{\mathbf{\Omega B}}{2\pi} \tag{3.50}$$

The total longitudinal current I flowing in a pulsar magnetosphere is,

† We do not touch upon the particular case when the magnetic axis is almost orthogonal to the rotation axis $\mathbf{B}_0 \perp \mathbf{\Omega}$, so that the product $\mathbf{\Omega B}_0$ may reverse sign within the polar cap.

therefore, conveniently represented in the form

$$I = i_0 I_c, \qquad I_c = \frac{B_0 \Omega^2 R^3}{2c} \tag{3.51}$$

where $i_0 < 1$ is the dimensionless current parameter and I_c is the total critical current in the polar cap.

Current jet near the light surface In order that the longitudinal current over the whole polar cap should flow only in one direction, it is necessary that there exist a reverse current onto the pulsar surface. This process of current closure is realized first of all by a current jet, J_R, flowing near the light surface S_c (Fig. 3.7).

The current jet J_R occurs due to electron and positron acceleration near the light surface. Indeed, the electric drift velocity $\mathbf{v}_{dr} = c[\mathbf{EB}]/B^2$ of particles on the light surface must be equal to the velocity of light.

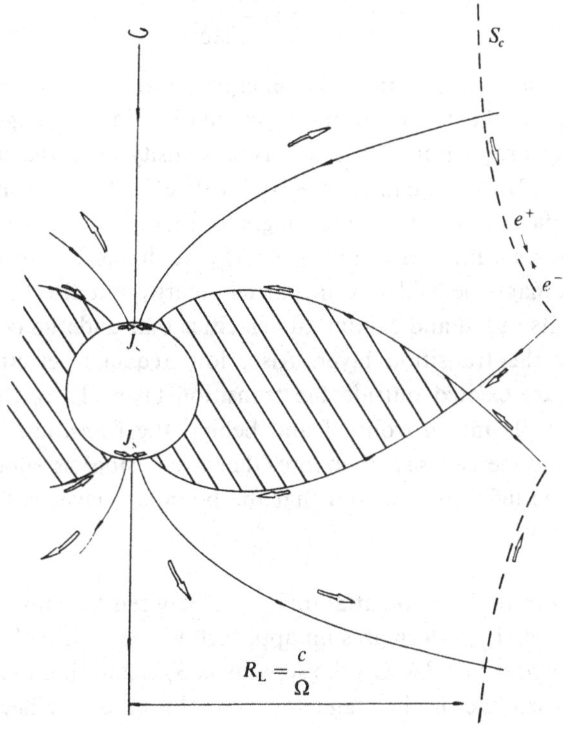

Fig. 3.7. The scheme of closure of currents in the magnetosphere of a pulsar.

Therefore, with approach to the light surface S_c the particle energy must increase sharply. As a result, the particles are accelerated, and therefore the state of equilibrium rotation for them is violated. The acceleration of particles arises from the fact that they not only drift along equipotential surface Φ (this drift is responsible for a uniform rotation) but also cross the equipotential surfaces, which provides them with energy from the electric field. Given this, the electrons must move in one direction (towards increasing Φ) and the positrons in the opposite direction (towards decreasing Φ), as shown in Fig. 3.7. As a result, in the boundary layer of thickness

$$\Delta \simeq \frac{c}{\Omega \lambda} \tag{3.52}$$

there forms a powerful jet of surface current J_R flowing along the light surface S_c perpendicular to the magnetic field lines.

In crossing the boundary layer, plasma electrons and positrons are accelerated up to energies

$$\mathscr{E} \simeq \frac{eI}{2c\lambda} = i_0 \frac{eBR^3\Omega^2}{2\lambda c^2} \tag{3.53}$$

From this one can see that the energy gained by the particles is proportional to the total current i_0 circulating in the magnetosphere and inversely proportional to the plasma density (λ is the multiplicity parameter (3.27)). The quantity $\mathscr{E} \sim 10^9$–10^{13} eV. Thus, flowing through the light surface and outside the magnetosphere, the electron–positron plasma acquires a rather substantial energy. It should be noted that it is not only the magnetic field and the particle energy, but also the density of plasma and its radial and azimuthal velocities that undergo considerable variation in the transition layer. Also, low-frequency electromagnetic oscillations are excited outside the transition layer. Thus, the physical properties of plasma in front of and behind the transition layer differ substantially. One can say, therefore, that a collisionless shock wave is excited on the light surface, and that the boundary layer is the front of this shock wave.

Closure of current The potential difference between the star surface and magnetosphere, Ψ_{max}, decreases on approach to the region of closed field lines, as shown in Fig. 3.6. On the separatrix S_f separating the regions of open and closed lines in the magnetosphere, this difference becomes zero:

$$\Psi_f = 0. \tag{3.54}$$

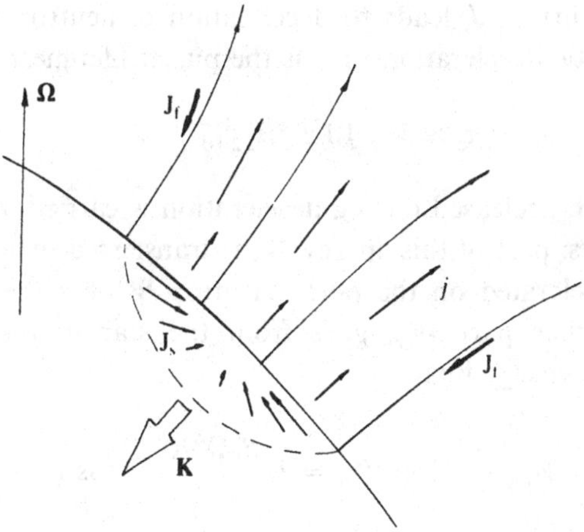

Fig. 3.8. Closure of currents near the surface of a neutron star. The ponderomotive action of the surface current leads to the appearance of the braking torque **K** directed opposite to the angular rotation velocity **Ω**.

This means that the current J_R flowing in the magnetosphere near the light surface can now return freely to the star surface along the magnetic field lines which form the separatrix S_f. The magnitude of the reverse current jet is

$$J_f = \frac{\Omega}{2\pi c} I \tag{3.55}$$

It compensates completely the direct longitudinal current flowing along the entire pulsar magnetosphere. This provides electric current circulation in the magnetosphere.

The complete closure of the current circuit is realized by the current jet J_s flowing along the neutron star surface, as shown in Figs 3.7 and 3.8.

3.5 Energetics of pulsars

As we have seen above, current circulation in a pulsar magnetosphere is closed by the current J_s flowing on the neutron star surface (see Fig. 3.8). Given this, the surface current $\mathbf{J}_s$ flows across the magnetic field lines. Because of this there arises a ponderomotive Ampère force

$$\mathbf{F}_A = \frac{1}{c}[\mathbf{J}_s\mathbf{B}] \tag{3.56}$$

in the direction opposite to the star rotation (see Fig. 3.8). Thus, the action of the surface current J_s leads to deceleration of neutron star rotation. The characteristic deceleration time is the pulsar lifetime τ_d:

$$\tau_d \simeq 3 \times 10^7 P^2 B_{12}^{-2} i_0^{-1} \quad \text{y} \tag{3.57}$$

The energy W_{tot} released during deceleration is carried away from the star in two ways: part of this energy W_p is transferred by a primary flux of particles accelerated on the potential drop Ψ near the neutron star surface. The other part W_{em} goes from the star in the form of an electromagnetic energy flux

$$\begin{aligned} W_{\text{tot}} &= W_p + W_{\text{em}} = L \frac{B_0^2 \Omega^4 R^6}{c^3} i_0 \cos \chi \\ W_p &= L \frac{B_0^2 \Omega^4 R^6}{c^3} i_0 \frac{\Psi}{\Psi_{\max}} \\ W_{\text{em}} &= L \frac{B_0^2 \Omega^4 R^6}{c^3} i_0 \left(\cos \chi - \frac{\Psi}{\Psi_{\max}} \right) \end{aligned} \tag{3.58}$$

Here χ is the angle between the magnetic and rotation axes of the pulsar, L is the numerical coefficient of the order of 0.3–0.5, and i_0 is a dimensional strength of current (3.51).

The energy W_p is expanded by a primary beam in the generation of electron–positron plasma, gamma-ray and radio-emission fluxes. The energy W_{em} is expended in electron–positron particle acceleration near the light surface and partly transferred to the low-frequency oscillations. The total flux of the energy W_{tot} carried away from the star is, naturally, equal to the energy released when star rotation is decelerated by current (3.58).

Thus, the energetics of active processes proceeding both in the internal pulsar magnetosphere and in the region of the light cylinder are provided at the expense of the energy released in deceleration of neutron star rotation by electric current.

The compatibility relation From (3.58) it is seen that the neutron star's rotation energy losses are proportional to the total longitudinal current i_0 (3.51) circulating in the pulsar magnetosphere. Is this current an arbitrary free parameter? A detailed study of magnetosphere electrodynamics has shown that the current i_0 is not arbitrary but is connected by a certain relation with the potential jump Ψ between the neutron star's surface and the magnetosphere (3.49). This relation is called the compatibility relation. It has the form

$$\Psi = \Psi_{max}(\chi)\left[1 - \left(1 - \frac{i_0^2}{i_{max}^2(\chi)}\right)^{1/2}\right]$$
$$i_{max} \simeq 1.5 \cos^2 \chi \qquad (3.59)$$

The compatibility relation (3.59) plays the role of a non-linear Ohm's law for the current flowing in the magnetosphere. It implies, in particular, that the dimensionless longitudinal current i_0 in the magnetosphere (3.51) cannot exceed the maximum value i_{max}, in fact, that the longitudinal current density $j_\parallel$ is less than j_c (3.7).

We now recall that the potential jump Ψ itself is uniquely determined by the vacuum 'breakdown' condition – the condition of a stationary generation of electron–positron plasma (3.17). Therefore, it is precisely the plasma generation process that determines both the magnitude of the potential jump Ψ and, by virtue of the compatibility relation (3.59), also the magnitude of the longitudinal current i_0 and, accordingly, the total energy losses induced by neutron star rotation deceleration (3.58).

It should be emphasized that from (3.59) we can immediately conclude that a rotating neutron star becomes a radio pulsar only if electron–positron plasma is generated from the vacuum near this star. Indeed, since

$$W_{tot} \propto i_0 \lesssim (\Psi/\Psi_{max})^{1/2} \qquad (\Psi_{max} \gtrsim 10^{13}\ \mathrm{V})$$

the energy losses W_{tot} due to star rotation can be large enough to provide observed emission of power $L_r \sim 10^{26}$–10^{30} erg s^{-1} only in the case $\Psi > 10^9$–10^{10} V. Such high values of the potential drop Ψ_0 near the star surface can occur only when plasma is generated from the vacuum.

Magnetodipole losses If a neutron star has a plasma magnetosphere, its energy losses due to rotation are determined by the expression (3.58). Let us compare them with the energy losses of a rotation magnetic dipole, W_{md} (Landau & Lifshitz, 1983):

$$W_{md} = \frac{1}{6}\frac{B_0^2\Omega^4R^6}{c^3}\sin^2\chi \qquad (3.60)$$

The dipole losses W_{md} (3.60) arise in vacuum, i.e. in the absence of a plasma-filled magnetosphere. Comparing the plasma W_{tot} and vacuum W_{md} losses, we can see first of all their different dependence on the inclination angle χ between the rotation axis and the magnetic axis. The dipole losses are maximal if the axes are orthogonal, $\chi = \pi/2$, and vanish completely if $\chi = 0$. The plasma losses W_{tot} are, in contrast, maximal in the axisymmetric case, $\chi = 0$, and decrease with increasing angle χ. Such a variation of plasma losses with χ seems quite natural from the physical point of view. The point is that the primary reason for the losses is the necessity of plasma generation on open field lines, but the amount of plasma proportional to n_c falls with increasing angle χ (see (3.5) and (3.27)).

The most important difference of plasma losses from dipole ones is that they are proportional to the total longitudinal current i_0 (3.51). The dimensionless current i_0 cannot exceed i_{max} (3.59). For $i_0 \sim i_{max}$ the plasma losses are of the same order of magnitude as the dipole losses, whereas for $i_0 \ll i_{max}$ the plasma losses are much smaller. In particular, in complete absence of longitudinal current ($i_0 = 0$) a rotating star surrounded by a plasma magnetosphere does not lose energy at all, irrespective of the inclination angle χ. The physical reason for this is that in the absence of longitudinal current a magnetosphere plasma is polarized in such a way that it completely suppresses the magnetodipole radiation. The Pointing vector on the light cylinder has in this case only the toroidal component, so that the energy flux from the star is identically zero.

Conclusion We can see that the physical ground of the whole complex of processes proceeding in a pulsar magnetosphere is exceedingly simple and natural: all this is simply the result of fast rotation in a vacuum of a strongly magnetized conducting body, i.e. a neutron star.

Rotation creates an electric field whose action, in the presence of the strong curvilinear magnetic field of the star, leads to an effective generation of electron–positron plasma, i.e. to the formation of a magnetosphere. In the magnetosphere of a rotating star, the plasma starts to corotate, inducing corotation currents, owing to which distant magnetic field lines open and go to infinity. Along these lines, plasma goes away from the star, and therefore it must be continuously generated in the vicinity of magnetic poles. This is just the cause of the lasting activity in the region of the magnetic poles, leading, in particular, to the appearance of a powerful directed flux of pulsar radio emission. The energetics of all the processes is provided by deceleration of the neutron star rotation, induced by electric currents. The leading role in the creation and maintenance of pulsar activity is played by the longitudinal current i_0 which circulates in the magnetosphere.

It is of importance that in this picture the role of the surrounding medium and the details of the surface structure of the magnetized rotating body are unimportant. This allows us to consider the whole problem on the basis of an autonomous physically closed theory involving, in effect, no model assumptions.

Parameters of pulsar magnetosphere

Size of the magnetosphere (light cylinder radius) (cm)	$R_L = \dfrac{c}{\Omega} = 4.77 \times 10^9 P$
Radius of the polar cap (cm)	$R_0 = R\left(\dfrac{\Omega R}{c} f^*\right)^{1/2} = 1.83 \times 10^4\left(R_6^3 \dfrac{f^*(\chi)}{f^*(0)}\right)^{1/2} P^{-1/2}$
Angle of the opened field lines' divergence on the level r from neutron star centre (deg)	$\Theta = 2\alpha; \quad \alpha = \dfrac{3}{2}\left(\dfrac{\Omega R}{c} f^*\right)^{1/2} \cdot \left(\dfrac{r}{R}\right)^{1/2}$ $= 1.57^\circ\left(R_6 \dfrac{f^*(\chi)}{f^*(0)}\right)^{1/2} P^{-1/2}\left(\dfrac{r}{R}\right)^{1/2}$
Magnetic field strength at the light cylinder (Gauss)	$B_c = B_0\left(\dfrac{\Omega R}{c}\right)^3 = 9.2 B_{12} P^{-3} R_6^3$
Density of the corotation (cm^{-3})	$n_c = -\dfrac{\mathbf{B}\Omega}{2\pi c e} = -6.94 \times 10^{10} B_{12} P^{-1} \cos\chi$
Current density of the corotation ($A\ cm^{-2}$)	$j_c = c e n_c = -3.33 \times 10^2 B_{12} P^{-1} \cos\chi$
Cylotron frequency of the non-relativistic electron (s^{-1})	$\omega_B = \dfrac{e B_0}{m_e c} = 1.76 \times 10^{19} B_{12}$
Cylotron frequency of the relativistic electron (s^{-1})	$\dfrac{\omega_B}{\gamma} = 1.76 \times 10^{17} B_{12} \gamma_{100}^{-1}$
Multiplicity of the electron–positron plasma	$\lambda = \dfrac{n_e}{n_c} = 10^3–10^5$ $\lambda_4 = \lambda \cdot 10^{-4}$
Plasma frequency (s^{-1})	$\omega_p = \left(\dfrac{4\pi n_e e^2}{m_e}\right)^{1/2} = 1.49 \times 10^{12}(B_{12} P^{-1} \lambda_4)^{1/2}$
Relativistic plasma frequency in the strong magnetic field (s^{-1})	$\Omega_p = \dfrac{\omega_p}{\gamma^{3/2}} = 1.49 \times 10^9 (B_{12} P^{-1} \lambda_4 \gamma_{100}^{-3})^{1/2}$

Debye length (cm)	$D_c = \frac{c}{\Omega_p} = 2.02 \times 10^1 (B_{12} P^{-1} \lambda_4 \gamma_{100}^{-3})^{-1/2}$
Curvature radius of the magnetic field lines (cm)	$\rho_l = \frac{4}{3} R \left(\frac{\Omega R}{c} f\right)^{-1/2} \left(\frac{r}{R}\right)^{1/2}$ $= 7.30 \times 10^7 P^{1/2} \left(\frac{r}{R}\right)^{1/2} \left(R_6 \frac{f^*(0)}{f}\right)^{1/2}$
Frequency of the curvature photon (s^{-1})	$\nu_c = \frac{3}{4\pi} \frac{c}{\rho_l} \gamma^3 = 9.81 \times 10^7 P^{-1/2} \gamma_{100}^3$ $\times \left(\frac{r}{R}\right)^{-1/2} \left(R_6 \frac{f^*(0)}{f}\right)^{-1/2}$

4

Electrodynamics of the pulsar magnetosphere

In Chapter 3 we considered the physical picture of the electrodynamics of currents and plasma in pulsar magnetosphere. The aim of the present chapter is to formulate a consistent theory of these processes. In Section 4.1 we derive the general equations of stationary electrodynamics in a uniformly rotating pulsar magnetosphere. In the derivation, we make no model assumptions. The equations obtained are solved in Section 4.2 for the cylindrically symmetric case when the star's rotation axis coincides with the magnetic dipole axis. In Section 4.3 we investigate the boundary layer appearing near the 'light' surface. The basic details of the magnetosphere structure in the case of an arbitrary inclination angle of the magnetic dipole axis to the rotation axis are established in Section 4.4. The deceleration of neutron star rotation and the associated energetic processes in the pulsar magnetosphere are discussed in Section 4.5

4.1 Basic equations

The dynamics of the pulsar magnetosphere is determined by the interaction of electron–positron plasma with the magnetic field of a rotating neutron star. It is described by the system of Maxwell equations for the electric **E** and magnetic **B** fields and by kinetic equations for the distribution functions of electrons F^- and positrons F^+:

$$\operatorname{div} \mathbf{E} = 4\pi\rho_e \tag{4.1}$$

$$\operatorname{rot} \mathbf{E} = -\frac{1}{c}\frac{\partial \mathbf{B}}{\partial t} \tag{4.2}$$

$$\operatorname{rot} \mathbf{B} = \frac{4\pi}{c}\mathbf{j} + \frac{1}{c}\frac{\partial \mathbf{E}}{\partial t}, \qquad \operatorname{div} \mathbf{B} = 0 \tag{4.3}$$

$$\frac{\partial F^{\pm}}{\partial t} + Bv_{\parallel}\frac{\partial}{\partial r_{\parallel}}\left(\frac{F^{\pm}}{B}\right) \pm \frac{e}{B}(\mathbf{BE})\frac{\partial F^{\pm}}{\partial p_{\parallel}} + \mathbf{v}_{\perp}\frac{\partial F^{\pm}}{\partial \mathbf{r}_{\perp}} = q \tag{4.4}$$

$$\rho_e = e(n^+ - n^-), \qquad \mathbf{j}_e = e(n^+\mathbf{v}^+ - n^-\mathbf{v}^-), \qquad v_{\parallel} = cp_{\parallel}/(p_{\parallel}^2 + m_e^2c^2)^{1/2}$$

Here $r_{\parallel}$ is the coordinate along the magnetic field line, $\mathbf{r}_{\perp}$ is the orthogonal coordinate, $F^{\pm}(p_{\parallel}, \mathbf{r}, t)$ is the distribution function of electrons and positrons over the longitudinal momenta, e is the electron charge, $q(\mathbf{r}, t)$ is the density of the source of electron–positron pairs, ρ_e is the charge density, $\mathbf{j}_e$ is the current density, $n^{\pm}$ is the particle concentration, and $\mathbf{v}^{\pm}$ is the mean velocity of particles.

$$n^{\pm} = \int F^{\pm}\,dp_{\parallel}, \qquad n^{\pm}v_{\parallel}^{\pm} = \frac{1}{m_e}\int \frac{p_{\parallel}F^{\pm}}{\gamma}\,dp_{\parallel}$$

$$\mathbf{p}_{\perp} = m_e\gamma\mathbf{v}_{\perp}, \qquad \gamma = \left(1 + \frac{p_{\parallel}^2 + p_{\perp}^2}{m_e^2c^2}\right)^{1/2}$$

Here m_e is the rest mass of an electron, γ is the Lorentz factor, and $\mathbf{p}_{\perp}$ is the transverse moment of particle described by the equation

$$\frac{\partial \mathbf{p}_{\perp}}{\partial t} + (\mathbf{v}_{\perp}\nabla)\mathbf{p}_{\perp} \mp e\left[\mathbf{E}_{\perp} + \frac{1}{m_e c\gamma}[\mathbf{p}_{\perp}\mathbf{B}]\right] = 0 \tag{4.5}$$

In eq. (4.4) we assumed no thermal dispersion difference in the transverse momenta $\mathbf{p}_{\perp}$. This corresponds to real conditions in a pulsar magnetosphere and is connected with a rapid loss of transverse momentum by an electron or a positron in a strong magnetic field due to synchrotron radiation. The duration of synchrotron radiation is

$$\tau_r \simeq \frac{4 \times 10^{-16}\gamma}{B_{12}^2}\ \text{s} \qquad (B_{12} = B \times 10^{-12}\ \text{G}^{-1}) \tag{4.6}$$

In contrast, particle collisions which tend to randomize spherically (symmetrize) the distribution function are very rare – the time of the free path is

$$\tau_l \simeq 30\gamma^2 n_{16}^{-1}\ \text{s} \qquad (n_{16} = n_e 10^{-16}\ \text{cm}^3) \tag{4.7}$$

Under the conditions of pulsar magnetosphere we always have $\tau_l \gg \tau_r$. Owing to this, the velocity components orthogonal to $\mathbf{B}$ are determined only by hydrodynamic motion in the fields $\mathbf{E}$ and $\mathbf{B}$ (4.5), and it is only the longitudinal component $p_{\parallel}$ that has a kinetic spread. As a result, the complete distribution functions of electrons and positrons over longitudinal and transverse momenta have the form

$$F^{\pm}(\mathbf{p}, \mathbf{r}, t) = F^{\pm}(p_{\parallel}, \mathbf{r}, t)\, \delta[\mathbf{p}_{\perp} - \mathbf{p}_{\perp}^{\pm}(\mathbf{r}, t)] \tag{4.8}$$

4.1.1 Simplification of the equations

Stationary state The initial equations can be simplified if we consider only the stationary state of a rotating magnetosphere and take into account that the amount of plasma in it is limited.

First, as mentioned in Chapter 1 (Fig. 1.10), the averaged picture of a periodic pulsar emission, which characterizes the state of the magnetosphere plasma, appears to be rather stable. This is not surprising – as mentioned above (see Section 3.5), the characteristic time of the evolution of neutron star rotation parameters is of the order of million years, whereas the rotation period P itself is only ~ 1 s. The star may therefore be assumed to rotate uniformly and the plasma generation to be stationary. The solution of equations (4.1)–(4.5) then depends on the time t and on the azimuthal angle ϕ in the combination $\phi - \Omega t$. This allows us to eliminate the time t from the equations via the change $\phi' = \phi - \Omega t$. Then for the arbitrary scalar function $a(\mathbf{r}_{\perp}, \phi')$ we have

$$\frac{\partial a}{\partial t} = -\Omega \frac{\partial a}{\partial \phi} = -c \frac{\Omega \rho_{\perp}}{c} \nabla_{\phi} a$$

The vector function $\mathbf{a}(\mathbf{r}_{\perp}, \phi')$ varies with time even when it is ϕ-independent. The point is that stationarity of the vector $\mathbf{a}$ in a rotating coordinate system implies its uniform rotation from point to point. Therefore,

$$\frac{\partial \mathbf{a}}{\partial t} = -c \frac{\Omega \rho_{\perp}}{c} \nabla_{\phi} \mathbf{a} + [\mathbf{\Omega} \mathbf{a}]$$

It is convenient to introduce the quantity

$$\boldsymbol{\beta}_R = \frac{1}{c} [\mathbf{\Omega} \mathbf{r}]$$

which is a dimensionless rotation velocity. Then, taking into account that $\operatorname{div} \boldsymbol{\beta}_R = 0$, we can represent the time derivatives of a and $\mathbf{a}$ in the form

$$\frac{\partial a}{\partial t} = -c \boldsymbol{\beta}_R \nabla a$$

$$\frac{\partial \mathbf{a}}{\partial t} = -c(\boldsymbol{\beta}_R \nabla)\mathbf{a} + c(\mathbf{a}\nabla)\boldsymbol{\beta}_R = c \operatorname{rot}[\boldsymbol{\beta}_R \mathbf{a}] - c\boldsymbol{\beta}_R \operatorname{div} \mathbf{a} \tag{4.9}$$

With allowance for the equality (4.9), equation (4.2) can be written as

$$\text{rot}\,\mathbf{E} = -\text{rot}[\boldsymbol{\beta}_R\mathbf{B}]$$

From this it follows that

$$\mathbf{E} = -[\boldsymbol{\beta}_R\mathbf{B}] - \nabla\Psi \tag{4.10}$$

where Ψ is the electric field potential in the system of coordinates rotating with an angular velocity $\boldsymbol{\Omega}$. As is clear from (4.10), the total electric potential in a rest frame is equal to

$$\Phi = \Phi_c + \Psi \tag{4.11}$$

Here $\Phi_c = \boldsymbol{\beta}_R \cdot \mathbf{A}$ is the potential of a uniformly rotating body and $\mathbf{A}$ is the vector potential of the magnetic field $\mathbf{B} = \text{rot}\,\mathbf{A}$. When $\beta_R \ll 1$, the potential Ψ shows how the actual motion of plasma in the magnetosphere differs from its joint rotation (corotation) with the pulsar; it reflects the interaction of the magnetic field and currents with plasma and is the most important characteristic of the magnetosphere. To distinguish between Ψ and Φ, we henceforth call Ψ a potential and Φ a total electric field potential (see (3.12)).

Taking into account (4.10), we obtain from (4.1) the expression for the charge density:

$$\rho_e = \rho_c + \frac{1}{4\pi}\boldsymbol{\beta}_R \cdot \text{rot}\,\mathbf{B} - \frac{1}{4\pi}\Delta\Psi \tag{4.12}$$

where ρ_c is the density of the Goldreich–Julian corotation charge (Goldreich & Julian, 1969) appearing in a plasma uniformly rotating in a magnetic field (3.5),

$$\rho_c = -\frac{\boldsymbol{\Omega}\mathbf{B}}{2\pi c}, \qquad n_c = \frac{\rho_c}{|e|} \tag{4.13}$$

It determines the characteristic value of the charge density ρ_e and electric current density $\mathbf{j}_e$ in a pulsar magnetosphere (3.7):

$$j_c = c\rho_c = -\frac{\boldsymbol{\Omega}\mathbf{B}}{2\pi} \tag{4.14}$$

Making use of (4.9) and (4.10), we obtain from (4.3)

$$\text{rot}\,\mathbf{B} = \frac{4\pi}{c}\mathbf{j}_e - [\boldsymbol{\beta}_R \cdot \text{rot}[\boldsymbol{\beta}_R\mathbf{B}]] + \nabla(\boldsymbol{\beta}_R\nabla\Psi) \tag{4.15}$$

$$\text{div}\,\mathbf{B} = 0 \tag{4.16}$$

The Maxwell equations thus acquire the stationary form (4.10)–(4.16) (Mestel, 1971; Beskin *et al.*, 1983).

The basic dimensionless parameters We now take into account the presence of electron–positron plasma. On open field lines, it is created by the source q (4.4) in immediate vicinity of a neutron star, and then moves from the star at a velocity close to c. We assume the plasma density n_e in magnetosphere to meet the requirement

$$n_c \ll n_e \ll n_B \tag{4.17}$$

Here n_c is the corotation density (4.13), and

$$n_B = \frac{B^2}{8\pi\langle\mathscr{E}\rangle} \tag{4.18}$$

where $\langle\mathscr{E}\rangle$ is the mean energy of plasma particles.

The condition $n_c \ll n_e$ implies that the plasma is quasi-neutral, i.e. that plasma polarization due to rotation leads only to a small charge separation: $|n^+ - n^-|/n_e \ll 1$. If the condition $n_e \ll n_B$ holds, the plasma energy density is always much lower than the magnetic field energy density, so that the plasma pressure is insignificant.

The inequalities (4.17) suggest that in the region of open field lines the electron–positron plasma source on the one hand is effective in filling the magnetosphere with a dense plasma, and on the other hand is not so powerful as to destroy the main structure of the magnetic field of the neutron star. These are just the conditions to be met by the plasma generation mechanism considered in Section 3.2.

Thus, in the theory of the magnetosphere, two large parameters, λ and μ, are singled out:

$$\lambda = \frac{n_e}{n_c} \gg 1 \tag{4.19}$$

$$\mu = \frac{n_B}{n_e} \gg 1 \tag{4.20}$$

The quantity λ is the multiplicity parameter describing the efficiency of plasma production in the pulsar magnetosphere. It plays the decisive role both for magnetosphere structure and for generation of radio emission. Usually, $\lambda \sim 10^3$–10^5 (see (3.27)). The parameter μ (4.20) shows the factor by which the magnetic field energy exceeds the plasma particle energy. As a rule, $\mu \sim 10^2$–10^4 (see (3.28)).

Simplification of the equations Making use of the parameters λ (4.19) and μ (4.20) we can expand the initial equations (4.1)–(4.5), in λ^{-1} and μ^{-1}. Owing to condition (4.19), the electric field $E_{\parallel}$ of polarization appears to be small since its magnitude is inversely proportional to the plasma density:

$$E_{\parallel} \propto n_e^{-1} \propto \lambda^{-1} \tag{4.21}$$

Therefore, in the zeroth approximation with respect to λ^{-1} we may assume that $E_{\parallel} = 0$, i.e. that the potential Ψ entering (4.10), (4.12) and (4.15) depends on $\mathbf{r}_{\perp}$ only

$$E_{\parallel} = 0, \qquad \Psi = \Psi(\mathbf{r}_{\perp}) \tag{4.22}$$

Now consider the electric current $\mathbf{j}$, more precisely, its components $\mathbf{j}_{\perp}$ orthogonal to the magnetic field $\mathbf{B}$:

$$\mathbf{j}_{\perp} = e(n^{+}\mathbf{v}_{\perp}^{+} - n^{-}\mathbf{v}_{\perp}^{-}). \tag{4.23}$$

The electron and positron velocities $\mathbf{v}_{\perp}^{\pm}$ are of the order of the rotation velocity $v_{\perp} \sim \Omega\rho_{\perp}$ of magnetic field lines ($\rho_{\perp}$ is the distance from the rotation axis). Consequently, in the entire region from the star surface to the 'light' surface, the condition $v_{\perp} < c$ holds. Under this condition the main role in (4.5) is played by the electron and positron drift in crossed fields:

$$\mathbf{v}_{\perp}^{+} = \mathbf{v}_{\perp}^{-} = \frac{c}{B^2}[\mathbf{EB}], \qquad \mathbf{j}_{\perp} = \frac{c}{B^2}[\mathbf{EB}]\rho_e \tag{4.24}$$

The corrections to $\mathbf{v}_{\perp}^{\pm}$ can be easily found from (4.5). They are of the order of $(\lambda\mu)^{-1}\mathbf{v}_{\perp}^{\pm}$ (for details see Beskin *et al.* (1983)). Thus, in the zeroth approximation with respect to $(\lambda\mu)^{-1}$ the drift approximation (4.24) holds, and the current $\mathbf{j}_{\perp}$ is given by the simple expressions (4.24) and (4.10).

As far as the longitudinal component $j_{\parallel}$ is concerned, it is free in the presence of collisionless plasma and the absence of longitudinal electric field (4.22), i.e. it is specified by the conditions on the boundaries and by the charge continuity equation

$$\frac{\partial \rho_e}{\partial t} + \operatorname{div} \mathbf{j}_e = 0 \tag{4.25}$$

which, under the stationarity conditions (4.9), acquires the form ($\operatorname{div} \boldsymbol{\beta}_R = 0$)

$$\operatorname{div}(\mathbf{j}_e - c\rho_e\boldsymbol{\beta}_R) = 0 \tag{4.26}$$

Thus, in the zeroth approximation with respect to the small parameters

λ^{-1} (4.19) and μ^{-1} (4.20), the quasi-stationary electrodynamic processes in the pulsar magnetosphere are described by the closed system of equations (4.10)–(4.15), (4.24), (4.25). Substituting the expression (4.10) for the electric field into (4.24) we obtain

$$\mathbf{j}_e = -\frac{c\rho_e}{B^2}[[\boldsymbol{\beta}_R\mathbf{B}]\mathbf{B}] - \frac{\rho_e c}{B^2}[\nabla\Psi\mathbf{B}] + i\mathbf{B} \tag{4.27}$$

or

$$\mathbf{j}_e = c\boldsymbol{\beta}_R\rho_e - c\frac{\rho_e}{B^2}[\nabla\Psi\mathbf{B}] + i_{\parallel}\mathbf{B} \tag{4.28}$$

$$i_{\parallel} = i - \frac{c\rho_e(\boldsymbol{\beta}_R\mathbf{B})}{B^2} \tag{4.29}$$

The quantity $i_{\parallel}$ is proportional to the longitudinal electric current and is described by the continuity equation (4.25).

We shall now find the equation specifying the magnetic field structure. To do so, we shall first express the plasma charge density ρ_e in terms of the magnetic field. We shall make use of eq. (4.12), whose second term on the right-hand side we find from (4.15) and (4.28):

$$\boldsymbol{\beta}_R \operatorname{rot}\mathbf{B} = \frac{4\pi}{c}\mathbf{j}_e\boldsymbol{\beta}_R + \boldsymbol{\beta}_R\nabla(\boldsymbol{\beta}_R\nabla\Psi)$$

$$= 4\pi\rho_e\beta_R^2 - \frac{4\pi\rho_e c}{B^2}\boldsymbol{\beta}_R[\nabla\Psi\mathbf{B}] + \frac{4\pi}{c}i_{\parallel}(\boldsymbol{\beta}_R\mathbf{B}) + \boldsymbol{\beta}_R\nabla(\boldsymbol{\beta}_R\nabla\Psi) \tag{4.30}$$

Substituting (4.30) into (4.12), we define ρ_e as a function of the magnetic field $\mathbf{B}$ and of the potential Ψ,

$$\rho_e = \frac{-\dfrac{\boldsymbol{\Omega}\mathbf{B}}{2\pi c} + \dfrac{i_{\parallel}}{c}(\boldsymbol{\beta}_R\mathbf{B}) - \dfrac{1}{4\pi}[\Delta\Psi - \boldsymbol{\beta}_R\nabla(\boldsymbol{\beta}_R\nabla\Psi)]}{1 - \beta_R^2 + \boldsymbol{\beta}_R[\nabla\Psi\mathbf{B}]/B^2} \tag{4.31}$$

Accordingly, the current $\mathbf{j}_e$ is described by the expression (4.28) with the density ρ_e specified by (4.31). Substituting it into (4.15), we finally arrive at the non-linear equation for $\mathbf{B}$ (Beskin *et al.*, 1983).

$$\operatorname{rot}\{\mathbf{B}(1 - \beta_R^2) + \boldsymbol{\beta}_R(\boldsymbol{\beta}_R\mathbf{B}) + [\boldsymbol{\beta}_R\nabla\Psi]\} = \frac{4\pi}{1 - \beta_R^2 + \boldsymbol{\beta}_R[\nabla\Psi\mathbf{B}]/B^2}$$
$$\left\{\frac{i_{\parallel}}{c}[(1 - \beta_R^2)\mathbf{B} + [\boldsymbol{\beta}_R\nabla\Psi]] + \frac{[\nabla\Psi\mathbf{B}]}{B^2}\left[\frac{\boldsymbol{\Omega}\mathbf{B}}{2\pi c} + \frac{1}{4\pi}(\Delta\Psi - \boldsymbol{\beta}_R\nabla(\boldsymbol{\beta}_R\nabla\Psi))\right]\right\} \tag{4.32}$$

To this we should add the equation

$$\operatorname{div} \mathbf{B} = 0 \tag{4.33}$$

The system of equations (4.32, 4.33) is the fundamental system of the theory of the pulsar magnetosphere completely filled with plasma. It describes the structure of the quasi-stationary magnetic field in the pulsar magnetosphere with allowance for the electric field $-\nabla\Psi$ and longitudinal current $i_{\|}$.

Note that since the charge continuity equation (4.25) follows from the Maxwell equations, it is not independent but is implied by (4.32). Therefore, the magnitude of longitudinal current $i_{\|}$ in the pulsar magnetosphere is also determined by eq. (4.32) and by boundary conditions. The potential $\Psi(\mathbf{r}_{\perp})$ is constant along magnetic field lines. In eq. (4.32) this potential and the longitudinal current $i_{\|}$ play the role of sources. It should be emphasized that the non-linear character of eq. (4.32) consists not only of the direct dependence of the coefficients on the magnetic field but also of the requirement of constancy of the potential Ψ on magnetic field lines, which themselves are determined by the solution of eqs. (4.32) and (4.33). This leads, in particular, to a complicated non-linear dependence of the solutions of eqs. (4.32) and (4.33) on the conditions on the magnetosphere boundaries.

4.1.2 Boundary conditions

Boundary conditions to eqs. (4.32), (4.33) have the following form. Near the star's surface, on the lower boundary of the magnetosphere, $S = S_0(\mathbf{r})$, the magnetic field of a pulsar is determined by the current flowing inside the star and along its surface

$$\mathbf{B}|_{S_0} = \mathbf{B}_0(\mathbf{r}) \tag{4.34}$$

Moreover, the longitudinal current

$$\mathbf{j}_e|_{S_0} = i_{\|}(\mathbf{r}_{\perp})\mathbf{B}_0 \tag{4.35}$$

flowing in and out of the magnetosphere, as well as the electric field potential

$$\Psi|_{S_0} = \Psi(\mathbf{r}_{\perp}) \tag{4.36}$$

are given here.

The regions of outflow are essentially different for closed and open (i.e.

going to infinity) magnetic field lines (see Section 3.1). Assuming the star's conductivity to be infinite (see Section 2.5), we may assume that in the region of closed field lines V_c there are no longitudinal currents and the magnetosphere corotates with the star. Therefore,

$$i_{\parallel}|_{V_c} = 0, \qquad \Psi|_{V_c} = 0 \tag{4.37}$$

This condition should be clarified. The point is that in a closed region of a magnetosphere the electric charge $\sim\rho_c$ (4.13) is necessary for the occurrence of corotation. To maintain the corotation, electric fields $\delta\Psi$ and currents $\delta i_{\parallel}$ are needed. However, $\delta\Psi \lesssim \langle\mathscr{E}\rangle/|e| \simeq 6 \times 10^7$ V (3.26), whereas a continuous generation of electron–positron plasma on open field lines requires the potentials $\Psi \simeq 10^{13}$ V (3.19). From this it follows that

$$\frac{\delta\Psi}{\Psi} \sim 10^{-5} \ll 1 \tag{4.38}$$

An estimate of the same order is valid for the currents. The condition (4.37) therefore holds everywhere in a closed magnetosphere in the zeroth approximation with respect to the parameter (4.38).

Another boundary condition arises on the surface S_d defined by the relation

$$1 - \beta_R^2 + \frac{\boldsymbol{\beta}_R[\nabla\Psi\mathbf{B}]}{B^2} = 0 \tag{4.39}$$

The right-hand side of eq. (4.32) has a singularity on this surface. The requirement that the magnetic field lines intersect the singular surface S_d, i.e. that the electric charge ρ_e (4.31) and the current $\mathbf{j}_e$ (4.28) remain finite on the surface, is a natural boundary condition of the problem. Note that the form of the singular surface is determined by the total electric field $\mathbf{E} = -[\boldsymbol{\beta}_R\mathbf{B}] - \nabla\Psi$. In the absence of the potential Ψ, i.e. for $\nabla\Psi = 0$, the whole of the magnetosphere rotates uniformly with an angular frequency $\boldsymbol{\Omega}$. Given this, the singular surface S_d (4.39) has the form of a cylinder of radius c/Ω. On this cylinder, the total electric field E becomes equal in magnitude to the magnetic field B, and the drift velocity of particle motion reaches the velocity of light. Consequently, the singular surface S_d coincides in this case with the *light surface* S_c. It is called the light cylinder. In the presence of the field $-\nabla\Psi$, the magnetosphere rotates non-uniformly, so that in the general case the light surface does not coincide with the light cylinder. It always lies farther from the pulsar than the light cylinder and than the singular surface S_d.

Note once again that near the light surface S_c the drift velocity of particles, $v_\perp^\pm$ (4.24), approaches the velocity of light. Therefore, near the light surface the particle energy (i.e. $\langle \mathscr{E} \rangle$) increases and, accordingly, the expansion parameter μ^{-1} (4.20, 4.18) also increases sharply. Owing to this fact, the conditions of applicability of eq. (4.32) are violated near S_c. A special boundary layer is formed here in which it is not enough to consider the drift approximation (4.24), but a more exact solution of the equation of electron and positron motion (4.5) is required. This problem is considered separately in Section 4.3.

In addition to the above-mentioned boundary conditions, it is also necessary to require the condition of disappearance of fields at infinity:

$$\mathbf{B} \to 0, \qquad \mathbf{r} \to \infty \tag{4.40}$$

At first glance this requirement may seem to give nothing new, since eqs. (4.32) and (4.33) hold only inside the light surface. But, as is readily seen (see Fig. 3.1), in the vicinity of the rotation axis of the star there exist field lines going to infinity inside the light surface. On these field lines, the field must vanish at infinity (4.40), and not merely vanish but fail to form separate closed structures not connected with the neutron star surface. Otherwise the boundary conditions (4.34)–(4.39) are not enough to determine the longitudinal current and the electric field potential, and there appears an uncertainty in the problem.

4.2 The axisymmetric case

4.2.1 Statement of the problem

The basic equation Let us first consider the simplest case. We assume an unperturbed magnetic field of a pulsar to be dipole and the dipole axis to be parallel to the rotation axis. In this case the problem is axisymmetric, i.e. all the functions depend only on two coordinates: z along the rotation axis and $\rho_\perp$ orthogonal to it. Note that the assumption concerning the dipole character of the unperturbed magnetic field of a pulsar in the problem of the structure of magnetosphere is quite natural. Indeed, a strong field perturbation by magnetospheric current shows up at distances which are always much larger than the star radius R; usually, $c/\Omega R \sim 10^4$ (see Section 3.1). The multipole fields of the currents flowing on the star surface or inside the star fall rapidly at large distances $r \gg R$. Therefore for $R \ll r \ll c/\Omega$ the dipole field always remains principal. This is confirmed by the observational data on pulsar radiation (see Chapter 1).

Suppose the relation $f(\rho_\perp, z) = \text{const}$ describes the magnetic surfaces where $f(\rho_\perp, z)$ is a scalar function. Then, by the definition of a magnetic surface,

$$\mathbf{B}\,\nabla f = 0 \tag{4.41}$$

Since, in addition, $\nabla f \cdot \nabla \Phi_0(\phi) = 0$, where $\Phi_0(\phi)$ is an arbitrary function of the azimuthal angle ϕ, the magnetic $\mathbf{B}$ can be represented in the general form as

$$\mathbf{B} = [\nabla f\ \nabla \Phi_0] + g\nabla\Phi_0$$

where g is an arbitrary function of the coordinates $\rho_\perp$, z: $g = g(\rho_\perp, z)$. Making use of eq. (4.33), div $\mathbf{B} = 0$, we find that $g\,\Delta\Phi_0 = 0$ or $\Phi_0 = \phi$. Taking into account that $\nabla\phi = (1/\rho_\perp)\mathbf{e}_\phi$, where $\mathbf{e}_\phi$ is the unit vector in the direction of the rotation angle ϕ, we represent $\mathbf{B}$ in the form

$$\mathbf{B} = \frac{1}{\rho_\perp}[\nabla f\ \mathbf{e}_\phi] + \frac{1}{\rho_\perp} g\mathbf{e}_\phi \tag{4.42}$$

By virtue of (4.22) the electric field potential Ψ is constant along magnetic field lines, and therefore

$$\Psi = \Psi(f) \tag{4.43}$$

Now substituting (4.42) and (4.43) into the main vector equation (4.32), we come to three equations for the scalar functions f, g and $i_\parallel$. Two of them, corresponding to the $\rho_\perp$ and z components of eq. (4.32), are represented as

$$\begin{gathered}\nabla g = \frac{1}{b}\left[\frac{4\pi i_\parallel}{c}\left(1 - \frac{\Omega^2\rho_\perp^2}{c^2} - \frac{\Omega\rho_\perp^2}{c}\frac{d\Psi}{df}\right) + \frac{\rho_\perp^2 g}{g^2 + (\nabla f)^2}\frac{d\Psi}{df}b_i\right]\nabla f \\ b = 1 - \frac{\Omega^2\rho_\perp^2}{c^2} - \frac{\Omega\rho_\perp^2}{c}\frac{d\Psi}{df}\frac{(\nabla f)^2}{g^2 + (\nabla f)^2}, \qquad b_1 = \frac{2\Omega}{c\rho_\perp}\frac{\partial f}{\partial \rho_\perp} + \Delta\Psi\end{gathered} \tag{4.44}$$

whence

$$g = g(f)$$

This relation can easily be verified by introducing the coordinate x^1 which lies on the magnetic surface and is orthogonal to $\mathbf{e}_\phi$. Then, by virtue of the axial symmetry $g = g(x^1, f)$, it follows from (4.44) that $\partial g/\partial x^1 = 0$, i.e. $g = g(f)$ and the current $i_\parallel$ is given by the expression

$$\frac{4\pi}{c}i_\parallel = \left(b\frac{dg}{df} - b_1\frac{\rho_\perp^2 g}{g^2 + (\nabla f)^2}\frac{d\Psi}{df}\right)\left(1 - \frac{\Omega^2\rho_\perp^2}{c^2} - \frac{\Omega\rho_\perp^2}{c}\frac{d\Psi}{df}\right)^{-1} \tag{4.45}$$

Now writing the equation corresponding to the component of eq. (4.32) with respect to $\mathbf{e}_\phi$ and discarding $i_\parallel$ according to (4.45), we bring it to the final form

$$-\Delta f\left[1-\frac{\rho_\perp^2\Omega^2}{c^2}\left(1+\frac{c}{\Omega}\frac{d\Psi}{df}\right)^2\right]+\frac{2}{\rho_\perp}\frac{df}{d\rho_\perp}-g\frac{dg}{df}$$
$$+\frac{\rho_\perp^2\Omega}{c}\left(1+\frac{c}{\Omega}\frac{d\Psi}{df}\right)\frac{d^2\Psi}{df^2}(\nabla f)^2=0 \quad (4.46)$$

For various particular cases, eq. (4.46) was obtained by Cohen & Rosenblum (1972); Mestel (1973); Scharlemann & Wagoner (1973); Ingraham (1973); Julian (1973); Michel (1973a); Salvati (1973); Endean (1974); Okamoto (1974); Hinata & Jackson (1974); Coppi & Pegoraro (1979). It is an extension of the Grad–Shafranov equation to the case $\Psi \neq 0$.

It is now convenient to go over to dimensionless variables and functions (shown as primed):

$$\rho_\perp=\frac{c}{\Omega}\rho'_\perp,\quad z=\frac{c}{\Omega}z',\quad f=\frac{M\Omega}{c}f',\quad \Psi=\frac{M\Omega^2}{c}\Psi'$$
$$g=\frac{M\Omega^2}{c^2}g',\quad \mathbf{B}=\frac{M\Omega^3}{c^3}\mathbf{B}',\quad i_\parallel=\frac{\Omega}{4\pi}i'_\parallel,\quad \rho_e=\frac{M\Omega^4}{4\pi c^4}\rho'_e \quad (4.47)$$

where M is the dipole magnetic moment. Writing eq. (4.46) in dimensionless variables and omitting the index primes, we obtain

$$-\Delta f\left[1-\rho_\perp^2\left(1+\frac{d\Psi}{df}\right)^2\right]+\frac{2}{\rho_\perp}\frac{\partial f}{\partial\rho_\perp}-g\frac{dg}{df}+\rho_\perp^2\left(1+\frac{d\Psi}{df}\right)\frac{d^2\Psi}{df^2}(\nabla f)^2=0 \quad (4.48)$$

Thus, we have reduced the problem to one equation (4.48) for the scalar function $f(\rho_\perp, z)$. The equation is non-linear, it depends on the functions $\Psi(f)$ and $g(f)$ which play the role of sources. In the absence of sources $\Psi=0$, $g=0$, eq. (4.48) assumes the simple form

$$-\Delta f[1-\rho_\perp^2]+\frac{2}{\rho_\perp}\frac{\partial f}{\partial\rho_\perp}=0 \quad (4.49)$$

Boundary conditions A concrete form of the functions $\Psi(f)$ and $g(f)$ in (4.48) is specified by boundary conditions. Indeed, the function $\Psi(f)$ is given directly by the condition (4.36) on the lower boundary of the magnetosphere S_0. As concerns $g(f)$, one should bear in mind that the boundary S_0 corresponds to the values $\rho_\perp \lesssim R$ and, accordingly, $\rho'_\perp = \Omega\rho_\perp/c \ll 1$. With allowance made for (4.46), the relation (4.45) for S_0 therefore becomes

$$i_{\|0} \equiv i_\| |_{S_0} = \frac{dg}{df} \tag{4.50}$$

This implies that the function $g(f)$ is defined directly by the longitudinal current, $i_\|$, given by the boundary condition (4.35). The total current I on the pulsar surface, when the surface conductivity is isotropic, is described by the function $g(f)$. Indeed,

$$I = 2\pi \int i_\| B_z \rho_\perp \, d\rho_\perp = \frac{M\Omega^2}{2c} g(f) \tag{4.51}$$

Recall that the field Ψ and longitudinal currents in the pulsar magnetosphere are given by the conditions (4.35) and (4.36) on open field lines only; on closed field lines, according to (4.37),

$$\Psi(f) = g(f) = 0 \qquad \text{at } f \geqslant f^* \tag{4.52}$$

Here $f = f^*$ is the separatrix between the regions of open and closed field lines. Bearing in mind that under stationary conditions the current flowing to the star surface from the magnetosphere should be equal to zero, we obtain

$$g(f^*) = g(0)$$

and taking into account (4.52), we have the condition on the pole

$$g(f)|_{f \to 0} = 0 \tag{4.53}$$

The boundary condition to eq. (4.48) is specified according to (4.34). As $\rho_\perp^2 + z^2 \to 0$, the magnetic field must be dipole. For the magnetic dipole,

$$\mathbf{B} = \operatorname{rot} \frac{1}{r^3} [\mathbf{Mr}] = \operatorname{rot} \frac{M}{r^3} \rho_\perp \mathbf{e}_\phi$$

On the other hand, in the absence of current ($g = 0$) the expression (4.42) yields

$$\mathbf{B} = \frac{1}{\rho_\perp} [\nabla f \, \mathbf{e}_\phi] = \operatorname{rot} f \frac{\mathbf{e}_\phi}{\rho_\perp}$$

From comparison of these two equations, with allowance for the definition of the dimensionless functions (4.47), we obtain

$$f|_{\rho_\perp^2+z^2\to 0} \to \frac{\rho_\perp^2}{(\rho_\perp^2+z^2)^{3/2}} \tag{4.54}$$

Moreover, the function $f(\rho_\perp, z)$ should not have divergences on the singular surface S_d,

$$\rho_\perp = \rho_\perp^*(f) = \left(1+\frac{d\Psi}{df}\right)^{-1} \tag{4.55}$$

on which the longitudinal electric current and the charge density go to infinity (see (4.45)). From (4.48) it follows then that

$$\frac{2}{\rho_\perp}\frac{\partial f}{\partial \rho_\perp}\bigg|_{\rho_\perp=\rho_\perp^*} = g\frac{dg}{df} - \left(1+\frac{d\Psi}{df}\right)^{-1}\frac{d^2\Psi}{df^2}(\nabla f)^2|_{\rho_\perp=\rho_\perp^*} \tag{4.56}$$

It is also obvious that $f(\rho_\perp, z)$ should be symmetric relative to the equatorial plane:

$$f(\rho_\perp, z) = f(\rho_\perp, -z)$$

It is also necessary that there hold the condition (4.40) at infinity in the neighbourhood of the rotation axis, $\rho_\perp \to 0$, $z \to \infty$, and it takes the form

$$\mathbf{B}|_{\rho_\perp\to 0, z\to\infty} = \frac{1}{\rho_\perp}([\nabla f\,\mathbf{e}_\phi] + g\mathbf{e}_\phi)|_{\rho_\perp\to 0, z\to\infty} \to 0 \tag{4.57}$$

The relations (4.52)–(4.47) are just the complete set of boundary conditions to eq. (4.48).

4.2.2 Structure of the magnetosphere

Functions of sources We now proceed to the solution of eq. (4.48). It should be noted that it is simplified by choosing the source functions $\Psi(f)$ and $g(f)$ to be

$$\Psi(f) = \Psi_0 - \beta_0 f, \qquad g = i_0 f \tag{4.58}$$

These relations have a simple physical meaning. According to the first relation, the total electric field in the magnetosphere (4.10) is $\mathbf{E} = -(1-\beta_0)\nabla f$. In such a field, plasma drifts with a constant angular frequency $\mathbf{\Omega}' = (1-\beta_0)\mathbf{\Omega}$. The parameter β_0 therefore describes the drift velocity deceleration by magnetospheric plasma (β_0 is always positive).

As is clear from (4.50), the second relation (4.58) corresponds to the existence of a constant longitudinal current i_0 near the pulsar surface. Equation (4.48), with allowance for (4.58), becomes

$$\Delta f[1 - \rho_\perp^2(1 - \beta_0)^2] - \frac{2}{\rho_\perp}\frac{\partial f}{\partial \rho_\perp} + i_0^2 f = 0$$

It obeys the following similarity law – in the variables

$$x_1 = \rho_\perp(1 - \beta_0), \qquad z_1 = z(1 - \beta_0) \tag{4.59}$$

it has the form

$$\Delta f(1 - x_1^2) - \frac{2}{x_1}\frac{\partial f}{\partial x_1} + \alpha_1 f = 0 \tag{4.60}$$

and, accordingly, depends on only one parameter:

$$\alpha_1 = \frac{i_0^2}{(1 - \beta_0)^2}$$

Equation (4.60) for the function f is linear. It is natural to seek its solution by expansion in a Fourier integral over z_1. Going over to the variables (4.59) and taking into account the symmetry condition $f(x_1, z_1) = f(x_1, -z_1)$, we represent the solution in the form

$$f(x_1, z_1) = \int_0^\infty dk\, R_k(x_1) \cos kz_1 \tag{4.61}$$

where the functions $R_k(x_1)$, as is clear from (4.60), satisfy the equation

$$\frac{d^2 R_k}{dx_1^2} - \frac{1 + x_1^2}{x_1(1 - x_1^2)}\frac{dR_k}{dx_1} + \left(\alpha_1 \frac{1 + x_1^2}{1 - x_1^2} - k^2\right) R_k = 0 \tag{4.62}$$

Asymptotic behaviour near the axis In the dimensionless variables (4.47), the radius of a neutron star $R' = \Omega R/c \to 0$. Hence, as $x_1 \to 0$, taking account of the condition (4.54), for the solutions of (4.62) we should choose the functions

$$\begin{aligned} R_k^0 &= \frac{2}{\pi}(1 - \beta_0)(k^2 - \alpha_1)^{1/2} x_1 K_1[(k^2 - \alpha_1)^{1/2} x_1], \qquad k^2 > \alpha_1 \\ R_k^0 &= -(1 - \beta_0)(\alpha_1 - k^2)^{1/2} x_1 Y_1[(\alpha_1 - k^2)^{1/2} x_1], \qquad k^2 < \alpha_1 \end{aligned} \tag{4.63}$$

which specify correctly the dipole magnetic field as $z \to 0$. Indeed, since

$$\int_0^\infty dk\, R_k^0 \cos kz_1 = -(1-\beta_0)\alpha_1\left(\frac{\pi}{2}\right)^{1/2}\frac{x_1^2}{r}\frac{Y_{3/2}(\alpha_1^{1/2}r)}{\alpha_1^{1/4}r^{1/2}}, \qquad r^2 = x_1^2 + z_1^2 \tag{4.64}$$

it follows that $f \to (1-\beta_0)\, x_1^2/r^3$ as $r \to 0$. Here K_1 and $Y_{1,3/2}$ are respectively Macdonald and Bessel functions of the second kind. However, when $r > \alpha_1^{-1/2}$, the function f given by the expression (4.64) starts oscillating, assuming both positive and negative values, whereas on the neutron star surface $f \geqslant 0$ always. This means that near the dipole axis ($x_1 \to 0$) at large distances from the star, the field has magnetic islands formed by the proper current and separated from the star. This contradicts the initial statement of the problem (see (4.40)). It is therefore necessary to require that f should be positive definite everywhere.

The general solution for f as $x_1 \to 0$ will be

$$f = \int_0^\infty dk\, R_k^0(x_1)\, \phi_1(\alpha_1, k) \cos kz_1 \tag{4.65}$$

where the function $\phi_1(\alpha_1, k)$ must satisfy the conditions

$$\phi_1(\alpha_1, k) \to 1 \qquad \text{as } k \to \infty \tag{4.66}$$

(this gives us the dipole magnetic field as $z_1 \to 0$), and

$$\phi_1(\alpha_1, k) \to 1 \qquad \text{as } \alpha_1 \to 0 \tag{4.67}$$

for all k (i.e. in the absence of current $f \to 0$ as $x_1 \to 0$). We find the form of the function $\phi_1(\alpha_1, k)$ from the condition of the absence of f oscillations in the region $z_1 \to \infty$. To do so, we consider the quantity

$$f^0 = \int_0^\infty dk\, R_k^0(x_1) \cos kz_1 = f_1^0 + f_2^0$$

$$f_1^0 = \frac{2}{\pi}(1-\beta_0)\int_{\alpha_1^{1/2}}^\infty dk(k^2-\alpha_1)^{1/2}x_1 K_1[(k^2-\alpha_1)^{1/2}x_1]\cos kz_1 \tag{4.68}$$

$$f_2^0 = -(1-\beta_0)\int_0^{\alpha_1^{1/2}} dk(\alpha_1-k^2)^{1/2}x_1 Y_1[(\alpha_1-k^2)^{1/2}x_1]\cos kz_1$$

We transform f_1^0 making use of the integral representation of the function K_1,

$$\xi K_1(\xi) = \int_0^\infty \frac{\cos \xi t}{(1+t^2)^{3/2}}\, dt$$

Then

$$f_1^0 = \frac{2}{\pi}(1-\beta_0)\int_0^\infty \frac{I(x_1, z_1, t, \alpha_1)}{(1+t^2)^{3/2}}\, dt$$

where

$$I = \int_0^\infty d\lambda \frac{\lambda}{(\lambda^2+\alpha_1)^{1/2}} \cos(\lambda x_1 t) \cos[(\lambda^2+\alpha_1)^{1/2} z_1] \qquad (4.69)$$

Let us single out the dipole component:

$$I_d = \int_0^\infty d\lambda \cos \lambda x_1 t \cos \lambda z_1 = \frac{\pi}{2}\delta(x_1 t - z_1)$$

$$\delta I = I - I_d = \delta I_1 + \delta I_2$$

$$\delta I_1 = \frac{1}{4}\int_{-\infty}^{\infty} d\lambda\, e^{i\lambda x_1 t}\left[\frac{\lambda}{\sigma(\lambda^2+\alpha_1)^{1/2}} e^{i\sigma(\lambda^2+\alpha_1)^{1/2} z_1} - e^{i\lambda z_1}\right] \qquad (4.70)$$

$$\delta I_2 = \frac{1}{4}\int_{-\infty}^{\infty} d\lambda\, e^{-i\lambda x_1 t}\left[\frac{\lambda}{\sigma(\lambda^2+\alpha_1)^{1/2}} e^{i\sigma(\lambda^2+\alpha_1)^{1/2} z_1} - e^{i\lambda z_1}\right]$$

$$\sigma = \mathrm{sign}(\lambda)$$

Integration in (4.70) will be performed in the complex region of λ, and the integration contour for δI_1 should be chosen as shown in Fig. 4.1.

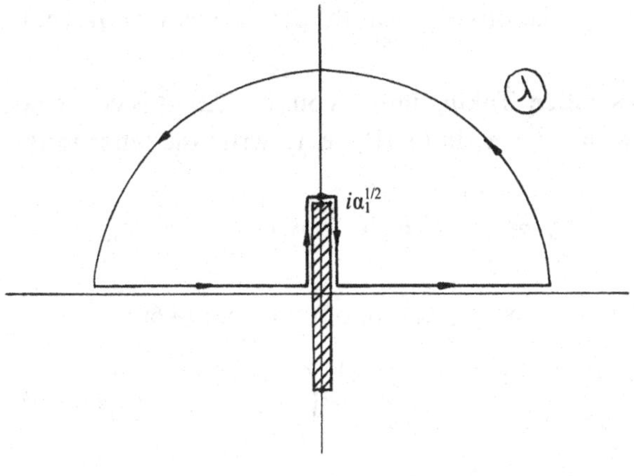

Fig. 4.1. Integration contour in formula (4.70).

Similarly, for δI_2 the integration contour will appear in the lower half-plane. As a result, we obtain

$$f^0 = \frac{x_1^2(1-\beta_0)}{(x_1^2+z_1^2)^{3/2}} + (1-\beta_0)\int_0^{\alpha_1^{1/2}} d\kappa \, \frac{x_1\kappa^2 H_{-1}(\kappa x_1)}{(\alpha_1-\kappa^2)^{1/2}} \cos[(\alpha_1-\kappa^2)^{1/2}z_1] \tag{4.71}$$

where $H_{-1}(\xi)$ is the Struve function. The second term in (4.71) describes the oscillations of f^0 with period $2\pi/\alpha_1^{1/2}$.

From the above consideration it follows that to get rid of the oscillations of the function f, we should choose $\phi_1(k, \alpha_1)$ such that

(a) $\phi_1 = 0$ for $k^2 < \alpha_1$,
(b) ϕ_1 must have a singularity in the region $k^2 > \alpha_1$,

so that the second term in the expression (4.71) vanishes. Besides, ϕ_1 must satisfy the conditions (4.66) and (4.67). All these conditions are satisfied by a single function:

$$\phi_1(k, \alpha_1) = \begin{cases} \dfrac{k}{(k^2-\alpha_1)^{1/2}}, & k^2 > \alpha_1 \\ 0, & k^2 < \alpha_1 \end{cases} \tag{4.72}$$

Calculating as above, we arrive at

$$f = \int_0^\infty dk \, R_k^0(x_1)\phi_1(k, \alpha_1)\cos kz_1 = \frac{(1-\beta_0)x_1^2}{(z_1^2+x_1^2)^{3/2}}, \qquad x_1 \to 0 \tag{4.73}$$

and therefore oscillations of f near the dipole axis are now actually absent.

The general solution Taking into account (4.72), it is convenient to make the change $k^2 = \lambda^2 + \alpha_1$ in (4.61), i.e. to write the latter in the form

$$f = \int_0^\infty d\lambda \, R_\lambda(x_1)\cos[(\lambda^2+\alpha_1)^{1/2}z_1]$$

Then, for the functions $R_\lambda(x_1)$ we obtain from (4.60)

$$\frac{d^2R_\lambda}{dx_1^2} - \frac{1+x_1^2}{x_1(1-x_1^2)}\frac{dR_\lambda}{dx_1} + \left(\alpha_1\frac{x_1^2}{1-x_1^2} - \lambda^2\right)R_\lambda = 0 \tag{4.74}$$

The boundary conditions (4.54) and (4.56) for $R_\lambda(x_1)$ take the following form: for $x_1 \to 0$,

$$R_\lambda \to \frac{2}{\pi}(1-\beta_0)\lambda x_1 K_1(\lambda x_1) \tag{4.75}$$

and for $x_1 \to 1$,

$$\left(\alpha_1 R_\lambda - 2\frac{dR_\lambda}{dx_1}\right) \to 0 \tag{4.76}$$

Here K_1 is the first-kind Macdonald function, and we have taken into account that as $x_1 \to 0$, eq. (4.74) becomes the Bessel equation and its solution should be chosen in accordance with the condition (4.76). Furthermore, from (4.74) and (4.76) we obtain the asymptotic expression for the functions $R_\lambda(x_1)$ as $x_1 \to 1$:

$$R_\lambda = D(\lambda) I_0\left[\left(\lambda^2 - \frac{\alpha_1^2}{4}\right)^{1/2}(1-x_1)\right]\exp[-\tfrac{1}{2}\alpha_1(1-x_1)] \tag{4.77}$$

where I_0 is the Bessel function of imaginary argument, and $D(\lambda)$ is a constant. Note also the asymptotic formulas for large values of $\lambda \gg 1$ which express the constant $D(\lambda)$ and the function $R_\lambda(x_1)$ in the region $\lambda^{-1} < x < 1 - \lambda^{-1}$:

$$D(\lambda) = \sqrt{2}\,\lambda(1-\beta_0)e^{-\lambda}$$

$$R_\lambda(x_1) = \left(\frac{2\lambda}{\pi}\right)^{1/2}(1-\beta_0)\left(\frac{x_1}{1-x_1^2}\right)^{1/2} e^{-\lambda x_1} \tag{4.78}$$

$$\times\left[1 + \frac{1}{\lambda}\left(\frac{3}{8}\frac{1}{x_1} + \frac{1}{4}\frac{x_1}{1-x_1^2} + \frac{1}{8}(1-2\alpha_1)\ln\frac{1-x_1}{1+x_1} - \frac{1}{2}\alpha_1 x_1\right) + O(\lambda^{-2})\right]$$

The expression (4.78) follows from the quasi-classical solution of eq. (4.74) when it is matched with formulas (4.75) and (4.77).

The solution of eq. (4.74) can be represented as a power series of $(1 - x_1^2)$

$$R_\lambda(x_1) = D(\lambda)\sum_{n=0}^{\infty} a_n(1-x_1^2)^n \tag{4.79}$$

where the coefficients a_n are connected by the recurrent relations

$$a_{n+1} = \frac{(4n^2 - \alpha_1)a_n + (\alpha_1^2 + \lambda^2)a_{n-1}}{4(n+1)^2}, \qquad a_0 = 1$$

Given this, according to (4.76) we have $a_1 = -\alpha_1/4$. Employing the expression (4.75) for finding $R_\lambda(x_1)$ at the point $x_1 = 0$, we obtain the expression for $D(\lambda)$:

$$D(\lambda) = \frac{2}{\pi}(1 - \beta_0)\left(\sum_{n=0}^{\infty} a_n\right)^{-1}$$

which, for $\lambda \gg 1$, takes the asymptotic value given by formula (4.78). As a result, the solution of eq. (4.74) is found numerically through the summation of series (4.79) and the integration (4.61) with the use of the asymptotics (4.77) and (4.78).

Complete corotation To begin with, we consider the simplest case of complete plasma corotation in the magnetosphere when the longitudinal current $i_\parallel$ and the electric field Ψ are absent, i.e. when $\alpha_1 = 0$, $\beta_0 = 0$. Equation (4.48) coincides in this case with (4.49), i.e. the magnetic field structure in the closed and open regions are described by one and the same equation. The results of the calculations in this case are shown in Fig. 4.2. The dashed line indicates the singular surface (4.55) and the coincident light surface, i.e. a cylinder of radius $\rho_\perp = 1$. Magnetic field lines, as can be seen from the figure, are divided into two classes: for $f > f^*$ they are closed and do not reach the singular surface, whereas for $f < f^*$ they are open, i.e. intersect the singular surface and go to infinity. The separatrix between these two classes corresponds to the value $f = f^*$,

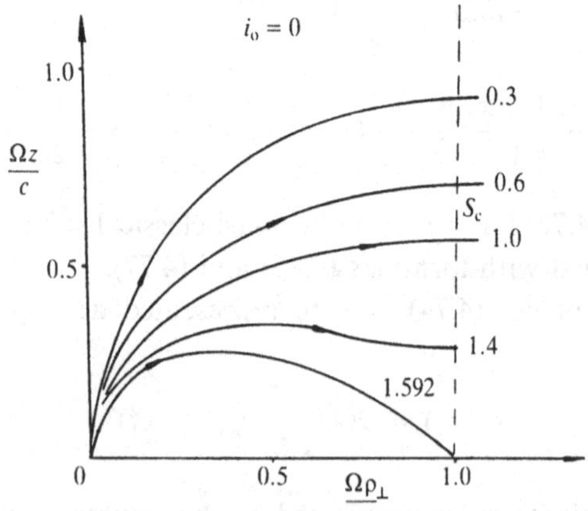

Fig. 4.2. Magnetic field lines for $\alpha_1 = \beta_0 = 0$. The dashed line indicates the light surface S_c coinciding in this case with the light cylinder $\rho_\perp = R_L$. The numbers are the values of f' on a given line.

where

$$f^* = \int_0^\infty d\lambda\, D(\lambda) = 1.592 \tag{4.80}$$

In the interaction with the light cylinder, i.e. when $z = 0$ and $\rho_\perp = 1$, there exists a singular point of the magnetic field, the so-called two-dimensional zero point $\mathbf{B} = 0$ (Syrovatskii, 1981) (in three dimensions we speak, of course, of a zero line which is a circle lying on the light cylinder $z = 0$, $\rho_\perp = 1, 0 \leqslant \phi \leqslant 2\pi$). For $z > 0$ and $z < 0$ the separatrices intersect at the zero point (see Fig. 4.2). The angle δ between intersecting separatrices and the quantity

$$\Delta^\circ f = \left[\frac{1}{\rho_\perp}\frac{\partial}{\partial \rho_\perp}\left(\rho_\perp \frac{\partial f}{\partial \rho_\perp}\right) + \frac{\partial^2 f}{\partial z^2}\right]_{z=0,\rho_\perp=1}$$

characterizing the run of field lines in the vicinity of the zero point

$$f - f^* = \Delta^\circ f\left[z^2 - \frac{1}{2}(1-\rho_\perp)^2\right] \tag{4.81}$$

are equal to

$$\tan\delta = 2\sqrt{2}, \qquad \Delta^\circ f = -\frac{1}{2}\int_0^\infty d\lambda\cdot\lambda^2 D(\lambda) = -4.006 \tag{4.82}$$

This follows directly from the expansion (4.79) as $z \to 0$, $\rho_\perp \to 1$.

The results represented in Fig. 4.2 were first obtained by Michel (1973b) and Mestel & Wang (1979) through direct numerical solution of eq. (4.49).

It should be noted that the polarization charge density (4.31) in the case of $i_\parallel = 0$ and $\Psi = 0$ has the form

$$\rho_e = -\frac{\mathbf{\Omega B}}{2\pi c(1-\beta_R^2)} \tag{4.83}$$

The singularity on the light cylinder $\beta_R = 1$ is absent here since, owing to the boundary condition (4.76), $\mathbf{\Omega B} = 0$ for $\beta_R = 1$.

We can see that in the case $i_\parallel = 0$, $\Psi = 0$ the polarization corotation charge density reverses sign on the surface $\mathbf{\Omega B} = 0$. At distances $r \ll R_L$, i.e. in those regions where the magnetic field is assumed to be dipole, this surface is a straight circular cone with a vertex angle $2\phi_c = 2\arccos 3^{-1/2} = 108°$ (Goldreich & Julian, 1969). At distances $r \lesssim R_L$, the cone surface becomes curved. It intersects the light cylinder $\beta_R = 1$ at height $z \simeq \pm 0.4 R_L$ from the equatorial plane (Michel, 1973b; Mestel & Wang, 1979).

4.2.3 Compatibility relation

We now proceed to the solution of the main problem – an account of longitudinal currents and electric field Ψ. Note that the expression (4.58) for $g(f)$ and $\Psi(f)$ for $i_0 \neq 0$, $\beta_0 \neq 0$ does not satisfy the condition (4.52). According to (4.52), in the region of closed field lines $f > f^*$, the field Ψ and longitudinal currents are absent, and eq. (4.49) is always valid there. On the boundary $f = f^*$, its solution must be matched with the solution of equation (4.60). Given this, by virtue of (4.52) and (4.58) the function of current $g(f)$ on the boundary $f = f^*$ must change from $g(f^* - 0) = i_0 f^*$ to $g(f^* + 0) = 0$, and the derivative dg/df increases sharply in magnitude, i.e. there occurs a jet of reverse longitudinal current:

$$i_{\parallel} = \left.\frac{dg}{df}\right|_{f=f^*} = -i_0 f^* \, \delta(f - f^*) \tag{4.84}$$

This corresponds to the following picture of currents in the pulsar magnetosphere: the current i_0 flows in one direction in the entire magnetosphere, while the reverse current (4.84) forms an intensive jet near the boundary $f = f^*$ (for details see Section 4.3). The electric field strength undergoes a similar rapid variation near the boundary $f = f^*$ in the case (4.58): $d\Psi/df = -\beta_0$ for $f = f^* - 0$ and $d\Psi/df = 0$ for $f = f^* + 0$. Accordingly, we observe here a sharp increase of the electric charge density:

$$\frac{d^2\Psi}{df^2}(\nabla f)^2 = -\beta_0(\nabla f)^2 \, \delta(f - f^*)$$

The electric field potential is always continuous as $f \to f^*$, and therefore from (4.58) we have

$$\Psi(f)|_{f \to f^*} \to 0, \qquad \Psi_0 = \beta_0 f^*$$

In matching the solutions in the closed and open regions, we assume according to (4.80) that the boundary between these regions is always the separatrix $f = f^* = 1.592$. In other words, we shall neglect the change in the shape of the closed region of the magnetosphere, which is strictly valid only in the case of a low value of current and electric field. The separatrix $f = f^*$ ends at a singular point $z = 0$, $\rho_\perp = 1$. Consequently, in our approximation, both the solution of (4.49) for the closed region and the solution of (4.48) for the open region must have a singular point for $z = 0$, $\rho_\perp = 1$. Taking account of this and also of the fact that near a singular point $(\nabla f)^2 \to 0$, we obtain from (4.48) and (4.81), as $z \to 0$, $\rho_\perp \to 1$, in the region of open field lines $f \to f^* - 0$:

$$\frac{1}{\Delta^\circ f} g \frac{dg}{df} = \left[2\frac{d\Psi}{df} + \left(\frac{d\Psi}{df}\right)^2 \right]\Bigg|_{f=f^*-0} \tag{4.85}$$

This is just the compatibility relation of solutions in closed and open regions (see Section 3.4). It establishes a connection between the electric field and electric current in pulsar magnetosphere. In the concrete case (4.58), it takes the form

$$\beta_0 = 1 - (1 - i_0^2/i_{\max}^2)^{1/2}, \qquad i_{\max}^2 = \frac{|\Delta^\circ f|}{f^*} \simeq 2.5 \tag{4.86}$$

It should be emphasized that all the quantities change most strongly near a singular point. Hence, matching near a singular point plays the leading role, and with allowance for fixation of the initial run of the separatrix near the pulsar surface $f = f^*$ is sufficient for a complete definition of the boundary. To clarify these words, we shall consider the exit $z_c(f_0)$ of the magnetic field line $f = f_0$ onto the light cylinder $\rho_\perp = 1$. It is described by the expression

$$z_c(f_0) = \int_0^1 \left(\frac{\partial z}{\partial \rho_\perp}\right)_{f_0} d\rho_\perp$$

Going over to integration over the entire volume $dV = 2\pi\rho_\perp\, d\rho_\perp\, dz$, we have

$$z_c(f_0) = \frac{1}{2\pi} \int_{\rho_\perp \leqslant 1} \frac{1}{\rho_\perp} \left(\frac{\partial z}{\partial \rho_\perp}\right)_{f_0} \delta[z - z(\rho_\perp, f_0)]\, dV$$

Since

$$\delta[z - z(\rho_\perp, f_0)] = \left|\frac{\partial f}{\partial z}\right|_{f_0} \delta(f - f_0)$$

and

$$\left(\frac{\partial z}{\partial \rho_\perp}\right)_{f_0} = -\frac{\partial f}{\partial \rho_\perp} \bigg/ \frac{\partial f}{\partial z}, \qquad \frac{\partial f}{\partial z} < 0$$

we finally come to

$$z_c(f_0) = \frac{1}{2\pi} \int_{\rho_\perp \leqslant 1} \delta(f - f_0) \frac{1}{\rho_\perp} \frac{\partial f}{\partial \rho_\perp}\, dV$$

We have taken into account here that the initial coordinates of field lines in the variables (4.47) are $\rho_\perp \to 0$, $z \to 0$. We now define $(1/\rho_\perp)(\partial f/\partial \rho_\perp)$

near the boundary $f = f^*$. Equation (4.48) implies that for $f_0 = f^* + 0$,

$$\frac{1}{\rho_\perp}\frac{\partial f}{\partial \rho_\perp} = \frac{1}{2}\Delta f(1-\rho_\perp^2)$$

and therefore

$$z_c^+ = \frac{1}{4\pi}\int_{\rho_\perp \leqslant 1} dV\,\Delta f(1-\rho_\perp^2)\,\delta[f-(f^*+0)]$$

Similarly, from eq. (4.60) for $f_0 = f^* - 0$ we have

$$\frac{1}{\rho_\perp}\frac{\partial f}{\partial \rho_\perp} = \frac{1}{2}\Delta f[1-\rho_\perp^2(1-\beta_0)^2] + \tfrac{1}{2}i_0^2 f$$

$$\begin{aligned} z_c^- &= \frac{1}{4\pi}\int_{\rho_\perp \leqslant 1} dV\,\Delta f[1-\rho_\perp^2(1-\beta_0)^2]\,\delta[f-(f^*-0)] \\ &\quad + \frac{i_0^2 f^*}{4\pi}\int_{\rho_\perp \leqslant 1} dV\,\delta[f-(f^*-0)] \qquad (4.87) \\ &= z_c^+ + \frac{1}{4\pi}\int_{\rho_\perp \leqslant 1} dV\,\{\Delta f\rho_\perp^2[1-(1-\beta_0)^2] + i_0^2 f^*\}\,\delta[f-(f^*-0)] \end{aligned}$$

Taking into account that, according to the solution in the closed region (Fig. 4.2), $z_c^+ = 0$, we find that the condition of coincidence of boundaries has the form $z_c^- = 0$. Next, taking into account that

$$\int_{\rho_\perp \leqslant 1} dV\,\delta(f-f^*) = -\int_0^1 \frac{\rho_\perp\,d\rho_\perp}{(\partial f/\partial z)_{f^*}}$$

in the case of coincident boundaries $f_0 = f^* \pm 0$,

$$\Delta f|_{f=f^*-0} = \Delta f|_{f=f^*+0} = \frac{2}{\rho_\perp(1-\rho_\perp^2)}\left(\frac{\partial f}{\partial \rho_\perp}\right)_{f^*}$$

we have from (4.87)

$$i_0^2 f^* C = [1-(1-\beta_0)^2]C_1 \qquad (4.88)$$

The constants C and C_1 are determined by integration over the displacement of z along the entire boundary field line:

$$C = \int_0^1 \frac{\rho_\perp\,d\rho_\perp}{(\partial f/\partial z)_{\rho_\perp, f=f^*}}, \qquad C_1 = \int_0^1 \frac{2\rho_\perp^2}{(1-\rho_\perp^2)}\left(\frac{\partial z}{\partial \rho_\perp}\right)_{f^*} d\rho_\perp \qquad (4.89)$$

However, both the integrals diverge logarithmically near the upper limit, i.e. near the singular point since here $\rho_\perp \to 1$ and $(\partial f/\partial z)_{f*} \to 0$. Indeed, as follows from (4.81),

$$\left(\frac{\partial f}{\partial z}\right)_{\rho_\perp} = -\sqrt{2}\,|\Delta f|\left[2\frac{(f-f^*)}{\Delta f} + (1-\rho_\perp)^2\right]^{1/2}$$

$$\left(\frac{\partial z}{\partial \rho_\perp}\right)_f = -\frac{1}{\sqrt{2}}(1-\rho_\perp)\left[2\frac{(f-f^*)}{\Delta f} + (1-\rho_\perp)^2\right]^{-1/2}$$

and as $f \to f^*$,

$$\left(\frac{\partial f}{\partial z}\right)_{\rho_\perp, f=f^*} = \sqrt{2}\,\Delta^\circ f(1-\rho_\perp), \qquad \left(\frac{\partial z}{\partial \rho_\perp}\right)_{f=f^*} = -\frac{1}{\sqrt{2}}$$

Substituting these expressions into (4.89), we make sure that the main contribution to the divergent integrals C and C_1 is made by the nearest neighbourhood of the point $\rho_\perp = 1$,

$$C \simeq \frac{1}{\sqrt{2}\,\Delta^\circ f}\ln\frac{1}{\varepsilon}, \qquad C_1 \simeq -\frac{1}{\sqrt{2}}\ln\frac{1}{\varepsilon}, \qquad \varepsilon = 1-\rho_\perp \to 0$$

Given this, the ratio C/C_1, however, remains a constant quantity

$$\frac{C}{C_1} = -\frac{1}{\Delta^\circ f}$$

With allowance for this relation, (4.88) coincides identically with (4.86). Note that to simplify calculations we have used (4.49) in the definition of Δf. An exhaustive analysis on the basis only of eq. (4.60), with account of the divergence of the integrals (4.89), yields the same results.

4.2.4 The structure of the magnetosphere in the presence of longitudinal current and the potential Ψ

In the presence of the longitudinal current $i_0 \neq 0$ and the potential $\beta_0 \neq 0$, we should seek the solution of eq. (4.60) for $f < f^*$ and match it with the solution of eq. (4.49) for $f > f^*$ with allowance made for the relation (4.86). The solution thus constructed is represented in Fig. 4.3 (Beskin *et al.*, 1983). The dashed lines in the figure indicate a singular S_d and light S_c surface. From the figure we can see that magnetic lines are as before separated into two groups. Closed lines are in the region $\rho_\perp \leqslant 1$ and do

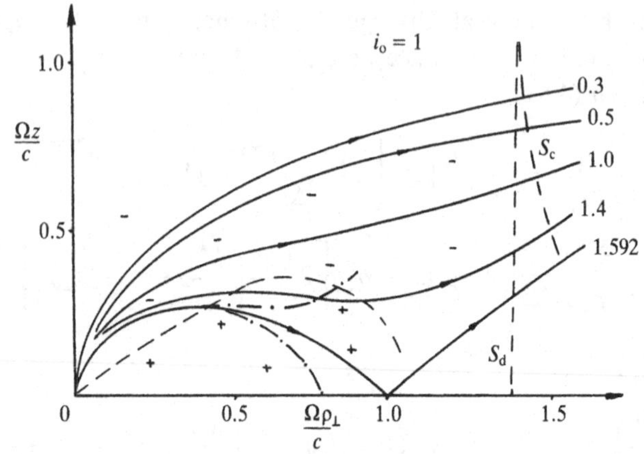

Fig. 4.3. Magnetic field lines for $i_0 = 1$; $\beta_0 = 0.275$. The dot-and-dash line indicates the field line $f = f^*$ in the case when the compatibility relation (4.86) does not hold. The sign of plasma charge density is also shown.

not reach the singular surface. Open lines traverse the singular and light surfaces. The zero point of the magnetic field is, as before, at $\rho_\perp = 1, z = 0$.

As is clear from (4.59), the singular surface S_d corresponds to $x_1 = 1$, i.e.

$$\rho_\perp = \frac{1}{1 - \beta_0}$$

On this surface, as is seen from (4.10), the electric field is

$$\mathbf{E} = \rho_\perp[\mathbf{B}\mathbf{e}_\phi] - \frac{d\Psi}{df}\nabla f$$

The quantity ∇f can be expressed in terms of the magnetic field $\mathbf{B}$. Indeed, from (4.42) it follows that $\nabla f = -\rho_\perp[\mathbf{B}\mathbf{e}_\phi]$, i.e.

$$\mathbf{E} = \rho_\perp\left(1 + \frac{d\Psi}{df}\right)[\mathbf{B}\mathbf{e}_\phi]$$

and

$$\mathbf{E}\left(\rho_\perp = \left(1 + \frac{d\psi}{df}\right)^{-1} = (1 - \beta_0)^{-1}\right) = [\mathbf{B}\mathbf{e}_\phi]$$

But since the longitudinal electric current is not equal to zero, the component $B_\phi \neq 0$ and, therefore, $E < B$. That is why the drift velocity of particles is less here than the velocity of light.

The surface where particles reach the light velocity (the light surface) is found from the condition (see (4.10))

$$E^2 = (1 - \beta_0)^2(\nabla f)^2 = B^2 = \frac{1}{\rho_\perp^2}[(\nabla f)^2 + g^2]$$

or

$$x_0^2 = 1 + \frac{g^2}{(\nabla f)^2} = 1 + \frac{\alpha_1(1 - \beta_0)^2 f^2(x_0, z_1)}{[\nabla f(x_0, z_1)]^2}, \qquad x_0 > 1 \tag{4.90}$$

For small currents

$$\alpha_1 \ll 1 \tag{4.91}$$

this surface is close to the singular surface $x_1 = 1$. To find the magnitude of the field near the light surface for small currents, we shall use the expansion of the function $f(x_1, z_1)$ in power series of $(1 - x_1^2)$ (4.61), (4.79). Keeping only first terms of the expansion, we represent the fields in the form

$$\begin{aligned}
B_x &= -\frac{1}{\rho_\perp}\frac{\partial f}{\partial z} = -\frac{(1 - \beta_0)^2}{x_1}\frac{\partial f}{\partial z_1} = (1 - \beta_0)^2 \int_0^\infty d\lambda\, \lambda D(\lambda) \sin \lambda z_1 \\
B_z &= \frac{1}{\rho_\perp}\frac{\partial f}{\partial \rho_\perp} = -\frac{(1 - \beta_0)^2}{x_1}\frac{\partial f}{\partial x_1} \\
&= \frac{\alpha_1(1 - \beta_0)^2}{2}\int_0^\infty d\lambda\, D(\lambda)\left[1 + \frac{\lambda^2}{2}\frac{(x_1^2 - 1)}{\alpha_1}\right]\cos \lambda z_1 \\
B_\phi &= \frac{1}{\rho_\perp} g = \frac{(1 - \beta_0)}{x_1} i_0 f = \alpha_1^{1/2}(1 - \beta_0)\int_0^\infty d\lambda\, D(\lambda) \cos \lambda z_1 \\
\mathbf{E} &= x_1[\mathbf{B}\mathbf{e}_\phi]
\end{aligned} \tag{4.92}$$

From (4.92) we can see that in the case under consideration $B_z \ll B_\phi \ll B_x$ ($z_1 \ll \alpha_1^{1/2}$). For the characteristic of magnetic field spirality in the vicinity of the light surface, it is convenient to introduce the parameter

$$\kappa_0(z_1) = \frac{B_\phi(z_1, x_0)}{B_x(z_1, x_0)} \tag{4.93}$$

which for small currents is equal to

$$\kappa_0(z_1) = \alpha_1^{1/2}\frac{\int_0^\infty d\lambda\, D(\lambda)\cos \lambda z_1}{\int_0^\infty d\lambda\, \lambda D(\lambda)\sin \lambda z_1} \tag{4.94}$$

The coordinate $x_0(z_1)$, according to (4.90) ($B_z \ll B_x$), is expressed via

$\kappa_0(z_1)$ by the relation

$$x_0^2 = 1 + \kappa_0^2 \tag{4.95}$$

For high values of z_1, all the fields fall exponentially. This is connected with the fact that the function

$$D^{-1}(\lambda) = \frac{\pi}{2(1-\beta_0)} \sum_{n=0}^{\infty} a_n$$

on the imaginary axis is oscillating. The first zero of the function $D^{-1}(ip)$ is responsible for the behaviour of the fields when $z_1 \gg 1$,

$$B_x \propto e^{-p_1 z_1}, \qquad z_1 \gg 1$$

In particular, $p_1 = 3.0$ for the case $i_0 = 0$ and $p_1 = 3.1$ for $i_0 = 1$. It then follows from (4.95) that κ_0 tends to a constant limit $\alpha_1^{1/2}/p_1$ when $z_1 \gg 1$. The dependence of the magnetic field components on z_1 is shown in Fig. 4.4 for $i_0 = 1$. Although the current is not small here, the qualitative picture of the fields is very close to the case of small currents considered above.

The polarization charge density ρ_e (4.31) in the region of open field lines will be written as

$$\rho_e = \frac{-\dfrac{\Omega\mathbf{B}}{2\pi c}(1-\beta_0) + \dfrac{\Omega}{8\pi c}(1-\beta_0)i_0^2\left(\dfrac{\Omega R}{c}\right)^3 B_0 f}{1 - \beta_R^2(1-\beta_0)^2}$$

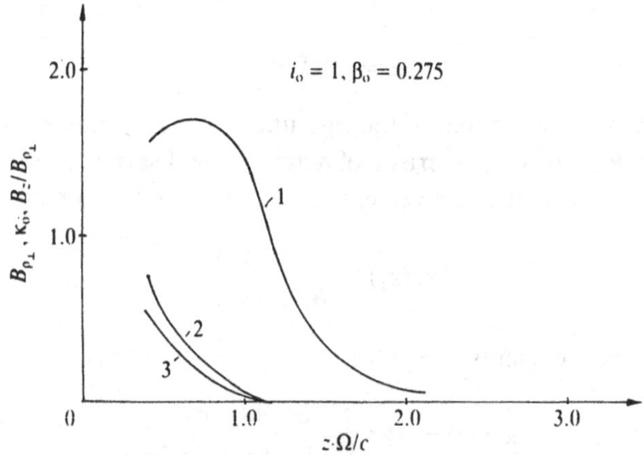

Fig. 4.4. The z-dependence of the magnetic field on the light surface ($i_0 = 1$): 1, $B_{\rho_\perp}$; 2, κ_0; 3, $B_z/B_{\rho_\perp}$.

As shown in Fig. 4.3, the charge density ρ_e for $i_0 \neq 0$, $\beta_0 \neq 0$ has on the whole the same structure as in the case $i_0 = 0$, $\beta_0 = 0$. In particular, the quantity ρ_e changes sign not only on closed but also on open field lines.

It should be noted that the compatibility relation (4.86) can be directly verified by numerical calculation. If it is not fulfilled, then in the solution of eq. (4.60) the boundary field line $f = f^*$ either intersects the boundary of a closed magnetosphere or goes far from it (the dot-and-dash lines in Fig. 4.3). So, the connection between the magnitudes of the electric field and longitudinal current for the solution of eq. (4.60) can be established only on the basis of numerical calculation by means of choosing i_0 and β_0 such that the boundaries of the closed and open regions coincide. The numerical dependence of β_0 on i_0 thus obtained coincides with (4.86) to sufficient accuracy.

We should emphasize the importance of the compatibility relation (4.86), which establishes the relation between the electric field and the longitudinal currents in the pulsar magnetosphere. It plays the role of a non-linear Ohm's law and, as we will see (Section 4.5.4) determines the pulsar energy losses. From (4.86) it follows, in addition, that the longitudinal current $i_0 = g(f^*)/f^*$ cannot exceed the critical value i_{max}. As is seen from (4.86), $i_{max} = 1.58$. One should bear in mind, however, that for high values of i_0 the relation (4.86) is itself approximate since it does not take into account the change in the boundary of the closed region of the magnetosphere, f^*, which possibly leads to a change in the value of the parameter i_{max}.

In dimensional quantities, taking into account (4.47) and (4.51), we obtain from this restriction on the density of the longitudinal and total current in pulsar magnetosphere,

$$|j_{\parallel}| \lesssim 0.79 \frac{\Omega B_0}{2\pi}, \qquad |I| \lesssim 3.95 R^3 \frac{\Omega}{c} \frac{B_0 \Omega}{2\pi} \tag{4.96}$$

We can see that the limiting value of the longitudinal current density is of the order of the corotation current $j_c = -\Omega B/2\pi$ (3.7), (4.14). From (4.86) it follows also that the electric field always decelerates the rotation of magnetosphere ($d\Psi/df < 0$). The deceleration increases with longitudinal current. As $i_0 \to i_{max}$, the total electric field $-\nabla(\Phi_c + \Psi) = -(1 + d\Psi/df)\nabla f$ vanishes, i.e. the magnetosphere plasma stops. Given this, the singular surface goes to infinity.

It is noteworthy that in the derivation of the compatibility relation we have used only the general conditions of the connection between the open and closed regions of the magnetosphere. A similar relation can be obtained for any magnetosphere in which there exist closed and open regions and eqs. (4.21) and (4.22) hold. Such conditions are usually satisfied, say, in the Earth and planetary magnetospheres. The necessity of the existence of the longitudinal potential difference Ψ between the body surface and magnetosphere in the presence of longitudinal current i_0, which follows from the compatibility relation, is a remarkable and, perhaps, the general property of magnetospheres. In particular, in the Earth and planetary magnetospheres the appearance of longitudinal potential difference in the regions where longitudinal currents flow leads to a continuous acceleration of electrons and ions, that is, serves as one of the sources of aurora polaris and radio emission of planets.

4.3 Boundary layer

4.3.1 The structure of the boundary layer

The solution constructed above is valid only before the light surface (4.90). On the light surfaces, the total electric and magnetic fields become equal. Therefore, near the light surface the drift velocity of particles (4.24) tends to the velocity of light, and the particle energy increases sharply. Given this, the conditions (4.17), (4.19) and (4.20) are violated, and eq. (4.32) becomes inapplicable. Near the light surface a special boundary layer is thus formed in which the drift approximation (4.24) turns out to be insufficient, and a more precise description of electron and positron motion is required.

Under the conditions (4.19), the boundary layer thickness $\Delta \simeq c/\Omega\lambda$ (3.52) is always small as compared with the magnetosphere scale $R_L \simeq c/\Omega$ (3.8) (this will be shown in detail in what follows). Therefore, under stationary conditions all the quantities in this layer undergo substantial variation only in the direction of the normal to the layer, $\rho_\perp$. Taking this fact into account, we write the Maxwell equation in the boundary layer as

$$\frac{\partial B_\phi}{\partial \rho_\perp} = \frac{4\pi}{c} j_z, \qquad \frac{\partial B_z}{\partial \rho_\perp} = -\frac{4\pi}{c} j_\phi, \qquad \frac{\partial}{\partial \rho_\perp}(\rho_\perp B_{\rho_\perp}) = 0$$
$$\frac{\partial^2 \Psi}{\partial \rho_\perp^2} = -2\frac{\Omega B_z}{c} + 4\pi\left(\frac{\Omega\rho_\perp}{c^2} j_\phi - \rho_e\right) \tag{4.97}$$

These equations follow directly from eqs. (4.15) and (4.12).

The electron and positron distribution functions are described by the kinetic equation (4.4, 4.5), where the source q is equal to zero (the plasma is created only in the vicinity of pulsar). We shall take into account that particles are strongly accelerated near the light surface. Their initial spread over longitudinal momenta is of little importance here, so we may assume

$$F^{\pm}(p_{\parallel}, \mathbf{r}) = n^{\pm}(\rho_{\perp})\,\delta[p_{\parallel} - p_{\parallel}^{\pm}(\rho_{\perp})] \tag{4.98}$$

Substituting (4.98) into the kinetic equation (4.4), leaving in it only derivatives with respect to the coordinate $\rho_{\perp}$ and integrating it over the momentum $p_{\parallel}$, we obtain as usual the hydrodynamic equations for the concentrations $n^{\pm}$ and mean momenta $\mathbf{p}^{\pm}$ of relativistic electrons and positrons:

$$\frac{d}{d\rho_{\perp}}\left(\frac{p_{\rho_{\perp}}^{\pm} n^{\pm}}{\gamma^{\pm}}\right) = 0$$

$$\begin{aligned} \frac{dp_{\rho_{\perp}}^{\pm}}{d\rho_{\perp}} &= \frac{p_{\phi}^{\pm 2}}{\rho_{\perp} p_{\rho_{\perp}}^{\pm}} \pm \frac{e\gamma^{\pm}}{p_{\rho_{\perp}}^{\pm}}\left\{m_e E_{\rho_{\perp}} + \frac{1}{c\gamma^{\pm}}[\mathbf{p}^{\pm}\mathbf{B}]_{\rho_{\perp}}\right\} \\ \frac{dp_{\phi}^{\pm}}{d\rho_{\perp}} &= -\frac{p_{\phi}^{\pm}}{\rho_{\perp}} \pm \frac{e\gamma^{\pm}}{p_{\rho_{\perp}}^{\pm}}\left\{m_e E_{\phi} + \frac{1}{c\gamma^{\pm}}[\mathbf{p}^{\pm}\mathbf{B}]_{\phi}\right\} \\ \frac{dp_z}{d\rho_{\perp}} &= \pm\frac{e\gamma^{\pm}}{p_{\rho_{\perp}}^{\pm}}\left\{m_e E_z + \frac{1}{c\gamma^{\pm}}[\mathbf{p}^{\pm}\mathbf{B}]_z\right\} \end{aligned} \tag{4.99}$$

Equations (4.97) and (4.99) make up a complete system describing plasma and field distribution in a thin boundary layer.

For simplicity we shall consider the case of a weak longitudinal current (4.91, 4.92), $\alpha_1 \ll 1$, or (4.94):

$$\kappa_0^2 \ll 1, \qquad \kappa_0 = \left.\frac{B_{\phi}}{B_{\rho_{\perp}}}\right|_{\rho_{\perp} = c/\Omega(1-\beta_0)-1} \tag{4.100}$$

In this, the component B_z on a singular surface (and a light surface close to it) is small, $B_z \simeq \kappa_0^2 B_{\rho_{\perp}}$ (4.92) and can therefore be neglected. Taking into account, in addition, that (4.97) implies $B_{\rho_{\perp}} \propto 1/\rho_{\perp}$, we present the magnetic field in the boundary layer in the form

$$B = B_c(z)\left\{\frac{1}{x}, \kappa, 0\right\}, \qquad \kappa = \kappa(x) = \frac{B_{\phi}(x)}{B_c} \tag{4.101}$$

where x is the dimensionless coordinate (4.47) $x = (\Omega\rho_{\perp}/c)(1 - \beta_0)$, and B_c is the magnitude of the radial magnetic field on the singular surface $x = 1$. The dependence of B_c on z is determined by the general solution before the boundary layer $x \gtrsim 1$ (see Fig. 4.3). According to (4.92), the electric field **E** in the boundary layer can be written with the same accuracy $\sim\kappa_0^2$ in the form

$$E = B_c(z)\{0, 0, 1\} \tag{4.102}$$

We shall make the energy and momentum of particles dimensionless by the value of initial energy $m_e c^2 \gamma_0$, i.e. the energy possessed by the particle before it enters the boundary layer for $x \lesssim 1$:

$$\gamma = \gamma_0 \gamma', \qquad \mathbf{p} = m_e c \gamma_0 \mathbf{p}'$$

We shall take into account (4.101) and (4.102) and introduce in eqs. (4.97) and (4.99) the total energy γ and the parameter κ (4.101) (primes omitted) instead of $\mathbf{p}$ and B_ϕ. Since for the energy $\mathscr{E}$ we have the equation $d\mathscr{E}/d\rho_\perp = e(\mathbf{Ep})/p_{\rho_\perp}$, and from the ϕ-component of eq. (4.99) it follows that $xp_\phi - \gamma = \text{const}$, the total system of equations acquires the form

$$\begin{aligned} &\frac{d\gamma}{dx} = \alpha\frac{p_z}{p_x}, \qquad p_\phi = \frac{\gamma - 1}{x} \\ &\frac{dp_z}{dx} = \alpha\left(\kappa + \frac{\gamma}{p_x} - \frac{\gamma - 1}{x^2 p_x}\right) \\ &p_x^2 + p_\phi^2 + p_z^2 + \frac{1}{\gamma_0^2} = \gamma^2 \\ &n = \gamma\frac{p_{x0}}{p_x}\lambda n_c, \qquad \frac{d\kappa}{dx} = 4\lambda\frac{p_z}{p_x} \end{aligned} \tag{4.103}$$

We have introduced here the dimensionless parameter

$$\alpha \equiv \lambda\mu = \frac{eB_c}{m_e c \gamma_0 \Omega(1 - \beta_0)} \tag{4.104}$$

Since $\lambda \gg 1$, $\mu \gg 1$, (4.19), (4.20), we always have $\alpha \gg 1$. Equations (4.103) hold, of course, only in a thin boundary layer Δx near the light surface $x = x_0$ (4.95):

$$\Delta x = x - x_0 \ll 1, \qquad x_0 \simeq 1 + \tfrac{1}{2}\kappa_0^2 \tag{4.105}$$

From the first and the last of equations (4.103) it follows that

$$\kappa(x) = \kappa_0 + \frac{4\lambda}{\alpha}(\gamma - 1) \tag{4.106}$$

i.e. that the ϕ-component of the magnetic field is expressed directly through the Lorentz factor γ of plasma particles. The relation (4.106) expresses conservation of the radial component of the energy flux:

$$\frac{c}{4\pi}[\mathbf{EB}]_{\rho_\perp} + (n^+ + n^-)m_e c^2 \gamma v_{\rho_\perp} = \text{const}$$

Indeed, since on the light surface $\mathbf{E} = B_c \mathbf{e}_z$ (4.102), the electromagnetic energy flux is proportional to B_ϕ. On the other hand, the particle flux is also conserved and is equal to $B_c\Omega(1 - \beta_0)\lambda/2\pi e$. As a result, we obtain the equality (4.106).

The drift approximation considered in the preceding section corresponds to the fact that in the equations of motion (4.99) the internal terms are small, i.e. $\mu \to \infty$ and, therefore, $\alpha \to \infty$ (4.104). In our case, the equations of drift approximation therefore take the simple form

$$\kappa = \kappa_0, \qquad p_z = 0, \qquad p_x = |\kappa_0|^{-1}\left(\gamma - \frac{\gamma - 1}{x^2}\right)$$

$$p_\phi = \frac{\gamma - 1}{x}, \qquad \gamma^2 = p_x^2 + p_\phi^2 \qquad (\gamma_0 \gg 1)$$

Solving these equations, we come to

$$\gamma = \frac{1 - |\kappa_0| x^2[1 - x^2(1 - \kappa_0^2)]^{-1/2}}{(1 - x^2)}$$

$$p_x = [1 - x^2(1 - \kappa_0^2)]^{-1/2}$$

From this we can see that in the drift approximation the energy and momentum of particles go to infinity on the light surface:

$$x = x_0 = (1 - \kappa_0^2)^{-1/2} \simeq 1 + \tfrac{1}{2}\kappa_0^2$$

$$\gamma \simeq \left(\frac{x_0}{2\kappa_0^2}\right)^2 (x_0 - x)^{-1/2}, \qquad p_x \simeq \left(\frac{x_0}{2}\right)^{1/2} (x_0 - x)^{-1/2} \tag{4.107}$$

Hence, as $x \to x_0$, the drift approximation is violated (4.20).

To find the exact solution of eqs. (4.103) near a singularity, we introduce new variables

$$s = \frac{4\lambda}{|\kappa_0|}(x - x_0), \qquad \Gamma_p = \frac{4}{\mu|\kappa_0|}\gamma, \qquad q_x = \frac{4}{\mu|\kappa_0|^2}p_x, \qquad q_z = \frac{4p_z}{\mu|\kappa_0|^2} \tag{4.108}$$

Then, under the condition that the longitudinal current is not very small ($|\kappa_0| > \mu^{-1/3}$), we have $\gamma \gg 1$, $p_x \gg 1$, $p_z \gg 1$, and eqs. (4.103) and (4.106) become

$$\frac{d\Gamma_p}{ds} = \frac{q_z}{q_x}$$

$$\frac{dq_z}{ds} = \Gamma_p\left(1 + \frac{1}{q_x}\right) - 1 \qquad (4.109)$$

$$q_x^2 + q_z^2 = \Gamma_p^2$$

These equations imply that

$$\frac{dq_x}{ds} = -\frac{q_z}{q_x}(\Gamma_p - 1) = -\frac{d\Gamma_p}{ds}(\Gamma_p - 1)$$

Consequently,

$$q_x = \Gamma_p - \tfrac{1}{2}\Gamma_p^2, \qquad q_x = \Gamma_p^{3/2}(1 - \Gamma_p/4)^{1/2}$$

From (4.109) we finally obtain

$$\Gamma_p = 2 - [4 - (s - s_0)^2]^{1/2} \qquad (4.110)$$

where s_0 is the value s for which $\Gamma_p \to 0$. Let us match the solution (4.110) with the solution of the drift approximation (4.107) which describes the behaviour of particle energy when approaching a singularity. Suppose s_1 is the matching point. Then, equating at this point the values of Γ_p and $d\Gamma_p/ds$ obtained according to (4.110) and (4.107), we obtain

$$s_1 = -a, \qquad s_0 = -5a, \qquad a = (2\lambda\mu^{-2}x_0)^{1/5}|\kappa_0|^{-1}$$

Since the solution (4.110) holds only for $(s_1 - s_0) \leqslant 2$, the indicated matching can be done only provided that $a < 1/2$ or $\mu|\kappa_0|^{5/2}\lambda^{-1/2} > 1$, which agrees with the approximation used in the derivation of eqs. (4.109).

The formulas obtained describe completely the distribution of particles and magnetic field in the boundary layer near the light surface.

From (4.107) and (4.110) it is first of all seen that in the boundary layer the electron and positron energy increases sharply:

$$\mathscr{E} = 2\mathscr{E}_m\left[1 - \left(1 - \frac{(s - s_0)^2}{4}\right)^{1/2}\right]$$

Here $\mathscr{E}_m$ is the limiting energy

$$\mathscr{E}_m = \frac{m_e e^2 \gamma_0 |\kappa_0| \mu}{4} = \frac{ecB_{c\phi}}{4\Omega\lambda(1-\beta_0)} = \frac{eI}{2c\lambda} \tag{4.111}$$

We have taken into account here that B_{c_ϕ} is proportional to the total longitudinal electric current I,

$$B_{c_\phi} = \frac{2I\Omega(1-\beta_0)}{c^2}$$

Thus, the limiting energy $\mathscr{E}_m$ is proportional to the total strength of longitudinal current I (4.51) in the pulsar magnetosphere and inversely proportional to the particle density $\lambda \propto n_e$ (4.19). Under optimal conditions, when the current I is maximal and reaches the value (4.96), particles are accelerated up to energies

$$\mathscr{E}_m \simeq 4 \times 10^6 B_{12} P^{-2} \lambda^{-1} \quad \text{MeV}$$

The variation of particle energy $\mathscr{E}$ in the transition layer is presented in Fig. 4.5. The limiting value $\mathscr{E} = \mathscr{E}_m$ is reached on the external boundary of the transition layer,

$$s = s_c = s_0 + \sqrt{3} \tag{4.112}$$

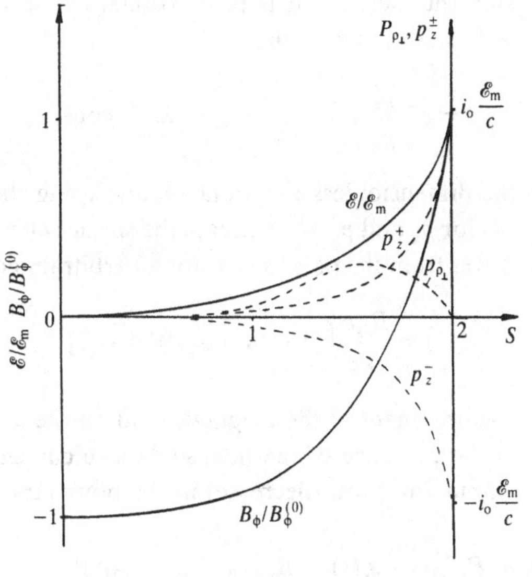

Fig. 4.5. The variation of the quantities $\mathbf{p}$, $\mathscr{E}$, B_ϕ in the transition layer.

The particle moment in the direction of rotation increases proportionally to the energy (4.103). The momentum in the radial direction

$$p_x = 2\frac{\mathscr{E}_m}{c}|\kappa_0|\left[1-\left(1-\frac{(s-s_0)^2}{4}\right)^{1/2}\right]\left[1-\frac{(s-s_0)^2}{4}\right]^{1/2} \tag{4.113}$$

also increases up to the value $p_{x_m} = \mathscr{E}_m|\kappa_0|/2c$ for $s = s_c$. Note that for $s = s_c$ the momentum p_x reaches its maximum: the derivative $(dp_x/ds)_{s=s_c} = 0$ (see Fig. 4.5).

The particle motion in the z direction is substantial only in the boundary layer. The momentum p_z increases here most sharply:

$$p_z^{\pm} = \pm\frac{\mathscr{E}_m}{c}|\kappa_0|(s-s_0)\left[1-\left(1-\frac{(s-s_0)^2}{4}\right)^{1/2}\right] \tag{4.114}$$

Its main specific feature is that particles of opposite charge sign – electrons and positrons – move along z in opposite directions (Fig. 4.5). This means that in a thin boundary layer there occurs a strong electric current and a jet of surface current is formed which flows along the light surface in the z direction.

The intensity of the current jet

$$J_s = \int (j_z^+ - j_z^-)\,d\rho_\perp$$

can be found using the fact that it is proportional to the radial current transferred by each plasma component:

$$j_z^{\pm} = j_x^+\frac{p_z^{\pm}}{p_x^{\pm}}, \qquad j_x^{\pm} = \pm e n_c \lambda c = \text{const}$$

Going over to the dimensionless coordinate s and using the expressions (4.113) and (4.114) for p_x and p_z, we estimate the surface current J_s flowing in the boundary layer from the point $s = s_0$ for an arbitrary coordinate s:

$$J_s(s) = \frac{B_{c\phi}c}{4\pi}[(4-(s-s_0)^2)^{1/2} - 2] \tag{4.115}$$

Here $B_{c\phi}$ is the ϕ-component of the magnetic field on the layer boundary $s = s_0$. Owing to the presence of the intense surface current J_s, the field B_ϕ undergoes a rapid variation (decrease) in the boundary layer:

$$B_\phi = B_{c\phi} + \frac{4\pi}{c}J_s(s) = B_{c\phi}[(4-(s-s_0)^2)^{1/2} - 1]$$

For $s = s_c$ the component B_ϕ vanishes (see Fig. 4.5). The vanishing of B_ϕ corresponds to a complete closure, by a current jet in the boundary layer, of the longitudinal current flowing in the magnetosphere.

4.3.2 The shock wave

The physical properties of plasma flowing behind the transition layer change radically. Indeed, before the transition layer the velocity of radial plasma flow is close to the velocity of light c, whereas behind the transition layer it is much lower:

$$v_{x_m} = \frac{cp_{x_m}}{m_e\gamma_m} = c\,\frac{|\kappa_0|}{4} \tag{4.116}$$

Correspondingly, the plasma density increases from $n_e = n_{e_0}$ to

$$n_{e_m} = 4n_{e_0}|\kappa_0|^{-1} \tag{4.117}$$

The azimuthal component of velocity increases sharply, $v_{\phi_m} \simeq c$, and the energy $\mathscr{E}$ of the azimuthal motion becomes the principal energy (4.111) $\mathscr{E} \simeq \mathscr{E}_m$. Furthermore, analysis shows that in the region behind the boundary s_c electromagnetic waves must be excited. It is natural to assume that the characteristic length of excited waves is of the order of the transition layer thickness. Accordingly, their frequency

$$\omega \simeq \frac{\lambda}{\kappa_0^2}\,\Omega \sim (10^3\text{–}10^7)\Omega \tag{4.118}$$

Waves are largely excited by the energy of particle motion along the z-axis (4.114) since the mean pulses of both electrons $\langle p_z^- \rangle$ and positrons $\langle p_z^+ \rangle$ for $s > s_c$ near the transition layer must be equal to zero:

$$\langle p_z^+ \rangle = \langle p_z^- \rangle = 0$$

Thus we can say that the light surface is the *collision-less shock front surface*. The front width is of the order of the transition layer thickness. On the wave front, the radial and azimuthal velocities of plasma, particle density and energy undergo a jump. Given this, in particular, while before the shock wave the plasma energy density was much lower than the energy density of the magnetic field (4.28), behind the shock wave this relation no longer holds ($B^2/8\pi \lesssim 2n_e\mathscr{E}_m$). This means that behind the shock front, i.e. in the external region of magnetosphere, the physical properties of plasma change significantly. Moreover, intensive electromagnetic oscillations (4.118) are excited here. At the distances $r > R_L$ from the neutron star the waves of MHD type should also be excited at the frequency of pulsar rotation. So, the general structure of plasma flow beyond the light surface is quite complicated. Some specific features of plasma flow in this region were investigated by Kennel & Coroniti (1984a,b). We should stress that the solution (4.110)–(4.114) obtained here holds only if the energy $\mathscr{E}_m$ (4.111) greatly exceeds the characteristic energy $m_e c^2 \gamma_0$ in magnetosphere plasma. If this is not the case, then there is no appreciable particle acceleration near the light surface, and the flux of particle leaving the shock front conserves energy of the order of $m_e c^2 \gamma_0$.

4.3.3 Closure of current

In a boundary layer, or in other words on a shock front, the longitudinal electric current is closed. The process of current closure and a simultaneous particle acceleration in the boundary layer can readily be understood physically. To this end, we compare the distribution of electric field potential (4.11) on the star surface $\Phi_c = \boldsymbol{\beta}_R \mathbf{A}|_{r=R}$ and in the magnetosphere $\Phi = \Phi_c + \Psi$. Since $\mathbf{A} = \mathbf{e}_\phi \rho_\perp M/r^3$ it follows that

$$\Phi_c = E_0 R f, \qquad E_0 = \frac{(\boldsymbol{\Omega}\mathbf{M})\Omega}{c^2 R} = \tfrac{1}{2} B_0 \left(\frac{\Omega R}{c}\right)^2 \operatorname{sign}(\boldsymbol{\Omega}\mathbf{M}) \qquad (4.119)$$

In the same dimensionless coordinates (4.47), the magnetosphere potential is

$$\Phi = E_0 R[f + \Psi(f)] \qquad (4.120)$$

The distribution of potentials Φ_c and Φ is shown in Fig. 4.6. The figure

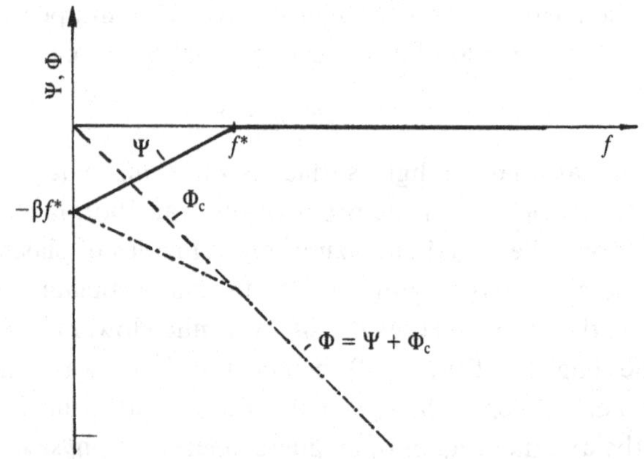

Fig. 4.6. The potentials of the magnetosphere Φ (dash-dot line), potential of the star Φ_c (dash-line) and the potential Ψ as functions of f.

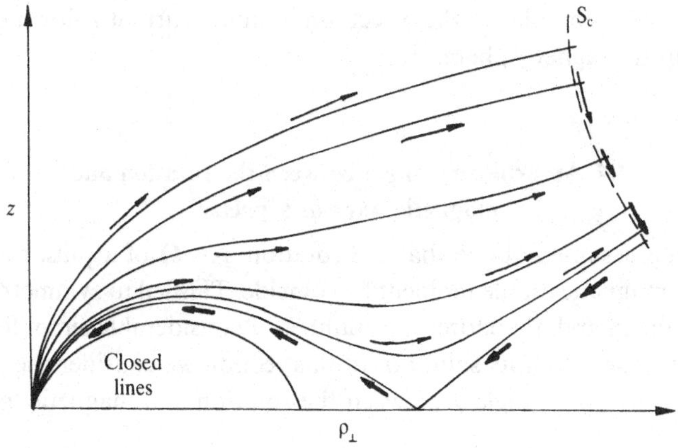

Fig. 4.7. Schematic representation of the electric currents in the pulsar magnetosphere.

presents a concrete case, discussed below, when the corotation charge ρ_c (4.13) in the polar region has a positive sign. In this case, on open field lines $f < f^*$ the body surface has an appreciable positive potential relative to the magnetosphere (the dashed line in Fig. 4.6). Therefore, only electrons can get onto the pulsar surface from the magnetosphere, so that the longitudinal current in the entire region of open lines in the magnetosphere always flows in one direction – from the pulsar, as shown in Fig. 4.7. In the boundary layer near the light surface the drift approximation (4.24) is violated, and particles are shifted across equipotential surfaces, gaining energy. As is clear from Fig. 4.6, to gain energy from the electric field, positrons must be displaced towards an increase of f and electrons in the opposite direction. As a result, an intensive current jet (4.115) which carries positive charges towards an increase of f (Fig. 4.7) appears in the boundary layer. As $f \to f^*$, the potential difference between the magnetosphere and the star's surface vanishes (see (4.52), we come to the closed field lines which have the same potential as the star surface (4.37)). This is shown in Fig. 4.6. Given this, positrons from the magnetosphere can get onto the star's surface. Therefore, at the place where the light surface intersects the field line $f = f^*$ the current jet turns to go back onto the pulsar surface along the separatrix (Fig. 4.7). The magnitude of the reverse current jet for $f = f^*$ is equal to (see (4.115))

$$J_f = -\frac{\Omega}{2\pi c} I \tag{4.121}$$

It compensates completely the direct longitudinal current I flowing in the entire pulsar magnetosphere.

4.4 An arbitrary angle between the rotation and magnetic axes of a pulsar

It has been assumed above that the rotation axis $\mathbf{\Omega}$ of a pulsar and the axis of its magnetic dipole moment $\mathbf{M}$ coincide. This led to symmetrization of the problem and, therefore, to a number of considerable simplifications in the construction of its solution. In this section we consider the general case of an arbitrary angle χ between the rotation and magnetic axes.

4.4.1 Complete corotation

As before, an important role is played by the complete corotation limit: longitudinal currents do not flow from the pulsar surface and an electric field is absent in the magnetosphere: $i_{\parallel} = 0$, $\Psi = 0$. It should be noted that, by virtue of (4.26),

$$\operatorname{div}\left\{i_{\parallel}\mathbf{B} - \frac{c\rho_e}{B^2}[\nabla\Psi\mathbf{B}]\right\} = 0 \tag{4.122}$$

In the absence of the electric field ($\Psi = 0$), from this relation it follows that

$$\frac{\partial}{\partial r_{\parallel}} i_{\parallel} = 0$$

i.e. the current $i_{\parallel}$ remains unchanged along the field line. Therefore, in the absence of the field and of the longitudinal current flowing down from the pulsar surface, $i_{\parallel}|_{S_0} = 0$, $\Psi|_{S_0} = 0$, the longitudinal current on the entire magnetosphere is equal to zero:

$$\Psi = 0, \qquad i_{\parallel} = 0 \tag{4.123}$$

We shall first consider the case of complete corotation (4.123). The system of equations (4.32), (4.33), with allowances made for (4.123), assumes the form

$$\begin{gathered} \operatorname{rot}\mathbf{G}_0 = 0, \qquad \mathbf{G}_0 = \mathbf{B}(1 - \beta_R^2) + \boldsymbol{\beta}_R(\mathbf{B}\boldsymbol{\beta}_R) \\ \operatorname{div}\mathbf{B} = 0 \\ \boldsymbol{\beta}_R = \frac{1}{c}[\mathbf{\Omega}\mathbf{r}] \end{gathered} \tag{4.124}$$

The solution (4.124) can obviously be represented as

$$\mathbf{B}(1-\beta_R^2)+\boldsymbol{\beta}_R(\mathbf{B}\boldsymbol{\beta}_R)=-\nabla\mathscr{H} \tag{4.125}$$

where $\mathscr{H}$ is a scalar function. In dimensionless cylindrical coordinates (4.47),

$$\rho'_\perp=\frac{\Omega\rho_\perp}{c},\qquad z'=\frac{\Omega z}{c},\qquad \phi$$

the dimensionless magnetic field $\mathbf{B}'=\mathbf{B}(c^3/M\Omega^3)$ is expressed in terms of the dimensionless effective potential $\mathscr{H}'=\mathscr{H}(c^2/M\Omega^2)$ by the relation

$$B_{\rho_\perp}=-\frac{\partial\mathscr{H}/\partial\rho_\perp}{1-\rho_\perp^2},\qquad B_\phi=-\frac{1}{\rho_\perp}\frac{\partial\mathscr{H}}{\partial\phi},\qquad B_z=-\frac{\partial\mathscr{H}/\partial z}{1-\rho_\perp^2} \tag{4.126}$$

The equation for $\mathscr{H}$ follows from (4.125), (4.33):

$$\frac{\partial^2\mathscr{H}}{\partial\rho_\perp^2}+\frac{1+\rho_\perp^2}{\rho_\perp(1-\rho_\perp^2)}\frac{\partial\mathscr{H}}{\partial\rho_\perp}+\frac{\partial^2\mathscr{H}}{\partial z^2}+\frac{1-\rho_\perp^2}{\rho_\perp^2}\frac{\partial^2\mathscr{H}}{\partial\phi^2}=0 \tag{4.127}$$

It was obtained by Henriksen and Norton (1975).

The linear equation (4.127) is valid in the entire magnetosphere. The total set of boundary conditions for the effective magnetic potential $\mathscr{H}$ is therefore specified according to (4.34)–(4.40). From the condition (4.34) it follows that as $\rho_\perp\to 0$, $z\to 0$, the magnetic field (4.126) must coincide with the dipole field $\mathscr{H}=(\mathbf{M}\cdot\mathbf{r})/Mr^3$, i.e.

$$\mathscr{H}|_{\rho_\perp\to 0,z\to 0}=\frac{z}{(z^2+\rho_\perp^2)^{3/2}}\cos\chi+\frac{\rho_\perp\cos\phi}{(z^2+\rho_\perp^2)^{3/2}}\sin\chi \tag{4.128}$$

The solution of the linear equation (4.127) with the boundary condition (4.128) is naturally sought in the form

$$\mathscr{H}(\rho_\perp,z,\phi)=\mathscr{H}_0(\rho_\perp,z)\cos\chi+\mathscr{H}_1(\rho_\perp,z)\cos\phi\sin\chi$$

The potential $\mathscr{H}_0$ describes the axisymmetric case $\chi=0$ and the potential $\mathscr{H}_1\cos\phi$ is the case of mutually orthogonal axes $\boldsymbol{\Omega}$ and $\mathbf{M}$ (i.e. $\chi=\pi/2$). The potentials $\mathscr{H}_0$ and $\mathscr{H}_1$ satisfy the equations

$$\frac{\partial^2\mathscr{H}_0}{\partial\rho_\perp^2}+\frac{1+\rho_\perp^2}{\rho_\perp(1-\rho_\perp^2)}\frac{\partial\mathscr{H}_0}{\partial\rho_\perp}+\frac{\partial^2\mathscr{H}_0}{\partial z^2}=0 \tag{4.129}$$

$$\frac{\partial^2\mathscr{H}_1}{\partial\rho_\perp^2}+\frac{1+\rho_\perp^2}{\rho_\perp(1-\rho_\perp^2)}\frac{\partial\mathscr{H}_1}{\partial\rho_\perp}+\frac{\partial^2\mathscr{H}_1}{\partial z^2}-\frac{1-\rho_\perp^2}{\rho_\perp^2}\mathscr{H}_1=0 \tag{4.130}$$

with the boundary conditions for $\rho_\perp \to 0$, $z \to 0$:

$$\mathscr{H}_0|_{\rho_\perp \to 0, z \to 0} = \frac{z}{(z^2 + \rho_\perp^2)^{3/2}}, \qquad \mathscr{H}_1|_{\rho_\perp \to 0, z \to 0} = \frac{\rho_\perp}{(z^2 + \rho_\perp^2)^{3/2}} \tag{4.131}$$

The second boundary condition (4.39) for $\mathscr{H}_0$ and $\mathscr{H}_1$ – the absence of singularities on the light surface – takes the form

$$\frac{\partial \mathscr{H}_0}{\partial \rho_\perp} = \frac{\partial \mathscr{H}_1}{\partial \rho_\perp} = 0 \qquad \text{at } \rho_\perp = 1 \tag{4.132}$$

Finally, the condition at infinity (4.40) will be written in the form $\mathscr{H}_0 \to 0$, $\mathscr{H}_1 \to 0$ as $z \to \infty$.

The solution of linear equations (4.129) and (4.130) is found by the method of separation of variables, similar to that considered in Section 4.2. Since from symmetry considerations it is clear that $\mathscr{H}_1$ is an even function of the coordinate z and $\mathscr{H}_0$ is an odd function, it follows that

$$\mathscr{H}_0(\rho_\perp, z) = \int_0^\infty d\lambda\, \xi_0(\rho_\perp, \lambda) \sin \lambda z \tag{4.133}$$

$$\mathscr{H}_1(\rho_\perp, z) = \int_0^\infty d\lambda\, \xi_1(\rho_\perp, \lambda) \cos \lambda z \tag{4.134}$$

and the functions $\xi_0(\rho_\perp, \lambda)$ and $\xi_1(\rho_\perp, \lambda)$ satisfy the equations

$$\begin{gathered} \frac{d^2\xi_0}{d\rho_\perp^2} + \frac{1 + \rho_\perp^2}{\rho_\perp(1 - \rho_\perp^2)} \frac{d\xi_0}{d\rho_\perp} - \lambda^2 \xi_0 = 0 \\ \frac{d^2\xi_1}{d\rho_\perp^2} + \frac{1 + \rho_\perp^2}{\rho_\perp(1 - \rho_\perp^2)} \frac{d\xi_1}{d\rho_\perp} - \left(\lambda^2 + \frac{1 - \rho_\perp^2}{\rho_\perp^2}\right)\xi_1 = 0 \end{gathered} \tag{4.135}$$

The boundary conditions follow from (4.131), (4.133) and (4.134): as $\rho_\perp \to 0$, according to the integral representation of the Macdonald functions (cf. (4.75)):

$$\begin{gathered} \lambda K_1(\lambda \rho_\perp) = \int_0^\infty \frac{\rho_\perp}{(z^2 + \rho_\perp^2)^{3/2}} \cos \lambda z\, dz \\ \xi_0(\rho_\perp, \lambda) = \frac{2}{\pi} \lambda K_0(\lambda \rho_\perp), \qquad \xi_1(\rho_\perp, \lambda) = \frac{2}{\pi} \lambda K_1(\lambda \rho_\perp) \end{gathered} \tag{4.136}$$

and as $\rho_\perp \to 1$ the solutions of eqs. (4.135) with allowance made for (4.132) come to asymptotics (cf. (4.77)):

$$\xi_{0,1}(\rho_\perp, \lambda) = \frac{8D_{0,1}(\lambda)}{\lambda}(1 - \rho_\perp)I_1[\lambda(1 - \rho_\perp)] \tag{4.137}$$

From this it is seen that on the light cylinder $\rho_\perp = 1$ not only $\partial\mathcal{H}/\partial\rho_\perp = 0$, but $\mathcal{H} = 0$, i.e. the magnetic field is directed only along $\rho_\perp$:

$$B_z = B_\phi = 0, \qquad B_{\rho_\perp} \neq 0 \tag{4.138}$$

For arbitrary values of $\rho_\perp$, the eigenfunctions ξ_0 and ξ_1 can be represented as the expansions

$$\begin{aligned} \xi_0(\rho_\perp, \lambda) &= D_0(\lambda) \sum_{n=2}^{\infty} b_n(\lambda)(1 - \rho_\perp^2)^n \\ \xi_1(\rho_\perp, \lambda) &= D_1(\lambda) \sum_{n=2}^{\infty} a_n(\lambda)(1 - \rho_\perp^2)^n \end{aligned} \tag{4.139}$$

where $a_2 = b_2 = 1$ and a_n and b_n satisfy the recurrence relations:

$$b_{n+1} = \frac{n}{n+1} b_n + \frac{\lambda^2}{4(n^2 - 1)} b_{n-1} \qquad (b_1 = 0)$$

$$\begin{aligned} a_{n+1} = {} & \frac{n(2n-3)}{n^2 - 1} a_n - \frac{4(n-1)(n-2) - \lambda^2}{4(n^2 - 1)} a_{n-1} \\ & - \frac{\lambda^2 - 1}{4(n^2 - 1)} a_{n-2} \qquad (a_0 = a_1 = 0) \end{aligned}$$

Next, the functions $D_0(\lambda)$ and $D_1(\lambda)$ in (4.139) should be so chosen that the expansion (4.139) satisfies the boundary conditions (4.136) as $\rho_\perp \to 0$. To this end, using the asymptotic expressions of the functions K_0 and K_1 as $\rho_\perp \to 0$, we find

$$\begin{aligned} D_0^{-1}(\lambda) &= \frac{\pi}{2\lambda} \lim_{\rho_\perp \to 0} \frac{1}{\ln \rho_\perp} \sum_{n=2}^{\infty} b_n(\lambda)(1 - \rho_\perp^2)^n \\ D_1^{-1}(\lambda) &= \frac{\pi}{2} \lim_{\rho_\perp \to 0} \rho_\perp \sum_{n=2}^{\infty} a_n(\lambda)(1 - \rho_\perp^2)^n \end{aligned} \tag{4.140}$$

The graphs of these functions are presented in Fig. 4.8. For $\lambda \gg 1$ the asymptotics of the functions $D_{0,1}(\lambda)$ can be obtained proceeding from the quasi-classical behaviour of the solution of eqs. (4.135) ($\lambda \gg 1$), when it is

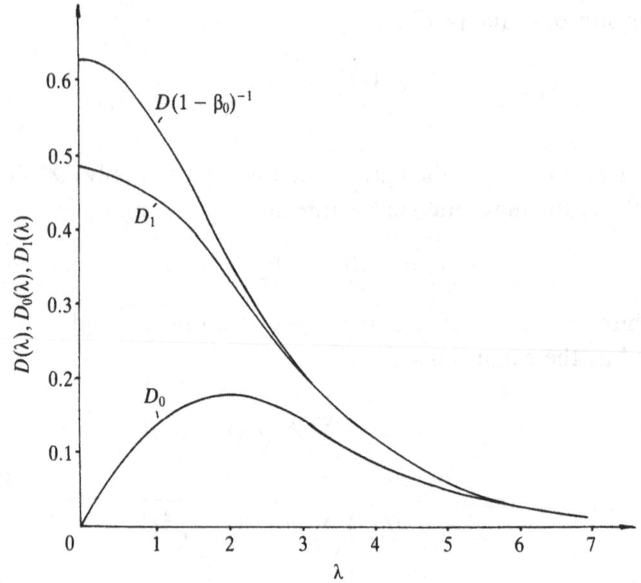

Fig. 4.8. Dependence of the functions $D(\lambda)$ (see (4.77)), $D_0(\lambda)$, $D_1(\lambda)$ on the parameter λ.

matched with (4.136) and (4.137)

$$D_{0,1}(\lambda) = \frac{\sqrt{2}}{4}\lambda^2 e^{-\lambda}$$

The equations obtained define completely the solutions of eq. (4.127). The magnetic field structure in the pulsar magnetosphere for different inclination angles χ is shown in Fig. 4.9. For $\chi = 0$, the solution, naturally, coincides with that obtained in Section 4.2. With increasing χ the region of closed magnetosphere is seen to increase and to decline from the rotation axis. In the plane where the closed magnetosphere touches the light cylinder, there lies a zero singular line, as in the axisymmetric case. Here the magnetic field $\mathbf{B} = 0$.

The equation specifying the zero line $z = z_0(\phi)$ of the magnetic field is found from the condition that $B_{\rho_\perp} = 0$ as soon as $\rho_\perp = 1$. Since (4.126), (4.133), (4.134) and (4.139) imply that

$$B_{\rho_\perp}|_{\rho_\perp=1} = 4\cos\chi \int_0^\infty D_0(\lambda)\sin(\lambda z)\, d\lambda + 4\sin\chi\cos\phi \int_0^\infty D_1(\lambda)\cos(\lambda z)\, d\lambda \tag{4.141}$$

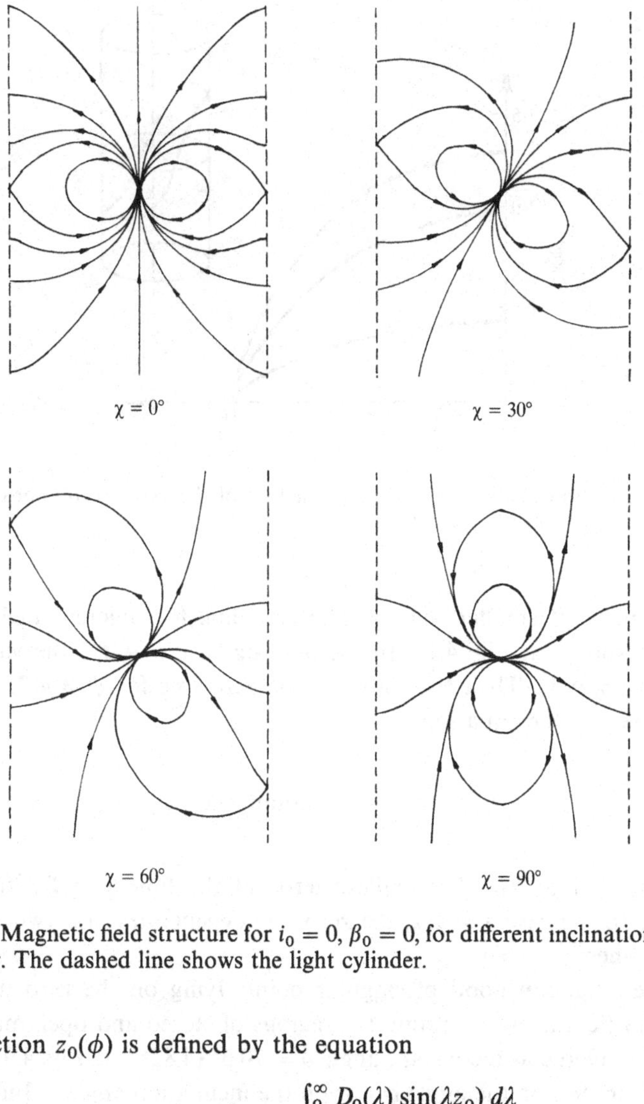

Fig. 4.9. Magnetic field structure for $i_0 = 0$, $\beta_0 = 0$, for different inclination angles of axes χ. The dashed line shows the light cylinder.

the function $z_0(\phi)$ is defined by the equation

$$\tan\chi\cos\phi = -\frac{\int_0^\infty D_0(\lambda)\sin(\lambda z_0)\,d\lambda}{\int_0^\infty D_1(\lambda)\cos(\lambda z_0)\,d\lambda} \tag{4.142}$$

For $z_0 < 1$ the zero line is close to an ellipse formed by the intersection of the light surface and the plane

$$z_0 = -\frac{\int_0^\infty D_1(\lambda)\,d\lambda}{\int_0^\infty D_0(\lambda)\lambda\,d\lambda}\tan\chi\cos\phi \tag{4.143}$$

Since $\int_0^\infty \lambda D_0(\lambda)\,d\lambda \gtrsim \int_0^\infty D_1(\lambda)\,d\lambda$, the inclination of this plane makes an

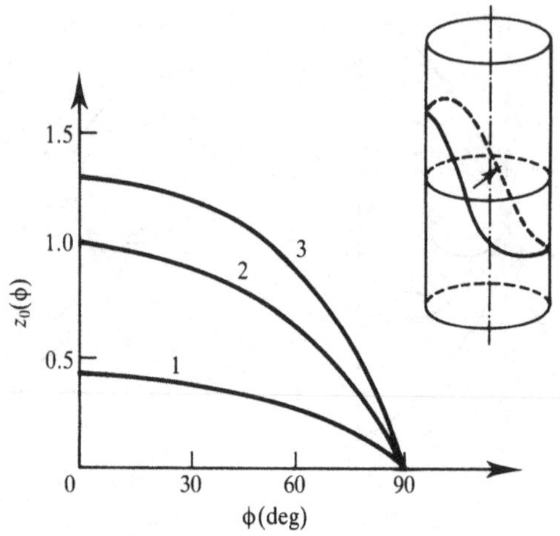

Fig. 4.10. Position of the zero line as a function of the cylindrical coordinate ϕ: 1, $\chi = 30°$; 2, $\chi = 60°$; 3, $\chi = 70°$.

angle somewhat smaller than χ. The zero line for different inclination angles is shown in Fig. 4.10. In the limiting case $\chi \to \pi/2$ the values z_0 heighten ($z_0 \gg 1$). Then the shape of the zero line for $\phi \neq \pi/2, 3\pi/2$ is determined by the equation

$$z_0(\phi) = \frac{1}{p_1 - p_0} \ln(|\tan \chi \cos \phi|) \tag{4.144}$$

where p_1 and p_0 are the smallest zeros of the functions $D_1^{-1}(ip)$ and $D_0^{-1}(ip)$. In the case $\chi = \pi/2$, the zero line degenerates into two vertical straight lines.

In the neighbourhood of singular points lying on the zero line, the boundary field lines separating the regions of closed and open magnetospheres converge at a constant angle $\delta = 70.5°$ (4.82) (see Fig. 4.9) which depends neither on the angle ϕ nor on the inclination angle χ. Indeed, as follows from (4.126), (4.139), the magnetic field near the zero line $z = z_0(\phi)$ is proportional to the declination from this line:

$$B_{\rho_\perp} \propto 2(z - z_0), \qquad B_z \propto -(1 - \rho_\perp), \qquad B_\phi \propto (1 - \rho_\perp)^2$$

Therefore, the equation of the field line $d\rho_\perp/dz = B_{\rho_\perp}/B_z$ has the solution $z - z_0(\phi) = \pm(1/\sqrt{2})(1 - \rho_\perp)$, which implies $\tan \delta = 2\sqrt{2}$.

On the light cylinder, only the z-component of the electric field and the

$\rho_\perp$-component of the magnetic field are non-zero, and according to (4.10) are equal in magnitude: $E_z = B_{\rho_\perp}$ (4.141). Thus in the case $i_0 = 0$, $\beta_0 = 0$ on the light cylinder, the Pointing vector $\mathbf{S} = (c/4\pi)[\mathbf{EB}]$ is tangent to the cylinder surface.

The charge density on the light cylinder is equal to

$$\rho_e = \frac{\Omega B_c}{2\pi c}\left[\cos\chi \int_0^\infty d\lambda\, D_0(\lambda)\lambda\cos(\lambda z) - \cos\phi\sin\chi \int_0^\infty d\lambda\, D_1(\lambda)\lambda\sin(\lambda z)\right]$$

$$= \frac{\Omega}{8\pi c}\frac{\partial B_{\rho_\perp}}{\partial z} \tag{4.145}$$

With distance along z from the zero line, the magnitudes of the fields first increase and then fall. At the extremum point of the magnetic field, the charge density reverses sign. For large z, both the magnetic field and the charge density fall exponentially. For example, for a magnetic field we obtain for $z \gg 1$,

$$B_{\rho_\perp}(\chi = 0) = 15e^{-3.0z}$$
$$B_{\rho_\perp}(\chi = \pi/2) = 30e^{-4.3z} \tag{4.146}$$

The corotation charge density in the region $\rho_\perp < R_L$ is described, as before, by the relation (4.83). Therefore, for $\chi \neq 0$ the charge reverses sign on the surface $\mathbf{\Omega B} = 0$.

The polar cap, i.e. the region on the star's surface where the open field lines come, in the axisymmetric case ($\chi = 0$) is a circle of radius

$$\rho'_\perp = R_0 = R\left(\frac{\Omega R}{c}f^*\right)^{1/2}$$

where $f^* = 1.592$ is the value of the parameter f for the separatrix between the regions of closed and open field lines.

For angles $\chi \neq 0$, the polar cap is deformed as shown in Fig. 4.11, preserving in the general case only the symmetry about the north–south meridian. Its boundary can therefore be represented as

$$\rho'_\perp = R\left[\frac{\Omega R}{c}f^*(\chi)\right]^{1/2}\eta(\cos\phi'), \qquad \int_0^{2\pi} d\phi'\,\eta^2(\cos\phi') = 2\pi \tag{4.147}$$

Here $\rho'_\perp$ and ϕ' are polar coordinates relative to the magnetic pole, and the angle ϕ' is counted from the direction towards the star's rotation angle.

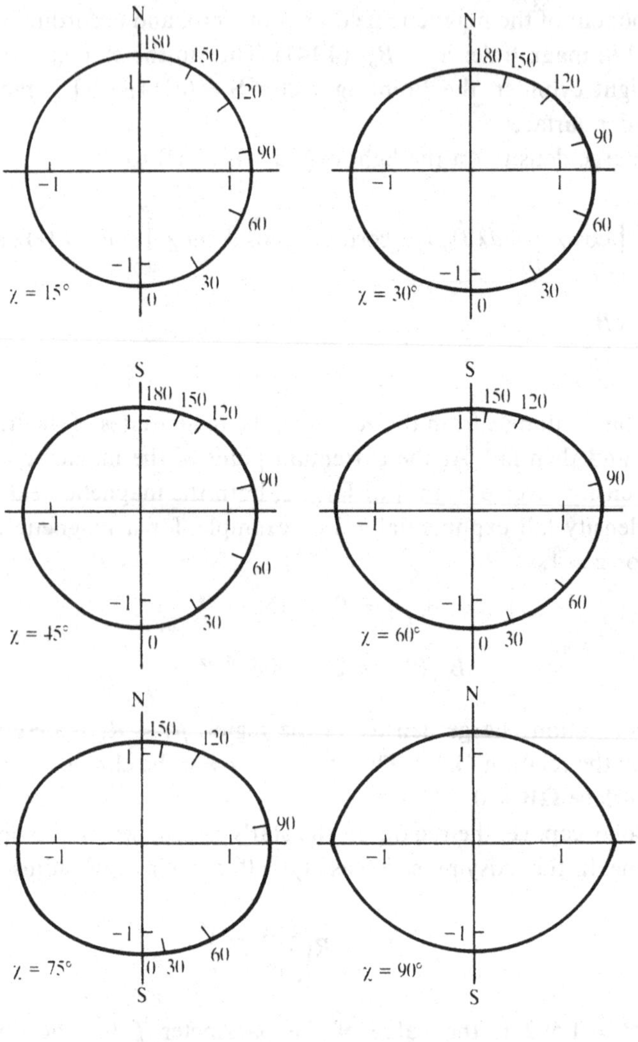

Fig. 4.11. The polar cap shape for different inclination angles of axes, χ. Numbers indicate the angle ϕ on the light cylinder surface made by the field line emanating from a given point on the star's surface. The unit scale corresponds to the distance $R(\Omega R/c)^{1/2}$.

The variation of the quantity $f^*(\chi)$ is shown in Fig. 4.12. As a result, the area of the polar cap

$$\sum(\chi) = \pi f^*(\chi)\Omega\frac{R^3}{c} \tag{4.148}$$

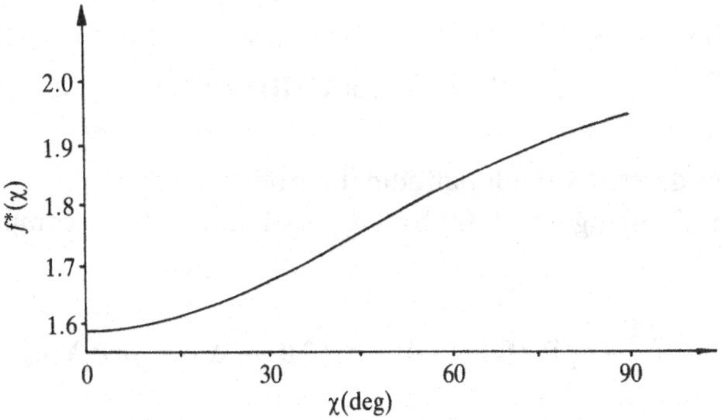

Fig. 4.12. Dependence of the parameter $f^*(\chi)$ on the angle χ.

ranges from $1.592\pi\Omega R^3/c$ for $\chi = 0$ to $1.95\pi\Omega R^3/c$ for $\chi = \pi/2$. The ratio of the principal scales of the polar cap varies between 1 and $\simeq 1.2$.

4.4.2 Magnetosphere in the presence of longitudinal current and the potential Ψ

We now proceed to the general case, when the longitudinal current $i_{\parallel} \neq 0$ and the potential $\Psi \neq 0$ exist in magnetosphere. Let $f(\rho_\perp, z, \phi) = \text{const}$ and $\theta(\rho_\perp, z, \phi) = \text{const}$ define two systems of magnetic surfaces, where $f(\rho_\perp, z, \phi)$ and $\theta(\rho_\perp, z, \phi)$ are scalar functions. The intersections of magnetic surfaces f and θ are magnetic field lines. Taking into account (4.33), we define the magnetic field $\mathbf{B}$ as

$$\mathbf{B} = [\nabla f\ \nabla\theta] \tag{4.149}$$

From this point on, we use dimensionless variables and functions (4.47).

Making allowance for (4.22), we can assume the electric field potential to be constant on the surface $f = \text{const}$ and therefore

$$\Psi = \Psi(f)$$

Since in this case the right-hand side of (4.32) is orthogonal to the vector ∇f, one of the components of the vector equation (4.32) takes the form

$$\begin{gathered} \operatorname{rot} \mathbf{G}\cdot\nabla f = \operatorname{div}[\mathbf{G}\ \nabla f] = 0 \\ \mathbf{G} = \mathbf{B}(1 - \beta_R^2) + \boldsymbol{\beta}_R(\boldsymbol{\beta}_R\mathbf{B}) + [\boldsymbol{\beta}_R\ \nabla\Psi] \end{gathered} \tag{4.150}$$

The second component can be obtained via multiplying (4.32) scalarly by **B**,

$$i_{\parallel} = \frac{1}{B^2}(\text{rot } \mathbf{G}\cdot\mathbf{B}) \tag{4.151}$$

This relation describes the longitudinal current $i_{\parallel}$.

Finally, multiplying eq. (4.32) by $\nabla\theta$, we obtain its third component in the form

$$\nabla\theta\cdot\text{rot } \mathbf{G} = \frac{\dfrac{d\Psi}{df}\{\rho_\perp B_\phi(\mathbf{B} \text{ rot } \mathbf{G}) - B^2[2B_z + \Delta\Psi - \rho_\perp^2\mathbf{e}_\phi\nabla(\mathbf{e}_\phi\,\nabla\Psi)]\}}{B^2(1-\rho_\perp^2) + \rho_\perp[\nabla\Psi\mathbf{B}]_\phi} \tag{4.152}$$

The current $i_{\parallel}$ is discarded here according to (4.151). Equations (4.150) and (4.152), along with (4.149), make up a complete system defining the functions $f(\rho_\perp, z, \phi)$ and $\theta(\rho_\perp, z, \phi)$.

In the axisymmetric case, the function f depends on two variables, $f = f(\rho_\perp, z)$, and the function θ is found using the following transformation of expression (4.42) for the magnetic field **B**:

$$\mathbf{B} = [\nabla f\cdot\nabla\phi] + g(f)\frac{1}{\rho_\perp\,\partial f/\partial z}[\nabla f\,\mathbf{e}_{\rho_\perp}]$$

$$= \left[\nabla f\cdot\nabla\left(\phi + g\int^{\rho_\perp}\frac{\partial z(f,\rho'_\perp)}{\partial f}\frac{d\rho'_\perp}{\rho'_\perp}\right)\right]$$

Then

$$\theta = \phi + g(f)\int^{\rho_\perp}\frac{\partial z(f,\rho'_\perp)}{\partial f}\frac{d\rho'_\perp}{\rho'_\perp}, \qquad \theta(\phi + 2\pi) = \theta(\phi) + 2\pi \tag{4.153}$$

For an arbitrary inclination angle χ, the coordinate θ retains its cyclic character and the natural normalization

$$\oint_{f=\text{const}} d\theta = 2\pi \tag{4.154}$$

Taking this into account, we obtain from (4.149)

$$\frac{1}{2\pi}\int \mathbf{B}\,d\mathbf{s} = \frac{1}{2\pi}\int[\nabla f\,\nabla\theta]\,d\mathbf{s}$$

$$= \frac{1}{2\pi}\int \text{rot}(f\,\nabla\theta)\,d\mathbf{s} = \frac{f}{2\pi}\oint_{f=\text{const}}\nabla\theta\,d\mathbf{l} = f \tag{4.155}$$

Here the integration is over the surface leaning upon the arbitrary contour C which lies on the surface $f(\rho_\perp, z, \phi) = \text{const}$. Since in the polar region of the star the magnetic field can be assumed constant, $\mathbf{B} = \mathbf{B}_0$, it follows from (4.155) that the value of f is proportional to the area, in the polar region, limited to the curve $f = \text{const}$. Introducing a dimensionless quantity $f'(f = M(\Omega/c)f')$ and making use of the equality (4.148), we find that in the region of open field lines f' varies within the range

$$0 \leqslant f' \leqslant f^*(\chi), \qquad f^*(\chi) = 1.592\frac{\Sigma(\chi)}{\Sigma_0} \tag{4.156}$$

where $\Sigma(\chi)$ is the area of the polar cap, $\Sigma_0 = \Sigma(\chi = 0)$ (see Fig. 4.12). The value of $f^*(\chi)$ varies from 1.592 for $\chi = 0$ to 1.95 for $\chi = \pi/2$.

In the axisymmetric case, the total arbitrary current in the region limited to the magnetic surface $f = \text{const}$ is constant and equal to $g(f)$ in the dimensionless variables (4.47). An analogous current conservation relation holds also for an arbitrary inclination angle χ of axes, which fact is seen directly from the expression (4.27) (the current cannot intersect the surface $f = \text{const}$). We shall find the total longitudinal current I flowing inside the surface limited to $f = \text{const}$:

$$I = \int \mathbf{j}_e\, d\mathbf{s} = \iint \left[i_\parallel \mathbf{B} - \frac{\rho_e c}{B^2}\frac{d\Psi}{df}[\nabla f \mathbf{B}] \right] d\mathbf{s}$$

Since

$$d\mathbf{s} = \left[\frac{d\mathbf{r}}{df}\frac{d\mathbf{r}}{d\theta}\right] df\, d\theta$$

then

$$I = \iint \left[i_\parallel - \frac{\rho_e c}{B^2}\frac{d\Psi}{df}\mathbf{B}\frac{d\mathbf{r}}{d\theta} \right] df\, d\theta \tag{4.157}$$

As in the axisymmetric case, we introduce the quantity $g_\chi(f) = I/2\pi$. Then (4.157) implies that

$$\frac{dg_\chi(f)}{df} = \frac{1}{2\pi}\int \left[i_\parallel - \frac{c\rho_e}{B^2}\frac{d\Psi}{df}\mathbf{B}\frac{d\mathbf{r}}{d\theta} \right] d\theta \tag{4.158}$$

Near the pulsar surface, the second term in (4.122) and (4.158) is small, and the current $i_\parallel$ is conserved here along the magnetic field line. In the axisymmetric case it does not depend on the angle ϕ and is equal to dg/df (4.50). For an arbitrary inclination angle χ, the longitudinal current near the star can depend on the coordinate θ. It is therefore convenient to distinguish in the boundary condition (4.35) between the symmetric and asymmetric parts of the current:

$$\mathbf{j}_{\parallel}|_{S_0} = [i_{\parallel_S}(f) + i_{\parallel_A}(f, \theta)]\mathbf{B}_0(f, \theta) \tag{4.159}$$

By virtue of the relation (4.158) and the definition of asymmetric current, there hold the equalities

$$i_{\parallel_S}(f) = \frac{dg_x(f)}{df}, \qquad \int_0^{2\pi} i_{\parallel_A}(f, \theta)\, d\theta = 0 \tag{4.160}$$

As we shall see below, the asymmetric part of the current is small as compared with the symmetric part.

4.4.3 The compatibility relation for $\chi \neq 0$

The system of equations (4.150), (4.152) for $i_{\parallel} \neq 0$ and $\Psi \neq 0$ holds only in the region of open field lines. On closed field lines, according to (4.37), we always have $i_{\parallel} = \Psi = 0$. Consequently, assuming, as in the axisymmetric case, the closed region of magnetosphere to be unchanged, we should necessarily match the solution of eqs. (4.150), (4.152) on the separatrix $f = f^*(\chi)$ with the solution in the closed region, obtained at the beginning of this section for an arbitrary inclination angle χ. It is essential that for almost all inclination angles, the region near the separatrix f^* is clearly distinguished, since it is only in this region that a reverse current flows, forming, as in the axisymmetric case, a narrow and strong current jet. Indeed, (4.13) implies that the charge density ρ_e near the magnetic pole has a constant sign (provided that $\mathbf{\Omega B}_0 \neq 0$, i.e. $\chi \neq \pi/2$). This implies that in the entire region of open field lines the potential Ψ has a constant sign and the longitudinal current has a constant direction. Consequently, the reverse current flows only near the separatrix, where Ψ vanishes (see Figs. 4.6 and 4.7).

To match the current region with the quiet region of magnetosphere, it is important to make compatible the magnetic surface near singular points lying on the 'zero' line, since even small variations of the quantity f which characterizes the magnetic surface lead to noticeable deviations in the position of the surface itself. Indeed, from the solution found at the beginning of this section it follows that the field line near the singular point $z = z_0(\phi)$ is described by the equation (cf. (4.81)):

$$f - f^* = \Delta_\perp^0 f(\phi)[(z - z_0)^2 - \tfrac{1}{2}(1 - \rho_\perp)^2] \tag{4.161}$$

where

$$\Delta_\perp^0 f = \left[\frac{1}{\rho_\perp}\frac{\partial}{\partial \rho_\perp}\left(\rho_\perp \frac{\partial f}{\partial \rho_\perp}\right) + \frac{\partial^2 f}{\partial z^2}\right]\Bigg|_{z=z_0(\phi),\rho_\perp=1} \tag{4.162}$$

and f^* is the value of f on the singular field line. From this we can see that the quantity

$$\frac{\partial z(f, \rho_\perp, \phi)}{\partial f} = (\Delta_\perp^0 f)^{-1}[2(1-\rho_\perp)^2 + 4(\Delta_\perp^0 f)^{-1}(f - f^*)]^{-1/2} \tag{4.163}$$

tends to infinity at the singular point $f = f^*$, $\rho_\perp \to 1$. Here the situation is quite similar to the axisymmetric case where almost all the deviation of the magnetic surface occurred in the neighbourhood of a singular surface, since it is just there that $(\partial f/\partial z)_{f^*} \to 0$, i.e. $\partial z/\partial f \to \infty$.

To perform matching, we first solve equation (4.150) in the neighbourhood of a singular point $z = z_0(\phi)$, $\rho_\perp = 1$. To do so, it is convenient to go over in (4.150) from the variables $\rho_\perp, z, \phi$ to $\rho_\perp, f, \phi$ and to consider $\theta(\rho_\perp, f, \phi)$ as an unknown function. Equation (4.150) in these variables ($z = z(\rho_\perp, f, \phi)$) assumes the form

$$\frac{\partial}{\partial \phi}\left[\frac{1-\rho_\perp^2}{\rho_\perp}\frac{\dfrac{\partial z}{\partial \rho_\perp}\dfrac{\partial z}{\partial \phi}}{\dfrac{\partial z}{\partial f}}\frac{\partial \theta}{\partial \rho_\perp} - \frac{1-\rho_\perp^2}{\rho_\perp}\frac{1+\left(\dfrac{\partial z}{\partial \rho_\perp}\right)^2}{\dfrac{\partial z}{\partial f}}\frac{\partial \theta}{\partial \phi} + \frac{d\Psi}{df}\rho_\perp\frac{1+\left(\dfrac{\partial z}{\partial \rho_\perp}\right)^2}{\dfrac{\partial z}{\partial f}}\right]$$

$$= \frac{\partial}{\partial \rho_\perp}\left[\frac{\dfrac{1-\rho_\perp^2}{\rho_\perp}\left(\dfrac{\partial z}{\partial \phi}\right)^2 + \rho_\perp}{\dfrac{\partial z}{\partial f}}\frac{\partial \theta}{\partial \rho_\perp} - \frac{1-\rho_\perp^2}{\rho_\perp}\frac{\dfrac{\partial z}{\partial \rho_\perp}\dfrac{\partial z}{\partial \phi}}{\dfrac{\partial z}{\partial f}}\frac{\partial \theta}{\partial \phi} + \frac{d\Psi}{df}\rho_\perp\frac{\dfrac{\partial z}{\partial \rho_\perp}\dfrac{\partial z}{\partial \phi}}{\dfrac{\partial z}{\partial f}}\right] \tag{4.164}$$

Taking account of the fact that in the neighbourhood of a singular point $\partial z/\partial f \to (1-\rho_\perp)^{-1}$, we represent the solution of eq. (4.164) as an expansion in power series of $\varepsilon = 1 - \rho_\perp$:

$$\frac{\partial \theta}{\partial \rho_\perp} = \frac{\partial z}{\partial f}\left[a^{(0)} + a^{(1)}\varepsilon + a^{(2)}\varepsilon^2 \int \frac{\partial z}{\partial f} d\rho_\perp + a^{(3)}\varepsilon^2 + \cdots\right] \tag{4.165}$$

In the expansion (4.165), in addition to integer powers of ε there also exist terms of the form $\varepsilon^n \ln \varepsilon$ for the reason that the right-hand side of eq. (4.164) is proportional to $d/d\varepsilon$. Substituting the series (4.165) into (4.164) and equating the coefficients at different powers of ε, we obtain

$$\frac{\partial \theta}{\partial \rho_\perp} = \frac{\partial z}{\partial f}\left[g_1 + \varepsilon\left[1 - 2\left(\frac{\partial z}{\partial \phi}\right)^2\right] g_1 - \frac{d\Psi}{df}\frac{\dfrac{\partial z}{\partial \phi}\dfrac{\partial z}{\partial \rho_\perp}}{\dfrac{\partial z}{\partial f}} + \cdots \right]$$

$$\frac{\partial \theta}{\partial \phi} = \frac{\partial}{\partial \phi}\left[g_1 \int \frac{\partial z}{\partial f}\, d\rho_\perp + g_2 + \cdots \right] \tag{4.166}$$

where $g_1(f, \phi)$ and $g_2(f, \phi)$ are arbitrary functions. An analogue of g_1 in the axisymmetric case is $g(f)$, and an analogue of g_2 is the angle ϕ (see the expression (4.153)).

The magnetic field $\mathbf{B} = [\nabla f\, \nabla\theta]$ on the light surface near a singular point in the zeroth order in ε is equal to

$$B_{\rho_\perp} = 0, \qquad B_\phi = g_1(f, \phi), \qquad B_z = g_1(f, \phi)\frac{\partial z_0}{\partial \phi}$$

From this we can see that the field **B** is parallel here to the 'zero line'. From eq. (4.163) and from the condition of the absence of divergences on the light surface, i.e. $\partial\theta/\partial\phi \neq \infty$ for $f = f^*$, $\rho_\perp = 1$, it follows that

$$g_1(f, \phi) = g_0(f)\, \Delta_\perp^0 f(\phi) \tag{4.167}$$

Substituting the solution (4.166), (4.167) into the equation (4.152) and leaving only first-order terms in ε, we obtain the following relation between the quantities $g_0(f)$ and $d\Psi/df$:

$$\frac{\partial}{\partial f}(\Delta_\perp^0 f \cdot g_0) = 2\frac{d\Psi}{df} g_0^{-1} + \frac{1}{2}\frac{d\Psi}{df}\left[4\left(\frac{dz_0}{d\phi}\right)^3 - \sqrt{2}\frac{d \ln \Delta_\perp^0 f}{d\phi}\right]$$
$$+ \left(\frac{d\Psi}{df}\right)^2 g_0^{-1}\left[1 + \left(\frac{dz_0}{d\phi}\right)^2\right] - \frac{3}{\sqrt{2}} g_0 \frac{d^2 z_0}{d\phi^2} + 2\frac{dz_0}{d\phi} \tag{4.168}$$

The quantity $g_0(f)$ can be expressed in terms of the total electric current flowing in the $\rho_\perp$-direction near the singular zero line of the magnetic field between two close magnetic surfaces f and f^*:

$$\delta I(f) = \int \mathbf{j}_e\, d\mathbf{s} = \int \operatorname{rot} \mathbf{G}\, d\mathbf{s} = \int \mathbf{G}\, d\mathbf{l}$$

Since, as is seen from (4.150), on the light cylinder $\rho_\perp = 1$ and $\mathbf{G} = B_\phi \mathbf{e}_\phi$, it follows that

$$\delta I(f) = \int_0^{2\pi} [g_1(f, \phi) - g_1(f^*, \phi)]\, d\phi \tag{4.169}$$

$$g_0(f) - g_0(f^*) = \delta I \left[\int_0^{2\pi} \Delta_\perp^0 f \, d\phi \right]^{-1} \tag{4.170}$$

On the other hand, the quantity δI is proportional to the difference $g_\chi(f) - g_\chi(f^*)$ (see (4.157) and (4.158)). Therefore,

$$g_0(f) = g_\chi(f) 2\pi \left[\int_0^{2\pi} |\Delta_\perp^0 f| \, d\phi \right]^{-1}$$

Integrating (4.168) over ϕ and expressing $d\Psi/df$ in terms of $g_\chi(f)$, we finally arrive at

$$\left(\frac{d\Psi}{df}\right)_{f=f^*} = -\beta_{\max}(\chi)\left[1 - \left(1 - \frac{g_\chi \, dg_\chi/df}{f^*(\chi) i_{\max}^2(\chi)}\right)^{1/2}_{f=f^*}\right] \tag{4.171}$$

Here

$$\beta_{\max}(\chi) = \left[1 + \frac{1}{2\pi}\int_0^{2\pi} \left(\frac{dz_0}{d\phi}\right)^2 d\phi\right]^{-1} \tag{4.172}$$

$$i_{\max}^2(\chi) = \frac{1}{2\pi}\int_0^{2\pi} |\Delta_\perp^0 f| \beta_{\max}(\chi) f^*(\chi)\, d\phi \tag{4.173}$$

Their values can as before (cf. Section 4.2) be found from the analysis of the solution of eqs. (4.32) and (4.33) for zero values of the current, $i_\parallel = 0$, and potential, $\Psi = 0$, obtained in Section 4.4.1.

The condition (4.171) is the 'compatibility relation', which relates in the inclined case the total current $g_\chi(f^*)$ and the electric field $d\Psi/df$ in the pulsar magnetosphere near the boundary $f = f^*$. In the case of linear dependence $d\Psi/df = -\beta_0$, $g_\chi(f) = i_0 f$ (4.58), it is fully analogous to the relation (4.86) for the axisymmetric case and can be presented in the form

$$\beta_0 = \beta_{\max}(\chi)\left[1 - \left(1 - \frac{i_0^2}{i_{\max}^2(\chi)}\right)^{1/2}\right] \tag{4.174}$$

It is only the coefficients $\beta_{\max}(\chi)$ and $i_{\max}(\chi)$ that vary as functions of the inclination angle χ. They are presented in Fig. 4.13. The relation (4.174) implies that for any inclination angle χ, as in the axisymmetric case, there exists a limiting current $i_{\max}(\chi)$ which decreases with increasing χ. The maximum value of the electric field $d\Psi/df$ is $\beta_{\max}(\chi)$.

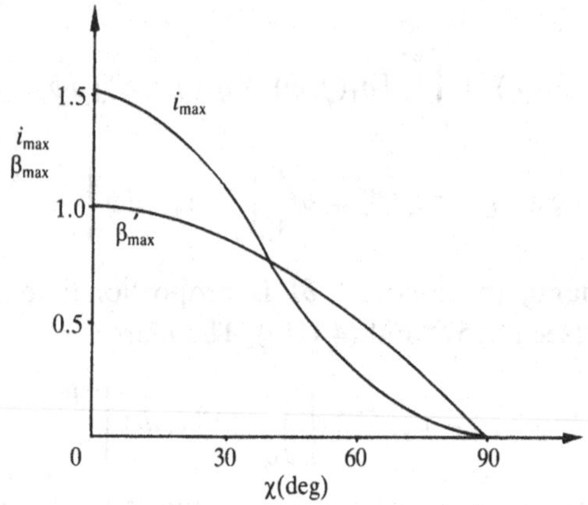

Fig. 4.13. The dimensionless current $i_{max}(\chi)$ and the potential $\beta_{max}(\chi)$ as functions of the angle χ.

It is important to note that the 'compatibility relation' (4.171) involves, in fact, only the symmetric current $i_{\|_s}$, in other words, only the value of the total current g_χ circulating in the magnetosphere.

The indicated matching of solutions suggested that the presence of the longitudinal current in the region of open field lines does not perturb fields in the region of closed field lines. To this end, the magnetic field created by longitudinal currents flowing in the magnetosphere must vanish on the separatrix $f = f^*$. In the axisymmetric case, $\chi = 0$, this is the case, since the magnetic perturbations are completely screened by the jet of symmetric reverse current flowing along a singular magnetic surface f^* (4.80). In the case of inclined axes, $\chi \neq 0$, the picture is not axisymmetric, and therefore the field compensation must be due to an additional asymmetric current flowing along the separatrix $f = f^*$.

Clearly, as the magnetic field created by longitudinal current should escape from the region of open field lines, the normal component **B** to the surface $f = f^*$ should vanish on the boundary $f = f^*$. Hence, to find the asymmetric current, it is necessary that the equations

$$\text{rot}\, \mathbf{B}_T = \frac{4\pi}{c} i_\| \mathbf{B}_0, \text{div}\, \mathbf{B}_T = 0 \tag{4.175}$$

which describes a perturbed magnetic field $\mathbf{B}_T$ (for $r \ll R_L$) be supplemented with the boundary condition

$$\mathbf{B}_{T_\perp}|_{f=f^*} = 0, \qquad \text{i.e. } \mathbf{B}_T \nabla f|_{f=f^*} = 0$$

Introducing the vector potential of magnetic current field $\mathbf{B}_T = \text{rot}(A_T\mathbf{B}_0/B_0)$, we bring (4.175) to the form

$$\Delta A_T = -\frac{4\pi}{c} i_\parallel B_0 \qquad (4.176)$$

Since the magnetic field line is given by the relation $A_T = \text{const}$, the condition of equality to zero of the normal component of the magnetic current field $\mathbf{B}_{T_\perp}$ on the boundary $f = f^*$ is equivalent to the condition

$$A_T[x_0(\phi'), \phi'] = \text{const}$$

where $x_0(\phi')$ is the polar cap boundary. Given this, the surface current flowing along the separatrix $f = f^*$ and necessary for screening the magnetic field $\mathbf{B}^T$, will be found from the condition

$$J_T = \frac{c}{4\pi}[\mathbf{B}_T|_{f=f^*}\mathbf{n}_\perp] = \frac{c}{4\pi}\frac{\mathbf{B}_0}{B_0}(\nabla A_T\mathbf{n}_\perp)|_{x=x_0(\phi')} \qquad (4.177)$$

where $\mathbf{n}_\perp$ is the normal to the surface $f = f^*$.

Thus, the construction, made above, in the neighbourhood of the separatrix $f = f^*$ proves the existence of the solution of eqs. (4.32) and (4.33) in the entire region of open field lines, which satisfies the indicated requirements and makes it possible to obtain the compatibility relation. The general structure of the magnetosphere and the picture of currents flowing there at an arbitrary inclination angle is not difficult to imagine on the basis of the solutions obtained in Sections 4.5.1 and 4.2.4 (Figs. 4.2 and 4.3).

4.5 The dynamics of neutron star rotation and the energy losses of a pulsar

4.5.1 Surface current

The longitudinal current flowing in pulsar magnetosphere is closed by the current $\mathbf{J}_s$ flowing along the neutron star surface (see Fig. 3.8). This current flows across the magnetic field $\mathbf{B}_0$ and is responsible for the star's rotation deceleration. The surface current $\mathbf{J}_s$, in the presence of a strong magnetic field, consists of two components: the current along the electric field $\mathbf{E}_s$ (the so-called Pedersen current) and the orthogonal Hall current:

$$\mathbf{J}_s = \hat{\Sigma}\mathbf{E}_s \qquad (4.178)$$

Here $\hat{\Sigma}$ is the surface conductivity tensor, which has two components $\Sigma_\perp$ (the Pederson conductivity) and $\Sigma_\wedge$ (the Hall conductivity). Accordingly (Gurevich *et al.*, 1976),

$$\mathbf{J}_s^{(1)} = \Sigma_\perp \mathbf{E}_s, \qquad \mathbf{J}_s^{(2)} = \Sigma_\wedge \left[\frac{\mathbf{B}_0}{B_0}\mathbf{E}_s\right] \tag{4.179}$$

$$\mathbf{J}_s = \mathbf{J}_s^{(1)} + \mathbf{J}_s^{(2)}$$

It is natural to assume the pulsar surface conductivity to be uniform and the electric field $\mathbf{E}_s$ to be potential. Then the component $\mathbf{J}_s^{(1)}$ of the surface current is potential, and the other component $\mathbf{J}_s^{(2)}$ is vortex:

$$\mathbf{J}_s^{(1)} = \nabla\xi, \qquad \mathbf{J}_s^{(2)} = \frac{\Sigma_\wedge}{\Sigma_\perp}\left[\frac{\mathbf{B}_0}{B_0}\nabla\xi\right] \tag{4.180}$$

where ξ is the potential of the surface current of the star. The equation for ξ follows from the continuity equation for current, $div\,\mathbf{j}_e = 0$. Taking into account the longitudinal current $i_\parallel \mathbf{B}_0$ flowing onto the star surface, we obtain

$$\Delta\xi = -i_\parallel B_0 \tag{4.181}$$

We introduce the quantity $x = R\sin\theta$ – the distance from the dipole axis to a given point in spherical coordinates R, θ, ϕ' – and taking into account the fact that the polar cap size is small ($\theta \lesssim (\Omega R/c)^{1/2} \ll 1$), rewrite (4.181) in the form

$$(1 - x^2)\frac{d^2\xi'}{dx^2} + \frac{1 - 2x^2}{x}\frac{d\xi'}{dx} + \frac{1}{x^2}\frac{d^2\xi'}{d\phi'^2} = i_0(x, \phi') \tag{4.182}$$

Here $\xi' = 4\pi\xi/B_0R^2\Omega$ is the dimensionless potential (the prime will be omitted below) and i_0 is the dimensionless longitudinal current (4.47). The boundary condition for eq. (4.182) is as follows. First, there must hold the condition

$$\xi[x = x_0(\phi'), \phi'] = \text{const} \tag{4.183}$$

where $x_0(\phi')$ is the polar cap boundary. The condition (4.183) is implied by the requirement of the absence of surface current on the star surface outside the polar region and, therefore, the absence of surface electric field. Note that eq. (4.181) for the current potential $\xi(x, \phi')$ with the boundary condition (4.183) is equivalent to eq. (4.176) for the potential A_T of the magnetic field $\mathbf{B}_T$ of the current. This implies that the quantities ξ and

A_T are proportional to each other:

$$\xi = \frac{c}{4\pi} A_T$$

The solution of eq. (4.182) for surface current depends on the distribution of longitudinal current $i_0(x, \phi')$ flowing from the magnetosphere into the polar cap. It is convenient to write it as

$$i_0(x, \phi') = i_S(x) + \left(\frac{\Omega R}{c}\right)^{1/2} i_A(x, \phi')$$
$$\int_0^{2\pi} d\phi' \int_0^{x_0(\phi')} x i_A(x, \phi')\, dx = 0 \tag{4.184}$$

where i_S and $i_A \lesssim 1$ are respectively symmetric and asymmetric currents (see Section 4.4.2). We have introduced here an additional factor $(\Omega R/c)^{1/2} \ll 1$ indicating that the asymmetric current amplitude is customarily $(\Omega R/c)^{1/2}$ times smaller than the symmetric current amplitude. This smallness is due to the small size of the polar cap. Indeed, the longitudinal current in the magnetosphere is proportional to the critical current j_c (4.14) which satisfies the relation

$$j_c \propto \frac{\mathbf{B}_0\mathbf{\Omega}}{B_0\Omega} \simeq \cos\chi - \sin\chi \cdot \theta$$

Here $\theta \lesssim (\Omega R/c)^{1/2}$ since $\theta \lesssim (\Sigma/\pi R^2)^{1/2}$, where Σ is the area of the polar cap (4.148). For inclination angles χ close to $\pi/2$, the condition of smallness of the current i_A as compared with i_S is, of course, violated.

As examples, to illustrate surface current distribution over a polar cap, we shall consider the solution of (4.182) for two limiting cases: $\chi = 0$ and $\chi = \pi/2$. In the former case $i_0 = i_S = \text{const}$, $i_A = 0$, $x_0 = \text{const}$ (i.e. $dx_0/d\phi' = 0$). From (4.182) and (4.183) we then obtain

$$\xi = \frac{i_0}{4} x^2$$

i.e. the current $\mathbf{J}_s = \nabla\xi$ is directed along the normal to the cap boundary and increases in magnitude proportionally to the distance from the dipole axis. In the latter case, symmetric current is absent completely and asymmetric current is an even function of the angle ϕ' counted from the direction towards the rotation axis. In the simplest case we may assume

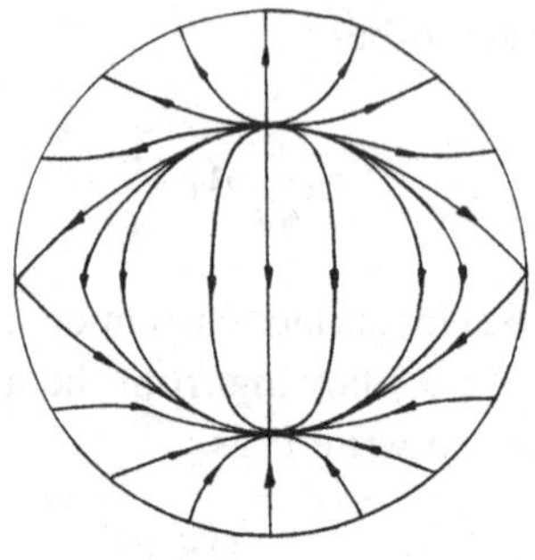

Fig. 4.14. Distribution of surface current for the asymmetric case $\chi = 90°$.

$i_A = i'_A\, x \cos\phi'$, and then

$$\xi = \tfrac{1}{8} i'_A\, x(x^2 - 1) \cos\phi' \tag{4.185}$$

The surface current distribution corresponding to (4.185) is shown in Fig. 4.14. The structure of current is seen to be sectorial, the direction of current being opposite in different sectors.

Note that a concrete structure of surface current for any inclination angle χ depends on the distribution of longitudinal current i_0 flowing in the magnetosphere and on the shape of the polar cap. But in all cases the current and the shape of the polar region remain symmetric about the north–south direction. This is connected with the fact that in the magnetosphere near a neutron star the magnetic field structure is symmetric about this direction.

4.5.2 Moment of forces

In the general case, the moment of forces acting upon a star is determined by the surface currents $\mathbf{J}_s$ following in the polar region,

$$\mathbf{K} = \frac{1}{c} \int [\mathbf{r}[\mathbf{J}_s(\mathbf{B}_0 + \mathbf{B}_T)]]\, ds \tag{4.186}$$

Here $\mathbf{B}_0$ is an unperturbed magnetic field and $\mathbf{B}_T$ is a perturbation due to the longitudinal current $i_\parallel$. The vortex part of the surface current $\mathbf{J}_s^{(2)}$ makes no direct contribution to the braking torque $\mathbf{K}$ but is only responsible for the perturbation of the magnetic moment of a star:

$$\delta\mathbf{M} = \frac{\Sigma_\wedge}{\Sigma_\perp} i_S \left(\frac{\Omega R}{c}\right)^3 \mathbf{M}$$

This is associated with the indicated symmetry of the polar cap and of the longitudinal current i_0. Under the condition $\Sigma_{\wedge}/\Sigma_{\perp} \ll (c/\Omega R)^3$, the perturbation of the magnetic moment of a star can be neglected.

The potential $\mathbf{J}_s^{(1)}$ is just responsible for deceleration. The expression (4.186) for braking torque will become

$$\mathbf{K} = \frac{K_{\|}\mathbf{M}}{M} + K_{\perp}\mathbf{n}_{\mathrm{ort}} \tag{4.187}$$

($\mathbf{n}_{\mathrm{ort}}$ is a unit vector orthogonal to $\mathbf{M}$ and lying in the plane $\boldsymbol{\Omega}\mathbf{M}$), where

$$K_{\|} = -\frac{1}{2\pi} B_0^2 \frac{\Omega R^4}{c} \int_0^{2\pi} d\phi' \int_0^{x_0(\phi')} dx\, x^2(1-x^2)^{1/2} \frac{\partial \xi}{\partial x} \tag{4.188}$$

and $K_{\perp} = K_{\perp}^{(0)} + K_{\perp}^{(1)}$, where

$$K_{\perp}^{(0)} = \frac{1}{2\pi} B_0^2 \frac{\Omega R^4}{c} \int_0^{2\pi} d\phi' \int_0^{x_0(\phi')} dx \left[x \frac{\partial \xi}{\partial x} \cos\phi' - \frac{\partial \xi}{\partial \phi'} \sin\phi' \right] \tag{4.189}$$

$$K_{\perp}^{(1)} = \frac{1}{2\pi} B_0^2 \frac{\Omega R^4}{c} \int_0^{2\pi} d\phi' \int_0^{x_0(\phi')} dx\, x^3 \frac{\partial \xi}{\partial x} \cos\phi' \tag{4.190}$$

We have taken into account here that both magnetic poles of a pulsar contribute to the star's deceleration.

We note first of all that the upper limit of integration over dx in (4.188)–(4.190) has the order of smallness $x_0 \simeq (\Omega R/c)^{1/2} \ll 1$. Therefore, the quantity $K_{\perp}^{(0)}$ could generally be $(\Omega R/c)^{-1/2}$ times as large as $K_{\|}$. However, owing to the boundary condition (4.183), the integral in (4.189)

$$\int_0^{x_0(\phi')} \left[x \frac{\partial \xi}{\partial x} \cos\phi' - \frac{\partial \xi}{\partial \phi'} \sin\phi' \right] dx$$

$$\equiv \frac{\partial}{\partial \phi'} \left\{ -\int_0^{x_0(\phi')} \xi \sin\phi'\, dx + \xi[x_0(\phi'), \phi'] x_0(\phi') \sin\phi' \right\}$$

is the total derivative with respect to ϕ'. We conclude, therefore, that $K_{\perp}^{(0)} = 0$, i.e. the higher-order terms in $(\Omega R/c)^{1/2}$, which formally occur in (4.189), are completely compensated (Beskin *et al.*, 1983; Good & Ng, 1985).

As a result, we have

$$K_{\parallel} = \frac{B_0^2\Omega^3R^6}{c^3}\left[-c_{\parallel}i_S + \mu_{\parallel}\left(\frac{\Omega R}{c}\right)^{1/2} i_A\right] \tag{4.191}$$

$$K_{\perp} = \frac{B_0^2\Omega^3R^6}{c^3}\left[\mu_{\perp}\left(\frac{\Omega R}{c}\right)^{1/2} i_S + c_{\perp}\frac{\Omega R}{c} i_A\right] \tag{4.192}$$

where $c_{\parallel} \sim c_{\perp} \simeq 1$ and the coefficients at the cross terms $\mu_{\parallel}$ and $\mu_{\perp}$ are connected only with asymmetry of the polar cap shape. But, as is seen from Fig. 4.11, the asymmetry is not large, and therefore $\mu_{\parallel} \ll 1$, $\mu_{\perp} \ll 1$ in particular

$$\mu_{\parallel}(0) = \mu_{\parallel}(\pi/2) = \mu_{\perp}(0) = \mu_{\perp}(\pi/2) = 0$$

The exact form of the coefficients $c_{\parallel}, c_{\perp}, \mu_{\parallel}, \mu_{\perp}$ can be obtained by solving the system (4.182), (4.183) and substituting the solution into (4.188), (4.190). In particular, in the axisymmetric case $\chi = 0$, $c_{\parallel} = f^{*2}(\chi = 0)/8$. The same expression holds also for an arbitrary angle χ not too close to $\pi/2$ (Beskin *et al.*, 1983):

$$c_{\parallel} = f^{*2}(\chi)/8 \tag{4.193}$$

i.e. $c_{\parallel}$ depends on the cap area only, and owing to its small asymmetry does not depend on the shape.

Thus, the main contribution to the braking torque **K** is made by symmetric current $K_{\parallel} \propto i_S$, while asymmetric current plays an essential part only for angles χ close to 90°, such that

$$\tan\chi > \left(\frac{\Omega R}{c}\right)^{-1/2}, \qquad |\chi - \pi/2| \lesssim \left(\frac{\Omega R}{c}\right)^{1/2} \tag{4.194}$$

when the amplitude of symmetric current i_S becomes sufficiently small.

4.5.3 Star rotation deceleration

The spherical case Let us now see what effect upon neutron star rotation is produced by the torque **K** (4.186), which is due to the electric current flowing on the polar cap surface. First we assume the star to be spherically symmetric: its inertia tensor $A_{r_{\alpha\beta}}$ is diagonal, and all three of its components are equal to A_r. In this case the equations of motion are

$$A_r \frac{d\Omega}{dt} = K_\parallel \cos\chi + K_\perp \sin\chi \tag{4.195}$$

$$A_r \Omega \frac{d\chi}{dt} = K_\perp \cos\chi - K_\parallel \sin\chi \tag{4.196}$$

As follows from (4.191), the torque **K** leads to a decrease in the angular frequency of pulsar rotation, i.e. to a deceleration. For angles χ not very close to 90°, we have, with allowance made for (4.191) and (4.193),

$$\frac{d\Omega}{dt} = -\frac{f^{*2}}{8}\frac{B_0^2 R^6 \Omega^3}{A_r c^3} i_S \cos\chi \tag{4.197}$$

Along with deceleration, the angle χ between the rotation axis and the direction of the magnetic field will increase:

$$\frac{d\chi}{dt} = \frac{f^{*2}}{8}\frac{B_0^2 \Omega^2 R^6}{A_r c^3} i_S \sin\chi \tag{4.198}$$

One can readily verify the invariance of the quantity (Heintzmann, 1981)

$$I_p = \Omega \sin\chi \tag{4.199}$$

From (4.197)–(4.199) it follows that as a star is decelerated by surface current (which closes up the currents circulating in the magnetosphere), the angle χ tends to 90° and the frequency Ω to $\mathbf{\Omega}_0 \sin\chi_0$, where $\mathbf{\Omega}_0$ and χ_0 are initial values of rotation frequency and inclination angle. The characteristic deceleration time τ_d from a frequency Ω_0 to $\Omega_0 \sin\chi_0$ can readily be estimated from the relations (4.197)–(4.199):

$$\tau_d \simeq \frac{c^3 A_r}{B_0^2 R^6 \Omega_0^2} \tag{4.200}$$

At the angle $\chi \simeq 90°$ when, as is seen from (4.195) and (4.196), the contribution of asymmetric current becomes substantial, the relations (4.197)–(4.199) are no longer valid. In this case, according to (4.193) and (4.195), we obtain

$$\frac{d\Omega}{dt} = -c_\perp \frac{B_0^2 R^7 \Omega^4}{A_r c^4} i_A \tag{4.201}$$

i.e. the star is still decelerated, although about $\Omega R/c$ times slower than for $\chi < 90°$. Note that for the simplest form of asymmetric current, $i_A \propto x \cos\phi'$, the constant $c_\perp = f^{*3}(\chi)/64$.

The effect of non-sphericity We now discuss the effect of deviation from spherical shape upon neutron star rotation, under the action of the moment of forces **K** (4.186). We shall assume a neutron star to be a symmetric top whose inertia tensor is diagonal and has the form

$$A_{r_{xx}} = A_{r_{yy}} = A_r, \quad A_{r_{zz}} = C_r, \quad C_r \neq A_r, \quad A_{r_{\alpha\beta}} = 0 \quad (\alpha \neq \beta) \tag{4.202}$$

i.e. the tensor component $A_{r_{\alpha\beta}}$ in the z-direction is distinct from the two others.

The Euler equations describing rotation of a non-spherical body with the moment of inertia (4.202) under the action of the moment of forces **K** have the form (Landau & Lifshitz, 1960a)

$$\begin{aligned} A_r \frac{d\Omega_x}{dt} + (C_r - A_r)\Omega_y\Omega_z &= K_x \\ A_r \frac{d\Omega_y}{dt} + (A_r - C_r)\Omega_x\Omega_z &= K_y \\ C_r \frac{d\Omega_z}{dt} &= K_z \end{aligned} \tag{4.203}$$

In the case of free motion ($\mathbf{K} = 0$), eqs. (4.203) describe nutation, i.e. rotation of the angular velocity $\mathbf{\Omega}$ vector relative to the z-axis. Their solutions have the form

$$\Omega_x = \Omega \sin\alpha \cos\Phi_n, \qquad \Omega_y = \Omega \sin\alpha \sin\Phi_n, \qquad \Omega_z = \Omega\cos\alpha$$

where α is the angle between the rotation axis and the z-axis, $\Phi_n = (C_r - A_r)\Omega t \cos\alpha / A_r + \Phi_{n0}$. The nutation period is

$$P_n = \frac{A_r}{C_r - A_r} P \cos^{-1}\alpha \tag{4.204}$$

It depends on the degree of non-sphericity of a rotating body, $(C_r - A_r)/A_r$.

If non-sphericity is caused by star deformation due to rotation, then (Pandharipande *et al.*, 1976)

$$\frac{C_r - A_r}{A_r} \simeq \frac{\Omega^2}{\rho_n G} \ll 1 \tag{4.205}$$

where ρ_n is the neutron star matter density (see Section 2.2), and G is the gravitational constant. From (4.204) and (4.205) it follows that the nutation period is

$$P_n \simeq 2 \times 10^6 P^3 \quad \text{s} \tag{4.206}$$

Another reason for non-sphericity is the influence of the magnetic field. In this case,

$$\frac{C_r - A_r}{A_r} \simeq \frac{B_0^2 R}{GM\rho_n}$$

and therefore

$$P_n \simeq (4\pi)^2 \frac{GM\rho_n}{B_0^2 R} \cos^{-1} \chi \tag{4.207}$$

From (4.206), (4.207), it is seen that the nutation period is seen to be much smaller than the characteristic time of pulsar deceleration under the action of the braking torque **K** (4.200). Therefore, taking into account the deceleration effect, we should average the Euler equations (4.203) over the nutation period P_n.

The averaging yields

$$\begin{aligned} \left\langle \frac{d\Omega}{dt} \right\rangle &= -\frac{f^{*2}}{8} \frac{B_0^2 \Omega^3 R^6}{C_r c^3} \cos\alpha \cos\theta \\ \left\langle \frac{d \sin\alpha}{dt} \right\rangle &= \frac{f^{*2}}{8} \frac{B_0^2 \Omega^2 R^6}{C_r c^3} \cos\alpha \sin\alpha \cos\theta \end{aligned} \tag{4.208}$$

where θ is the angle between the z-axis and the direction of magnetic moment **M**. From (4.208) it follows that $\langle d\Omega/dt \rangle = -\langle d \sin\alpha/dt \rangle \Omega/\sin\alpha$. Replacing $\langle d\Omega/dt \rangle$ by $d\langle\Omega\rangle/dt$ and $\sin\alpha$ by $\langle \sin\alpha \rangle$, we come to the relation

$$\langle\Omega\rangle\langle\sin\alpha\rangle = \text{const} \tag{4.209}$$

which is equivalent to the expression (4.199) obtained in the case of strict spherical symmetry. The presence of nutations, therefore, does not change the conclusion that the angle χ between the rotation axis and the magnetic dipole axis tends to 90°. Indeed, as follows from (4.209), the angle α tends to 90° as

$$\chi = \alpha + \theta \cos(2\pi t/P_n)$$

and, accordingly, $\langle\chi\rangle \to \pi/2$.

Thus we can see that the weak asymmetry of a neutron star does not change the character of the evolution of its rotation, caused by electric currents. In particular, the angle χ between the rotation axis and the magnetic moment always tends to 90°.

Note that in the case of a magnetic dipole rotating in vacuum, when the moment of forces is due only to magnetodipole radiation (Landau & Lifshitz, 1983),

$$\mathbf{K} = \frac{2}{3}\frac{\Omega^2}{c^3}[\mathbf{M}[\mathbf{M}\boldsymbol{\Omega}]]$$

Nutations lead to a significant change of rotation dynamics: the angle $\langle\chi\rangle$ tends to 55°, whereas for a spherically symmetric body, $\chi \to 0$ (Goldreich, 1970).

4.5.4 Energy losses of a pulsar

The energy flux To determine the energy losses of a neutron star, we transform (4.195) and (4.196) as follows:

$$W_{\text{tot}} = \frac{d\mathscr{E}_{\text{kin}}}{dt} = \frac{c}{4\pi}\int (\boldsymbol{\beta}_R\mathbf{B})(\mathbf{B}\,d\mathbf{s}) \tag{4.120}$$

On the other hand, using the expression (4.10) for the electric field **E**, we obtain for the Pointing vector flux

$$W_{\text{em}} = \frac{c}{4\pi}\int [\mathbf{EB}]\,d\mathbf{s} = \frac{c}{4\pi}\int (\boldsymbol{\beta}_R\mathbf{B})(\mathbf{B}\,d\mathbf{s}) + \frac{c}{4\pi}\int [\nabla\Psi\mathbf{B}_T]\,d\mathbf{s} \tag{4.211}$$

the last summand being the energy flux, transferred by particles, taken with a reversed sign. Indeed,

$$W_p = \int \Psi\mathbf{j}_e\,d\mathbf{s} = \frac{c}{4\pi}\int \Psi \operatorname{rot}\mathbf{B}_T\,d\mathbf{s} = -\frac{c}{4\pi}\int [\nabla\Psi\mathbf{B}_T]\,d\mathbf{s} \tag{4.212}$$

In the derivation of (4.212) we have used the fact that $\Psi = 0$ on the boundary of the outflow region.

Thus, we see that the energy carried away from the star consists of two parts – the energy transferred by an electromagnetic field and the energy (4.212) transferred by a current, i.e. by particles. This energy goes to generation of electron–positron plasma (see Section 3.2). Given this, the particles generating electron–positron plasma gain energy on the potential drop Ψ in the double layer at the pulsar surface. In the case of linear dependence of Ψ on f (which we considered in Section 4.2, formula (4.58)), we have

$$W_p = \frac{f^{*2}(\chi)}{8}\frac{B_0^2\Omega^4R^6}{c^3}i_0\beta_0 \tag{4.213}$$

irrespective of the shape of the polar cap. The energy transferred by an electromagnetic field is

$$W_{\rm em} = \frac{f^{*2}(\chi)}{8} \frac{B_0^2 \Omega^4 R^6}{c^3} i_0(\cos\chi - \beta_0) \tag{4.214}$$

Finally, the total rotational energy losses of a pulsar make up

$$W_{\rm tot} = \frac{f^{*2}(\chi)}{8} \frac{B_0^2 \Omega^4 R^6}{c^3} i_0 \cos\chi \tag{4.215}$$

Note that since, according to the 'compatibility relation' (4.174), the dimensionless potential $\beta_0 \leqslant \beta_{\max}(\chi) \simeq \cos\chi$, the energy transferred by the electromagnetic field is always positive.

We see that particles in the pulsar magnetosphere carry away only part of the energy, this part not being large for $\beta_0 \ll 1$. However, as shown in Section 4.3, in the boundary layer near the light surface, electrons and positrons are accelerated and gain the energy $\mathscr{E}$ (4.111). Then the power expended to their acceleration is

$$W'_p = \int (n^+ + n^-) v_{p_\perp} \mathscr{E}\, ds$$

Integration is carried out here over the light surface. Taking into account that when intersecting the boundary layer near the light surface, particles gain the energy $\mathscr{E} = \mathscr{E}_m = el/2c\lambda$ (4.111), and that owing to particle flux conservation the quantity $(n^+ + n^-)v_{p_\perp}\, ds = 2\lambda B_0 \Omega \cos\chi / 2\pi e\, ds_0$ (where ds_0 is the element of the surface near the star in the polar cap region), we obtain

$$\begin{aligned} W'_p &= \frac{B_0 \Omega \cos\chi}{2\pi c} j_\parallel \int s_0\, ds_0 \\ &= \frac{\pi B_0 \Omega \cos\chi}{2c} j_\parallel \frac{f^{*2} R^6 \Omega^2}{c^3} = \frac{f^{*2}}{8} \frac{B_0^2 \Omega^4 R^6}{c^3} i_0 \cos\chi \end{aligned} \tag{4.216}$$

This implies that in the region of the light surface the energy transferred by an electromagnetic field is given to accelerated particles.

Thus, we are led to the following picture of energy transfer in the magnetosphere of a neutron star. The ponderomotive action of surface currents (4.178) induces pulsar deceleration. The energy released in this process is transferred along open field lines owing both to the electromagnetic energy flux (4.211) and to the energy of charged particles accelerated in the double layer near the star's surface (4.212). Then, in the region of the light surface, plasma is accelerated additionally owing to the electromagnetic energy flux (4.211).

Relations between longitudinal current and potential drop Ψ The expression (4.216) for the total energy gained by particles as they cross the boundary layer near the light surface was obtained under conditions of a small longitudinal current i_0 (4.91, 4.100) in the first order in $i_0 \to 0$. Its comparison with the total energy losses of a pulsar (4.215) and with the energy transferred by an electromagnetic field (4.214), with allowance for the fact that only this energy can be used for particle acceleration near the light surface, suggests an important conclusion: the equality of energies is possible only under condition $\beta_0 \to 0$ as $i_0 \to 0$.

In other words, the values of the parameter β_0 (i.e. the mean potential drop of the field $\bar{\Psi}$ near the pulsar surface) and the longitudinal current i_0 are not independent – there is a relation

$$\bar{\Psi}(i_0) \to 0 \qquad \text{at } i_0 \to 0 \tag{4.217}$$

between them. Directly from (4.48), and from the corrections (4.105) and (4.107) to the solution considered in the layer, it follows that as $i_0 \to 0$ this relation must look like

$$\bar{\Psi} = c_0 i_0^2 \tag{4.218}$$

where c_0 is a constant. In a similar way, the possibility of closure of the longitudinal current in the layer restricts its limiting value $i_0 \lesssim i_c$, where $i_c \simeq 1$.

The same relation between $\bar{\Psi}$ and i_0, of course, follows directly from the compatibility relation (4.86). But the compatibility relation was obtained in Section 4.2 as the condition of matching solutions at the boundary of closed and open regions of the pulsar magnetosphere for a concrete model of field and current distribution (4.58). The expressions (4.215), (4.216) and formulas (4.217), (4.218) which follow from them are of a more general character – they were derived without an investigation of this model.

This gives a new understanding of the physical meaning of the 'compatibility relation'. For arbitrary values of i_0 and Ψ, equations (4.32) with the boundary conditions (4.34)–(4.37) describe an arbitrary system of currents and energy fluxes in magnetosphere plasma. In the stationary state we additionally have to satisfy the natural physical requirements. First, currents must be closed, the closure proceeding in a thin boundary layer near the light surface. Second, all the energy released must leave the magnetosphere, i.e. the energy flux in magnetosphere must always be directed away from the star. These general requirements are responsible for the existence of a compatibility relation between the potential and the current.

4.5.5 Vacuum losses

We now consider the situation in which plasma is completely absent, so that the energy losses of a rotating star are due to generation of magnetodipole radiation. We shall show that in this vacuum case the star rotation deceleration is also due to a ponderomotive action of currents flowing along the star surface. We shall also show that the known expression for magnetodipole energy losses (3.60) can be obtained in the framework of a quasi-stationary formalism (4.10)–(4.15).

We consider a uniformly magnetized sphere of radius R which rotates in a vacuum with an angular velocity $\mathbf{\Omega}$ and possesses a total magnetic moment $\mathbf{M}$. The electric and magnetic fields outside the sphere will be determined by the Maxwell equations (4.10)–(4.15), in which we put $\mathbf{j}_e = 0$ and $\rho_e = 0$.

In this case, (4.10) and (4.15) become

$$\mathbf{E} + [\boldsymbol{\beta}_R \mathbf{B}] = -\nabla\Psi \tag{4.219}$$

$$\mathbf{B} - [\boldsymbol{\beta}_R \mathbf{E}] = -\nabla\mathscr{H} \tag{4.220}$$

Here $\Psi(\rho_\perp, \phi - \Omega t, z)$ and $\mathscr{H}(\rho_\perp, \phi - \Omega t, z)$ are scalar functions which completely determine the fields $\mathbf{E}$ and $\mathbf{B}$:

$$\mathbf{E}_p = \frac{1}{1-\beta_R^2}(-\nabla\Psi + [\boldsymbol{\beta}_R \nabla\mathscr{H}]) \tag{4.221}$$

$$E_\phi = -\frac{1}{\rho_\perp}\frac{\partial\Psi}{\partial\phi} \tag{4.222}$$

$$\mathbf{B}_p = \frac{1}{1-\beta_R^2}(-\nabla\mathscr{H} - [\boldsymbol{\beta}_R \nabla\Psi]) \tag{4.223}$$

$$B_\phi = -\frac{1}{\rho_\perp}\frac{\partial\mathscr{H}}{\partial\phi} \tag{4.224}$$

where $\mathbf{E}_p$ and $\mathbf{B}_p$ are poloidal fields.

Now substituting the fields $\mathbf{E}$ and $\mathbf{B}$, defined in terms of the potentials

Ψ and $\mathscr{H}$, into (4.1) ($\rho_e = 0$) and (4.3), we obtain

$$\hat{M}_r \Psi - 2\frac{\partial}{\partial z}\mathscr{H} = 0 \tag{4.225}$$

$$\hat{M}_r \mathscr{H} + 2\frac{\partial}{\partial z}\Psi = 0 \tag{4.226}$$

where

$$\hat{M}_r = \Delta - x^2\Delta + 2x\frac{\partial}{\partial x} - \frac{\partial^2}{\partial \phi^2} + x^2\frac{\partial^2}{\partial \phi^2}$$

$x = \Omega\rho_\perp/c$ and Δ is the Laplace operator.

The boundary conditions to eqs. (4.225) and (4.226) should be imposed both when $\mathbf{r} \to 0$ and when $\mathbf{r} \to \infty$. When $\mathbf{r} \to 0$, the boundary conditions are:

(a) The dipole magnetic field of a magnetized sphere, such that (in spherical coordinates r, θ, ϕ)

$$\mathscr{H}_{r\to 0} \to M\frac{\cos\theta}{r^2}\cos\chi + M\frac{\sin\theta}{r^2}\sin\chi\cos\phi \tag{4.227}$$

(b) Assuming that a magnetized sphere is an ideal conductor, such that $\mathbf{E}_{\text{in}} + [\beta_R \mathbf{B}_{\text{in}}] = 0$, we obtain, according to (4.219),

$$\Psi|_{r=R} = 0 \tag{4.228}$$

The conditions at infinity can be formulated as the absence of an energy flux in the direction to the sphere. This boundary condition corresponds to the choice of retarding potentials only (see Landau & Lifshitz, 1983). Besides, all fields must naturally vanish at infinity (see (4.40)).

We now proceed to the solution of eqs. (4.225) and (4.226). First, for $r \ll c/\Omega$ eq. (4.225) gives

$$\Psi = \frac{\Omega M}{c}\left[-\frac{\sin^2\theta}{r} - \frac{2}{3}\frac{R^2}{r^3}P_2(\cos\theta) + \frac{2}{3R}\right]\cos\chi$$
$$+\frac{\Omega M}{c}\left[\frac{\sin\theta\cos\theta}{r} - \frac{R^2}{r^3}\sin\theta\cos\theta\right]\sin\chi \tag{4.229}$$

where $P_2(\cos\theta) = \frac{1}{2}(3\cos^2\theta - 1)$ is a Legendre polynomial. It can be easily verified that the terms proportional to R^2/r^3 (defined by the boundary condition (4.228)) correspond to the quadrupole electric field induced by star rotation (Deutsch, 1955). The magnitude of the quadrupole moment $\mathscr{D}_{\alpha\beta}$ connected with charge redistribution inside the sphere, which is responsible for fulfilment of the condition $\mathbf{E}_{\text{in}} + [\boldsymbol{\beta}_R \mathbf{B}_{\text{in}}] = 0$, is equal to

$$\mathscr{D}_{\alpha\beta} = \frac{R^2}{c}[M_\alpha \Omega_\beta + M_\beta \Omega_\alpha - \tfrac{2}{3}(\mathbf{M\Omega})\delta_{\alpha\beta}] \tag{4.230}$$

Hence, the quadrupole radiation of the sphere must be of the order of

$$W_q \simeq \frac{\Omega^6 \mathscr{D}^2}{c^5} \sim \left(\frac{\Omega R}{c}\right)^4 W_{\text{md}}$$

and, therefore, it makes up only a small fraction compared to magneto-dipole losses $W_{\text{md}} \sim M^2\Omega^4/c^3$. In the limit $R \to 0$ there remain only fields associated with the magnetic dipole M, to the determination of which we shall restrict our further consideration.

As has been mentioned above, the boundary condition at infinity requires that the potentials Ψ and $\mathscr{H}$ contain only a departing wave, and that the electric and magnetic fields tend to zero as $r \to \infty$. Then, as can be obtained directly from equations (4.225) and (4.226), the potentials Ψ and $\mathscr{H}$ must have the form

$$\begin{aligned} \Psi &= M \sin\theta \cos\theta \sin\chi \, \mathrm{Re}\left(\frac{\Omega}{c}\frac{D_4}{r} + \frac{\Omega^2}{c^2} D_5\right) e^{i(\phi + \Omega r/c)} \\ \mathscr{H} &= M \sin\theta \sin\chi \, \mathrm{Re}\left(\frac{D_1}{r^2} + \frac{\Omega}{c}\frac{D_2}{r} + \frac{\Omega^2}{c^2} D_3\right) e^{i(\phi + \Omega r/c)} \end{aligned} \tag{4.231}$$

with constant coefficients $D_{1,2,3,4,5}$. Substituting the potentials (4.231) into (4.225, 4.226), we find these coefficients

$$D_1 = 1, \qquad D_2 = -i, \qquad D_3 = -1, \qquad D_4 = 1, \qquad D_5 = -i \tag{4.232}$$

A direct substitution shows that the fields **E** and **B**, defined using the relations (4.221)–(4.224) and (4.231)–(4.232), coincide with the well-known solutions of the Maxwell equations for a magnetic dipole rotating in a vacuum (see e.g. Landau & Lifshitz, 1983).

As a result, substituting (4.231), (4.232) into the equation for a Pointing vector flux,

$$W_{\text{em}} = \frac{c}{4\pi}\int [\mathbf{EB}]\, ds \tag{4.233}$$

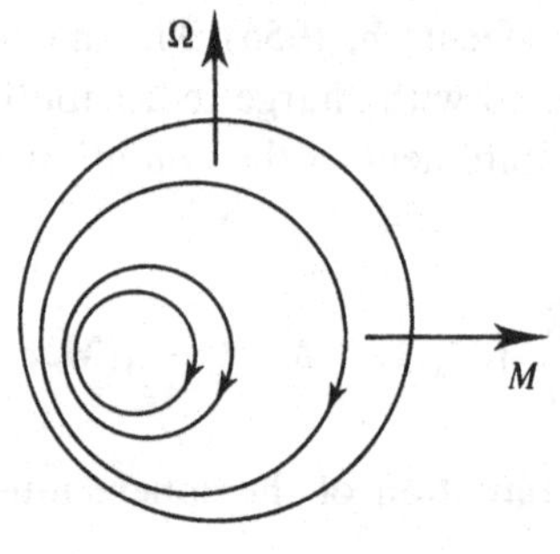

Fig. 4.15. Surface currents slowing down a neutron star in the vacuum case for $\chi = 90°$.

we obtain

$$W_{\text{em}} = \frac{\Omega}{4\pi}\int \frac{\partial \mathscr{H}}{\partial \phi}\frac{\partial \mathscr{H}}{\partial r}\,ds = \frac{1}{6}\frac{\Omega^4 R^6 B_0^2}{c^3}\sin^2\chi \tag{4.234}$$

It is important to emphasize that the relation (4.234) can be obtained directly from the analysis of ponderomotive forces of sphere deceleration induced by currents flowing on the sphere surface. Indeed, if we assume all the currents to be concentrated on the surface, the quantity W_{em} (4.234) can be brought to the form (cf. (4.197, 4.199))

$$W_{\text{em}} = \mathbf{\Omega}\cdot\mathbf{K}_s \tag{4.235}$$

where

$$\mathbf{K}_s = \frac{1}{c}\int [\mathbf{r}[\mathbf{J}_s\mathbf{B}]]\,ds, \qquad \mathbf{J}_s = \frac{c}{4\pi}[\mathbf{nB}]$$

Given this, the deceleration is both due to the interaction of the dipole magnetic field $\mathbf{B}_d = [3\mathbf{n}(\mathbf{Mn}) - \mathbf{M}]/r^3$ with the current induced by the vortex component of the magnetic field $\delta\mathbf{B} = \mathbf{B} - \mathbf{B}_d$,

$$\mathbf{J}_s = \frac{c}{4\pi}[\mathbf{n}\,\delta\mathbf{B}]$$

(see Fig. 4.15), and due to the interaction of surface currents of 'zero approximation' creating a dipole magnetic field

$$\mathbf{J}_s^d = \frac{c}{4\pi}[\mathbf{nB}_d]$$

with the vortex magnetic field $\delta\mathbf{B}$.

Comparison of plasma and vacuum losses We now compare the energy losses of a rotating magnetosphere filled with plasma and the losses of a magnetic dipole rotating in a vacuum (4.234), which was earlier used for estimating pulsar deceleration (Manchester & Taylor, 1977). First we can see a different dependence W_{tot} on the inclination angle χ: the dipole losses are maximal when the axes are orthogonal and disappear completely as $\chi \to 0$. In contrast, the plasma losses (4.215) are maximal in the axisymmetric case and decrease with increasing χ. Such a decrease of plasma losses seems quite natural. The point is that the initial reason for losses is the necessity of continuous plasma generation on open field lines, whereas the amount of plasma, proportional to ρ_c, falls with increasing χ (see (4.13)).

The most essential difference of plasma losses from dipole losses is that they are proportional to the longitudinal current flowing from the pulsar surface, which is expressed in formula (4.215) by the dimensionless factor i_0:

$$i_0 = 2 \cos \chi(j_e/j_c)$$

where j_e is the current density and j_c is the characteristic current determined according to (3.7), (4.14), $j_c \propto \cos \chi$.

The factor i_0 cannot exceed the critical value $i_{max}(\chi)$ (4.173). Therefore, for $i_0 \simeq i_{max}$ the plasma losses are of the same order as the dipole ones, whereas for $i_0 \ll 1$ the plasma losses are much smaller. In particular, in a complete absence of current, $i_0 = 0$, a star surrounded by a plasma magnetosphere loses no energy at all, irrespective of the inclination angle χ. The physical meaning of this can readily be comprehended by considering the energy flux for the solution of complete corotation obtained in Section 4.4.1. In this case, the Pointing vector in the magnetosphere and on the light cylinder, as mentioned in Section 4.4.1, has only a ϕ-component; therefore, the energy flux going away from the star is equal to zero. It means that in complete absence of a longitudinal current, the magnetosphere plasma is so polarized as to completely suppress magnetodipole radiation.

It should be stressed that, formally, the absence of radiation is determined by the boundary condition at infinity. The point is that equation (4.130) admits another solution (Endean, 1976, 1983),

$$\mathscr{H}_E = a_E[xJ_0(x) - J_1(x)] \sin \chi \sin \phi$$

where J_0 and J_1 are Bessel functions and a_E is an arbitrary constant. It can be readily verified that it is precisely the potential $\mathscr{H}_E$ that would be responsible for the energy flux through the light cylinder surface, since it is only the potential $\mathscr{H}_E$ that leads to a non-zero azimuthal magnetic field $B_\phi(R_L) \propto J_1'(1) \cos\phi \sin\chi$ for $\rho_\perp = R_L$. However, in spite of the fact that the potential $\mathscr{H}_E$ does not lead to infinite fields at the light cylinder $\beta_R = 1$, the field **B** associated with $\mathscr{H}_E$ does not depend on z and, therefore, does not vanish at infinity. Thus, it is just the condition at infinity that suggests the equality $a_E = 0$ and, hence, $W_{em} = 0$ for $i_0 = 0$.

It is also noteworthy that in the vacuum case the two second-order equations (4.225) and (4.226) for the potentials Ψ and $\mathscr{H}$ can be written as a fourth-order one for the potential Ψ or for $\mathscr{H}$. It is just for this reason that in the vacuum case two independent solutions are possible which correspond to retarded and advanced potentials – the choice in (4.231) of retarded potentials only is connected with the additional physical assumption expressed by the condition at infinity (4.40). In the case of a plasma-filled magnetosphere with $\beta_0 = 0$, $i_0 = 0$, equation (4.127) is of the second order only (by the way, it coincides with (4.226) provided that $\Psi \equiv 0$) and so it has a unique solution in the form of a standing wave not taking energy away to infinity.

5

Generation of electron–positron plasma in the pulsar magnetosphere

In Section 3.2 we presented the physical picture of electron–positron plasma generation in the pulsar magnetosphere. In this chapter we shall construct a consistent theory of this process. In Section 5.1 we derive the general system of kinetic equations for electrons, positrons and gamma-ray quanta in curvilinear magnetic and inhomogeneous electric fields. We take into account the production of electron–positron pairs and radiation of curvature photons and synchrophotons by the particles produced in these processes. In Section 5.2 we establish the conditions of breakdown and find the threshold value Ψ_0 of the potential drop near the star's surface. In Section 5.3 we investigate the process of particle multiplication and determine the distribution functions of electrons and positrons, as well as the spectrum, direction and intensity of the gamma-ray emission generated in these processes.

5.1 Basic equations

Electron–positron plasma is generated in the vicinity of a neutron star by a flux of charged particles accelerated to high energies in a potential electric field in the double layer near the star's surface. The equation for the electric field potential near a neutron star rotating with a constant angular velocity can readily be derived using (4.1) and taking into account the quasi-stationarity conditions (4.10). We have

$$\Delta\varphi = -\frac{4\pi e^2}{m_e c^2}(n^+ - n^- - n_c) \tag{5.1}$$

Here n_c is the corotation density (3.5) and φ is the dimensionless electric field potential

$$\varphi = \frac{e\Psi}{m_e c^2} \tag{5.2}$$

Note that the magnetic field perturbations induced by longitudinal currents near the star's surface are not large, $\delta\mathbf{B}/\mathbf{B} \sim (R\Omega/c)^{3/2} i_0 \ll 1$. The magnetic field in formula (3.5) is therefore the proper field of the star, and we can think of it as given.

In the kinetic equations for electrons and positrons (4.4) it is convenient to go over from the longitudinal particle momentum $p_{\parallel}$ to the Lorentz factor

$$\gamma = \left(1 + \frac{p_{\parallel}^2}{m_e^2 c^2}\right)^{1/2} \tag{5.3}$$

Then equations (4.4) become

$$\frac{\partial F_{\sigma_p}^{\pm}}{\partial t} + \sigma_p c\mathbf{B}\left(\frac{\partial}{\partial \mathbf{r}} \mp \nabla\phi \frac{\partial}{\partial \gamma}\right)\frac{(\gamma^2 - 1)^{1/2}}{\gamma B} F_{\sigma_p}^{\pm} = q \tag{5.4}$$

We have disregarded here the transverse drift $\mathbf{v}_{\perp}$; the factor and the index $\sigma_p = \text{sign}(\mathbf{pB}) = \pm 1$ characterize the direction of longitudinal momentum $p_{\parallel}$ with respect to the magnetic field.

Next, the operator

$$q = Q_N - S_S \tag{5.5}$$

describes the processes of production Q_N and scattering S_S of electrons and positrons. Since, as shown in Section 3.2, these processes are connected with radiation of high-energy photons, it follows that (5.4) should be supplemented with another equation for the distribution function of photons $N_{\text{ph}}(\mathbf{k}, \mathbf{z}, t)$ over the momenta $\mathbf{k}$,

$$\frac{\partial N_{\text{ph}}}{\partial t} + c\frac{\mathbf{k}}{k}\frac{\partial N_{\text{ph}}}{\mathbf{dr}} = Q_F + Q_S - D \tag{5.6}$$

where the operator Q_F describes photon production by fast electrons and positrons, Q_S describes the generation of synchrophotons, i.e. photons of synchrotron radiation, and D the death of photons due to pair production.

We now derive concrete expressions for the scattering operators S_S, production and decay Q_N, Q_F, Q_S, D in eqs. (5.4)–(5.6).

Photon production The radiation of curvature photons by electrons and positrons is induced by particle motion along a curvilinear magnetic field.

To determine the number of photons, it is convenient to employ the theory of synchrotron radiation represented, for instance, by Landau and Lifshitz (1983). In spite of the fact that the field line curvature radius is not constant in space, the theory of synchrotron radiation is quite fit for the description of curvature radiation. The point is that photon radiation proceeds from a small region of the trajectory of the order of $\rho_l/\gamma \ll \rho_l$ on which the quantity ρ_l can be thought of as constant. According to Landau & Lifshitz (1983), the probability of photon radiation with momentum $\mathbf{k}$ by a charged particle of energy γ is expressed by

$$P(k, \gamma)\, dk = \frac{3^{1/2}}{2\pi} \frac{\alpha_\hbar c}{\rho_l} \frac{\gamma}{k} \Phi_S\left(\frac{k}{k_c}\right) dk \tag{5.7}$$

From here on it is convenient to understand k is a wave number, dimensionless with respect to the Compton wavelength $\lambda\!\!\!^{-} = \hbar/m_e c$ (i.e. $k' = k\lambda\!\!\!^{-}$). Next, $\alpha_\hbar = 1/137$ is the fine structure constant, the function

$$\Phi_S(z) = z \int_z^\infty K_{5/3}(x)\, dx, \qquad k_c = \frac{3}{2} \frac{\lambda\!\!\!^{-}}{\rho_l} \gamma^3 \tag{5.8}$$

and $K_p(x)$ is the Macdonald function. We shall present the asymptotics of the function $\Phi_S(z)$ in a form useful for what follows:

$$\begin{aligned} \Phi_S(z) &= 3\Gamma(5/3)(z/2)^{1/3}\left[1 + \frac{3}{4}\left(\frac{z}{2}\right)^2 + \cdots\right] - \frac{\pi}{\sqrt{3}} z, \qquad z \ll 1 \\ \Phi_S(z) &= (\pi/2)^{1/2} z^{1/2} e^{-z}\left(1 + \frac{55}{72}\frac{1}{z} + \cdots\right), \qquad z \gg 1 \end{aligned} \tag{5.9}$$

The complete form of this function is presented in Landau & Lifshitz (1983). Integrating the probability density $P(k, \gamma)$ (5.7), we derive the expression for the general number of photons with a given momentum k produced per unit time by fast electrons and positrons moving in the direction σ_p:

$$Q_F = \int_1^\infty d\gamma [F^+_{\sigma_p}(\gamma) + F^-_{\sigma_p}(\gamma)] \int_0^\infty P(k, \gamma) \delta\left(\mathbf{k} - \frac{\mathbf{B}}{B} \sigma_p k\right) dk \tag{5.10}$$

We have taken into account here that a curvature photon moves in the same direction as the parent particle.

Particle scattering The radiation of curvature photons leads to particle scattering. Since the motion is one-dimensional, the scattering operator

S_S has the general form

$$S_S = \int_0^\infty [F(\gamma)P(\gamma, k) - F(\gamma + k)P(\gamma + k, k)]\, dk \tag{5.11}$$

Assuming the energy of the emitted photons to be low compared with the particle energy $k \ll \gamma$, we can make a power series expansion of k in the integrand of (5.11). Using for formula (5.7) $P(\gamma, k)$ and restricting ourselves, as is customary in the Fokker–Planck method, to the leading terms of expansion, we obtain

$$S_S = -\tfrac{2}{3}\alpha_\hbar \frac{c\lambda\!\!\!^{-}}{\rho_l^2}\frac{\partial}{\partial\gamma}\left[\gamma^4 F + \frac{55}{32\sqrt{3}}\frac{\lambda\!\!\!^{-}}{\rho_l}\frac{\partial}{\partial\gamma}(\gamma^7 F)\right] \tag{5.12}$$

One can see that the expansion (5.12) is, in fact, in powers of $\lambda\!\!\!^{-}\gamma^2/\rho_l$. In the pulsar magnetosphere the condition $\lambda\!\!\!^{-}\gamma^2/\rho_l \ll 1$ always holds well, so that the particle scattering is described by the differential operator (5.12) to sufficient accuracy.

Pair production The probability of production, per unit time, of an electron–positron pair by a photon with momentum **k** moving in a magnetic field **B** is given by the expression (Berestetskii *et al.*, 1971)

$$W(\mathbf{k}) = \frac{3^{3/2}}{2^{9/2}}\frac{\alpha_\hbar c}{\lambda\!\!\!^{-}} b_0|\sin\beta_{\text{ph}}|\exp\left\{-\frac{8}{3kb_0|\sin\beta_{\text{ph}}|}\right\}\Theta[k|\sin\beta_{\text{ph}}| - k_{\perp 0}] \tag{5.13}$$

Here β_{ph} is the angle between **k** and **B**. $\Theta(x)$ is a step theta-function. $b_0 = B/B_\hbar$, $B_\hbar = m_e^2c^3/e\hbar \simeq 4.4 \times 10^{13}$ Gauss. Formula (5.13) is valid for

$$kb_0|\sin\beta_{\text{ph}}| \ll 1 \tag{5.14}$$

when the production probability is exponentially small. Next, $k_{\perp 0}$ in (5.13) is the minimal value of the photon momentum component orthogonal to **B**, for which pair production is still possible. It can easily be determined from the law of conservation of energy and the longitudinal component of momentum. Indeed, when production proceeds in not very strong magnetic fields, $b_0 \lesssim 0.1$, and when electrons and positrons are produced with almost equal longitudinal momenta (Beskin, 1982a; Dougherty & Harding, 1983), the energy γ_0 of the particles produced and the angle β_p between the direction of their momentum and the direction of the magnetic field are equal to

$$2\gamma_0 = k, \qquad 2(\gamma_0^2 - 1)^{1/2}\cos\beta_p = k\cos\beta_{ph} \tag{5.15}$$

or

$$\gamma_0 = \frac{k}{2}, \qquad \sin^2\beta_p = \frac{k^2\sin^2\beta_{ph} - 4}{k^2 - 4}$$

We can see that production is possible only for $k^2\sin^2\beta_{ph} > 4$, which just determines the value of $k_{\perp_o}$ in (5.13)

$$k_{\perp_o} = (k\sin\beta_{ph})_0 = 2 \tag{5.16}$$

Knowing the pair production probability (5.13), we can find the free path l of a photon with momentum k moving in a curvilinear magnetic field which has a constant magnitude b_0 and a constant curvature radius ρ_l. Then the angle β_{ph} between the direction of photon motion and the direction of the magnetic field is proportional to the path s passed by the photon, $\beta_{ph} = s/\rho_l$. The free path l is found from the condition

$$\frac{1}{c}\int_0^l W(\mathbf{k})\,ds \simeq 1$$

or

$$\frac{3^{3/2}}{2^{9/2}}\frac{\alpha_\hbar b_0}{\lambda\!\!\!^-\rho_l}\int_{2\rho_l/k}^{l} s\exp\left[-\frac{8\rho_l}{3kb_0 s}\right]ds = 1$$

The lower integration limit $2\rho_l/k$ is determined by the pair production threshold (5.16) $s_0 = \rho_l\beta_{ph_0} = 2\rho_l/k$. Integrating by parts we obtain from this equality

$$l = 8\rho_l/3kb_0\left[\ln\left[\frac{\alpha_\hbar\rho_l}{2\sqrt{6}\lambda\!\!\!^-k^2b_0}\right] - 3\ln\left[\tfrac{1}{2}\ln\left[\frac{\alpha_\hbar\rho_l}{2\sqrt{6}\lambda\!\!\!^-k^2b_0}\right]\right]\right] \tag{5.17}$$

For formula (5.13) and, correspondingly, the inequality (5.14) to hold, the condition

$$k^2 \ll \frac{\alpha_\hbar\rho_l}{\lambda\!\!\!^-b_0}$$

should be met.

When $(k_\perp - 2) \gg b_0$, the electron and positron produced find themselves at high Landau levels. Emitting synchrotron radiation, they go over to a lower level. The radiating time τ_r is very small (4.6) under the conditions of pulsar magnetosphere, so that the synchrotron production proceeds practically simultaneously with pair production and at the same point of the space. Given this, the particle does not change its longitudinal velocity $v_\parallel$ during synchrophoton radiation, i.e.

$$v_\parallel^2 = c^2 \cos^2 \beta_p \frac{(\gamma_0^2 - 1)}{\gamma_0^2} = \text{const} = c^2(\gamma^2 - 1)/\gamma^2 \tag{5.18}$$

where γ is the longitudinal energy of the particle after radiation. Expressing $\cos^2 \beta_p$ and γ_0 in (5.18) by means of (5.15), we find that γ is determined directly by the momentum of the parent photon:

$$\gamma = \frac{1}{|\sin \beta_{\mathrm{ph}}|} \tag{5.19}$$

The general number of electron–positron pairs produced by photons per unit time is therefore given by the expression

$$Q_N = \int W(\mathbf{k}) N_{\mathrm{ph}}(\mathbf{k}) \delta\left(\gamma - \frac{1}{|\sin \beta_{\mathrm{ph}}|}\right) d\mathbf{k} \tag{5.20}$$

where the production probability $W(\mathbf{k})$ is determined according to (5.13) and (5.16).

Synchrophoton production We shall now determine the number of generated synchrophotons. We shall use the fact that under the condition $b_0 \lesssim 0.1$ the radiation process can be considered classically (Berestetskii *et al.*, 1971). Following, as before, the theory of synchrotron radiation (Landau & Lifshitz, 1983), we find that the power radiated within the interval of wave numbers dk is

$$\begin{aligned} dI &= \frac{\sqrt{3}}{2\pi} \alpha_\hbar \frac{m_e c^3}{\lambda\!\!\!^-} b_0 |\sin \beta_{\mathrm{ph}}| \Phi_S(k/k_B)\, dk \\ k_B &= \tfrac{3}{2} b_0 \gamma_0^2 |\sin \beta_{\mathrm{ph}}| \end{aligned} \tag{5.21}$$

where the function $\Phi_S(z)$ is defined according to (5.8) and (5.9). Note that quantum corrections to the classical formula (5.21) become substantial for high photon energies comparable with the particle energy $k \simeq \gamma$ and for small $k \lesssim b_0$ when the photon energy becomes of the order of the cyclotron energy. Within the intermediate energy range $b_0 \ll k \ll \gamma$, which we are interested in, the difference of exact expressions from the classical expression (5.21) is insignificant (see Harding & Preece, 1987).

The distribution of generated synchrophotons over $\mathbf{k}$ is determined from the condition of equality between the energy lost by the particle and the energy expended to synchrophoton generation: $G(\mathbf{k})km_ec^2\,d\mathbf{k} = dI\,dt$. Substituting here dI from (5.21), we obtain

$$\begin{aligned} G(k) &= -\frac{1}{2\pi|\sin\beta_{\mathrm{ph}}|k^3m_ec^2}\frac{dI}{dk}\frac{dt}{d\beta_{\mathrm{ph}}} \\ &= -\frac{\sqrt{3}\alpha_\hbar}{(2\pi)^2}\frac{cb_0}{\lambda\!\!\!^{-} k^3}\Phi_S(k/k_B)\frac{dt}{d\beta_{\mathrm{ph}}} \end{aligned} \tag{5.22}$$

The quantity $dt/d\beta_{\mathrm{ph}}$ can be found from the dynamics of variation of the energy γ_0 and the angle β_p of the particle in the course of radiation. Indeed, from (5.21) and (5.17) it follows that

$$\begin{aligned} \frac{d\gamma_0}{dt} &= -\tfrac{2}{3}\alpha_\hbar\frac{c}{\lambda\!\!\!^{-}}\sin^2\beta_p(\gamma_0^2-1)b_0^2 \\ \frac{d\beta_p}{dt} &= \cot\beta_p\frac{1}{\gamma_0(\gamma_0^2-1)}\frac{d\gamma_0}{dt} \end{aligned} \tag{5.23}$$

Since radiation is concentrated near the 'velocity cone' of a relativistic particle, we have $\beta_{\mathrm{ph}} \simeq \beta_p$. If we take into account this fact and substitute $(d\beta_p/dt)^{-1}$ from (5.23) into (5.22), we obtain

$$G(\mathbf{k}) = \frac{3^{3/2}}{8\pi^2}\frac{\Phi_S(k/k_{B_0})}{k^3b_0|\sin\beta_{\mathrm{ph}}|\left[\dfrac{1}{\gamma^2}-\sin^2\beta_{\mathrm{ph}}\right]^{1/2}}$$

$$k_{B_0} = \tfrac{3}{2}b_0|\sin\beta_{\mathrm{ph}}|\cos^2\beta_{\mathrm{ph}}\Big/\left(\frac{1}{\gamma^2}-\sin^2\beta_{\mathrm{ph}}\right)$$

We have also made allowance for the fact that, by virtue of (5.15) and (5.19), the distribution of radiated synchrophotons depends on the wave vector $\mathbf{k}'$ of the parent photon:

$$\frac{1}{\gamma} = |\sin\beta'_{\mathrm{ph}}|, \qquad k_{B_0} = \tfrac{3}{2}b_0\frac{|\sin\beta_{\mathrm{ph}}|\cos^2\beta_{\mathrm{ph}}}{\sin^2\beta'_{\mathrm{ph}}-\sin^2\beta_{\mathrm{ph}}}$$

$$\sin^2\beta'_{\mathrm{ph}}-\sin^2\beta_{\mathrm{ph}} \geqslant \frac{4\cos^2\beta_{\mathrm{ph}}}{k'^2}$$

Finally, the number of synchrophotons generated per unit time is given by the expression

$$Q_S = 2\int W(\mathbf{k}')N_{\rm ph}(\mathbf{k}')G(\mathbf{k})\Theta[\sin^2\beta'_{\rm ph} - \sin^2\beta_{\rm ph} - 4\cos^2\beta_{\rm ph}|\mathbf{k}'^2]\,d\mathbf{k}'$$

$$= \frac{3^{3/2}}{(2\pi)^2}\frac{1}{k^3 b_0|\sin\beta_{\rm ph}|}\int\frac{W(\mathbf{k}')N_{\rm ph}(\mathbf{k}')\Phi_S(k/k_{B_0})}{(\sin^2\beta'_{\rm ph} - \sin^2\beta_{\rm ph})^{1/2}}$$

$$\times\ \Theta[\sin^2\beta'_{\rm ph} - \sin^2\beta_{\rm ph} - 4\cos^2\beta_{\rm ph}/k'^2]\,d\mathbf{k}' \tag{5.24}$$

Photon decay The quantity $\mathscr{D}$ describing the photon decay upon electron–positron pair production is clearly

$$\mathscr{D} = W(\mathbf{k})N_{\rm ph}(\mathbf{k}) \tag{5.25}$$

Thus, the system of kinetic equations (5.4), (5.6), with allowance made for equations (5.10), (5.12), (5.20), (5.24) and (5.25) describing the scattering and production of electrons, positrons and photons, is closed. The equations hold under the conditions

$$b_0 \lesssim 0.1, \qquad \lambda\!\!\!^{-}\gamma^2/\rho_l \ll 1, \qquad k^2 \ll \frac{\alpha_\hbar \rho_l}{\lambda\!\!\!^{-} b_0} \tag{5.26}$$

which are always met in the pulsar magnetosphere where

$$\rho_l \simeq 7\times 10^7 P^{1/2}\ \text{cm}, \qquad P \lesssim 1\ \text{s}, \qquad \gamma \simeq 10^7, k \simeq 10^4 \tag{5.27}$$

Boundary conditions The boundary conditions to eqs. (5.1), (5.4) and (5.6) are given on the star's surface for $h = 0$ and for sufficiently high values of $h \gtrsim R$, where the plasma generation practically ceases (h is the height above the star surface in the polar cap region).

For $h = 0$,

$$\begin{aligned} F_1^{\pm}(\gamma, h = 0) &= K^{\pm}(F_{-1}^{+}, F_{-1}^{-}, N_{{\rm ph}_{-1}})|_{h=0} + F_{1_0}^{\pm} \\ N_{{\rm ph}_1}(k, h = 0) &= K_N(F_{-1}^{+}, F_{-1}^{-}, N_{{\rm ph}_{-1}})|_{h=0} + N_{1_0} \end{aligned} \tag{5.28}$$

For definiteness we have assumed the vector **B** to be directed from the star surface. The functions $F_1^{\pm}$ and $N_{{\rm ph}_1}$ therefore correspond to positrons, electrons and photons flying from the surface, while the functions $F_{-1}^{\pm}$ and $N_{{\rm ph}_{-1}}$ correspond to those flying towards the surface. The coefficients $K^{\pm}$ and K_N describe the electrons, positrons and photons of energy higher than 1 MeV which are knocked out, i.e. emitted from the star surface by the impact of accelerated particles falling onto the star from the plasma;

in the general case, these coefficients are linear operators. The coefficients $F^{\pm}_{1_0}$ and N_{1_0} describe an independent emission of the same particles from the star surface due to other processes (thermal emission, cold emission induced by the electric field).

At large distance $h \gtrsim R$, all particles of the generated plasma are moving away from the star, i.e.

$$F^{+}_{-1} = F^{-}_{-1} = N_{\mathrm{ph}-1} = 0, \qquad h > R \tag{5.29}$$

At the same heights, owing to an intensive production of secondary particles, plasma screens the longitudinal electric field, so that the potential φ tends to a constant value independent of the height. According to (5.1), it is specified by the condition

$$n^{+} - n^{-} = n_c - \frac{m_e c^2}{4\pi e^2} \Delta_{\perp}\varphi \tag{5.30}$$

where $\Delta_{\perp}$ is the transverse part of the Laplacian. Assuming axial symmetry in the vicinity of the polar cap, we can represent $\Delta_{\perp}$ as

$$\Delta_{\perp} = \frac{1}{\rho'_1}\frac{\partial}{\partial \rho'_{\perp}}\left(\rho'_{\perp}\frac{\partial}{\partial \rho'_{\perp}}\right)$$

where $\rho'_{\perp}$ is the radial distance in the polar cap measured from the magnetic dipole axis. Since, according to (4.58),

$$\varphi = \frac{Me\Omega^2\beta_0}{m_e c^4}\left(f^* - \frac{c}{\Omega R^3}\rho'^2_{\perp}\right)$$

the condition (5.30), with allowance made for (4.174), can be written as

$$n^{+} - n^{-} = \int_1^{\infty} (F^{+}_1 - F^{-}_1)\, d\gamma = n_c(1 - i_0^2)^{1/2}, \qquad i_0 = j_{\parallel_e}/j_c \tag{5.31}$$

5.2 The double layer

The double layer is a region, near the star's surface, with a strong electric field (see Section 3.2). In this field, charged particles are accelerated to high energies and radiate curvature γ-ray quanta which generate plasma when absorbed in the strong magnetic field of a pulsar. On this basis, we can readily formulate a qualitative theory of the double layer.

The Ruderman–Sutherland qualitative theory According to the Poisson equation (5.1), the height H of the double layer and the maximal value of the electric field potential φ_0 are related as

$$\frac{H^2}{\varphi_0} \simeq \left(\frac{4\pi e^2}{m_e c^2} n_c\right)^{-1} \simeq \frac{\lambdabar c}{\Omega b_0 \cos\chi} \tag{5.32}$$

On the other hand, the layer height H must be of the order of the free path length l of quanta with respect to the electron–positron pair production. From (5.17) we have

$$H \simeq l = \frac{\rho_l}{k b_0 \ln \Lambda}, \qquad \Lambda = \frac{\alpha_\hbar \rho_l}{2\sqrt{6}\lambdabar b_0 k^2} \tag{5.33}$$

where k is the energy of a curvature photon emitted by an accelerated particle of energy $\gamma \simeq \varphi_0$. The quantity k corresponds to the energy of a photon processing the maximal emission probability $k \simeq k_c$ (5.8). Thus, $k \simeq \lambdabar\varphi_0^3/\rho_l$ which, with allowance for (5.33), gives another equation relating H and φ_0:

$$H\varphi_0^3 \simeq \frac{\rho_l^2}{\lambdabar b_0 \ln \Lambda} \tag{5.34}$$

So, from (5.32) and (5.34) we find the values of H and φ_0 (Ruderman & Sutherland, 1975):

$$\begin{aligned} H &\simeq (\lambdabar^2 \rho_l^2 c^3 \Omega^{-3} b_0^{-4} \cos^{-3}\chi \ln^{-1}\Lambda)^{1/7} \\ \varphi_0 &\simeq (\lambdabar^{-3} \rho_l^4 c^{-1} \Omega b_0^{-1} \cos\chi \ln^{-2}\Lambda)^{1/7} \end{aligned} \tag{5.35}$$

as well as the characteristic photon energy k (5.33):

$$k \simeq \frac{\rho_l}{H b_0 \ln \Lambda}$$

This simple theory of the double layer has played an important role in the development of ideas concerning the mechanism of plasma generation. It allowed us to determine the characteristic values of the principal quantities: the double layer height $H \simeq 10^4$ cm, the potential drop $\varphi_0 \simeq 10^7$, as well as their dependence on the magnitudes of the magnetic field $B = b_0 B_\hbar$, of the neutron star rotation period $P = 2\pi/\Omega$, of the magnetic field curvature radius ρ_l, and of the inclination angle χ between the axes.

At the same time, in the framework of this theory it remains unclear how the conditions for stationary plasma generation occur. Indeed, conserving the longitudinal momentum of the parent photon, the electron and positron produced move away from the neutron star. Under what conditions will one of these particles be trapped by the field and accelerated towards the star? Will it be also able to emit photons which will produce pairs before they reach the star surface in order that the plasma generation process will be stationary? What is the role of particles knocked out from the star surface? What is the role of the longitudinal electric current flowing in the double layer? To answer all these questions, we should construct a more consistent quantitative theory, which is done below in line with Gurevich & Istomin (1985).

5.2.1 The distribution function of curvature photons

In the polar region of the pulsar magnetosphere it is natural to choose a cylindric reference frame characterized by three coordinates: the height h above the star surface, the transverse coordinate f related to the distance from the dipole axis $\rho'_\perp$ as $\rho'^2_\perp = (\Omega R^3/c)f$, and the coordinate ϕ in the azimuthal direction.

Owing to azimuthal symmetry, the magnetic field $\mathbf{B}$ in the double layer has two components:

$$B_h = B_0 = \text{const}, \qquad B_f = \frac{B_0 h}{\rho_l}, \qquad B_\phi = 0 \tag{5.36}$$

Since $H \ll \rho_l$, R, the curvature radius of magnetic field lines, does not depend on the height h above the surface but is a function of the coordinate $\rho'_\perp$. For the dipole magnetic field $\rho_l = \frac{4}{3}R^2/\rho'_\perp$, i.e. in the variable f (4.47, 4.54),

$$\rho_l = \left(\frac{16Rc}{9\Omega f}\right)^{1/2} \sim 10^8 \text{ cm} \tag{5.37}$$

Under the double layer conditions, the distribution function of photons $N_{\text{ph}}(\mathbf{k}, \mathbf{r}, t)$ is $N_{\text{ph}}(k, k_f, h, f, \text{sign}\, k_h, t)$. We single out here the magnitude of the wave vector $k = |k_h|$ and the component k_f. Note that $k_f \simeq \beta_{\text{ph}}$ and in the cases of interest we always have $\sin\beta_{\text{ph}} \ll 1$ (5.14). In the particle production and photon emission process, it is just the component of the photon wave vector that is the most important characteristic. Owing to the deviation of the field line from the vertical, the magnitude of this component will depend on the height h. Therefore, instead of k_f it is convenient to introduce the quantity $\kappa_\perp$ proportional to the transverse (to the direction of the magnetic field) component of the wave vector:

$$\kappa_\perp = \tfrac{3}{4} b_0 k \left(\frac{k}{\rho_l} \operatorname{sign} k_h - \frac{k_f}{k} \right) \tag{5.38}$$

In the variables k, $\kappa_\perp$, h, f, σ_{ph}, t, the equation for the distribution function of photons (5.6), with allowance for (5.10) and (5.25) for Q_F and D, can be represented as

$$\frac{\partial N_{ph}}{\partial t} + c\sigma_{ph} \frac{\partial N_{ph}}{\partial h} + \tfrac{3}{4} c b_0 \frac{k}{\rho_l} \frac{\partial N_{ph}}{\partial \kappa_\perp} = -W(\mathbf{k}) N_{ph} + \sum_{+,-} \int_1^\infty F^\pm_{\sigma_p} P(k, \gamma) \delta(\kappa_\perp) \cdot \delta_{\sigma_p, \sigma_{ph}}\, d\gamma \tag{5.39}$$

We have here made renormalization of the distribution function of photons conditioned by the introduction of the new variable $\kappa_\perp$, i.e. we have gone over to the new function $N'_{ph}(k, \kappa_\perp) = 2\pi(\tfrac{3}{4} b_0)^{-2} \kappa_\perp N_{ph}$, so that the quantity $N'\, dk\, d\kappa_\perp$ is the photon density in the coordinate space (the prime is hereafter omitted). Under the condition $H \ll \rho_l$, the displacement in f is insignificant, so that the quantity f enters eq. (5.39) only as the parameter $\rho_l = \rho_l(f)$. We have also taken into account that the free path of pair production by synchrophotons is much larger than in the case of the main photons; therefore, the role of synchrophotons in the double layer is negligible and the term Q_S is disregarded. The boundary conditions to (5.39) were given earlier (see (5.28) and (5.29)).

We shall first neglect the flux of gamma-ray quanta knocked out of the star surface, i.e. assume that in the boundary conditions (5.28)

$$K_N = 0, \qquad N_{1_0} = 0$$

The stationary solution of the linear equation (5.39) can readily be obtained by integration along the characteristics. After integration over $\kappa'_\perp$ it is written as

$$N_{ph}(k, \kappa_\perp, h, \sigma_{ph}) = \frac{4\rho_l}{3cb_0 k} \sum_{+,-} \int_1^\infty d\gamma\, F^\pm\!\left(\gamma, \sigma_p, h - \frac{4}{3} \frac{\rho_l \kappa_\perp}{k b_0} \sigma_p \right) P(k, \gamma) \times I(\kappa_\perp, k) c(h) \Theta(\kappa) \delta_{\sigma_p, \sigma_{ph}} \tag{5.40}$$

where

$$I(\kappa_\perp, k) = \begin{cases} \exp\left[-\dfrac{\alpha_\hbar}{\sqrt{6}} \dfrac{\rho_l}{\lambda\!\!\!^{-} k^2 b_0} \displaystyle\int_{3/2b_0}^{\kappa_\perp} \kappa'_\perp \, e^{-2/\kappa'_\perp} \, d\kappa'_\perp \right], & \kappa_\perp \geqslant \tfrac{3}{2} b_0 \\ 1, & \kappa_\perp < \tfrac{3}{2} b_0 \end{cases} \tag{5.41}$$

and

$$c(h) = \begin{cases} \Theta\left(h - \dfrac{4}{3} \dfrac{\rho_l \kappa_\perp}{k b_0} \right), & \sigma_{\mathrm{ph}} > 0 \\ \Theta\left(h_0 - h - \dfrac{4}{3} \dfrac{\rho_l \kappa_\perp}{k b_0} \right), & \sigma_{\mathrm{ph}} < 0 \end{cases} \tag{5.42}$$

Here $h_0 > H$ is the value of the coordinate h for which there holds the boundary condition (5.29). We can thus see that the distribution function of photons at a given point h is determined by positron and electron distribution at the point $h' = h - \frac{4}{3}\rho_l \kappa_\perp \sigma_{\mathrm{ph}}/k b_0$, at which the photons were emitted. Moreover, from the point of emission h' to h we observe a monotonic extension of the distribution function of photons in transverse momenta $\kappa_\perp$, which is described by the factor $c(h)$ (5.42). As soon as $\kappa_\perp > \frac{3}{2} b_0$, photons are absorbed with the production of electron–positron pairs, and their distribution function starts falling, which is described by the function $I(\kappa_\perp, k)$ (5.41, 5.40). The energy distribution of photons (i.e. the distribution in k), is mainly determined by the probability of radiation of a 'curvature' photon by an accelerated charged particle $P(k, \gamma)$ (5.7).

5.2.2 The distribution function of electrons and positrons

Substituting the function N_{ph} (5.40) into (5.4), we obtain a closed equation for the distribution function of particles, which under stationary conditions has the form

$$c\sigma_p \left[\frac{\partial}{\partial h} \left(\frac{(\gamma^2 - 1)^{1/2}}{\gamma} F^\pm \right) \mp \nabla \varphi \frac{\partial}{\partial \gamma} \left(\frac{(\gamma^2 - 1)^{1/2}}{\gamma} F^\pm \right) \right] = -S_S + Q_N$$

$$Q_N = \frac{2^{1/2}}{3\pi} \frac{c \alpha_\hbar^2}{\lambda\!\!\!^{-} b_0^2} \int_{3/2b_0}^{\infty} d\kappa_\perp \int_0^\infty \frac{\kappa_\perp^2}{k^3} I(\kappa_\perp, k) \, e^{-2/\kappa_\perp} c(h) \delta\left(k - \frac{4}{3b_0} \gamma \kappa_\perp \right) dk$$

$$\times \sum_{+,-} \int_1^\infty d\gamma' \, F^\pm\left(\gamma', \sigma_p, h - \frac{\rho_l}{\gamma} \sigma_p \right) \gamma' \Phi_S\left[\frac{2}{3} \frac{\rho_l k}{\lambda\!\!\!^{-} \gamma'^3} \right]. \tag{5.43}$$

The boundary conditions to (5.43) are specified according to (5.28) and (5.29). In the sequel we disregard the free particle outflow from the surface, putting $F^{\pm}_{1_0} = 0$.

The two-flow approximation Equations (5.43) can be substantially simplified. First of all, the diffusion smearing of the distribution of particles over the energies $\Delta\gamma$, which is described by the second term in the expression for S_S (5.12), depends on the distance h from the surface

$$\Delta\gamma \simeq \gamma\varepsilon_1, \qquad \varepsilon_1 = \left(\frac{2}{3}\frac{\alpha_\hbar\gamma}{\rho_l}\right)^{1/2}\frac{\lambdabar}{\rho_l}\gamma^2 h^{1/2}, \qquad \varepsilon_1 \ll 1$$

and is negligibly small on scales of the order of the size of the double layer H (5.35) under the usual conditions for pulsars (5.27). The energy distribution $\Delta\gamma$ due to particle production is also insignificant:

$$\Delta\gamma \simeq \gamma\varepsilon_2, \qquad \varepsilon_2 = \frac{3}{4}\frac{\lambdabar}{\rho_l}\gamma^2, \qquad \varepsilon_2 \ll 1$$

Thus, in the zero approximation with respect to small parameters ε_1 and ε_2, the particle energy in the double layer is completely determined by the action of the dynamical forces, namely, by the electric field and the deceleration (the first term in S_S). This means that in this approximation the energy distribution functions of particles have the form of δ-functions:

$$F^{\pm} = n^{\pm}(h)\delta(\gamma - \gamma_1^{\pm}(h)) \tag{5.44}$$

In other words, in a double layer there exist, so to say, two hydrodynamic particle flows moving towards each other. Taking into account (5.44) and (5.12), we find from the kinetic equation (5.43)

$$n^{+}(h) = \frac{n^{+}(0)\gamma_1^{+}}{((\gamma_1^{+})^2 - 1)^{1/2}}, \qquad n^{-}(h) = \frac{n^{-}(h_m)\gamma_1^{-}}{((\gamma_1^{-})^2 - 1)^{1/2}} \tag{5.45}$$

$$\frac{d\gamma_1^{\pm}}{dh} = \mp\frac{d\varphi}{dh} - \frac{2}{3}\frac{\alpha_\hbar\lambdabar}{\rho_l^2}\gamma_1^{\pm 4}, \qquad \gamma_1^{+}(h=0) = \gamma_1^{-}(h=h_m) \simeq 1 \tag{5.46}$$

Note that, as shown below, under the conditions practically realized in pulsars the role of the second term in (5.46) proves to be negligible.

In (5.46) we have assumed, for definiteness, that the star's surface potential is positive relative to the plasma $\varphi_0 > 0$; this corresponds to $n_c > 0$, i.e. $\mathbf{B}_0\boldsymbol{\Omega} < 0$. In this case, positrons are accelerated by the field towards the magnetosphere and electrons towards the star, which is allowed in (5.45). In the opposite case, $\mathbf{B}_0\mathbf{\Omega} > 0$, we have

$$n^-(h) = \frac{n^-(0)\gamma_1^-}{((\gamma_1^-)^2 - 1)^{1/2}}, \qquad n^+(h) = \frac{\gamma_1^+ n^+(h_m)}{((\gamma_1^+)^2 - 1)^{1/2}} \tag{5.47}$$

Reflected particles It is noteworthy that the particle density $n^-(h_m)$ in (5.45) (and, respectively, $n^+(h_m)$ in (5.47)) is the density of particles produced near the point h_m and reflected by the electric field in the same region – the region near the potential maximum (for them $\sigma_p < 0$). To find this quantity we should know the form of the distribution function of particles produced up to $\Delta\gamma/\phi_0 \simeq \varepsilon_2$. We shall bear in mind that the function $I(\kappa_\perp, k = 4\gamma\kappa_\perp/3b_0)$ (5.41) describing the angular distribution of curvature photons can be approximated with a sufficient accuracy by the Θ-function

$$I(\kappa_\perp, k = 4\gamma\kappa_\perp/3b_0) = \Theta(\Lambda^{-1} - \kappa_\perp), \qquad \Lambda = \tfrac{1}{2}\ln\left(\frac{3^{3/2}\alpha_\hbar\rho_l b_0}{2^{11/2}\lambda\!\!\!^{-}\gamma^2}\right)$$

Substituting the distribution of fast particles in the form (5.44) and (5.45) into the expression (5.43) for Q_N and going over to the trajectory of particle motion $\gamma_1^\pm(h)$ described by equation (5.46) for the distribution functions of produced particles for $h \sim h_m$, $F^\pm(\gamma, h, \sigma_p = 1)$ and $F^\pm(\gamma, h, \sigma_p = -1)$, we have

$$F^\pm(\gamma, h, \sigma_p = \pm 1) = \frac{\sqrt{3}}{2\pi}\frac{\alpha_\hbar n^+(0)}{\rho_l}\frac{\gamma}{(\gamma^2-1)^{1/2}}\int_0^h \frac{\gamma_1^+\left(h' - \frac{\rho_l}{\gamma'}\right)}{\gamma'}$$
$$\times\, \Phi_S\left[\frac{4k}{3}\frac{\rho_l}{\lambda\!\!\!^{-}}\gamma'\left(\gamma_1^+\left(h' - \frac{\rho_l}{\gamma'}\right)\right)^{-3}\right]dh'$$
$$+\, n^+(0)\,\delta(\gamma - \gamma_1^+(h))$$
$$\gamma' = \gamma + \gamma_1^\pm(h') - \gamma_1^\pm(h), \qquad k = 2/(3b_0\Lambda) \tag{5.48}$$

For particles produced for $h \sim 0$ we have, respectively,

$$F^\pm(\gamma, h, \sigma_p = \pm 1) = \frac{\sqrt{3}}{2\pi}\frac{\alpha_\hbar n^-(h_m)}{\rho_l}\frac{\gamma}{(\gamma^2-1)^{1/2}}\int_h^{h_m} \frac{\gamma_1^-\left(h' + \frac{\rho_l}{\gamma'}\right)}{\gamma'}$$
$$\times\, \Phi_S\left[\frac{4k}{3}\frac{\rho_l}{\lambda\!\!\!^{-}}\gamma'\left(\gamma_1^-\left(h' + \frac{\rho_l}{\gamma'}\right)\right)^{-3}\right]dh'$$
$$+\, n^-(h_m)\delta(\gamma - \gamma_1^-(h)) \tag{5.49}$$

From (5.48) and (5.49) we can see that the effective production of electron–positron pairs proceeds near the upper, $h = h_m$, and lower, $h = 0$, boundary layer on the scale Δh small compared to the entire thickness of the double layer $H \simeq h_m$. Indeed, the production becomes effective only if the argument of the function Φ_S entering formulas (5.48) and (5.49) becomes of the order of unity (recall that for a high value of the argument the function Φ_S is exponentially small). From this condition we have

$$\frac{4k}{3}\frac{\rho_l^2}{\lambda\!\!\!^{-}}\frac{1}{\gamma^5}\left(\frac{\partial\gamma_1}{\partial h}\right)^2 \Delta h \simeq 1$$

but since $\partial\gamma_1/\partial h \sim \gamma/H$, it follows that

$$\frac{\Delta h}{H} \simeq b_0 \Lambda \frac{H}{\rho_0} \gamma \varepsilon_2 \sim 10^{-1} \tag{5.50}$$

The condition (5.50) implies that the plasma multiplication region is small and the whole process has a simple character: to each accelerated positron there correspond K_m electrons produced and reflected back by the electric field near the point $h = h_m$:

$$n^-(h_m) = K_m n^+(0) \tag{5.51}$$

Similarly, near the star surface an electron accelerated by the field generates K_1 positrons. Then, taking account of the boundary condition (5.28), we have

$$n^+(0) = K_0 n^-(h_m), \qquad K_0 = K_1 + K, \qquad K = K^+ + K_N \tag{5.52}$$

where K^+ is the coefficient of particle multiplication due to knocking out of the star's surface of particles of opposite sign (5.28). It is assumed in (5.52) that the energy of knocked out particles is small as compared to the energy of particles incident onto the surface. It is also assumed that there is no independent emission of particles from the star's surface: $F^+_{1_0} = 0$.

The reflection factors Substituting (5.51) into (5.52), we obtain the following 'breakdown' condition, i.e. the condition of the existence of a stationary plasma generation in a double layer:

$$K_0 K_m = 1, \qquad K_m = \frac{1}{K_0} \tag{5.53}$$

The expressions for the factors K_m and K_1 can be derived by integrating the distribution function of produced particles (5.48) and (5.49) over the energy. Changing the integration order will lead us to the following expressions:

$$K_1 = \frac{\sqrt{3}}{2\pi}\frac{\alpha_\hbar}{\rho_l}\int_{h_m - t_m}^{h_m} dt\, \gamma_1^-(t) \times \left\{ Y\left[\frac{4k\rho_l^2}{3(\gamma_1^-(t))^3 \lambda\!\!\!^{-}(h_2 - t)}\right] - Y\left[\frac{4k\rho_l^2}{3(\gamma_1^-(t))^3 \lambda\!\!\!^{-}(h_1 - t)}\right]\right\} \tag{5.54}$$

$$K_m = \frac{\sqrt{3}}{2\pi}\frac{\alpha_\hbar}{\rho_l}\int_0^{t_m} dt\, \gamma_1^+(t) \times \left\{ Y\left[\frac{4k\rho_l^2}{3(\gamma_1^+(t))^3 \lambda\!\!\!^{-}(h_2 - t)}\right] - Y\left[\frac{4k\rho_l^2}{3(\gamma_1^+(t))^3 \lambda\!\!\!^{-}(h_1 - t)}\right]\right\} \tag{5.55}$$

where $h_{1,2}(h, t)$ are roots of the equations

$$h_{1,2} - \rho_l/[1 + \gamma_1^+(h) - \gamma_1^+(h_{1,2})] = t$$

$$h_{1,2} - \rho_l/[1 + \gamma_1^-(h_m - h) - \gamma_1^-(h_{1,2})] = t$$

and t_m is the value of t for which $h_1 = h_2$. The function $Y(x)$ entering (5.54) and (5.55) is equal to

$$Y(x) = \int_x^\infty (\eta - x)K_{5/3}(\eta)\, d\eta = \begin{cases} \dfrac{5\pi}{3} - 9\Gamma(5/3)x^{1/3}/2^{1/3} + \pi x/\sqrt{3} + \cdots & x \ll 1 \\ (\pi/2x)^{1/2}\, e^{-x} & x \gg 1 \end{cases} \tag{5.56}$$

Table 5.1 shows the values of Y versus x.

The factors K_1 and K_m depend on the Lorentz-factor $\gamma_1^\pm(h)$ of particles accelerated in the potential $\varphi(h)$ according to (5.54) and (5.55). The dependence $\gamma_1^\pm(h)$ is given by the expression (5.46).

Table 5.1. *The function* $Y(x)$

x	0.0	0.01	0.05	0.1	0.2	0.5	1.0	1.5	2.0	3.0	4.0
$Y(x)$	5.24	3.87	2.95	2.42	1.82	0.99	0.45	0.23	0.12	0.037	0.012

5.2.3 The 'breakdown' potential

Equations (5.46) should be solved together with the Poisson equation (5.1) which, in the assumption of a thin double layer (i.e. H – the layer height – is smaller than the polar cap size $R_0 = R((\Omega R/c)f^*)^{1/2}$; $H \ll R_0$) takes the form

$$\frac{d^2\varphi}{dh'^2} = 1 - (n^+ - n^-), \qquad h' = \frac{h}{D_c}, \qquad D_c = \left(\frac{m_e c^2}{4\pi n_c e^2}\right)^{1/2} \tag{5.57}$$

The concentrations n^+ and n^- are expressed here in the units n_c (3.5). According to (5.28) and (5.31), the boundary conditions to eq. (5.57) are

$$\varphi(0) = 0, \qquad \varphi(h'_m) = \varphi_0, \qquad d\varphi/dh'|_{h'=h'_m} = 0, \qquad h'_m = \frac{h_m}{D_c} \tag{5.58}$$

The last condition follows from the necessity of a smooth matching of the potential in the double layer $h' < h'_m$ and in the quasi-neutral region $h' > h'_m$ (5.31). It should be noted that this condition strictly corresponds to the hydrodynamic approximation (5.44) only when the spread of the distribution function is neglected. Indeed, as can be seen from (5.48) and (5.49), a small spread of particles over energy is observed. This leads to the appearance near the point h'_m of the 'hump' in the potential $\Delta\phi$ of the order of

$$\frac{\Delta\varphi}{\varphi_0} \sim \varepsilon_2 \ll 1$$

We neglect this small quantity below, i.e. we assume that the longitudinal electric field in the zero order in ε_2 for $h' > h'_m$ vanishes.

Bearing in mind that by virtue of (5.50) the concentrations n^+ and n^- in the layer are constant, we easily find the solution of eq. (5.57) with the boundary conditions (5.58):

$$\varphi = \tfrac{1}{2}(1 - n^+ + n^-)(h' - h'_m)^2 + \varphi_0, \qquad h' < h'_m$$

Here

$$h'_m = (2\varphi_0)^{1/2}[1 - (n^+ - n^-)]^{-1/2}$$

φ_0 is the difference of potentials between the star and the magnetosphere. The concentration of electrons n^- and positrons n^+ in a layer is conveniently expressed in terms of the current density $i_0 = j_\parallel/j_c$:

$$n^+ = \frac{i_0}{1 + K_m}, \qquad n^- = \frac{i_0 K_m}{1 + K_m}$$

For the thickness of a double layer we then obtain

$$h_m = D_c(2\varphi_0)^{1/2}(1 - pi_0)^{-1/2}, \qquad p = \frac{1 - K_m}{1 + K_m}$$

Substituting this expression into (5.51), (5.55), we obtain from (5.53), with allowance made for (5.52), the integral expression which determines the threshold value of the field potential φ_0.

The result of the solution of this equation is conveniently represented in the form (Gurevich & Istomin, 1985)

$$\varphi_0 = \frac{2^{5/7}}{3^{4/7}} \frac{\rho_l^{4/7}(b_0\Lambda)^{-2/7}}{\lambda\!\!\!^{-2/7} D_c^{2/7}} (1 - pi_0)^{1/7} b^{-2/7} \tag{5.59}$$

where b is a function of the dimensionless parameter d:

$$d = \frac{2^{4/7}\alpha_\hbar}{3^{5/14}\pi} \rho_l^{-1/7} \lambda\!\!\!^{-3/7} D_c^{4/7}(1 - pi_0)^{-2/7}(b_0\Lambda)^{-3/7} \tag{5.60}$$

The qualitative character of the dependences (5.59) and (5.60) can be understood if we take into account that the main contribution to the integrals (5.54) and (5.55) is made by the region where the arguments of the function Y are of the order of unity. The exact solution was determined numerically.

The dependence of b on d for different values of the multiplication coefficient $K = K^{\pm} + K_N$ is represented in Figs. 5.1 and 5.2. We can see that in the range of variation $10 \lesssim d \lesssim 30$, which is interesting for pulsars, the quantity $b(d, K)$ changes little and does not differ greatly from its initial value for $d \gg 1$. The dependence of b on K for $d \gg 1$ is also shown. The numerical expressions for φ_0 and H have the form

$$\begin{aligned} \varphi_0 &= 1.1 \times 10^7 \rho_{l_7}^{4/7} P^{-1/7} B_{12}^{-1/7} \cos^{1/7}\chi(1 - pi_0)^{1/7} b^{-2/7}(\Lambda/8)^{-2/7} \\ H &= 9.5 \times 10^3 \rho_{l_7}^{2/7} P^{3/7} B_{12}^{-4/7} \cos^{-3/7}\chi(1 - pi_0)^{-3/7} b^{-1/7}(\Lambda/8)^{-1/7}\ \text{cm} \end{aligned} \tag{5.61}$$

Here $\varphi_0 = e\Psi_0/m_e c^2$ is expressed in dimensionless units:

$$\Lambda = 7.66 - \tfrac{3}{14}\ln\rho_{l_7} + \tfrac{3}{7}\ln\left[\frac{P}{\cos\chi(1 - pi_0)}\right] - \tfrac{1}{14}\ln B_{12} - \tfrac{1}{7}\ln b$$

The exact solution obtained agrees rather well with the approximate estimation of φ_0 and H (5.35). It should be emphasized that (5.60) and (5.61) (see Fig. 5.1) permit an account of the influence upon the breakdown conditions of the interaction of fast particles with the neutron star

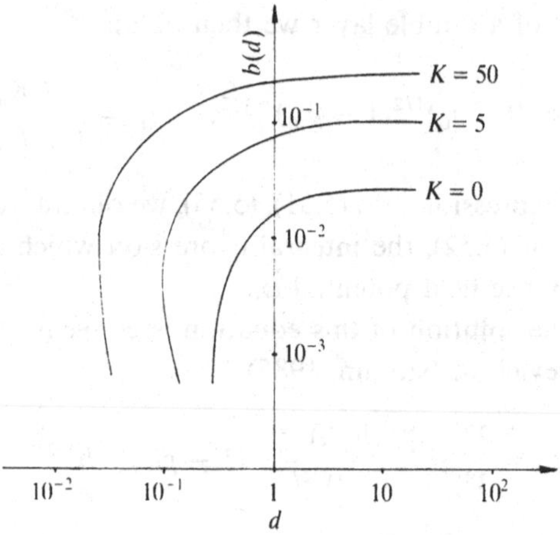

Fig. 5.1. Behaviour of the function $b(d)$.

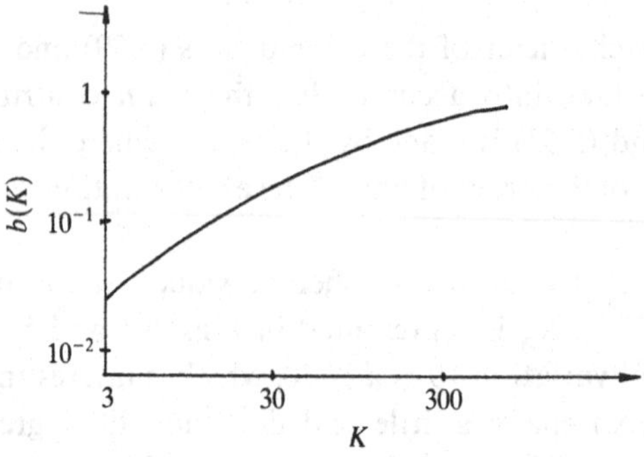

Fig. 5.2. Dependence of the parameter b on the multiplication coefficient K for $d \gg 1$.

surface – it is only necessary to know the number of particles and/or gamma-ray quanta knocked out of the surface. Getting onto the neutron star surface, a fast particle induces there an inward-going electromagnetic cascade accompanied by an excitation of iron nuclei (Jones, 1978). As a result, particles do not leave the surface for the magnetosphere while gamma-ray quanta, which have a sufficient energy to produce an electron–positron pair in the magnetic field, can go out of the star's surface into the magnetosphere. The exact calculation carried out by Bogovalov & Kotov (1989) showed that for incident particle energy $\gamma \sim 10^7$ (and for fields $B_0 \simeq 10^{12}$ G) the K value reaches 10–30. The dependence of K on

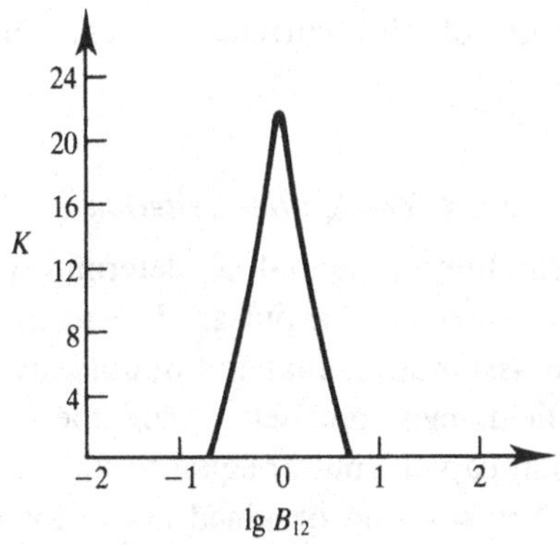

Fig. 5.3. Dependence of the multiplicity of birth K on the magnetic field B_0 for a primary particle with energy 10^7 MeV. From Bogovalov & Kotov (1989).

the magnetic field B_0 is shown in Fig. 5.3. Taking account of K leads to a noticeable variation of the magnitude of the breakdown potential. Note that taking account of some other processes near the surface (considered in papers included in the list of additional references) leads to smaller effects and can be effectively included by the renormalization of the quantity K.

Note also that the expressions (5.61) is applicable in the case of large magnetic fields, when the condition $b_0 \leqslant 0.1$ (5.26) does not hold. We need only replace Λ by $\frac{2}{3}b_0^{-1}$ (i.e. $k = 1$) in the formulas, which will change the dependence of the quantity φ_0 on the magnetic field $\propto B_{12}^{1/7}$ as compared to (5.35). This is connected with the fact that in large magnetic fields $\frac{2}{3}b_0^{-1} < \Lambda$ the free path of quanta relative to the production of electron–positron pairs is no longer dependent on the magnitude of the magnetic field: $l \simeq 2\rho_l/k$ (Beskin, 1982a; Daugherty & Harding, 1983).

It is important that the breakdown potential φ_0 depends, according to (5.61), on the magnitude of the longitudinal current i_0. This means that, strictly speaking, formula (5.61) establishes not the value of the potential but the relation between φ_0 and i_0. Another relation, at between φ_0 and i_0, is found from the dynamics of current in the magnetosphere of a neutron star and has the form (4.174). Thus, the true values of the breakdown potential φ_0 and current i_0 are found from a simultaneous solutions of eqs. (4.174) and (5.61). For the current $i_0(f)$ we then have

$$\frac{1-(1-i_0^2)^{1/2}}{(1-pi_0)^{1/7}} = 0.85\rho_{l_7}^{4/7}P^{13/7}B_{12}^{-8/7}(\cos\chi)^{-6/7}b^{-2/7}\left(1-\frac{f}{f^*}\right)^{-1}\left(\frac{\Lambda}{8}\right)^{-2/7} \qquad (5.62)$$

Next, from the value of the current we can find the potential $\varphi_0(f)$.

5.2.4 The ignition criterion

The value of the breakdown potential φ_0 determined above cannot be realized for all the parameters of a pulsar. The restriction is connected with violation of the assumption that the double layer is thin. Indeed, the region of open field lines, precisely where the double layer exists, embraces only the polar cap of a pulsar equal to $R_0 = R(2\pi R f^*(\chi)/Pc)^{1/2}$, $1.59 \leqslant f^*(\chi) \leqslant 1.95$. The solution obtained above for the potential φ is valid only for $H \ll R_0$. If H approaches R_0 in magnitude, we should solve not equation (5.57), but the total Poisson equation (5.1). Expressing, as before, the quantity $n^+ - n^-$ in terms of the electric current i_0 flowing in the polar magnetosphere, we write (5.1) in the form

$$\frac{1}{\rho'_\perp}\frac{\partial}{\partial\rho'_\perp}\left(\rho'_\perp\frac{\partial\varphi}{\partial\rho'_\perp}\right)+\frac{\partial^2\varphi}{\partial h^2}=-\frac{4\pi e^2 n_c}{m_e c^2}(1-pi_0) \tag{5.63}$$

Here $\rho'_\perp$ is the distance to the dipole axis. The boundary conditions to eq. (5.63) are as follows: on the star surface $h = 0$, while on the lateral surface of the 'cylinder' of radius $\rho'_\perp = R_0$ (i.e. in the boundary of closed field line region) the potential φ is equal to zero (see (4.58)). The corresponding solution of (5.63) is

$$\varphi=-\sum_{n=1}^{\infty}\frac{8\pi e^2 n_c(1-pi_0)R_0^2}{m_e c^2\mu_n^3 J_1(\mu_n)}J_0\left(\mu_n\frac{\rho'_\perp}{R_0}\right)(1-e^{-\mu_n(h/R_0)}) \tag{5.64}$$

$J_0(x)$ and $J_1(x)$ are Bessel functions, μ_n are the roots of the zeroth-order Bessel function. If the height of the double layer H, which is determined by the conditions of the electron–positron multiplication pairs (5.61), is smaller than the polar cap size $H < R_0/\mu_1$, the solution (5.64) coincides with (5.57) and (5.58). But as is seen from (5.64), the potential φ does not grow with the height of the double layer if $H > R_0/\mu_1$. So H cannot exceed R_0, and the potential φ cannot exceed its maximal value φ_m:

$$\varphi_m=-\frac{\pi e^2 n_c(1-pi_0)R_0^2}{m_e c^2}$$

Substituting the expressions (5.61) for φ_0 and H into these inequalities and employing the relation

$$\rho_l \simeq \frac{4}{3}\frac{R^2}{R_0}=\frac{4}{3}R\left(\frac{\Omega R}{c}\right)^{-1/2} \tag{5.65}$$

which holds in the case of a dipole magnetic field, we obtain the condition determining the possibility of electron–positron plasma generation in the polar region of the star:

$$PB_{12}^{-8/15} < 1.25b^{2/15}(K)\cos^{2/5}\chi \tag{5.66}$$

The relation (5.66) is the condition of the existence of stationary cascade processes of particle generation in a magnetosphere. It is in fact the criterion of pulsar 'ignition'.

5.3 Plasma multiplication

5.3.1 Deceleration of a primary beam

We have considered electron and positron production in a double layer near the neutron star surface $h < H$. In the quasi-neutrality region $h > H$ an intensive multiplication of the electron–positron plasma is prolonged. The source of multiplication is a beam of fast positrons (or electrons) accelerated in the double layer to energies $\gamma \simeq 10^7$. Moving along the field lines of a curvilinear magnetic field, these primary particles radiate plasma-generating curvature photons.

Therefore, let us first consider the distribution function of primary particles $F_0(\gamma, h, f, t)$. Its variation due to scattering upon radiation of curvature photons is described under stationary conditions by (5.4) and (5.12):

$$\frac{\partial F_0}{\partial h} = \frac{2\alpha_\hbar}{3}\frac{\lambda\!\!\!^{-}}{\rho_l^2}\frac{\partial}{\partial\gamma}\left[\gamma^4 F_0 + \frac{55}{32\sqrt{3}}\frac{\lambda\!\!\!^{-}}{\rho_l}\frac{\partial}{\partial\gamma}(\gamma^7 F_0)\right] \tag{5.67}$$

According to (5.44), the boundary conditions to eq. (5.67) have the form

$$F_0|_{h=H} = n_0\delta(\gamma - \gamma_0), \qquad n_0 = \frac{i_0 n_c K_m}{1 + K_m}$$

where the quantity γ_0 depends on the potential drop $|\varphi_0|$ according to (5.46), so that we usually have $\gamma_0 = |\varphi_0|$.

It is convenient to introduce the dimensionless variables

$$\Gamma_p = \frac{\gamma}{\gamma_0}, \qquad \eta = \tfrac{2}{3}\alpha_\hbar \lambda\!\!\!^{-}\gamma_0^3 \int_H^h \frac{dh'}{\rho_l^2(h')} \tag{5.68}$$

We take into account here that the curvature radius $\rho_l(h)$ varies with h

and when $h/R \ll (c/R\Omega)^{1/3}$ we have

$$\rho_l(h) = \rho_{l_0}\left(1 + \frac{h}{R}\right)^{1/2}, \qquad \rho_{l_0} = \frac{4}{3}\left(\frac{cR}{\Omega f}\right)^{1/2} \qquad (0 < f < f^*)$$

Then

$$\frac{\partial F_0}{\partial \eta} = \frac{\partial}{\partial \Gamma_p}\left[\Gamma_p^4 F_0 + \varepsilon_d \frac{\rho_{l_0}}{\rho_l(\eta)} \frac{\partial}{\partial \Gamma_p}(\Gamma_p^7 F_0)\right]$$
$$\varepsilon_d = \frac{55}{32\sqrt{3}} \frac{\lambda\!\!\!^{-} \gamma_0^2}{\rho_{l_0}} \tag{5.69}$$

Note that under the conditions of a pulsar (5.27) the parameter ε_d is always small, $\varepsilon_d \simeq 10^{-4}$. This means that the broadening Δ of the distribution function is always small too, $\Delta \propto \varepsilon_d^{1/2}$, so that the solution of (5.69) can be sought in the form

$$F_0 = \frac{n_0}{(\pi \Delta^2(\eta))^{1/2}} \exp\left[-\frac{(\Gamma_p - \bar{\Gamma}_p(\eta))^2}{\Delta^2(\eta)}\right] \tag{5.70}$$

Substituting (5.70) into (5.69), making a power series expansion of $\varepsilon_d^{1/2}$ and keeping the leading terms ε_d^{-1} and $\varepsilon_d^{-1/2}$, we obtain

$$\bar{\Gamma}_p = (1 + 3\eta)^{-1/3}$$
$$\Delta^2(\eta) = \frac{4\varepsilon_d}{(1 + 3\eta)^{8/3}} \int_0^\eta (1 + 3\eta')^{1/3}\left(1 - \frac{\eta'}{\eta_m}\right) d\eta' \tag{5.71}$$

where η_m is the maximal value of η:

$$\eta_m = \frac{2}{3} \frac{\alpha_h \lambda\!\!\!^{-} R \gamma_0^3}{\rho_{l_0}^2} \ln\left(\frac{c}{\Omega R}\right)$$
$$= \frac{3}{8} \frac{\alpha_h \lambda\!\!\!^{-} \gamma_0^3 \Omega}{c} f \ln\left(\frac{c}{\Omega R}\right)$$

The function $\Delta(\eta)$ describing the broadening of the distribution function F_0 is shown in Fig. 5.4 for different values of the parameter η_m. We can see that at first the broadening increases with η (and therefore with h proportionally to $h^{1/2}$), then reaches its maximum Δ_m and starts falling. The maximal value Δ_m is equal to

$$\Delta_m = \varepsilon_d^{1/2}\begin{cases}(12\eta_m/5)^{1/2} & \text{at } \eta = \eta_m - (24/5)^{3/2}\eta_m^{5/2}, \quad 3\eta_m \ll 1 \\ 1/2 & \text{at } \eta = 2^{3/4} - 1, \quad 3\eta_m \gg 1\end{cases}$$

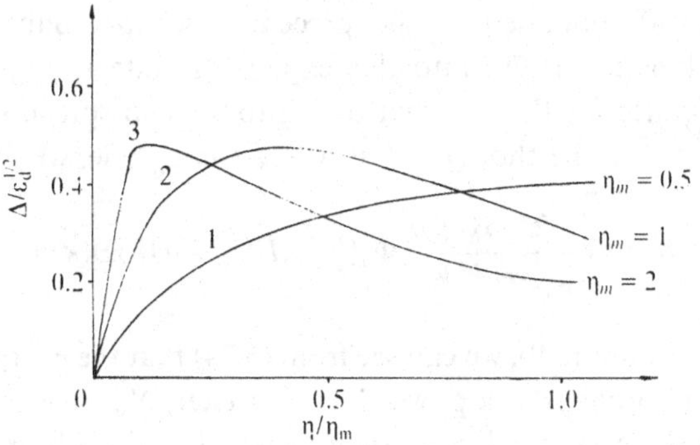

Fig. 5.4. The beam broadening function $\Delta(\eta)$.

It is important that for high h values, when the parameter η reaches its limiting value η_m, particle deceleration ceases and the distribution function acquires its stationary form (5.70) with the parameters $\bar{\Gamma}(\eta_m)$, $\Delta(\eta_m)$ (5.71). This implies that the total energy expended by the primary particles for radiation of curvature photons,

$$\mathscr{E} = \gamma_0 m_e c^2 [1 - (1 + 3\eta_m)^{-1/3}] \tag{5.72}$$

depends strongly on the parameter η_m. For $\eta_m \gtrsim 1$ the quantity $\mathscr{E} \simeq \gamma_0 m_e c^2$. If $\eta_m \ll 1$, then $\mathscr{E} \simeq \gamma_0 m_e c^2 \eta_m \propto \gamma_0^4$. From this we can readily see that the quantity γ_0, i.e. the value of the potential φ_0 between the star's surface and the magnetosphere, has the critical value φ_{0_k} determined by the condition $\eta_m \simeq 1/3$:

$$\varphi_{0_k} \simeq 3 \times 10^7 P^{1/3} \left(\frac{f^*}{f} \right)^{1/3} \tag{5.73}$$

If $|\varphi_0| > \varphi_{0_k}$, then a substantial energy equal to about the whole energy of the primary beam goes to plasma generation. Provided that $|\varphi_0| < \varphi_{0_k}$, the energy going to plasma generation falls very rapidly $\propto (\varphi_0/\varphi_{0_k})^4$, and due to this fact the generation process must be strongly suppressed already for $|\varphi_0|/\varphi_{0_k} \lesssim 0.3$.

5.3.2 *Curvature photons*

We now consider photon multiplication. A flux of primary particles yields a curvature radiation whose spectrum N_{pho} in the magnetosphere region not very far from the neutron star surface $h \ll R$ is described by the expression (5.40) obtained in the preceding section. Substituting the distribution function (5.70) into this expression, integrating over γ and making allowance for the fact that owing to a small spread of F_0 over γ the distribution can be thought of as a low-energy one, we find

$$N_{\text{ph}_0} = \frac{2}{\pi\sqrt{3}} \frac{\alpha_\hbar n_0 \gamma_0}{b_0 k^2} \Phi_S(k/k_c) I(\kappa_\perp, k) c(h) \Theta(\kappa_\perp) \tag{5.74}$$

Taking into account (5.9), we can see from (5.74) that the energy spectrum of curvature photons is of a power-law character, $N_{\text{ph}_0} \propto k^{-5/3}$ up to the values $k \simeq 0.3k_c$ (5.8). For $k > k_c$, the function $N_{\text{ph}_0}(k)$ falls exponentially. The distribution of curvature photons over angles, more precisely over the variable $\kappa_\perp$ (5.38), proves to be constant for $\kappa_\perp < \kappa_\perp^*(k)$ and to decrease rapidly for $\kappa_\perp > \kappa_\perp^*$ with the width $\Delta\kappa_\perp \simeq \kappa_\perp^{*2}$, where

$$\begin{aligned} \kappa_\perp^*(k) &= \frac{1}{\Lambda - \frac{3}{2}\ln\Lambda} \\ \Lambda(k) &= \frac{1}{2}\ln\left(\frac{\alpha_\hbar \rho_l}{2\sqrt{6}\lambda\!\!\!^{-} b_0 k^2}\right), \qquad \Lambda \gg 1 \end{aligned} \tag{5.75}$$

We have taken into account here that under the conditions (5.26) we always have $b_0 < \frac{2}{3}\Lambda^{-1}$.

5.3.3 The cascade process of synchrophoton generation

Curvature photons generate electron–positron pairs and simultaneously synchrophotons. The free path l of a photon, with respect to pair production and, therefore, to the production of synchrophotons, is determined by the expression (5.17).

The synchrophoton generation process is of cascade character (Tademaru, 1973). Curvature radiation quanta generate pairs and synchrophotons of the first generation N_{ph_1}. Their energy is Λ times less than the energy of curvature photons, and the free path l (5.17) is accordingly Λ times greater. Synchrophotons of the first generation produce pairs and, simultaneously, synchrophotons of the second generation N_{ph_2}, etc. The general distribution function of photons N_{ph} can thus be represented in the form of a sum:

$$N_{\text{ph}} = N_{\text{ph}_0} + N_{\text{ph}_1} + N_{\text{ph}_2} + \cdots \tag{5.76}$$

Substituting the expansion (5.76) into (5.6), (5.10), (5.24) and (5.25) and making allowance for (5.39) and (5.19), we arrive at the following closed chain of stationary equations for synchrophotons:

$$\frac{\partial N_{\mathrm{ph}_i}}{\partial h}+\frac{3}{4}\frac{kb_0}{\rho_l\kappa_\perp}\frac{\partial}{\partial\kappa_\perp}(\sqrt{\kappa_\perp^2-\xi^2}\,N_{\mathrm{ph}_i})$$

$$=-\frac{3^{1/2}}{2^{5/2}}\frac{\alpha_\hbar}{\lambda\!\!\!^{-} k}\kappa_\perp e^{-2/\kappa_\perp}N_{\mathrm{ph}_i}\Theta\left(\kappa_\perp-\frac{3}{2}b_0\right)+Q_S(N_{\mathrm{ph}_{i-1}}) \qquad (5.77)$$

$$Q_S=\frac{3^2}{\pi^2 2^{9/2}}\frac{\alpha_\hbar}{\lambda\!\!\!^{-} b_0}\frac{1}{k^2\kappa_\perp(\kappa_\perp^2-\xi^2)^{1/2}}\int\frac{\kappa_\perp'^2e^{-2/\kappa_\perp'}}{\left[\kappa_\perp'^2-\dfrac{k'^2}{k^2}\kappa_\perp^2\right]^{1/2}}$$

$$\times\Phi_S\left\{\frac{8}{9b_0^2}\frac{k^2}{k'^2\kappa_\perp}\left[\kappa_\perp'^2-\frac{k'^2}{k^2}\kappa_\perp^2\right]\right\}\Theta\left[\kappa_\perp'^2-\frac{k'^2}{k^2}\kappa_\perp^2-\frac{9}{4}b_0^2\right]$$

$$\times N_{\mathrm{ph}}(\kappa_\perp',k',\xi')\,d\kappa_\perp'\,dk'\,d\xi' \qquad (5.78)$$

Here, as in (5.39), we have passed over to the angular variable $\kappa_\perp$ (5.38) (we have assumed that sign $\sigma_{\mathrm{ph}}>0$). Moreover, since synchrophotons are produced equiprobably with all angles ϕ_{ph} relative to the direction of the magnetic field, the distribution function now also depends on the angle ϕ_{ph}. This dependence is allowed for in (5.77) through the quantity $\xi=\kappa_\perp\sin\phi_{\mathrm{ph}}$ conserved under the photon motion and therefore entering (5.77) as a parameter.

The first term on the right-hand side of (5.77) describes the decay of photons due to pair production, while the second term is the source of synchrophotons and is determined by photons of a preceding generation. The source of first-generation synchrophotons is photons of curvature radiation N_{ph_0} (5.74).

We can easily write the general solution of the linear equation (5.77). Doing so, we shall take into account that, as is seen from (5.77), photon absorption occurs only when $\kappa_\perp>\frac{3}{2}b_0$ and they are produced with the same angles β_{ph} as their parent photons, i.e. $\kappa_\perp\simeq\Lambda^{-2}$. Then, for not very small fields $b_0>\frac{2}{3}\Lambda^{-2}$, the solution of (5.77) for $\kappa_1>\frac{3}{2}b_0$ has the form ($Q_S=0$)

$$N_{\mathrm{ph}_i}=N_i^0\left(h-\frac{4}{3}\frac{\rho_l}{kb_0}\sqrt{\kappa_\perp^2-\xi^2},k\right)\times(\kappa_\perp^2-\xi^2)^{-1/2}$$

$$\times\exp\left\{-\frac{\alpha_\hbar\rho_l}{\sqrt{6}\,\lambda\!\!\!^{-} k^2b_0}\int_{3/2b_0}^{\kappa_\perp}\frac{\kappa_\perp'^2e^{-2/\kappa_\perp'}\,d\kappa_\perp'}{\sqrt{\kappa_\perp'^2-\xi^2}}\right\}\Theta\left[\frac{3}{2}b_0-|\xi|\right] \qquad (5.79)$$

Substituting (5.79) into the expression for Q_S (5.78) and integrating it over $\kappa'_\perp$ and ξ' with allowance made for the fact that the exponent contains a rapidly varying function, we have

$$Q_S = \frac{3^{5/2}}{4\pi^2 \rho_l} \frac{1}{k^2 \kappa_\perp (\kappa_\perp^2 - \xi^2)^{1/2}} \int_0^\infty k'^2 \Phi_S\left(\frac{2k^2}{k'^2 \kappa_\perp}\right)$$
$$\times \exp\left\{-\frac{\alpha_\hbar \rho_l}{\sqrt{6}\,\lambda\!\!\!^{-} k'^2 b_0} \int_{(3/2)b_0}^{[(k'^2/k^2)\kappa_\perp^2 + (9/4)b_0^2]^{1/2}} \kappa'_\perp\, e^{-2/\kappa'_\perp}\, d\kappa'_\perp\right\}$$
$$\times N^0\left(h - \frac{4}{3}\frac{\rho_l}{kb_0}\kappa_\perp, k'\right) dk'$$

Next, integrating (5.77) over $\kappa_\perp$ for a constant quantity

$$h - \frac{4}{3}\frac{\rho_l}{kb_0}\sqrt{\kappa_\perp^2 - \xi^2}$$

we shall find the distribution function of synchrophotons for $\kappa_\perp < \frac{3}{2}b_0$:

$$N_{\mathrm{ph}_i} = \frac{4\rho_l}{3kb_0}(\kappa_\perp^2 - \xi^2)^{-1/2} \int_\xi^{\kappa_\perp} \kappa'_\perp Q_S\left(\kappa', h + \frac{4}{3}\frac{\rho_l}{kb_0}\sqrt{\kappa_\perp'^2 - \xi^2}\right) d\kappa'_\perp \tag{5.80}$$

The matching of the functions (5.79) and (5.80) for $\kappa_\perp = \frac{3}{2}b_0$ determines the function N_i^0 in (5.79):

$$N_i^0 = \frac{3^{3/2}}{\pi^2 b_0} \frac{1}{k^3} \int_{4k/3b_0 a_m}^\infty dk'\, k'^2 N_{i-1}^0(k')\, Y\left(\frac{k}{k'} a_m\right) \tag{5.81}$$

where the expression $Y(x)$ is specified in the preceding section by formula (5.56), see Table 5.1, and

$$a_m = \frac{2}{3b_0}\left[\left(\frac{2}{3b_0 \Lambda}\right)^2 - 1\right]^{-1/2} \simeq \Lambda \gg 1$$

Since $Y(x)$ decreases exponentially for $x > 1$, it follows that in the zeroth approximation with respect to the parameter $a_m^{-1} \simeq \Lambda^{-1}$ the solution (5.81) can be represented in a step-like form

$$N_i^0 = \frac{\Gamma(5/3)}{\pi \cdot 3^{5/6}} \frac{\alpha_\hbar n_0}{b_0^2} \left(\frac{\rho_l}{\lambda\!\!\!^{-}}\right)^{1/3} k_c^{4/3} \left(\frac{3^{3/2}}{\pi b_0}\right)^i \frac{1}{k^3} M_i^0\left(\frac{k a_m^i}{k_c}\right)$$
$$\times [\Theta(k - k_c a_m^{-i-1}) - \Theta(k - k_c a_m^{-i})] \tag{5.82}$$

Here the function $M_i^0(x)$ is represented in terms of the previous one, M_{i-1}^0, by the integral relation:

$$M_i^0(x) = \int_x^1 \frac{dy}{y} M_{i-1}^0(y) Y\left(\frac{x}{y}\right), \qquad x > a_m^{-1}$$

and M_0^0 corresponds to the distribution function of curvature photons, $M_0^0 = y^{4/3}$. The quantity $k_c = \frac{3}{2}\lambda\!\!\!^{-}\gamma_0^3/\rho_l$. The functions $M_i^0(x)/x^3$ are plotted in Fig. 5.5. We can see that with increasing i the functions M_i^0 become ever steeper and are concentrated near small x. This means that the synchrophoton spectrum becomes increasingly soft.

The dependence $N_{\text{ph}}(k, \kappa_\perp, h)$ on the angle $\kappa_\perp$ and on the coordinate h for $\kappa_\perp < \frac{3}{2}b_0$ is given by the expression, following from (5.80),

$$N_{\text{ph}_i}(k, \kappa_\perp, h, \xi) = \frac{\Theta\left(h - \dfrac{4}{3}\dfrac{\rho_l}{kb_0}\sqrt{\kappa_\perp^2 - \xi^2}\right)}{(\kappa_\perp^2 - \xi^2)^{1/2}} \times \begin{cases} N_i^0(k), & \kappa_\perp > 2a_m^{-2} \\ \displaystyle\int_{k(2/\kappa_\perp)^{1/2}}^{2k/\kappa_\perp a_m} dk'\, k'^2 N_i^0(k') Y\left(\frac{2k^2}{k'^2\kappa_\perp}\right) \\ \displaystyle\quad + \int_{2k/\kappa_\perp a_m}^{\infty} dk'\, k'^2 N_i^0(k') Y\left(\frac{ka_m}{k'}\right), & \kappa_\perp < 2a_m^{-2} \end{cases} \tag{5.83}$$

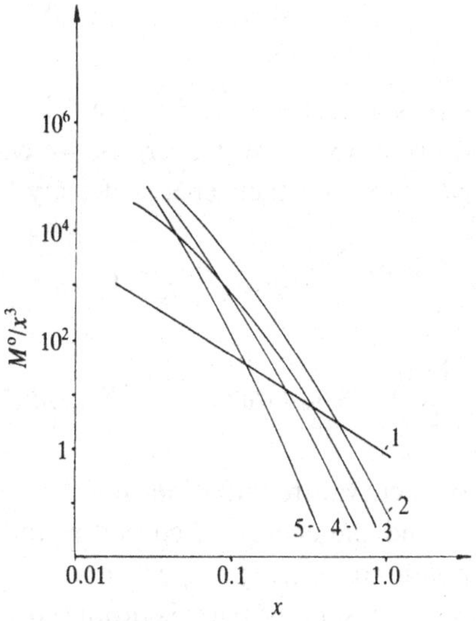

Fig. 5.5. Spectra of several generations of photons.

Thus, N_{ph} changes smoothly within the range $0 < \kappa_1 < 2a_m^{-2}$, remains constant for $2a_m^{-2} < \kappa_\perp < \kappa_\perp^*$ (where $\kappa_\perp^*$ is determined according to (5.75)), and is then sharply cut off for $\kappa_\perp > \kappa_\perp^*$, since an intensive photon absorption starts here due to pair production.

5.3.4 The properties of outflowing plasma

The energy spectrum We shall now determine the energy spectrum of electrons and positrons produced. From (5.4) and (5.20) it follows that

$$\frac{\partial F^\pm}{\partial h} = \frac{3^{1/2}}{2^{5/2}}\frac{\alpha_\hbar}{\lambdabar}\int \frac{\kappa_\perp^2 e^{-2/\kappa_\perp}}{k} N_{ph}(k, \kappa_\perp, \xi)\Theta(\kappa_\perp - \tfrac{3}{2}b_0)\delta\left(\gamma - \frac{3}{4}\frac{kb_0}{\kappa_\perp}\right) dk\, d\kappa_\perp\, d\xi$$

Substituting N_{ph} from (5.79) and integrating, we obtain

$$\frac{\partial F^\pm}{\partial h} = \frac{4}{\rho_l \Lambda^2}\gamma N^0\left(k = \frac{4\gamma}{3b_0\Lambda}, h - \frac{\rho_l}{\gamma}\right)\Theta\left(\gamma - \frac{\rho_l}{h}\right) \tag{5.84}$$

This implies that the energy spectrum of particles is similar to the photon spectrum, with the only difference that their indices differ by unity – the particle spectrum is harder. The energy of first-generation particles varies within the range

$$\gamma_{max} a_m^{-1} < \gamma < \gamma_{max} = \frac{9}{8}\frac{\lambdabar}{\rho_l}\gamma_0^3 b_0 \Lambda \tag{5.85}$$

They have a power-law spectrum $\propto \gamma^{-2/3}$ (since $N_0^0 \propto k^{-5/3}$, see (5.74)). Integrating the distribution (5.84) over energies, we can readily find the density $n_i^\pm$ of these particles and their energy density $\tilde{\mathscr{E}}_1^\pm$:

$$\begin{aligned} n_1^\pm &= \frac{3^{3/2}\Gamma(5/3)2^{2/3}}{\pi}\alpha_\hbar n_0 a_m \frac{\rho_l}{\lambdabar}\gamma_0^{-2}(1 - a_m^{-1/3}) \\ \tilde{\mathscr{E}}_1^\pm &= \frac{3^{5/2}\Gamma(5/3)}{2^{7/3}\pi}\alpha_\hbar a_m n_0 \gamma_0 m_e c^2 = 0.9\alpha_\hbar a_m n_0 \gamma_0 m_e c^2 \end{aligned} \tag{5.86}$$

The photon density of curvature radiation is of the order of the electron and positron density, and the energy of curvature radiation is a_m times lower, which corresponds to its steeper spectrum.

The next generation of plasma particles is produced by synchrophotons radiated in the course of first-generation production. Their energy varies within the range

$$\gamma_{\max} a_m^{-2} < \gamma < \gamma_{\max} a_m^{-1}$$

Their density and energy are respectively equal to

$$n_2^\pm = \frac{3^2 \cdot 2^{8/3}\Gamma(5/3)}{\pi^3} \frac{\rho_l}{\lambda\!\!\!^{-}} \gamma_0^{-2} \frac{\alpha_{\hbar} n_0}{b_0}$$
$$\tilde{\mathscr{E}}_2^\pm = \frac{3^3 \cdot 2^{5/3}\Gamma(5/3)}{\pi^3} \frac{\alpha_\hbar n_0 \gamma_0 m_e c^2}{b_0} \ln a_m \tag{5.87}$$

This cascade process could be continued. But owing to the fact that the magnetic field decreases in magnitude, $B = B_0[(h + R)/R]^{-3}$, and its curvature radius increases,

$$\rho_l = \rho_{l_0}[(h + R)/R]^{1/2}, \qquad \rho_{l_0} = 9.2 \times 10^7 P^{1/2} f^{-1/2}\ \mathrm{cm}$$

(5.37) at large distances from the star, the magnetosphere becomes transparent for gamma-ray quanta of rather low energy $k < k_{\min}$. The quantity $k_{\min}$ is found from the condition

$$\frac{1}{c}\int_0^\infty W(k_{\min})\, dh \simeq 1 \tag{5.88}$$

where $W(k)$ is the probability of electron–positron pair production by a gamma-ray quantum; it is given by the expression (5.13). Taking into account the dependence of B and ρ_l on h, as well as the fact that the angle β_{ph} between the wave vector $\mathbf{k}$ of a photon and the direction of the magnetic field varies with h ($d\beta_{\mathrm{ph}} = dh/\rho_l(h)$), i.e.

$$\beta_{\mathrm{ph}} = \frac{2R}{\rho_{l_0}}\left(\frac{h}{R}\right)^{1/2}$$

we have

$$W(k) = \frac{3^{3/2}}{2^{7/2}} \frac{\alpha_\hbar c}{\lambda\!\!\!^{-}} \frac{b_0 R}{\rho_{l_0}} \left(\frac{h}{R}\right)^{-5/2} \exp\left\{-\frac{4}{3}\frac{\rho_{l_0}}{kRb_0}\left(\frac{h}{R}\right)^{5/2}\right\} \Theta\left[h - \frac{\rho_{l_0}^2}{Rk^2}\right]$$

From this we can see that the quantity W falls rapidly with distance from the star. Substituting this expression into (5.88), we find

$$k_{\min} = \frac{\rho_{l_0}}{R}\left[\frac{3}{4} b_0 \ln\left(\frac{3^{4/3}}{5 \cdot 2^{13/6}} \frac{\alpha_\hbar}{\lambda\!\!\!^{-}} b_0^{5/6} \frac{R^2}{\rho_{l_0}}\right)\right]^{-1/6} \tag{5.89}$$

or

$$k_{\min} = 10^2 P^{1/2} B_{12}^{-1/6} f^{-1/2} \left(\frac{L}{24}\right)^{-1/6}$$

$$L = 24 + \tfrac{5}{6} \ln B_{12} - \tfrac{1}{2} \ln P$$

Thus, gamma-ray quanta of energy lower than minimal $k < k_{\min}$ go freely through the magnetosphere without being absorbed in it. As a result, the spectrum of particles generated in the magnetosphere must be cut off for $\gamma < \gamma_{\min}$. Since

$$\gamma_{\min} = \beta_{\rm ph}^{-1} = \frac{\rho_{l_0}}{2R}\left(\frac{h_{\max}}{R}\right)^{-1/2} \quad \text{and} \quad \left(\frac{h_{\max}}{R}\right)^{1/2} = \frac{\rho_{l_0}}{Rk_{\min}}$$

then

$$\gamma_{\min} = \frac{k_{\min}}{2} \simeq 5 \times 10^1 P^{1/2} B_{12}^{-1/6} f^{-1/2} \simeq 10^2 \tag{5.90}$$

On the other hand, according to (5.85) and (5.8), the maximal energy of produced particles and photons is

$$\gamma_{\max} = \frac{9}{8}\frac{\lambda\!\!\!^{-}}{\rho_{l_0}} \gamma_0^3 b_0 \Lambda(k_{\max}), \qquad k_{\max} = \frac{3}{2}\frac{\lambda\!\!\!^{-}}{\rho_{l_0}} \gamma_0^3$$

Substituting the quantity $\gamma_0 = |\phi_0|$ (5.61) into this expression for $\gamma_{\max}$, we obtain

$$\begin{aligned} \gamma_{\max} &= 5.6 \times 10^3 P^{-1/14} B_{12}^{4/7} \cos^{3/7} \chi(\Lambda_k/11) \\ \Lambda_k &= 10.8 + \tfrac{9}{28} \ln P - \tfrac{1}{14} \ln B_{12} - \tfrac{3}{7} \ln \cos \chi \end{aligned} \tag{5.91}$$

The plasma multiplication parameter We shall take into account the fact that each next generation of particles has the energy a_m times lower than that of the previous generation, where $a_m \simeq \Lambda(k_{\max}) = \Lambda_k$. Comparing the expressions (5.91) and (5.90), we have

$$\frac{\gamma_{\max}}{\gamma_{\min}} \simeq 10^2 P^{-4/7} B_{12}^{31/42} \simeq \Lambda_k^2$$

This means that the cascade process stops mainly at a second generation; a third generation of particles is generated completely only for very small P and large B_0. Therefore, for the density of electron–positron plasma generated in a magnetosphere we can use the expression (5.87), approximately, which implies that in the same approximation the plasma multiplication parameter λ important for pulsar magnetosphere is specified by a simple expression (Beskin, Gurevich & Istomin, 1988):

$$\lambda = \frac{n^{\pm}}{n_c} = \frac{3^2 \cdot 2^{8/3}}{\pi^3} \Gamma(5/3) \frac{\rho_{l0}}{\lambda\!\!\!^{-}} \gamma_0^{-2} \frac{\alpha_\hbar i_0}{b_0}$$

$$\approx 6 \times 10^3 P^{3/14} B_{12}^{-5/7} \cos^{-2/7} \chi \qquad (5.92)$$

It is noteworthy that formula (5.92) is, in fact, the lower boundary for the quantity λ. For example, not only two generations of secondary plasma but also some part of the third generation can actually be produced. The maximal value of λ can be estimated from energy considerations. Indeed, according to (5.72), the total energy density transferred by a primary beam to plasma particles produced is equal to

$$\tilde{\mathscr{E}} = \frac{i_0 n_c K_m}{1 + K_m} (1 - a_\gamma)[1 - (1 + 3\eta_m)^{-1/3}] \gamma_0 m_e c^2 \qquad (5.93)$$

The factor $1/(1 + K_m)$ here allows for the energy which goes to heating of the pulsar surface, the factor $(1 + 3\eta_m)^{-1/3}$ the energy remaining with fast particles (5.72), and a_γ for the fraction of energy transferred to the photon flux. Equating (5.93) to the energy density of the plasma produced,

$$\tilde{\mathscr{E}}_p = 2 n_e \bar{\gamma} m_e c^2$$

we obtain

$$\lambda = \frac{n_e}{n_c} = i_0 \frac{K_m(1 - a_\gamma)}{2(1 + K_m)} [1 - (1 + 3\eta_m)^{-1/3}] \frac{\gamma_0}{\bar{\gamma}}$$

From this, with account taken of the fact that $\gamma_0 = |\varphi_0|$ (5.61) and $\bar{\gamma} \gtrsim \gamma_{\min}$ (5.90), we find that, under the conditions $\eta_m > 1/3$ (5.73),

$$\lambda \simeq 2 \times 10^5 i_0 P^{-5/14} B_{12}^{1/42} \left(\frac{f}{f^*}\right)^{3/14} \cos^{1/7} \chi \qquad (5.94)$$

Thus, the plasma multiplication parameter can actually take the values $\lambda \sim 10^3$–10^5.

The expressions (5.87), (5.89) and (5.82) describe the concentration, energy and distribution function of particles and gamma-ray quanta in the electron positron plasma generated in a pulsar magnetosphere. Note that the analytical theory presented above is in agreement with the numerical estimates of plasma multiplication obtained using the Monte Carlo method (Daugherty & Harding, 1982). An example of the results reported by these authors is presented in Fig. 5.6. The particle spectrum

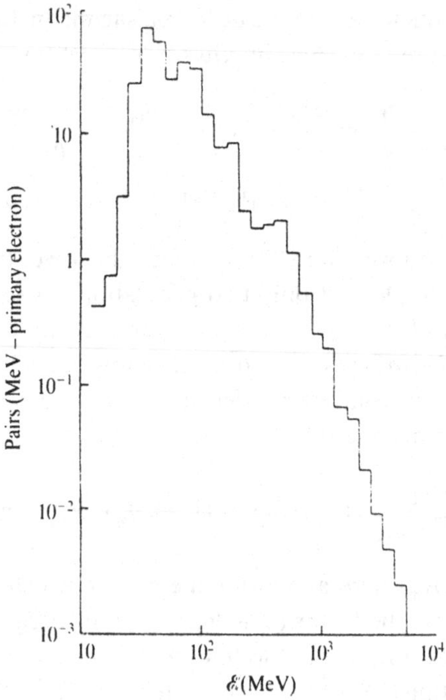

Fig. 5.6. The spectrum of secondary e^+e^- pairs created by a primary particle of energy 10^{13} eV in a magnetic field $B_0 = 10^{12}$ G. From Daugherty & Harding (1982).

is on average proportional to γ^{-2}, which does not contradict the two dependences $\gamma^{-2/3}$ and $\gamma^{-5/2}$ obtained for the first and second generations of multiplied particles. The values $\gamma_{max} \sim 10^4$, $\gamma_{min} \sim 10^2$ and $\lambda \sim 10^4$ are also in rather close agreement with formulas (5.90), (5.91) and (5.92).

The differences between the distribution functions It should be emphasized that, as mentioned in Section 3.2, although electrons and positrons in plasma are quite equivalent, their final distribution functions established in a pulsar magnetosphere are substantially different. The reason is that in the quasi-neutrality region $h > H$ there appears a weak longitudinal electric field, owing to the action of which the electric charge density becomes equal to the corotation density ρ_c (3.5). Under the action of the field one of the components of the plasma produced is decelerated, which slightly increases its density, while the other is accelerated and its density falls. The potential difference $|\Delta\varphi|$ responsible for this field is

$$|\Delta\varphi| \lesssim \gamma_{min}$$

It is appreciably smaller than the potential φ_0 (5.61) in the 'double layer' region.

We shall consider this process in more detail. From the condition of conservation of the total electric current flowing within a given field line tube and from the condition of equality of the plasma charge density to the critical density ρ_c, we have

$$n^+ - n^- = (1 - i_0)\frac{\rho_c}{e}$$

$$j^+ + j^- = 0, \qquad j^\pm = \pm ec\int\left(1 - \frac{1}{2\gamma^2}\right)F^\pm(\gamma)\,d\gamma$$

The charge density and the current of the initial beam of accelerated particles is taken into account here. Taking into consideration the fact that the distribution functions of plasma particles are power-law, $F^\pm \propto \gamma^{-\bar{\nu}}$, we obtain from the latter expressions two equations specifying the quantities $\gamma_{\min}^+$ and $\gamma_{\min}^-$:

$$\begin{aligned} n^+ - n^- &= (1 - i_0)\frac{\rho_c}{e} \\ \frac{n^+}{(\gamma_{\min}^+)^2} - \frac{n^-}{(\gamma_{\min}^-)^2} &= 2\,\frac{\bar{\nu}+1}{\bar{\nu}-1}(1 - i_0)\frac{\rho_c}{e} \end{aligned} \qquad (5.95)$$

The solution of equations (5.95) depends on the quantity ρ_c, which is proportional to the cosine of the angle θ_m between the direction of the magnetic field and the rotation axis. Owing to the change in the direction of the magnetic field line, the quantity can change significantly along this very field line, even traversing zero (4.83, 4.145). In the solution of equations (5.95) we therefore distinguish between two cases: (a) $\gamma_{\min}^2 > \lambda \cos\chi/\cos\theta_m$ and (b) $\gamma_{\min}^2 < \lambda\cos\chi/\cos\theta_m$.

In case (a), the solution of (5.95) yields

$$\begin{aligned} \gamma_{\min}^+ &= \left[\frac{(\bar{\nu}-1)}{2(\bar{\nu}+1)(1-i_0)}\lambda\frac{\cos\chi}{\cos\theta_m}\right]^{1/2} \simeq \lambda^{1/2} \\ \gamma_{\min}^- &\simeq \gamma_{\min} \qquad \text{at } \rho_c > 0 \\ \gamma_{\min}^+ &= \gamma_{\min} \\ \gamma_{\min}^- &= \left[\frac{(\bar{\nu}-1)}{2(\bar{\nu}+1)(1-i_0)}\lambda\frac{\cos\chi}{\cos\theta_m}\right]^{1/2} \simeq \lambda^{1/2} \qquad \text{at } \rho_c < 0 \end{aligned} \qquad (5.96)$$

In case (b):

$$\gamma_{\min}^{\pm} = \gamma_{\min}\left[1 \mp \frac{(\bar{\nu}+1)}{2(\bar{\nu}-1)}\frac{\gamma_{\min}^{2}\cos\theta_m}{\lambda\cos\chi}(1-i_0)\right] \quad \text{at } \rho_c > 0$$

$$\gamma_{\min}^{\pm} = \gamma_{\min}\left[1 \pm \frac{(\bar{\nu}-1)}{2(\bar{\nu}-1)}\frac{\gamma_{\min}^{2}\cos\theta_m}{\lambda\cos\chi}(1-i_0)\right] \quad \text{at } \rho_c < 0 \tag{5.97}$$

We can see that the sign of the charge of the slow component coincides with the sign of the corotation charge, i.e. with the sign of the primary beam. In case (a), where the deviation of the field line from the initial direction is not large ($\cos\theta_m \simeq \cos\chi$, we usually have $\gamma_{\min}^2 \gg \lambda$), the distinction in the minimal energies of the distribution functions of electrons and positrons is appreciable. Given this, the characteristic energy of one of the components is several times less than that of the other (Cheng & Ruderman, 1977). Owing to this, the effective particle mass is substantially different: slow particles have a smaller effective mass and play, as it were, the role of electrons; while fast particles play the role of ions in ordinary plasma. This fact, as will be seen below, is essential in the electrodynamics of waves in magnetosphere plasma.

6
Pulsar radio emission

We have determined the properties of electron–positron plasma in the pulsar magnetosphere. It is generated in the polar cap region and moves along a curvilinear magnetic field at a velocity close to the velocity of light. Particles of this plasma must give curvature radiation whose basic characteristics – the frequency range (3.31) and the direction (3.32) – are close to the observed pulsar radio emission. However, the wavelength of curvature radiation (3.33) is much larger than the distance between particles in a plasma flux. In this case an important role is played by a collective interaction between the particles and the field. In other words, the radiation must be coherent (Ginzburg *et al.*, 1969), i.e. must be a fundamental mode of plasma oscillations. To find and to describe quantitatively the radio emission generation mechanism, it is therefore necessary to investigate the excitation of natural high-frequency oscillations of a relativistic plasma flux moving in the polar region of a magnetosphere. This chapter is aimed at solving this problem.

In Section 6.1 we investigate the dielectric properties of a relativistic electron–positron plasma placed in a homogeneous magnetic field and find the fundamental modes of electromagnetic oscillations. In Section 6.2 we take into account the effect of magnetic field inhomogeneity, i.e. the curvature which determines mode instability. The non-linear stabilization and mode transformation effects are studied in Section 6.3.

6.1 Electrodynamics of relativistic electron–positron plasma in a homogeneous magnetic field

6.1.1 Dielectric permittivity of a homogeneous plasma

We shall consider a relativistic electron–positron plasma placed in a strong homogeneous magnetic field B. Owing to the high strength of the external field B, all the particles are in an unexcited state at a lower Landau level (see Chapter 5). The unperturbed distribution functions of electrons and positrons of a homogeneous plasma should, therefore, be written as (4.8)

$$F^{\pm}(\mathbf{p}) = n^{\pm} F^{\pm}_{\parallel}(p_{\parallel})\delta(\mathbf{p}_{\perp}) \tag{6.1}$$

where $p_{\parallel}$ and $\mathbf{p}_{\perp}$ are the longitudinal and transverse components of particle momentum and $n^{\pm}$ is the particle density. From (6.1) it follows that $\int F^{\pm}_{\parallel}(p_{\parallel})\, dp_{\parallel} = 1$. We shall introduce a unit vector $\mathbf{b}$ in the direction of the magnetic field $\mathbf{B}_0$. Then the distribution function of particles is conveniently represented as

$$F^{\pm}(\mathbf{p}) = n^{\pm} \int F^{\pm}_{\parallel}(p_{\parallel})\delta(\mathbf{p} - \mathbf{b}\cdot p_{\parallel})\, dp_{\parallel} \tag{6.2}$$

Small perturbations of distribution functions are described by the linearized kinetic equation

$$\frac{\partial f^{\pm}}{\partial t} + \mathbf{v}\frac{\partial f^{\pm}}{\partial \mathbf{r}} \pm \frac{e}{c}[\mathbf{v}\mathbf{B}_0]\frac{\partial f^{\pm}}{\partial \mathbf{p}} = \mp\frac{e}{c}\{c\mathbf{E} + [\mathbf{v}\mathbf{B}]\}\frac{\partial F^{\pm}}{\partial \mathbf{p}} \tag{6.3}$$

We now consider the perturbations of the distribution functions $f^{\pm}$ induced by the field of a plane electromagnetic wave:

$$\mathbf{E} \sim \exp\{-i\omega t + i\mathbf{k}\mathbf{r}\}, \qquad \mathbf{B} = \frac{c}{\omega}[\mathbf{k}\mathbf{E}] \tag{6.4}$$

The solution of eq. (6.3), (6.4) can be obtained by the method of integration over trajectories (Shafranov, 1967). Going over to the unperturbed trajectory of particle motion, $\mathbf{r}(t)$, $\mathbf{p}(t)$, we obtain

$$f^{\pm} = \mp\frac{en^{\pm}}{\omega} e^{-i\omega t + i\mathbf{k}\mathbf{r}} \int_{-\infty}^{t} dt' \exp\{i\omega(t - t') - i\mathbf{k}(\mathbf{r} - \mathbf{r}(t'))\}$$
$$\times [(\omega - \mathbf{k}\mathbf{v}')\delta_{\beta\sigma} + k_{\sigma}v'_{\beta}]\frac{\partial F^{\pm}}{\partial p'_{\sigma}} E_{\beta} \tag{6.5}$$

Here $\mathbf{r}' = \mathbf{r}(t')$, $\mathbf{v}' = \mathbf{v}(t')$, $\mathbf{p}' = \mathbf{p}(t')$ are the coordinate, the velocity and the momentum of particle at an instant of time t' provided that at a current instant t at a point $\mathbf{r}$ the velocity and momentum are equal to $\mathbf{v}$ and $\mathbf{p}$. The trajectory of particle motion in the homogeneous magnetic field $\mathbf{B}_0$ is given, as is known (Landau & Lifshitz, 1983), by the expressions

$$p_{\parallel} = p'_{\parallel}, \qquad v_{\parallel} = v'_{\parallel}, \qquad z - z' = v'_{\parallel}(t - t')$$

$$\bar{p} = \bar{p}' \exp\left[-i\frac{\omega_B}{\gamma}(t - t')\right], \qquad \bar{v} = \bar{v}' \exp\left[-i\frac{\omega_B}{\gamma}(t - t')\right] \tag{6.6}$$

$$\bar{r} - \bar{r}' = \frac{i}{m_e\omega_B}\bar{p}'\left\{\exp\left[-i\frac{\omega_B}{\gamma}(t - t')\right] - 1\right\}$$

Here z is the coordinate along the magnetic field $\mathbf{B}_0$, x and y across the magnetic field,

$$\bar{p} = p_x + ip_y, \qquad \bar{v} = v_x + iv_y, \qquad \bar{r} = r_x + ir_y$$

$\omega_B = eB_0/m_ec$ is the cyclotron frequency, and γ is the Lorentz-factor.

Knowing the perturbation of the distribution function $f^{\pm}$ we find the electric current $j^{\pm}$ in the plasma, induced by the electromagnetic wave

$$\mathbf{j}^{\pm} = \pm e\int f^{\pm}\mathbf{v}\,dp$$

$$\begin{aligned} j_{\alpha}^{\pm} = e^{-i\omega t + i\mathbf{kr}}E_{\beta}\frac{e^2n^{\pm}}{\omega}\int_{-\infty}^{t} dt' \int \mathbf{dp}\int F_{\parallel}^{\pm}(p_{\parallel})\,dp_{\parallel} \\ \times\ \delta(\mathbf{p}' - \mathbf{b}p_{\parallel})\frac{\partial}{\partial p'_{\sigma}}v_{\alpha}\exp\{i\omega(t - t') - i\mathbf{k}(\mathbf{r} - \mathbf{r}')\} \\ \times\ [(\omega - \mathbf{kv}')\delta_{\beta\sigma} + k_{\sigma}v'_{\beta}] \end{aligned} \tag{6.7}$$

We have taken into account here that with allowance for (6.2) the derivative

$$\frac{\partial F^{\pm}}{\partial p'_{\sigma}} = n^{\pm}\int F_{\parallel}^{\pm}(p_{\parallel})\frac{\partial}{\partial p'_{\sigma}}\delta(\mathbf{p}' - \mathbf{b}p_{\parallel})\,dp_{\parallel}$$

The expression (6.7) determines the plasma conductivity $\sigma_{\alpha\beta}$ and, accordingly, the dielectric permittivity tensor $\varepsilon_{\alpha\beta}$

$$\varepsilon_{\alpha\beta} = \delta_{\alpha\beta} + \frac{4\pi i}{\omega}\sigma_{\alpha\beta}$$

$$\begin{aligned} \sigma_{\alpha\beta} = \frac{e^2}{\omega}\left\langle n^{\pm}\int_{-\infty}^{t} dt'\int \mathbf{dp}\,\delta(\mathbf{p}' - \mathbf{b}p_{\parallel})\frac{\partial}{\partial p'_{\sigma}}m_ev_{\alpha}\right. \\ \left.\times\ \exp\{i\omega(t - t') - i\mathbf{k}(\mathbf{r} - \mathbf{r}')\}[(\omega - \mathbf{kv}')\delta_{\beta\sigma} + k_{\sigma}v'_{\beta}]\right\rangle \end{aligned} \tag{6.8}$$

From here on, the angular brackets imply averaging over the unperturbed longitudinal distribution function $F_{\parallel}^{\pm}(p_{\parallel})$ and summation over the types of particles:

$$\langle \cdots \rangle = \sum_{+,-} \int_{-\infty}^{\infty} dp_{\parallel} \cdots F_{\parallel}^{\pm}(p_{\parallel}) \tag{6.9}$$

Substituting the expression for the trajectory of particle motion (6.6) into (6.8) and taking into account the relation $d\mathbf{p} = d\mathbf{p}'$,

$$\frac{\partial v_\lambda}{\partial p_\mu} = \frac{1}{m_e \gamma}\left(\delta_{\lambda\mu} - \frac{p_\lambda p_\mu}{\gamma^2 m_e^2 c^2}\right)$$

we obtain the general expression for the dielectric permittivity tensor $\varepsilon_{\alpha\beta}(\omega, \mathbf{k})$ in the form (Godfray *et al.*, 1975; Suvorov & Chugunov, 1975; Hardee & Rose, 1976):

$$\varepsilon_{\alpha\beta}(\omega, \mathbf{k}) =$$
$$\begin{pmatrix} 1 + \left\langle \dfrac{\omega_p^2 \gamma \tilde{\omega}^2}{\omega^2(\omega_B^2 - \gamma^2 \tilde{\omega}^2)} \right\rangle & i\left\langle \dfrac{\omega_p^2 \omega_B \tilde{\omega}}{\omega^2(\omega_B^2 - \gamma^2 \tilde{\omega}^2)} \right\rangle & \left\langle \dfrac{\omega_p^2 \gamma k_x v_{\parallel} \tilde{\omega}}{\omega^2(\omega_B^2 - \gamma^2 \tilde{\omega}^2)} \right\rangle \\ -i\left\langle \dfrac{\omega_p^2 \omega_B \tilde{\omega}}{\omega^2(\omega_B^2 - \gamma^2 \tilde{\omega}^2)} \right\rangle & 1 + \left\langle \dfrac{\omega_p^2 \gamma \tilde{\omega}^2}{\omega^2(\omega_B^2 - \gamma^2 \tilde{\omega}^2)} \right\rangle & -i\left\langle \dfrac{\omega_p^2 \omega_B k_x v_{\parallel}}{\omega^2(\omega_B^2 - \gamma^2 \tilde{\omega}^2)} \right\rangle \\ \left\langle \dfrac{\omega_p^2 \gamma k_x v_{\parallel} \tilde{\omega}}{\omega^2(\omega_B^2 - \gamma^2 \tilde{\omega}^2)} \right\rangle & i\left\langle \dfrac{\omega_p^2 \omega_B k_x v_{\parallel}}{\omega^2(\omega_B^2 - \gamma^2 \tilde{\omega}^2)} \right\rangle & 1 - \left\langle \dfrac{\omega_p^2}{\gamma^3 \tilde{\omega}^2} \right\rangle + \left\langle \dfrac{\omega_p^2 \gamma k_x^2 v_{\parallel}^2}{\omega^2(\omega_B^2 - \gamma^2 \tilde{\omega}^2)} \right\rangle \end{pmatrix} \tag{6.10}$$

Here

$$\omega_p^2 = \frac{4\pi n^{\pm} e^2}{m_e}, \qquad \tilde{\omega} = \omega - k_z v_{\parallel}, \qquad \gamma = \left(1 - \frac{v_{\parallel}^2}{c^2}\right)^{-1/2}$$

It is assumed here that the x-axis lies in the $\mathbf{kB}_0$-plane and, therefore, the wave vector $\mathbf{k}$ has two components k_z and k_x. Since the magnetic field of a pulsar is actually very strong, the cyclotron frequency ω_B in a rather large region near the star $r \lesssim 10^3 R$ exceeds greatly the frequency ω of observed radio waves ($\omega_{B_0} \simeq 1.8 \times 10^{19} B_{12}$ Hz, $\omega \lesssim 10^{11}$ Hz). The basic electrodynamic properties of pulsar magnetosphere plasma can therefore be described in the approximation of infinitely strong magnetic field $\omega_B \to \infty$. Given this, the dielectric permittivity tensor $\varepsilon_{\alpha\beta}$ (6.10) takes a particularly simple form

$$\varepsilon_{\alpha\beta} = \begin{pmatrix} 1 & 0 & 0 \\ 0 & 1 & 0 \\ 0 & 0 & 1 - \left\langle \dfrac{\omega_p^2}{\gamma^3 \tilde{\omega}^2} \right\rangle \end{pmatrix} \tag{6.11}$$

The expression (6.11) implies that plasma conductivity in an infinitely strong magnetic field is non-zero only in the direction of the external field.

6.1.2 Normal modes of electromagnetic oscillations

We are first of all interested in the natural modes whose frequencies are close to those of the radiofrequency band. For the plane waves (6.4), the Maxwell equations (4.1)–(4.3) take the form (Landau & Lifshitz, 1960b)

$$\mathscr{D} - \frac{c}{\omega}[\mathbf{kB}] = 0, \qquad \mathbf{B} = \frac{c}{\omega}[\mathbf{kE}], \qquad \mathscr{D}_\alpha = \varepsilon_{\alpha\beta} E_\beta \tag{6.12}$$

From (6.12) there follow ordinary equations for the components of the vector of the electric field of a wave:

$$\left(k^2 \delta_{\alpha\beta} - k_\alpha k_\beta - \frac{\omega^2}{c^2} \varepsilon_{\alpha\beta} \right) E_\beta = 0 \tag{6.13}$$

The condition of solvability (6.13) yields the dispersion equation

$$\det(n^2 \delta_{\alpha\beta} - n_\alpha n_\beta - \varepsilon_{\alpha\beta}) = 0, \qquad \mathbf{n} = \frac{\mathbf{k}c}{\omega} \tag{6.14}$$

which determines the refractive indices n of the natural modes of electromagnetic oscillations. Substituting the expression (6.11) for $\varepsilon_{\alpha\beta}$ into the dispersion equation and denoting $\cos\theta = \mathbf{kB}_0/kB_0$ we obtain

$$(1 - n^2)\left[1 - n^2 - (1 - n^2 \cos^2\theta)\left\langle \frac{\omega_p^2}{\gamma^3 \tilde{\omega}^2} \right\rangle \right] = 0 \tag{6.15}$$

Here the quantity $\tilde{\omega} = \omega(1 - n\cos\theta(v_\parallel/c))$ is proportional to the wave frequency $\omega' = \gamma\tilde{\omega}$ in a coordinate system moving at a velocity $v_\parallel$ relative to the laboratory system, i.e. a system related to the neutron star. Note that it is convenient for us to work precisely in the laboratory system of coordinates, since it is only in this case that we can go correctly (without involving non-inertial coordinate systems) to the case of a curvilinear magnetic field.

As is seen from (6.15), from the general dispersion equation there splits one mode:

$$n_1^2 = 1 \tag{6.16}$$

As follows from (6.13), (6.11) and (6.10), the equality (6.16) corresponds to a linearly polarized transverse wave t_1 whose electric vector is perpendicular to the $\mathbf{kB}_0$-plane. As we can see, the refractive index n of this wave is identically unity. This is not surprising since in an infinite magnetic field free particle motion across the magnetic field lines is impossible. The response of such a medium to any wave whose electric vector is always perpendicular to the external magnetic field is therefore also equal to zero. Such a wave will be called 'extraordinary'.

The dispersion equation for the rest of the natural modes has the form, according to (6.15),

$$n^2 = 1 - (1 - n^2 \cos^2 \theta) \frac{\omega_p^2}{\omega^2} \left\langle \gamma^{-3} \left(1 - n \cos \theta \frac{v_\parallel}{c}\right)^{-2} \right\rangle \tag{6.17}$$

Equation (6.17) is a fourth-power equation with respect to n, i.e. there generally exist another four modes of electromagnetic oscillations. The case of small angles θ between the vectors $\mathbf{k}$ and $\mathbf{B}_0$ is of interest for us, and we henceforth give special attention to it. Given this, one mode – the one propagating opposite to the plasma flow direction $\cos \theta \simeq -1$ – falls out of consideration. Thus, there remain three modes. As is seen from (6.17),

$$n - 1 \simeq \left(\frac{\omega_p^2}{\omega^2} \langle \gamma^{-3} \rangle \right)^{1/2}$$

and since in the pulsar magnetosphere

$$\omega_p \gamma_c^{-3/2} = 1.5 \times 10^9 (\lambda_4 B_{12} P^{-1} \gamma_{100}^{-3})^{1/2} < \omega \tag{6.18}$$

the refractive indices of all the modes are close to unity: $n \simeq 1$. Here we have again used the notation

$$\gamma_c = \langle \gamma^{-3} \rangle^{-1/3}, \qquad \lambda_4 = \lambda \times 10^{-4},$$
$$B_{12} = B_0 \times 10^{-12}\,\mathrm{G}^{-1}, \qquad \gamma_{100} = \gamma_c \times 10^{-2} \tag{6.19}$$

As a result, for small angles $\theta \ll 1$, the dispersion equation (6.17) can be represented as ($v_\parallel/c = 1 - 1/2\gamma^2$, $\gamma \gg 1$):

$$1 - n = \left(1 - n + \frac{\theta^2}{2}\right)\frac{\omega_p^2}{\omega^2}\langle\gamma^{-3}[1 - n + \tfrac{1}{2}(\theta^2 + \gamma^{-2})]^{-2}\rangle \quad (6.20)$$

In the solution of eq. (6.20) we single out two cases: (1) $|1 - n| \ll \gamma^{-2}$ and (2) $|1 - n| \gg \gamma^{-2}$. Since $1 - n \simeq \omega_p/(\omega\gamma_c^{3/2})$, case (1) corresponds to the inequality

$$A_p \equiv \frac{\omega_p^2\gamma_c}{\omega^2} \ll 1 \quad (6.21)$$

The dependence of the refractive indices n_j for the three modes (6.20) and for the 'extraordinary' mode (6.16) on the angle θ for $A_p \ll 1$ is shown in Fig. 6.1a.

In the inverse limit,

$$A_p \gg 1 \quad (6.22)$$

in the expression for $\tilde{\omega}$ we can neglect the difference of $v_\parallel$ from c, which will make eq. (6.20) much simpler:

$$1 - n = \left(1 - n + \frac{\theta^2}{2}\right)\frac{\omega_p^2}{\omega^2}\langle\gamma^{-3}\rangle\left(1 - n + \frac{\theta^2}{2}\right)^{-2} \quad (6.23)$$

We can see that in the region $A_p \gg 1$ the refractive indices n of modes do not, in fact, depend on the form of the distribution functions of particles, i.e. normal waves in the radiofrequency band are here hydrodynamic modes. This case is of particular interest for us since the condition $A_p \gg 1$ corresponds to the dense plasma located in the internal regions of the magnetosphere, where plasma effects are most substantial.

From (6.23) we obtain the following expressions for the refractive indices of normal modes:

$$n_2 = 1 + \frac{\theta^2}{4} - \left(\frac{\omega_p^2}{\omega^2}\langle\gamma^{-3}\rangle + \frac{\theta^4}{16}\right)^{1/2} \quad (6.24a)$$

$$n_3 = 1 + \frac{\theta^2}{4} + \left(\frac{\omega_p^2}{\omega^2}\langle\gamma^{-3}\rangle + \frac{\theta^4}{16}\right)^{1/2} \quad (6.24b)$$

$$n_4 = 1 + \frac{\theta^2}{2} = \frac{1}{\cos\theta} \quad (6.24c)$$

The dependence of the refractive indices for the four modes (6.16) and (6.20) for $A_p \gg 1$ on the angle θ is shown in Fig. 6.1b. From the figure

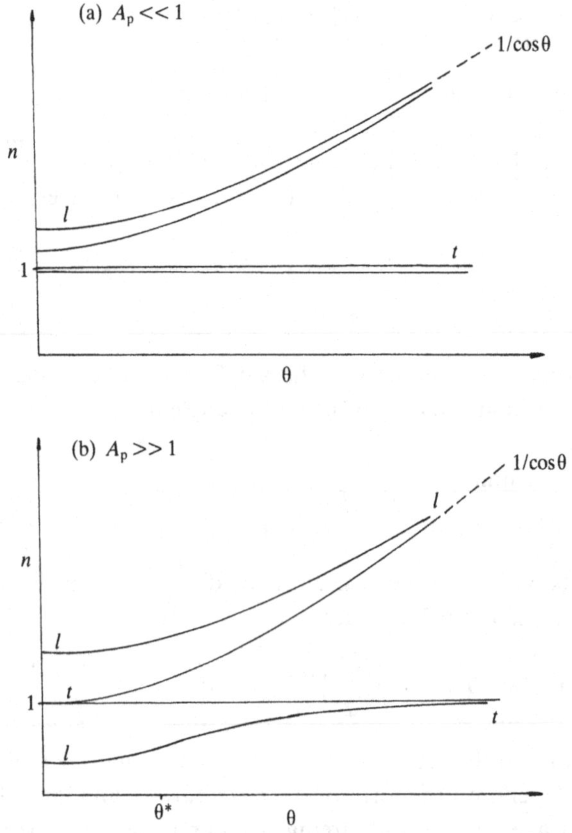

Fig. 6.1. Dependence of the refractive index n_j on the angle θ. (*a*) $A_p \ll 1$; (*b*) $A_p \gg 1$.

we can see that for both $\theta \ll \theta^*$ and $\theta \gg \theta^*$ where

$$\theta^* = \left(\frac{\omega_p^2}{\omega^2}\langle\gamma^{-3}\rangle\right)^{1/4} \tag{6.25}$$

in the magnetosphere plasma there may propagate four normal waves of the radiofrequency band: two plasma waves (*l*-modes) and two transverse modes (*t*-modes).

Wave polarization is determined by eq. (6.13) and depends on the refractive indices of the waves

$$\frac{E_x}{E_z} = \frac{n_x n_z}{n_z^2 - 1} \tag{6.26}$$

Substituting the expressions for n (6.16) and (6.20) in (6.26), we obtain the following polarization relations in the case $A_p \gg 1$, $\theta \ll 1$:

$$\left(\frac{E_x}{E_z}\right)_1 = -\theta \tag{6.27a}$$

$$\left(\frac{E_x}{E_z}\right)_2 = -\frac{\theta}{2}\left[\left(\frac{\omega_p^2}{\omega^2}\langle\gamma^{-3}\rangle + \frac{\theta^4}{16}\right)^{1/2} + \frac{\theta^2}{4}\right]^{-1} \tag{6.27b}$$

$$\left(\frac{E_x}{E_z}\right)_3 = \begin{cases} \dfrac{\theta}{2}\left[\left(\dfrac{\omega_p^2}{\omega^2}\langle\gamma^{-3}\rangle + \dfrac{\theta^4}{16}\right)^{1/2} - \dfrac{\theta^2}{4}\right]^{-1}, & \theta < \theta^* A_p^{1/4} \\ \theta\gamma_c^2 \gg 1, & \theta > \theta^* A_p^{1/4} \end{cases} \tag{6.27c}$$

$$\left(\frac{E_x}{E_z}\right)_4 = \begin{cases} 4A_p/\theta \gg 1, & \theta < \theta^* A_p^{1/4} \\ \theta\gamma_c^2 \gg 1, & \theta > \theta^* A_p^{1/4} \end{cases} \tag{6.27d}$$

The corrections to the relations (6.24) and (6.27) can be found, for instance, in Lominadze & Pataraya (1982). We can see that for the angles $\theta \ll \theta^*$ the normal modes $j = 2, 3$ are purely longitudinal ($E_x/E_z \ll 1$) and (Lominadze *et al.*, 1979)

$$n_{2,3} = 1 \mp \left(\frac{\omega_p^2}{\omega^2}\langle\gamma^{-3}\rangle\right)^{1/2} \tag{6.28}$$

But, as is seen from Fig. 6.1b, as the angle θ increases, the modes behave differently. Thus, the normal mode $j = 2$ for $\theta \gg \theta^*$ goes over to a transverse electromagnetic wave with $n_2 \simeq 1$ whose electric vector lies in the plane which contains the vectors **k** and $\mathbf{B}_0$. We shall call it an 'ordinary' mode. For the normal wave $j = 3$, for large angles θ its refractive index tends to the value (6.24d), which corresponds to the 'drift' or Alfven type dispersion $\omega = (\mathbf{kv})$. Note that the energy of this wave is negative since in a reference frame moving with a particle flux it has a negative frequency $\omega' = \gamma\tilde{\omega} < 0$. Finally, the fourth normal mode $j = 4$ in the entire region of angles is a purely drift or Alfven wave (6.24c).

6.1.3 Cyclotron resonance

We have considered only the limit of an infinitely strong magnetic field (6.11). We now proceed to gyromagnetic correlations which occur due to a final magnitude of the field. They are particularly significant in the vicinity of cyclotron resonance, when the wave frequency in a moving coordinate system coincides with the gyrorotation frequency of electrons and positrons:

$$\frac{|\omega_B|}{\gamma} = \omega - k_z v_{\|} \tag{6.29}$$

Note that in view of strong relativistic effects ($\gamma \gg 1$), for the waves which propagate at small angles to the direction of the magnetic field ($\theta \ll 1$) and whose refractive index is close to unity ($|1 - n| \ll 1$), the resonance condition (6.29) assumes the form

$$\frac{|\omega_B|}{\gamma} = \omega[(1 - n) + \tfrac{1}{2}(\theta^2 + \gamma^{-2})] \tag{6.30}$$

The condition (6.30) implies that resonance can be realized only at large distances from the star, $r > 10^3 R$, when the magnetic field falls noticeably. For fast pulsars, the resonance condition (6.30) cannot be fulfilled in the magnetosphere (i.e. inside the light cylinder, see Gedalin & Machabeli (1983)).

To find the eigenmodes, it is necessary to solve the dispersion equation with the general dielectric permittivity tensor (6.10). It should be noted that, although the components of an electron–positron plasma have the charge-to-mass ratios equal in magnitude and opposite in sign, the non-diagonal elements of the tensor ε_{xy}, ε_{yx}, ε_{yz} and ε_{zy} odd in the gyrofrequency ω_B turn out to be non-zero. The point is that the energy distribution functions of electrons and positrons in a pulsar magnetosphere differ substantially (see Section 5.3.4). Therefore, the effective particle masses are different too, and owing to this the gyrotropic terms in $\varepsilon_{\alpha\beta}$ do not vanish. In particular, when $|1 - n| \ll \theta^2, \gamma^{-2}$ and $|\omega_B|/\gamma > \tilde{\omega}$ (which is valid at large distances from the star), we have

$$\varepsilon_{xy} = -\varepsilon_{yx} = i\left\langle \frac{\omega_p^2 \omega_B}{2\omega\, \omega_B^2}(\gamma^{-2} + \theta^2) \right\rangle = \frac{i}{2}\frac{\omega_p^2}{\omega|\omega_B|}\left\langle \frac{1}{\gamma^2} \right\rangle_{\pm} \tag{6.31}$$

where

$$\left\langle \frac{1}{\gamma^2} \right\rangle_{\pm} = \left\langle \frac{\text{sign}(e)}{\gamma^2} \right\rangle \equiv \left\langle \frac{1}{\gamma^2} \right\rangle_{+} - \left\langle \frac{1}{\gamma^2} \right\rangle_{-}$$

Given this, the quantity ε_{xy} proves generally not to be small compared with $\delta\varepsilon_{xx} = \varepsilon_{xx} - 1$.

The dispersion equation becomes much simpler for waves for which $\theta \ll 1$ and $|1 - n| \ll 1$. Keeping in the dispersion equation only terms quadratic in (n^2-1),

$$\delta\varepsilon_{\perp\perp} = \varepsilon_{xx} - 1 = \varepsilon_{yy} - 1, \quad \delta\varepsilon_{zz} = \varepsilon_{zz} - 1, \quad \varepsilon_{xz} = -\varepsilon_{zx}, \quad \varepsilon_{yz} = -\varepsilon_{zy}$$

we obtain the following solution

$$n_{1,2}^2 = 1 + \delta\varepsilon_{\perp\perp} - \theta\varepsilon_{xz} + \frac{\theta^2}{2}\delta\varepsilon_{zz}$$
$$\pm \tfrac{1}{2}[(\theta^2\delta\varepsilon_{zz} - 2\theta\varepsilon_{xz})^2 - 4(\varepsilon_{xy} + \theta\varepsilon_{yz})^2]^{1/2} \tag{6.32}$$

The polarization of these two modes is given by (6.13), and with allowance for (6.32) can be represented as

$$\left(\frac{E_x}{E_y}\right)_{1,2} = -\frac{\theta^2\delta\varepsilon_{zz} - 2\theta\varepsilon_{xz} \pm [(\theta^2\delta\varepsilon_{zz} - 2\theta\varepsilon_{xz})^2 - 4(\varepsilon_{xy} + \theta\varepsilon_{yz})^2]^{1/2}}{2i|\varepsilon_{xy}|} \tag{6.33}$$

The leading terms in (6.32) and (6.33) are, in fact, $\delta\varepsilon_{\perp\perp}$, ε_{xy} and $\theta^2\delta\varepsilon_{zz}$. In (6.32) and (6.33) the 'plus' sign corresponds to an extraordinary wave and the 'minus' sign to an ordinary wave. When $\omega_B \gg \omega$ (that is, for small distances from the star), the leading term in (6.32) and (6.33) is $\theta^2\delta\varepsilon_{zz}$. Accordingly, the refractive indices $n_{1,2}$ go over to the expressions given by formulas (6.16) and (6.17).

When approaching the cyclotron resonance $\gamma\tilde{\omega} \to \omega_B$ the quantities $\delta\varepsilon_{\perp\perp}$ and ε_{xy} become appreciable. In this region the expressions (6.32) and (6.33) can be considerably simplified. Assuming electromagnetic waves to be generated in the lower regions of the magnetosphere (this is grounded in Section 6.2.2 below) we see that by the moment of cyclotron resonance they have considerable angles θ due to magnetic field curvature. Therefore, by virtue of the relation $A_p \ll 1$ (6.21) there hold the inequalities

$$|1-n| \ll \frac{1}{\gamma_c^2} \ll \theta^2 \tag{6.34}$$

Given this, $\tilde{\omega} = \frac{1}{2}\omega\theta^2$. Substituting this relation into the expressions for $\varepsilon_{\alpha\beta}$ (6.10), we make sure that

$$\theta^2\delta\varepsilon_{zz} - 2\theta\varepsilon_{xz} = -\theta^2\left\langle\frac{\omega_p^2}{\gamma^3\tilde{\omega}^2}\right\rangle, \qquad \varepsilon_{xy} \gg \theta\varepsilon_{yz}$$

The ratio of $\theta^2\delta\varepsilon_{zz}$ to ε_{xy} is naturally characterized by the parameter μ_c:

$$\mu_c = \frac{|\varepsilon_{xy}|}{\theta^2\delta\varepsilon_{zz}} = 2\langle\gamma^{-3}\rangle^{-1}\left\langle\frac{\tilde{\omega}\omega_B}{\gamma^2(\omega_B^2 - \gamma^2\tilde{\omega}^2)}\right\rangle$$

Then the expression (6.33) for polarization can be represented as

$$\left(\frac{E_x}{E_y}\right)_{1,2} = \frac{1 \pm (1 + 4\mu_c^2)^{1/2}}{2i\mu_c} \tag{6.35}$$

In the region $\omega_B \gg \gamma\tilde{\omega}$,

$$\mu_c = 2\frac{\tilde{\omega}}{\omega_B}\langle\gamma^{-2}\rangle_{\pm}\langle\gamma^{-3}\rangle^{-1} \ll 1$$

so that polarization of normal waves is linear:

$$\left(\frac{E_x}{E_y}\right)_1 = -i\mu_c, \qquad \left(\frac{E_x}{E_y}\right)_2 = \frac{i}{\mu_c} \tag{6.36}$$

However, in the region of cyclotron resonance $\omega_B \simeq \gamma\tilde{\omega} = \frac{1}{2}\gamma\omega\theta^2$, $\mu_c \gg 1$, and the waves become circularly polarized:

$$\left(\frac{E_x}{E_y}\right)_{1,2} = \mp i \tag{6.37}$$

The refractive indices (6.32) in the resonance region take the form

$$n_{1,2}^2 = 1 + \delta\varepsilon_{\perp\perp} \pm |\varepsilon_{xy}| = 1 \pm \frac{\omega_p^2\tilde{\omega}}{\omega^2}\left\langle\frac{1}{\omega_B \mp \gamma\tilde{\omega}}\right\rangle \tag{6.38}$$

From (6.38) it follows that a wave with a refractive index n_1 (the upper sign) undergoes resonance on positrons since the rotation direction of the electric field vector of this wave coincides with gyrorotation of positrons, whereas the wave n_2 undergoes resonance on electrons.

In conclusion we should note that in studying in this section the behaviour of normal modes of relativistic electron–positron plasma we have discussed only the refractive indices n_j of waves which are determined by the Hermitian part of the dielectric permittivity tensor $\varepsilon_{\alpha\beta}^{H}(\omega, \mathbf{k})$ (6.10), (6.11). The coefficients of absorption or wave amplification, κ_j, describing the wave amplitude variation, are determined by the anti-Hermitian part of the tensor $\varepsilon_{\alpha\beta}^{AH}(\omega, \mathbf{k})$. Since in the plasma in question particle collisions do not play any role (4.7), the anti-Hermitian part $\varepsilon_{\alpha\beta}^{AH}(\omega, \mathbf{k})$ is conditioned by the Landau interaction only, i.e. by resonance interaction between waves and particles. The latter depends essentially on the form of the distribution function of particles. The thorough analysis carried out for magnetosphere plasma has shown that under real conditions of pulsars the role of resonance processes in the homogeneous magnetic field approximation is insignificant (see Lominadze & Pataraya (1982)). The picture changes drastically if we take into account the magnetic field inhomogeneity (curvature). The follow-up section is devoted to the study of this problem.

6.2 Electrodynamics of relativistic electron–positron plasma in a curvilinear magnetic field

We have considered the electrodynamic properties of relativistic plasma in a homogeneous magnetic field. But for the curvature mechanism of electromagnetic wave excitation the most important fact is that plasma particles move along a curvilinear magnetic field, since only in this case does curvature radiation occur.

6.2.1 Dielectric permittivity

We now proceed to the discussion of dielectric properties of a relativistic electron–positron plasma placed in a curved magnetic field. In this case, the medium is not homogeneous in space – the characteristic scale of inhomogeneity $\mathscr{L}$ is the curvature radius of magnetic field lines ρ_l.

The dielectric permittivity of an inhomogeneous plasma The conditions of wave propagation in an inhomogeneous plasma differ strongly from those in a homogeneous plasma. Indeed, in a homogeneous medium, waves of small amplitude are plane (6.4): they have constant phase and group velocities. Their properties are completely determined by the dielectric permittivity tensor of a medium, i.e. by its Hermitian $\varepsilon^{H}_{\alpha\beta}(\omega, \mathbf{k})$ and anti-Hermitian $\varepsilon^{\mathrm{AH}}_{\alpha\beta}(\omega, \mathbf{k})$ parts. In an inhomogeneous medium, plane waves are not eigenfunctions of linearized equations, the phase and group velocities of waves are not constant in space. The wave amplitude can change at the expense of its group velocity variation.

An especially strong variation even in a weakly inhomogeneous medium

$$k\mathscr{L} \gg 1, \qquad \frac{\kappa}{k} \simeq \frac{|\varepsilon^{\mathrm{AH}}|}{|\varepsilon^{H}|} \ll 1, \qquad \kappa = \operatorname{Im} k \tag{6.39}$$

is undergone by the resonance interaction between particles and a wave. Indeed, the resonance interaction is connected with particles moving at a velocity coinciding with the phase velocity of the wave, i.e. with phase matching between the wave and the particles. At the same time, in an inhomogeneous medium neither the phase velocity of waves nor the velocities of free motion of particles are constant. This must lead to a violation of the phase matching and can completely change the character of phase interaction.

A consistent solution of the linear problem that provided a correct description of the interaction between waves and plasma particles in an inhomogeneous medium was obtained by Beskin *et al.* (1987b). Below, without giving detailed reasoning, we present the general result of this paper and then apply it to the case of a relativistic electron–positron plasma moving in a curvilinear magnetic field.

We shall consider a system of linearized Maxwell equations and kinetic equations for particles in an inhomogeneous collisionless plasma. We shall represent the field of a wave of frequency ω (as we did with (6.4) above), in the form of the Fourier integral

$$\mathbf{E}(\mathbf{k}) = \frac{1}{(2\pi)^3}\int \mathbf{E}(\mathbf{r})e^{-i\mathbf{kr}}\,d\mathbf{r}, \qquad \mathbf{B}(\mathbf{k}) = \frac{c}{\omega}\,[\mathbf{kE}(\mathbf{k})] \tag{6.40}$$

The response of the medium to the Fourier-harmonic of the field is found from the linearized kinetic equations (6.3) and has, as before, the general form (6.5). The only difference is that the functions $\mathbf{r}(t')$, $\mathbf{v}(t')$, $\mathbf{p}(t')$ determining the particle trajectory should now be calculated with allowance for the inhomogeneity of the medium (in particular, in the case of a curvilinear magnetic field, they are not described by simple formulae (6.6)). As a result, the response of the medium now depends on the point $\mathbf{r}$ of the space. For the conductivity $\sigma^0_{\alpha\beta}$, according to (6.5) and (6.7),

$$\sigma^0_{\alpha\beta}(\omega, \mathbf{k}, \mathbf{r}) = -e^2\int d\mathbf{p}\,v_\alpha \int_{-\infty}^{t} dt'\,\exp[i\omega(t-t') - i\mathbf{k}(\mathbf{r}-\mathbf{r}')]$$
$$\times\left[\left(1 - \frac{\mathbf{kv}'}{\omega}\right)\delta_{\beta\sigma} + \frac{k_\sigma v'_\beta}{\omega}\right]\frac{\partial F}{\partial p'_\sigma} \tag{6.41}$$

Taking into account (6.41), we then obtain from the Maxwell equations the following linear integro-differential equation which describes the wave field evolution:

$$\Delta\mathbf{E} - \nabla\,\mathrm{div}\,\mathbf{E} + \frac{\omega^2}{c^2}\mathbf{E} = -\frac{4\pi i\omega}{c^2}\mathbf{j}(\mathbf{r}, \omega)$$
$$j_\alpha = \frac{1}{(2\pi)^3}\int\!\!\int d\mathbf{k}\,d\mathbf{r}'\,e^{i\mathbf{k}(\mathbf{r}-\mathbf{r}')}\sigma^0_{\alpha\beta}(\omega, \mathbf{k}, \mathbf{r})E_\beta(\mathbf{r}') \tag{6.42}$$

The solution of this equation under the conditions (6.39) is naturally sought in the framework of geometrical optics, when the field can be represented as a wave packet with a slowly varying amplitude:

$$\mathbf{E}(\mathbf{r}) = \mathbf{E}^0(\mathbf{r})e^{i\psi_p(\mathbf{r})} \tag{6.43}$$

By means of a special expansion (see Beskin *et al.*, 1987b) the equation for the electromagnetic field (6.42) can be reduced to the form (6.13)

$$\left(k^2\delta_{\alpha\beta} - k_\alpha k_\beta - \frac{\omega^2}{c^2}\varepsilon_{\alpha\beta}\right)E^0_\beta = 0 \tag{6.44}$$

The quantity $\varepsilon_{\alpha\beta}(\omega, \mathbf{k}, \mathbf{r})$ involved here is just the dielectric permittivity tensor of an inhomogeneous collisionless plasma. It is described by an expression similar to (6.8) and (6.10):

$$\varepsilon_{\alpha\beta}(\omega, \mathbf{k}, \mathbf{r}) = \delta_{\alpha\beta} - \frac{4\pi i e^2}{\omega^2}\int d\mathbf{p}\, v_\alpha \int_{-\infty}^{t} dt' \exp[i\omega(t-t') - i\mathbf{k}\mathbf{R}^*]$$
$$\times \det{}^{-1}\left(\delta_{\mu\nu} - \frac{1}{2}\frac{\partial L_\mu}{\partial r_\nu}\right)$$
$$\times \left[(\omega - \mathbf{k}\mathbf{v}')\delta_{\beta\sigma} + k_\sigma v'_\beta + \frac{i}{2}\frac{\partial}{\partial r_\sigma}v'_\beta - \frac{i}{2}\delta_{\beta\sigma}\frac{\partial}{\partial r_\chi}v'_\chi\right]\frac{\partial F^\pm}{\partial p'_\sigma}\bigg|_{\mathbf{r}+\mathbf{R}^*/2} \tag{6.45}$$

Here, as in (6.9), we carry out summation over repeated indices and over all sorts of particles. The quantity $\mathbf{R}^*$ in (6.45) is related to the trajectory of particle motion in an inhomogeneous medium as follows. We take into account the fact that the quantity $\mathbf{r} - \mathbf{r}'$ is a function of the coordinate $\mathbf{r}$, the momentum $\mathbf{p}$ and the time differences $t - t'$

$$\mathbf{r} - \mathbf{r}' = \mathbf{L}(\mathbf{r}, \mathbf{p}, t - t')$$

Then the vector $\mathbf{R}^*(\mathbf{r}, \mathbf{p}, t - t')$ is the solution of the equation

$$\mathbf{R}^* = \mathbf{L}(\mathbf{r} + \mathbf{R}^*/2, \mathbf{p}, t - t') \tag{6.46}$$

It should be emphasized that the operator $\partial/\partial\mathbf{r}$ in (6.45) acts also upon the derivative $\partial F^\pm/\partial\mathbf{p}'$.

The expression (6.45) holds for any inhomogeneous medium provided that the conditions (6.39) are satisfied. The normal modes are determined using the dispersion equation (6.44) (cf. (6.13)–(6.17)). The electric field of waves is described after this by the usual formulas of geometrical optics.

We shall point out some general properties of the dielectric permittivity tensor (6.45) of an inhomogeneous plasma.

1. The law of conservation of the energy of waves and particles is fulfilled in all orders of small parameters (6.39). An exact fulfilment of the energy conservation law just underlines the construction of the solution of (6.44), (6.45).
2. Under time reversal $t \to -t$, a wave with a reversed front $\omega \to -\omega$, $\mathbf{k} \to -\mathbf{k}$, i.e. a wave running in the opposite direction, gives the same response in the medium as the initial wave (the permutation of indices

$$\varepsilon_{\alpha\beta}(t \to -t, -\omega, -\mathbf{k}, -\mathbf{B}, \mathbf{r}) = \varepsilon_{\beta\alpha}(\omega, \mathbf{k}, \mathbf{B}, \mathbf{r})$$

should be taken into account). Exactly the same symmetry is inherent in the dielectric permittivity tensor in a homogeneous medium.
3. The dielectric permittivity tensor (6.45) can be represented in a symmetrized form (Bernstein & Friedland, 1983):

$$\varepsilon_{\alpha\beta}(\omega, \mathbf{k}, \mathbf{r}) = \frac{1}{(2\pi)^3} \int \varepsilon^0_{\alpha\beta}(\omega, \mathbf{k}', \mathbf{r} + \boldsymbol{\eta}/2) e^{i(\mathbf{k}'-\mathbf{k})\boldsymbol{\eta}} \, d\mathbf{k}' \, d\boldsymbol{\eta} \qquad (6.47)$$

This relation establishes, in addition, the connection between the tensor $\varepsilon_{\alpha\beta}(\omega, \mathbf{k}, \mathbf{r})$ and a direct response to the Fourier component of the field (6.40) in an inhomogeneous plasma

$$\varepsilon^0_{\alpha\beta}(\omega, \mathbf{k}, \mathbf{r}) = \delta_{\alpha\beta} + \frac{4\pi i}{\omega} \sigma^0_{\alpha\beta}(\omega, \mathbf{k}, \mathbf{r})$$

where the conductivity $\sigma^0_{\alpha\beta}$ is determined according to (6.41).

The dielectric permittivity of plasma in a curvilinear magnetic field We now proceed concretely to the case of relativistic electron–positron plasma moving in a curvilinear magnetic field. At each point we introduce three vectors: **b** along the direction of the magnetic field, **n** along the normal and **l** along the binormal. To these vectors, as shown in Fig. 6.2, there corresponds an orthogonal coordinate system z, x, y.

Owing to the magnetic field curvature, the distribution function $F^{\pm}(\mathbf{p}, \mathbf{r})$ of particles is represented in the form (see (6.2))

$$F^{\pm}(\mathbf{p}, \mathbf{r}) = n^{\pm} \int F^{\pm}_{\parallel}(p_{\parallel}) \delta[\mathbf{p} - p_{\parallel} \mathbf{b}(\mathbf{r}) + \mathbf{p}^{\pm}_{\mathrm{dr}}] \, dp_{\parallel} \qquad (6.48)$$

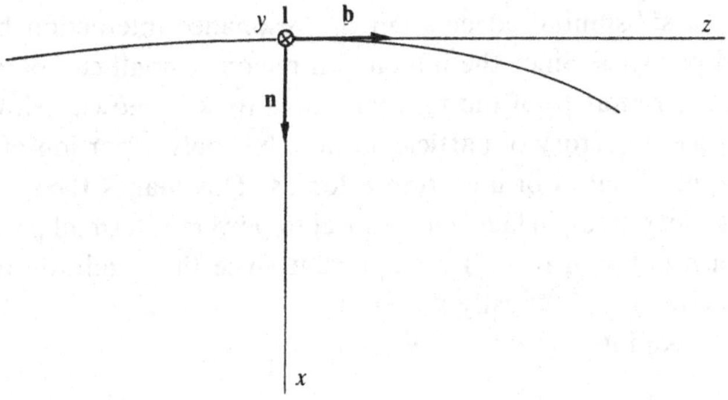

Fig. 6.2. Local coordinate system connected with a curved magnetic field line.

The momentum $\mathbf{p}_{dr}^{\pm}$ describes the drift of charged particles in a curved magnetic field:

$$\mathbf{p}_{dr}^{\pm} = \pm \frac{v_{\|}}{|\omega_B| \rho_l} p_{\|} \mathbf{l}$$

As before, we go over to the limit of an infinitely strong magnetic field ($B \to \infty$). This can be done if $\omega_B \gg \omega$ and $p_{dr}/m_e c \ll \gamma_c^{-1}$. The transverse motion is completely 'frozen' here, so that

$$\frac{\partial F^{\pm}}{\partial p'_{\sigma}} = n^{\pm} \int \frac{\partial F_{\|}^{\pm}}{\partial p_{\|}} \frac{p'_{\sigma}}{p_{\|}} \delta(\mathbf{p} - p_{\|}\mathbf{b})\, dp_{\|} \tag{6.49}$$

For what follows it is necessary to know the trajectory of particle motion (cf. (6.6)):

$$\mathbf{r} - \mathbf{r}' = \mathbf{v}(t - t') - \tfrac{1}{2}\mathbf{a}(t - t')^2 + \tfrac{1}{6}\dot{\mathbf{a}}(t - t')^3 + \cdots \tag{6.50}$$

Here the quantities $\mathbf{v}$, $\mathbf{a}$ and $\dot{\mathbf{a}}$ are the velocity, acceleration and a derivative of the acceleration of a particle when it moves along a magnetic field of curvature radius ρ_l:

$$\mathbf{v} = v_{\|}\mathbf{b}, \qquad \mathbf{a} = \frac{v_{\|}^2}{\rho_l}\mathbf{n}, \qquad \dot{\mathbf{a}} = -\frac{v_{\|}^3}{\rho_l^2}\left(\mathbf{b} + \mathbf{n}\left(\mathbf{b}\frac{d\rho_l}{d\mathbf{r}}\right) + \frac{\rho_l}{\rho_\tau}\mathbf{l}\right) \tag{6.51}$$

Here ρ_τ is the torsion radius of the magnetic field line.

We can see that during the motion of a particle in a curvilinear field, it is not only the velocity but also the acceleration that changes. All this must have a substantial effect upon the resonance interaction between waves and particles. Since the interaction region is small compared with the curvature radius ρ_l of the magnetic field (this is shown below), it is not the whole trajectory of particle motion, but only a portion of it near a given point **r** that is of importance for us. This makes the expansion (6.50) possible. It goes, in fact, with respect to powers of a small parameter, but the term cubic in $(t - t')$ is important since the quadratic term, as we shall see below, practically vanishes.

Next we take into account that, since

$$\begin{aligned}\mathbf{b}(\mathbf{r} + \mathbf{R}^*/2) &= \mathbf{b}(\mathbf{r}) + \mathbf{n}(\mathbf{bR}^*)/2\rho_l - \mathbf{b}(\mathbf{bR}^*)^2/8\rho_l^2 + \cdots \\ \mathbf{n}(\mathbf{r} + \mathbf{R}^*/2) &= \mathbf{n}(\mathbf{r}) - \mathbf{b}(\mathbf{bR}^*)/2\rho_l + \cdots\end{aligned} \tag{6.52}$$

the quantity $\mathbf{R}^*$ entering into expression (6.46) is equal to

$$\mathbf{R}^* = \mathbf{b}v_{\parallel}(t - t') - \frac{\left[\mathbf{b} + \mathbf{n}\left(\mathbf{b}\dfrac{d\rho_l}{d\mathbf{r}}\right)\right]v_{\parallel}^3(t - t')^3}{24\rho_l^2} \tag{6.53}$$

To the same accuracy in (6.45),

$$\det\left|\delta_{\mu\nu} - \frac{\partial L_\mu(\mathbf{r} + \mathbf{R}^*/2)}{2\,\partial r_\nu}\right| = 1$$

We have assumed here that $\rho_\tau \to \infty$, since in the internal regions of pulsar magnetosphere $\rho_\tau \gg \rho_l$. It should be stressed that the quantity $\mathbf{R}^*$ in (6.52) (and accordingly $\omega(t - t') - \mathbf{kR}^*$, which enters into the exponent in (6.45)) contains only a linear and cubic terms and is, therefore, an odd function of the difference $t - t'$. This corresponds to the above-mentioned general property of tensor symmetry (6.45) under time reversal.

Substituting (6.48), (6.52) and (6.53), as well as the relation

$$\mathbf{p}' = p_{\parallel}\left[\mathbf{b} - \frac{\mathbf{n}v_{\parallel}(t - t')}{\rho_l} - \frac{\mathbf{b}v_{\parallel}^2(t - t')^2}{2\rho_l^2}\right]$$

into (6.45), we write the expression for the dielectric permittivity tensor of plasma in the form

$$\varepsilon_{\alpha\beta} = \delta_{\alpha\beta} - i \sum_{+,-} \frac{\omega_p^2}{\omega} \int dp_{\parallel} m_e v_{\parallel} \frac{\partial F_{\parallel}^{\pm}}{\partial p_{\parallel}} \int_0^{\infty} d\tau \, \mathrm{Ex}(\omega, \mathbf{k}, p_{\parallel}, \tau)$$

$$\times \left[b_\alpha b_\beta \left(1 - \frac{v_{\parallel}^2 \tau^2}{4\rho_l^2} \right) + (n_\alpha b_\beta - n_\beta b_\alpha) \frac{v_{\parallel}}{2\rho_l} \tau \right.$$

$$\left. - (n_\alpha b_\beta + n_\beta b_\alpha) \frac{v_{\parallel}^2}{8\rho_l^2} \left(\mathbf{b} \frac{d\rho_l}{d\mathbf{r}} \right) \tau^2 - n_\alpha n_\beta \frac{v_{\parallel}^2}{4\rho_l^2} \tau^2 \right]$$

$$\mathrm{Ex}(\omega, \mathbf{k}, p_{\parallel}, \tau) = \exp\left[i(\omega - k_z v_{\parallel})\tau + \frac{i}{24} \frac{v_{\parallel}^3 \tau^3}{\rho_l^2} \left[k_z + k_x \left(\mathbf{b} \frac{d\rho_l}{d\mathbf{r}} \right) \right] \right]$$

Since curvature radiation, just as synchrotron radiation, is directed towards a narrow cone of angle $\theta \lesssim \gamma^{-1}$, we are most interested in the waves propagating at a rather small angle to the magnetic field direction:

$$\theta \lesssim \gamma_c^{-1} \tag{6.54}$$

Given this, we can neglect the quantities containing the derivative $d\rho_l/d\mathbf{r}$ in the expression for $\varepsilon_{\alpha\beta}$ provided that $\mathbf{b}\, d\rho_l/d\mathbf{r} < \gamma$ (which, in fact, always holds).

Integrating over $d\tau$, we have

$$\varepsilon_{\alpha\beta}(\omega, \mathbf{k}, \mathbf{r}) = \delta_{\alpha\beta} - 2\pi i \frac{\rho_l^{2/3}}{k_z^{1/3}} \sum_{+,-} \frac{\omega_p^2}{\omega} \int dp_{\parallel} \frac{\partial F_{\parallel}^{\pm}}{\partial p_{\parallel}} \times \begin{pmatrix} \frac{\mathscr{F}''(\xi)}{(k_z\rho_l)^{2/3}} & 0 & -i\frac{\mathscr{F}'(\xi)}{(k_z\rho_l)^{1/3}} \\ 0 & 0 & 0 \\ i\frac{\mathscr{F}'(\xi)}{(k_z\rho_l)^{1/3}} & 0 & \mathscr{F}(\xi) \end{pmatrix} \tag{6.55}$$

where

$$\mathscr{F}(\xi) = \mathrm{Ai}(\xi) + i\,\mathrm{Gi}(\xi) = \frac{1}{\pi} \int_0^{\infty} d\tau \exp(i\tau\xi + i\tau^3/3) \tag{6.56}$$

and primes indicate derivatives with respect to the variable

$$\xi = 2(\omega - k_z v_{\parallel}) \frac{\rho_l^{2/3}}{k_z^{1/3} v_{\parallel}} \tag{6.57}$$

Integrating in (6.55) by parts, we can write the tensor $\varepsilon_{\alpha\beta}(\omega, \mathbf{k}, \mathbf{r})$ in the final form

$$\varepsilon_{\alpha\beta} = \begin{pmatrix} 1 + 4\pi \dfrac{\rho_l^{2/3}}{k_z^{4/3}} \left\langle \omega_p^2 \dfrac{\mathrm{Gi}'''(\xi) - i\,\mathrm{Ai}'''(\xi)}{\gamma^3 v_\parallel^2} \right\rangle & 0 & -4\pi \dfrac{\rho_l}{k_z} \left\langle \omega_p^2 \dfrac{\mathrm{Ai}''(\xi) + i\,\mathrm{Gi}''(\xi)}{\gamma^3 v_\parallel^2} \right\rangle \\ 0 & 1 & 0 \\ 4\pi \dfrac{\rho_l}{k_z} \left\langle \omega_p^2 \dfrac{\mathrm{Ai}''(\xi) + i\,\mathrm{Gi}''(\xi)}{\gamma^3 v_\parallel^2} \right\rangle & 0 & 1 + 4\pi \dfrac{\rho_l^{4/3}}{k_z^{2/3}} \left\langle \omega_p^2 \dfrac{\mathrm{Gi}'(\xi) - i\,\mathrm{Ai}'(\xi)}{\gamma^3 v_\parallel^2} \right\rangle \end{pmatrix} \tag{6.58}$$

Here the notation $\langle\ \rangle$ stands for the averaging of (6.9) over the longitudinal distribution function of electrons and positrons.

For future purposes, let us present asymptotical expansions of the Airy function $\mathrm{Ai}(z)$ and the relative function $\mathrm{Gi}(z)$ entering in (6.56) (Abramowitz & Stegun, 1964; Nikishov and Ritus, 1986). For $|z| \gg 1$

$$\mathscr{F}(z) = \mathrm{Ai}(z) + i\,\mathrm{Gi}(z)$$

$$= \begin{cases} \dfrac{i}{\pi z}\left(1 + \dfrac{1\cdot 2}{z^3} + \dfrac{1\cdot 2\cdot 4\cdot 5}{z^6} + \cdots\right), & \pi > \arg z > -\dfrac{\pi}{3} \quad (6.59\text{a}) \\ -\pi^{1/2} z^{-1/4} \exp(-\tfrac{2}{3} z^{3/2}) + \dfrac{i}{\pi z}\left(1 + \dfrac{1\cdot 2}{z^3} + \dfrac{1\cdot 2\cdot 4\cdot 5}{z^6} + \cdots\right), & \\ & -\dfrac{\pi}{3} > \arg z \geqslant -\pi \quad (6.59\text{b}) \end{cases}$$

Note that for a real argument ξ, the Hermitian part of the tensor $\varepsilon_{\alpha\beta}$ is related to the function $\mathrm{Gi}(\xi)$ and the non-Hermitian part to the function $\mathrm{Ai}(\xi)$. In particular, for $\xi \gg 1$, the anti-Hermitian part of the tensor $\varepsilon_{\alpha\beta}$ is exponentially small since

$$\mathrm{Ai}(\xi) = \tfrac{1}{2}\pi^{-1/2}\xi^{-1/4}\exp(-\tfrac{2}{3}\xi^{3/2}), \qquad \mathrm{Im}\,\xi = 0, \mathrm{Re}\,\xi > 0 \tag{6.60}$$

It is easy to show that, for $n \simeq 1$,

$$\xi = \left(\frac{\rho_l \omega}{c\gamma^3}\right)^{2/3}(1 + \gamma^2\theta^2)$$

and the condition $\xi \gg 1$ just corresponds to the frequencies

$$\omega \gg \omega_c \simeq \frac{c}{\rho_l}\gamma^3$$

for which the intensity of curvature radiation is also small.

It is noteworthy that the argument ξ is, in fact, a complex quantity.

For the natural modes propagating at small angles to the direction of motion of relativistic particles (6.54), i.e. Re $k_z > 0$, attenuation corresponds to the condition Im $k_z > 0$ and amplification to Im $k_z < 0$. Henceforth we consider the case Re $\mathbf{k} \gg$ Im $\mathbf{k}$, and the imaginary part $\mathbf{k}$ should be taken into account in the numerator (6.57). As concerns the frequency ω, in our statement of the problem it should be considered as a real quantity. As a result, to wave amplification there corresponds the inequality Im $\xi > 0$.

Let us show how the limiting transition of the dielectric permittivity tensor (6.58) to the case of a direct magnetic field (6.11) proceeds. As is seen from (6.57), for large ρ_l the condition $|\xi| \gg 1$ holds. Using the asymptotics (6.59), we obtain

$$\varepsilon_{\alpha\beta} = \begin{pmatrix} 1 - \dfrac{3}{2}\left\langle \dfrac{\omega_p^2 v_\parallel^2}{\gamma^3 \tilde{\omega}^4 \rho_l^2} \right\rangle & 0 & -i\left\langle \dfrac{\omega_p^2 v_\parallel}{\gamma^3 \tilde{\omega}^3 \rho_l} \right\rangle \\ 0 & 1 & 0 \\ i\left\langle \dfrac{\omega_p^2 v_\parallel}{\gamma^3 \tilde{\omega}^3 \rho_l} \right\rangle & 0 & 1 - \left\langle \dfrac{\omega_p^2}{\gamma^3 \tilde{\omega}^2} \right\rangle \end{pmatrix} \tag{6.61}$$

As might be expected, within the limit $\rho_l \to \infty$ this tensor tends to the dielectric permittivity tensor of a homogeneous plasma (6.11).

6.2.2 Normal modes of electromagnetic oscillations

The dispersion equation We shall now discuss the natural modes of electromagnetic oscillations. The dispersion equation (6.13) for the tensor (6.58) takes the form

$$\begin{aligned}(1 - n^2)^2 &+ (1 - n^2)(1 - n^2 \cos^2\theta)\delta\varepsilon_{zz} \\ &+ (1 - n^2)(1 - n^2 \sin^2\theta \cos^2\phi)\delta\varepsilon_{\perp\perp} \\ &+ (1 - n^2 + n^2 \sin^2\theta \sin^2\phi)(\delta\varepsilon_{zz}\cdot\delta\varepsilon_{\perp\perp} - \varepsilon_{xz}\cdot\varepsilon_{zx}) = 0 \end{aligned} \tag{6.62}$$

where $\delta\varepsilon_{\perp\perp} = \varepsilon_{xx} - 1$, $\delta\varepsilon_{zz} = \varepsilon_{zz} - 1$, $k_z = k\cos\theta$, $k_x = k\sin\theta\cos\phi$. The properties of solutions of eq. (6.62) depend essentially on the quantity $\delta\varepsilon_{zz}$. In particular, if $\delta\varepsilon_{zz} \ll 1$ as is seen from (6.58), the quantities $\delta\varepsilon_{\perp\perp}$ and ε_{xz} are also small here since $\varepsilon_{xz} \simeq (k_z\rho_l)^{-1/3}\delta\varepsilon_{zz}$, $\delta\varepsilon_{\perp\perp} \simeq (k_z\rho_l)^{-2/3}\delta\varepsilon_{zz}$, the refractive indices n_j will be equal to

$$n_{t_1}^2 = 1 \tag{6.63}$$

$$n_{t_2}^2 = 1 + \delta\varepsilon_{\perp\perp} + \delta\varepsilon_{zz}\sin^2\theta \tag{6.64}$$

both these waves corresponding to transverse oscillations. For the case $\delta\varepsilon_{zz} \gg 1$, we shall first write the expressions for n_j^2 only for the vector **k** which lies in the plane of the curved magnetic field, i.e. when $k_y = k \sin\theta \sin\phi = 0$. Taking into account that

$$\delta\varepsilon_{\perp\perp} \ll \delta\varepsilon_{zz}$$

we obtain

$$n_{t_1}^2 = 1 \tag{6.65}$$

$$n^2 = 1 + \frac{\delta\varepsilon_{zz}\sin^2\theta + \delta\varepsilon_{\perp\perp} + (\delta\varepsilon_{zz}\cdot\delta\varepsilon_{\perp\perp} - \varepsilon_{xz}\cdot\varepsilon_{zx})}{1 + \delta\varepsilon_{zz}\cdot\cos^2\theta} \tag{6.66}$$

It should be emphasized that equation (6.66), the same as (6.17), determines the refractive index implicitly, so that its solution describes both transverse and plasma waves. On the other hand, as in the case of a homogeneous magnetic field, the normal waves (6.63) and (6.65) correspond to a polarization under which the electric wave vector is perpendicular to the plane where the magnetic field line lies. For $k_y = 0$ this wave (which we shall also refer to as extraordinary) does not interact with plasma placed in an infinitely strong magnetic field and therefore cannot be amplified.

Classification of solutions We shall now classify the solutions (6.63)–(6.66) depending on physical parameters, namely, the plasma density $n^{\pm}$ (and, therefore, the Langmuir frequency ω_p), the curvature radius ρ_l of magnetic field lines and the oscillation frequency ω. In Fig. 6.3 on the plane $\omega - \rho_l$ there are regions in which the properties of normal waves are essentially different. The coordinates of the 'knot' point

$$\omega^* = \omega_p \gamma_c^{1/2}, \qquad \rho_l^* = \frac{c}{\omega_p}\gamma_c^{5/2} \qquad (\gamma_c = \langle\gamma^{-3}\rangle^{-1/3}) \tag{6.67}$$

depend only on the quantity γ_c and on the electron–positron plasma density.

Region I ($\delta\varepsilon_{zz} \gg 1$) corresponds to the case $|1 - n| \gg 1/\gamma_c^2$ when the quantity γ^{-2} can be neglected in the expression for

$$\tilde{\omega} = \omega - k_z v_{\parallel} = \tfrac{1}{2}\omega[\gamma^{-2} + \theta^2 + 2(1 - n)]$$

Owing to this fact, the condition $\delta\varepsilon_{zz} \gg 1$ can be written in the form $a \gg 1$,

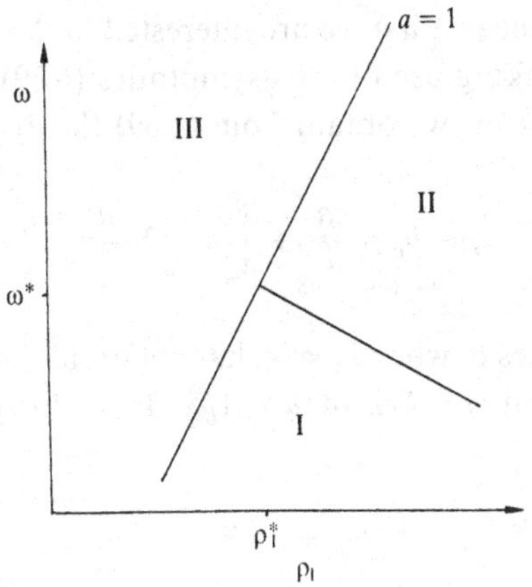

Fig. 6.3. Three regions of parameters on the ω–ρ_l plane. Unstable curvature-plasma waves exist in region I. In region III there may exist only two transverse waves.

where

$$a = 4\pi \frac{\rho_l^{4/3}}{k_z^{2/3}c^2}\left\langle\frac{\omega_p^2}{\gamma^3}\right\rangle \tag{6.68}$$

An important dimensionless parameter a expresses the relation between the eigenfrequency of oscillations of a relativistic plasma $\langle\omega_p^2/\gamma^3\rangle^{1/2}$ and the characteristic time of wave–plasma interaction $\rho_l^{2/3}k_z^{-1/3}/c$ determined by the inhomogeneity. The parameter a determines the influence of inhomogeneity upon plasma oscillations.

For $a \gg 1$, the dispersion equation (6.66) takes the form

$$\xi - b_\theta = ia\xi\mathscr{F}'(\xi) + ia\mathscr{F}'''(\xi) + a^2[\mathscr{F}'''(\xi)\cdot\mathscr{F}'(\xi) - \mathscr{F}''^2(\xi)] \tag{6.69}$$

where

$$b_\theta = (k_z\rho_l)^{2/3}\theta^2 \tag{6.70}$$

Given this, the refractive index n_j is related, according to (6.57), to the root ξ_j of the dispersion equation (6.69) in a simple way:

$$n_j^2 = 1 + \theta^2 - \frac{\xi_j}{(k_z\rho_l)^{2/3}} \tag{6.71}$$

Note that the parameter a separates regions I and III ($a = 1$). For the boundary between the regions I and II, the corresponding condition will be formulated below.

Curvature plasma modes First we are interested in the natural modes, for which $\mathrm{Im}\,\xi \geqslant 0$. Making use of the asymptotics (6.59) containing all the modes of interest for us, we obtain from (6.69) the dispersion equation

$$\xi - b_\theta = \frac{a}{\pi\xi} + \frac{6a}{4\xi^4} - 2\frac{a^2}{\pi^2\xi^6} \tag{6.72}$$

which for small angles θ, when $b_\theta \ll \xi$, has solutions $\xi \propto a^u$ ($a \gg 1$). From (6.72) we obtain that $u = 1/2$ or $u = 1/5$. Thus, for $\theta < \theta_\parallel$ ($b_\theta < a^{4/5}$), where

$$\theta_\parallel = a^{3/20}\theta^* \tag{6.73}$$

and θ^* is given by (6.25), equation (6.72) has five branches of undamped oscillations:

$$\xi_{2,3} = \pm\left(\frac{a}{\pi}\right)^{1/2} \tag{6.74}$$

$$\xi_{4,5,6} = \exp\left(\frac{2\pi}{5}im\right)\left(\frac{2a}{\pi}\right)^{1/5}, \qquad m = 0, 1, 2 \tag{6.75}$$

If $\theta \gg \theta_\parallel$ (which is, of course, possible only if $a^{3/20}\theta^* \ll 1$), then here

$$\xi_2 = b_\theta = (k_z\rho_l)^{2/3}\theta^2 \tag{6.76}$$

$$\xi_{3,4,5,6} = \exp\left(\frac{\pi i}{3}m\right)\frac{2^{1/6}}{\pi^{1/3}}\frac{a^{1/3}}{(k_z\rho_l)^{1/9}\theta^{1/3}}, \qquad m = 3, 0, 1, 2 \tag{6.77}$$

Number 1 we have left for the extraordinary mode $n_1^2 \equiv 1$ (see (6.63) and (6.65)). The asymptotics (6.74)–(6.77) hold only for $|\xi| \gg 1$.

It can easily be verified, using the definition of the quantity a (6.68), that roots 2 and 3 in formulas (6.74), (6.76) and (6.77) correspond exactly to the normal modes $j = 2, 3$ for a homogeneous magnetic field. Therefore, for them there hold the relations (6.24a) and (6.24b) for the refractive indices n_j, and (6.27b) and (6.27c) for their polarization. This is not surprising, since according to (6.68) the asymptotic $a \to \infty$ just corresponds to the limit $\rho_l \to \infty$.

On the other hand, for a finite curvature radius ρ_l there occurs splitting of the t–l mode $j = 4$ (which exists in the direct magnetic field) into three branches of oscillations (6.75), (6.77), as shown in Fig. 6.4. Two of them,

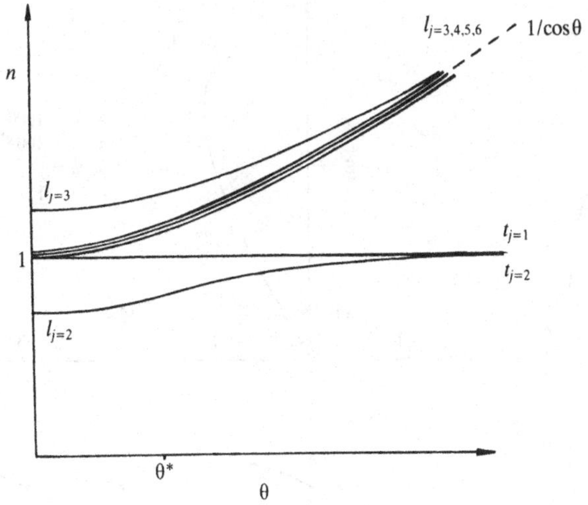

Fig. 6.4. Dependence of the refractive index n_j on the angle θ in a curved magnetic field (region I). Unstable curvature-plasma waves correspond to Alfven oscillations with $n \simeq 1/\cos\theta$.

as is seen from the relations (6.75) and (6.77) are unstable. For these modes,

$$\operatorname{Im} n_{5,6} = \begin{cases} -2^{2/5} \sin\left(\frac{2\pi}{5} l\right) \left\langle \frac{\omega_p^2}{\gamma^3} \right\rangle^{1/5} (k_z^2 \rho_l c)^{-2/5}, & l = 1, 2, \theta \ll \theta_\parallel \\ -\frac{3^{1/2}}{2^{7/6}} \left\langle \frac{\omega_p^2}{\gamma^3} \right\rangle^{1/3} (k_z^3 \rho_l c^2 \theta)^{-1/3}, & \theta \gg \theta_\parallel \end{cases} \tag{6.78}$$

We see that Im n depends in a power-law manner on the curvature radius ρ_l. The polarization of such waves is close to transverse. Indeed, (6.13) and (6.61) imply

$$\left(n_z^2 - 1 + \frac{3}{2}\left\langle \frac{\omega_p^2 c^2}{\gamma^3 \tilde{\omega}^4 \rho_l^2} \right\rangle\right) E_x + i\left\langle \frac{\omega_p^2 c}{\gamma^3 \tilde{\omega}^3 \rho_l} \right\rangle E_z = 0$$

and for the dispersion $n^2 = 1 + \theta^2$ (6.71) we obtain

$$\left(\frac{E_z}{E_x}\right)_{4,5,6} \simeq i \frac{c}{\rho_l \tilde{\omega}_{4,5,6}}, \quad \left|\frac{E_z}{E_x}\right| \ll 1 \tag{6.79}$$

It should be stressed, however, that, as distinct from the case of a homogeneous magnetic field, the longitudinal component of the electric field E_z in normal modes $j = 4, 5, 6$ is non-zero even in the case $\theta = 0$.

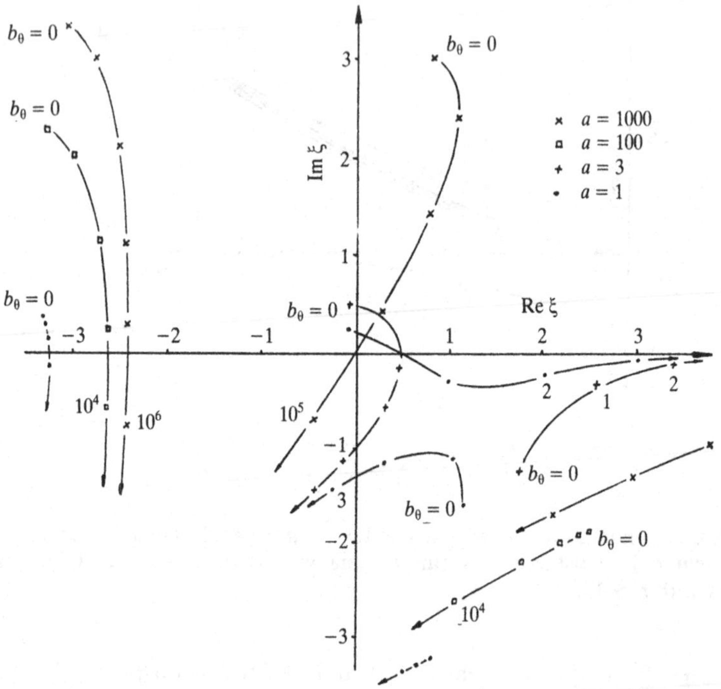

Fig. 6.5. Motion of the roots ξ_j of the dispersion equation (6.69) depending on the parameter b_θ (6.70).

The new modes which we call curvature-plasma modes, are unstable. The existence of these modes, as we shall see below, leads to the appearance of the powerful radio emission of pulsars.

For $a \gg 1$, as the angle θ increases, four of the five roots ξ_j pass to the lower half-plane, so that the corresponding normal waves become damping. The motion of the roots is shown in Fig. 6.5 (see also formula (6.77)). For instance, the roots corresponding to increasing waves with $m = 1$ ($j = 5$) and $m = 2$ ($j = 6$) in formulas (6.75) and (6.77) go over to the lower half-plane for $\theta = \theta^{\text{out}}_{\parallel 5,6}$ where

$$\theta^{\text{out}}_{\parallel 5,6} = u_{5,6} a^{3/4} \theta^*, \qquad u_5 = 0.28. \qquad u_6 = 0.77 \qquad (6.80)$$

The only undamped normal wave here is the mode ξ_2 (6.76).

Note also that for $a \simeq 1$ there occurs mutual transformation of an unstable wave with $m = 1$ ($j = 5$) and a transverse mode ξ_2 (6.75). This is shown in detail in Fig. 6.6. We can see that for $a > 1.15$ the root ξ_5 of the unstable mode $m = 1$ with increasing angle θ is moving to the lower

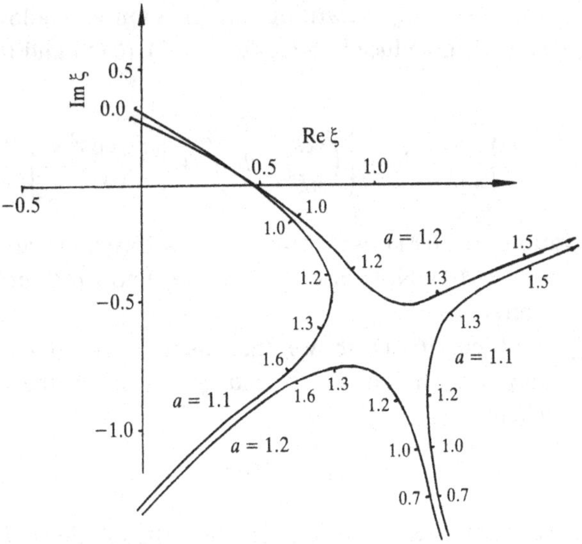

Fig. 6.6. The domain of transformation of two normal modes. Numbers indicate b_θ-values.

half-plane (i.e. such a wave damped for large θ). In contrast, for $a < 1.15$ this mode comes to the asymptotics $\xi = b_\theta = (k_z \rho_l)^{2/3}\theta^2$ and, therefore, propagates freely for $(k_z \rho_l)^{2/3}\theta^2 \gg 1$ as the transverse mode.

We note that the increasing modes found above correspond to a hydrodynamic instability, exactly as might be expected because the limit $a \gg 1$ corresponds to a high particle density (cf. the case $A_p \gg 1$ in a homogeneous field (6.22)). In particular, the solutions (6.74)–(6.77) can be obtained directly from the asymptotic expression (6.61) for the dielectric permittivity tensor through the change $\tilde{\omega} \to \omega - k_z c$, $v_\parallel \to c$. The hydrodynamic character of the instability is also indicated by the fact that the imaginary parts of the refractive index (6.78) depend on the particle density $n^\pm$ in a power-law manner.

Finally, we shall present exact expressions for the case when the vector **k** has a component perpendicular to the plane of the magnetic field line, i.e. when $k_y = k \sin\theta \sin\phi \neq 0$. In this case, the solution of the dispersion equation (6.62) has the form

$$n^2 = 1 + \frac{\delta\varepsilon_{zz}\theta^2 + \delta\varepsilon_{\perp\perp} + D_\varepsilon \pm [(\delta\varepsilon_{zz}\theta^2 + \delta\varepsilon_{\perp\perp} + D_\varepsilon)^2 - 4\theta^2\cos^2\phi D_\varepsilon(1 + \delta\varepsilon_{zz})]^{1/2}}{2(1 + \delta\varepsilon_{zz})}$$

where $D_\varepsilon = \delta\varepsilon_{zz}\delta\varepsilon_{\perp\perp} - \varepsilon_{xz}\varepsilon_{zx}$. Rewriting this equation with allowance for the quantities a, b_θ, ξ introduced above (see (6.57), (6.68) and (6.70)), we obtain

$$\xi - b_\theta = \frac{1}{2}\left\{\frac{2a}{\pi\xi^4} - b_\theta \pm \left[\left(\frac{2a}{\pi\xi^4} - b_\theta\right)^2 + \frac{8ab_\theta \cos^2\phi}{\pi\xi^4}\right]^{1/2}\right\} \tag{6.81}$$

In the derivation of (6.81) we have used the asymptotic dielectric permittivity tensor (6.61). Note also that eq. (6.81) now includes also the 'extraordinary' mode $j = 1$.

The analysis of eq. (6.81) shows that its solutions coincide with (6.74)–(6.77) only for the angles θ and ϕ satisfying the condition $\theta < \theta_\perp/\cos\phi$, where

$$\theta_\perp = a^{-3/20}\theta^* \tag{6.82}$$

If the inverse inequality $\theta > \theta_\perp/\cos\phi$ holds, then the solutions of eq. (6.81) have the form

$$\begin{aligned} \xi_4 &= b_\theta \\ \xi_{1,5,6} &= \exp\left(\frac{\pi i l}{2}\right)\left(\frac{2a}{\pi b_\theta}\right)^{1/4}\cos^{1/2}\phi, \qquad l = 0, 1, 2 \end{aligned} \tag{6.83}$$

The dispersion properties of the normal waves $j = 2, 3$ are determined, as before, by the relations (6.24a) and (6.24b).

Thus, we can see that the 'amplification cone' is, in fact, strongly elongated along the x-axis, as shown in Fig. 6.7. Hence, the ratio of the angles $\theta_\parallel$ and $\theta_\perp$ (formulas (6.73) and (6.82)) within which the

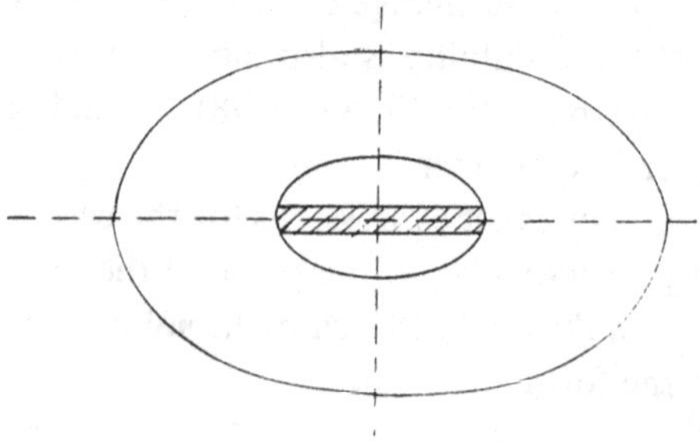

Fig. 6.7. Structure of the amplification cone. The internal region corresponds to the largest increments. In the shaded region the linear transformation of waves $j = 1$ and $j = 4$ is ineffective.

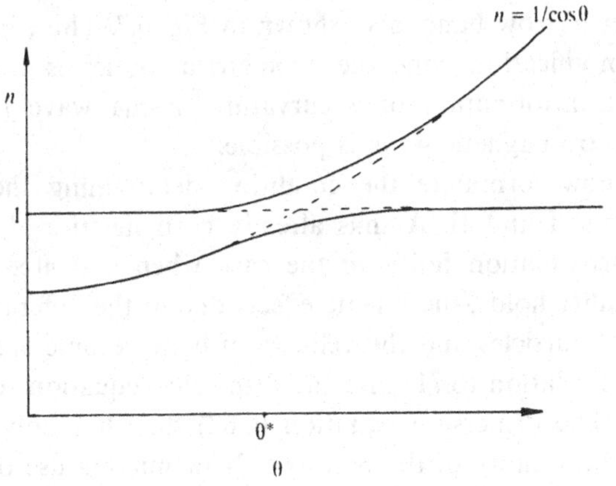

Fig. 6.8. Linear transformation of two normal waves $j = 1$ and $j = 4$ within the range of angles $\theta \simeq \theta_\perp$ (6.82).

curvature-plasma wave amplification is most efficient is

$$\frac{\theta_\parallel}{\theta_\perp} \simeq a^{3/10} \gg 1 \tag{6.84}$$

The same applies to the whole of the amplification cone. Indeed, according to (6.80), (6.81), the external size of the amplification cone is

$$\theta_\perp^{\text{out}} \simeq a^{1/4}\theta^*, \qquad \theta_\parallel^{\text{out}} \simeq a^{3/4}\theta^* \tag{6.85}$$

so that $\theta_\perp^{\text{out}} \ll \theta_\parallel^{\text{out}}$. The spatial dimension of the region that contributes to amplification is of the order of $\rho_l \theta_\parallel^{\text{out}}$. Consequently, this size is much smaller than ρ_l, which was, in fact, used in the expansion (6.50).

Let us pay attention to another important fact connected with the possibility of mode transformation within the range of angles $\theta \simeq \theta_\perp$, shown in Fig. 6.8. We can see that with increasing angle θ for $k_y \neq 0$ the curvature-plasma mode $j = 4$ goes over to the transverse wave $j = 1$ with $n_j = 1$, and the transverse wave $j = 1$ to the curvature-plasma mode. Given this, the condition

$$g_0 = \frac{1}{k_z} \frac{\partial n/\partial l}{|n_1 - n_2|} \simeq a^{-3/10} \cos^{-3}\phi \ll 1 \tag{6.86}$$

determining, as is known, the validity of the approximation of geometrical optics (Ginzburg, 1970), fails only at angles

$$\cos\phi < a^{-1/10} \ll 1 \tag{6.87}$$

i.e. in a rather narrow band, also shown in Fig. 6.7. Thus, in practically the entire amplification cone the geometrical optics is valid, so that an effective transformation of a curvature-plasma wave $j = 4$ into a transverse electromagnetic wave is possible.

We shall now formulate the condition determining the boundary between regions I and II. As has already been mentioned, the hydrodynamic approximation holds in the case when $|1 - n| \gg 1/\gamma_c^2$. If the inverse inequality holds, the kinetic effects due to the difference between the velocity of particles and the velocity of light become substantial. In particular, the relation (6.71) and the dispersion equation (6.69) are no longer valid. The dispersion equation (6.62) then has only four roots, which lie in the vicinity of the real axis. Now making use of (6.71), the condition $|1 - n| = 1/\gamma_c^2$ can be written in the form

$$\omega = \frac{\omega_p^{1/2} c^{1/2}}{\rho_l^{1/2}} \langle \gamma^{-3} \rangle^{-7/12} \tag{6.88}$$

This relation just determines the second boundary of region I (Fig. 6.3).

Thus, in region I there exist six undamped normal waves. At sufficiently small angles θ, two of them (curvature-plasma modes) have negative reabsorption coefficient $\mu_j = 2\omega/c \operatorname{Im} n_j$: these waves can be effectively amplified.

Region II occupies a sector $a \gg 1$, $|1 - n| \ll \gamma_c^{-2}$. In this sector, the dispersion equation (6.62) has four roots which practically coincide with the corresponding solutions for a homogeneous magnetic field, just as might be expected since region II corresponds to the limit $\rho_l \to \infty$, as is seen in Fig. 6.3. Consequently, in this region there exist two longitudinal and two transverse normal waves, shown in Fig. 6.1. There are no unstable modes here, so that an effective amplification is impossible in region II. Therefore, we shall not dwell on this case.

In *region III* as is already clear from the previous analysis, along with an ordinary mode there exists a single transverse mode t_2. Indeed, as we can see from the dispersion equation (6.69), for $a \gg 1$ we can in the first approximation neglect the terms in the right-hand side, and thereby

$$\xi_2 = (k_z \rho_l)^{2/3} (\theta^2 + \gamma^{-2}) \tag{6.89}$$

which just corresponds to a transverse mode with $n_2^2 \simeq 1$. The right-hand side of eq. (6.69) determines the imaginary part of the refractive index.

As a result, we obtain

$$n_{t_2}^2 = 1 + \frac{a}{(k_z\rho_l)^{2/3}}[\mathcal{F}'''(\xi) + (k_z\rho_l)^{2/3}\theta^2\mathcal{F}'(\xi)] \tag{6.90}$$

All the other roots of the dispersion equation (6.69) for $a \ll 1$ are in the region $-\pi < \arg\xi < -\pi/3$, where for $|\xi| \gg 1$

$$\mathcal{F}(\xi) = \frac{1}{\pi^{1/2}\xi^{1/4}}\exp(-\tfrac{2}{3}\xi^{3/2}) \tag{6.91}$$

i.e. all roots are in the lower half-plane. Consequently, normal waves corresponding to such roots will damp rapidly.

Thus, in region III there may propagate only two transverse modes (6.63), (6.64), with the refraction indices n_j close to unity. Note that this conclusion remains valid in the case $\omega \gg \omega_c$. The point is that due to the asymptotics (6.59a), in the upper half-plane the function $\mathcal{F}(\xi)$ along with the derivatives tends modulo to zero as $|\xi| \to \infty$, so that the corrections to unity in the equation (6.64) will be small irrespective of the form of the argument ξ.

The disappearance of plasma waves other than the transverse modes t_1 and t_2 is due to the fact that, as mentioned above (6.68), in the region of parameters $a \ll 1$ the plasma correction to the dielectric constant is $\ll 1$, which means that the time of interaction between waves and plasma particles $\tau = \rho_l^{2/3}k_z^{-1/3}/v_\parallel$ is essentially less than the characteristic time of proper oscillations, which is equal to the inverse plasma frequency $(\omega_p^2\langle\gamma^{-3}\rangle)^{-1/2}$. Under these conditions the waves do not practically fill plasma, so that for $a \ll 1$ there may exist only two transverse modes, as in vacuum.

6.2.3 The region of cyclotron resonance

Concluding this section, we present without derivation the expression for the component ε_{yy} of the dielectric permittivity tensor in the case of a finite magnetic field. As has already been said, special attention should be given here to the region near the cyclotron resonance. As can be seen from the calculation (Beskin *et al.*, 1988),

$$\varepsilon_{yy} = 1 + \pi\left\langle \frac{\omega_p^2}{\omega^2}\frac{\tilde{\omega}}{\gamma}\frac{\rho_l^{2/3}}{k_z^{1/3}v_\parallel}[-\mathrm{Gi}(x^+) - \mathrm{Gi}(x^-) + i\,\mathrm{Ai}(x^+) + i\,\mathrm{Ai}(x^-)]\right\rangle \tag{6.92}$$

where

$$x^{\pm} = 2\left(\tilde{\omega} \pm \frac{\omega_B}{\gamma}\right)\frac{\rho_l^{2/3}}{k_z^{1/3}v_{\parallel}}$$

Formula (6.92) allows us to extend the expressions (6.10) and (6.31) for 'cyclotron' corrections to the case of a curved magnetic field.

As might be expected, allowance for the finite curvature of magnetic field lines leads to cyclotron resonance broadening,

$$2\pi\rho_l^{2/3}k_z^{-1/3}v_{\parallel}^{-1}\,\mathrm{Ai}(x^{\pm}) \qquad \text{instead of } \pi\delta(\tilde{\omega} \pm \omega_B/\gamma)$$

But it can readily be verified that in the range of parameters in question $\rho_l \simeq 10^9$ cm, $\omega \simeq 10^9\ \mathrm{s}^{-1}$, and in the cyclotron resonance region $\omega_B/\gamma = \tilde{\omega}$, we have

$$\frac{\omega_B}{\gamma}\frac{\rho_l^{2/3}}{k_z^{1/3}v_{\parallel}} \gg 1 \tag{6.93}$$

Owing to this, the energy range of particles where they are in resonance with the wave remains sufficiently narrow. In particular, if the width of the distribution function is

$$\frac{\Delta\mathscr{E}}{\mathscr{E}} \gg \frac{\omega^{1/3}c^{2/3}\gamma_c}{\omega_B\rho_l^{2/3}} = 10^{-6}\left(\frac{\nu}{1\ \mathrm{GHz}}\right)^{-2/3}\left(\frac{\rho_l}{10^9\ \mathrm{cm}}\right)^{-2/3}\left(\frac{\tilde{\omega}}{\omega}\right)^{-1} \tag{6.94}$$

then formula (6.92) will coincide with the corresponding expression (6.10) for a homogeneous magnetic field.

Indeed, in this case the arguments x^+ and x^- are modulo much larger than unity for all particle energies. Hence, we can use the asymptotics (6.59). For a sufficiently wide distribution function, the oscillating factors appear to be 'smeared out', and the asymptotics $1/\pi x^{\pm}$ yield a result exactly coinciding with the limit of a homogeneous magnetic field. Only when the inverse inequality holds in (6.93) is the influence of magnetic field line curvatures appreciable. In particular, in the energy range where $x^- < 0$, $x^+ < 0$, the imaginary part will oscillate (Shaposhnikov, 1981; Beskin *et al.*, 1988).

Clearly, however, the condition (6.94) is rather rigid, in any case for the main electron plasma. The only exception is the primary beam, for which resonance occurs at small distances from the star's surface, where $\tilde{\omega}/\omega \ll 1$ and the condition (6.94) becomes less rigid. But even in this case, the natural beam width $\Delta\mathscr{E}/\mathscr{E} \sim 10^{-2}$ appears to be large enough (see (5.71) and Fig. 5.4). We may therefore conclude that the influence of magnetic field line curvature upon cyclotron resonance in the pulsar magnetosphere is insignificant, so in the consideration of cyclotron effects it suffices to use the homogeneous field approximation (Section 6.1.3).

6.3 Non-linear processes in inhomogeneous relativistic electron–positron plasma

6.3.1 Specificities of non-linear interaction in a curved magnetic field

In the preceding section we have investigated the linear properties of plasma flowing in pulsar magnetosphere along open field lines of a magnetic field. It has been shown that in a curved magnetic field there arise unstable curvature-plasma modes which have substantial increments. For real conditions in the magnetosphere of a neutron star, the amplification of these waves is large, so that it is necessary to take into account non-linear effects. Also, the effects of plasma inhomogeneity are of importance here. Inhomogeneity changes appreciably the electromagnetic properties of relativistic electron–positron plasma – both its linear and non-linear characteristics. Moreover, the field curvature leads to the interaction not only of waves having electric field components along a strong magnetic field, as in the homogeneous case, but also of waves whose electric vector is perpendicular to the magnetic field line, as is the case with curvature-plasma waves (6.79).

We consider here the non-linear processes which restrict the growth of unstable modes, as well as the non-linear interaction between different modes. As shown below, the energy of excited waves is much lower than the plasma energy, and therefore we can consider the non-linear processes in the framework of the theory of weak turbulence. But the effects of plasma inhomogeneity are considerable here, and strictly speaking we should use the equations allowing for these effects in the non-linear processes. These equations were obtained by Istomin (1988).

As shown in the preceding section, an effective excitation of curvature-plasma modes processes only in a narrow cone of angles θ near the direction of the magnetic field (see Fig. 6.7). Outside this cone, unstable modes fade away. It is therefore natural to expect that the instability will be most rapidly compensated at the expense of the non-linear processes which lead to small turns of the wave vectors, thus providing the escape of waves from the instability region. The three-wave interaction proves to be the most effective in this respect. Note that the three-wave interaction is cubic in charge, so that the contributions of electrons and positrons in this interaction are opposite in sign. But they do not fully compensate each other in the pulsar magnetosphere, since the distribution functions of electrons and positrons are essentially distinct (see Section 5.3.4).

6.3.2 An approximate theory of non-linear stabilization

We shall first consider stabilization of curvature-plasma waves in the simplest quasihomogeneous approximation. Namely, we neglect the influence of magnetic field curvature upon non-linear wave interaction. This allows a direct use of the general methods of the theory of non-linear wave interaction (Tsytovich, 1971).

As is well known, to investigate non-linear wave interaction we should calculate the non-linear conductivity of an electron–positron plasma. Since a strong instability of curvature-plasma oscillations exists under conditions in which the velocity spread of particles can be neglected (see the condition (6.88)), we calculate the matrix element of the three-wave interaction in the hydrodynamic limit.

Suppose there exist three waves whose longitudinal electric fields are E_{z_1}, E_{z_2} and E_{z_3}. The frequencies and wave vectors obey the conservation laws $\omega_1 = \omega_2 + \omega_3$, $k_{z_1} = k_{z_2} + k_{z_3}$. The magnitude of the non-linear plasma response at a frequency ω_1 is found from the equations of continuity and momentum for electron and positron liquids:

$$\frac{\partial n}{\partial t} + \frac{\partial}{\partial \mathbf{r}}(n\mathbf{v}) = 0$$
$$\frac{\partial \mathbf{p}}{\partial t} + (\mathbf{v}\nabla)\mathbf{p} = \frac{e}{c}[\mathbf{v}\mathbf{B}_0] + e\left\{\mathbf{E} + \frac{1}{\omega}[\mathbf{v}[\mathbf{k}\mathbf{E}]]\right\} \tag{6.95}$$

Within the limit of an infinitely strong magnetic field $B_0 \to \infty$, the equation of motion implies that the transverse velocities of particle motion $\mathbf{v}_\perp$ tend to zero in any order of perturbations with respect to the amplitudes of field waves $\mathbf{E}$. The density and longitudinal velocity of particles can be represented as a power series expansion of the wave amplitudes:

$$n = n_0 + \delta n^{(1)} + \delta n^{(2)} + \cdots$$
$$v = v_\parallel + \delta v_\parallel^{(1)} + \delta v_\parallel^{(2)} + \cdots \tag{6.96}$$

where n_0 and $v_\parallel$ are the unperturbed density and the particle velocity parallel to $\mathbf{B}_0$; the quantities $\delta n^{(1)}$ and $\delta v_\parallel^{(1)}$ are proportional to field amplitudes

$$\delta n^{(1)} = n_0 \frac{k_z}{\tilde{\omega}} \delta v_{\parallel}^{(1)}, \qquad \delta v_{\parallel}^{(1)} = \frac{\delta p_{\parallel}^{(1)}}{m_e \gamma^3}, \qquad \delta p_{\parallel}^{(1)} = \frac{ieE_z}{\tilde{\omega}}, \qquad \tilde{\omega} = \omega - k_z v_{\parallel} \tag{6.97}$$

In the second order we have, respectively,

$$\begin{aligned} \delta n_1^{(2)} &= \frac{n_0 k_{z_1}}{\tilde{\omega}_1} \left[\delta v_{\parallel_1}^{(2)} + \frac{1}{2} \left(\frac{k_{z_2}}{\tilde{\omega}_2} + \frac{k_{z_3}}{\tilde{\omega}_3} \right) \delta v_{\parallel_2}^{(1)} \cdot \delta v_{\parallel_3}^{(1)} \right] \\ \delta v_{\parallel}^{(2)} &= \frac{1}{2} \left(\frac{k_{z_1}}{\tilde{\omega}_1} - 3\gamma^2 \frac{v_{\parallel}}{c^2} \right) \delta v_{\parallel_2}^{(1)} \cdot \delta v_{\parallel_3}^{(1)} \end{aligned} \tag{6.98}$$

The second summand in (6.98), which is proportional to γ^2, is a consequence of the non-linear relation between $\delta v_{\parallel}$ and $\delta p_{\parallel}$. Namely,

$$\delta v_{\parallel} = \frac{\delta p_{\parallel}}{m_e \gamma^3} - \frac{3}{2} \frac{v_{\parallel}}{m_e^2 c^2 \gamma^4} (\delta p_{\parallel})^2 + \cdots$$

Using (6.97) and (6.98), we can represent the non-linear current $j_z^{(2)}$ as $j_z^{(2)} = (e\omega/k_z)\delta n^{(2)}$ since

$$j_z^{(2)} = -\frac{1}{2} \frac{n_0 e^3}{m_e^2 \gamma^6} \frac{\omega_1}{\tilde{\omega}_1 \tilde{\omega}_2 \tilde{\omega}_3} \frac{1}{v_{\parallel}} \left(\frac{\omega_1}{\tilde{\omega}_1} + \frac{\omega_2}{\tilde{\omega}_2} + \frac{\omega_3}{\tilde{\omega}_3} - 3\gamma^2 \right) E_{z_2} E_{z_3} \tag{6.99}$$

Within the non-relativistic limit ($v_{\parallel} \to 0$) formula (6.99) corresponds to the well-known expression for the longitudinal non-linear conductivity of a cold plasma in a strong magnetic field (Tsytovich, 1971):

$$\sigma_z^N = -\frac{1}{2} \frac{n_0 e^3}{m_e^2} \frac{1}{\omega_2 \omega_3} \left(\frac{k_{z_1}}{\omega_1} + \frac{k_{z_2}}{\omega_2} + \frac{k_{z_3}}{\omega_3} \right) \tag{6.100}$$

In our case of the interaction among curvature-plasma waves, the shift of the frequency $\tilde{\omega}$ is so large that in the expression (6.99) the relativistic term proportional to γ^2 becomes more significant ($\tilde{\omega}/\omega \gg \gamma^{-2}$):

$$\sigma_z^N = \frac{3}{2} \frac{n_0 e^3}{m_e^2 \gamma^4 c} \frac{\omega_1}{\tilde{\omega}_1 \tilde{\omega}_2 \tilde{\omega}_3} \tag{6.101}$$

From (6.99) it follows that it is only the longitudinal components of the electric fields of waves that interact. Unstable modes are mainly polarized across the external magnetic field (see (6.88)), and therefore in the homogeneous field approximation their interaction (6.101) is strongly suppressed.

Let us now determine the stationary spectrum obtained at the expense of transfer of the energy of unstable modes from the amplification cone as a result of the non-linear interaction. The equation that takes into account the non-linear current looks like

$$(n_1^2\delta_{\alpha\beta} - n_{1_\alpha}n_{1_\beta} - \varepsilon_{\alpha\beta})E_{1_\beta} = \frac{4\pi i}{\omega_1}\sigma_\alpha^N E_{z_2}E_{z_3} \tag{6.102}$$

where on the right-hand side we have the non-linear conductivity, which has only one longitudinal component.

On the left-hand side of (6.102) it is necessary to take into account the imaginary part of the frequency shift $\tilde{\omega}_1$ responsible for the instability of curvature-plasma waves. The dielectric permittivity tensor $\varepsilon_{\alpha\beta}$ should therefore be taken with allowance for the magnetic field curvature. Substituting $\varepsilon_{\alpha\beta}$ from (6.61), we reduce (6.102) to the form

$$\left(\frac{4\tilde{\omega}_1^5\rho_l^2}{c^2\omega_p^2\omega_1}\langle\gamma^{-3}\rangle^{-1} - 1\right)E_{z_1} = \tfrac{3}{2}i\frac{e}{m_e c}\langle\gamma^{-3}\rangle^{-1}\langle\gamma^{-4}\rangle\frac{\tilde{\omega}_1}{\tilde{\omega}_2\tilde{\omega}_3}E_{z_2}E_{z_3} \tag{6.103}$$

In the linear approximation we observe a wave amplification with the increment $\tilde{\omega}_1$ (6.78). To solve the non-linear problem, we go over to the spectral decomposition of the field:

$$\mathbf{E} = \int \mathbf{E}(\mathbf{k}, \mathbf{r}, t)\exp\{-i\omega t + i\mathbf{kr})\,d\mathbf{k}$$

Then using the relation $\tilde{\omega} \simeq [\omega_p^2c^2\omega\langle\gamma^{-3}\rangle\rho_l^{-2}]^{1/5}$ we have from (6.103) for spectral density $E_z(\omega, \theta)$

$$E_z(\omega_1, \theta) = \frac{e}{m_e c}\langle\gamma^{-3}\rangle^{-1}\langle\gamma^{-4}\rangle\omega_1^{6/5}\left(\frac{\omega_p^2c^2}{\rho_l^2}\langle\gamma^{-3}\rangle\right)^{-1/5}$$
$$\times\int\frac{E_z(\omega_2, \theta_2)E_z(\omega_1 - \omega_2, \theta_1 - \theta_2)}{\omega_2^{1/5}(\omega_1 - \omega_2)^{6/5}}\,d\omega_2\,d\theta_2 \tag{6.104}$$

Now assuming for simplicity that inside the amplification cone the wave amplitude does not depend on the angle θ, and taking into account that the integration limits in (6.104) do not depend on ω_1, we make sure that this equation can have power-like solutions of the form $E_z \propto \omega^{6/5}$. Then from (6.104) we obtain

$$E_z(\omega, \theta) = \frac{m_e c}{e} \langle \gamma^{-4} \rangle^{-1} \langle \gamma^{-3} \rangle^{4/5} \omega_p^{-2/5} \left(\frac{c}{\rho_l} \right)^{3/5} \frac{\omega^{6/5}}{\omega_{\max}^{7/5}} \tag{6.105}$$

where $\omega_{\max}$ is the maximum frequency of excited curvature-plasma waves, i.e. the value of the frequency for which the hydrodynamic instability disappears:

$$\frac{\tilde{\omega}}{\omega} = \tfrac{1}{2}\gamma_{\min}^{-2}$$

According to (6.78),

$$\omega_{\max} \simeq \gamma_{\min}^{7/4} \left(\frac{\omega_p c}{\rho_l} \right)^{1/2} \tag{6.106}$$

The condition (6.106) coincides naturally with the boundary (6.88) between regions I and II.

Thus, the three-wave interaction actually leads to stabilization of the instability and brings the spectrum to the stationary level. The energy density of stabilized oscillations is small compared with the energy density of the plasma (for more details see Section 6.3.4). The weak turbulence approximation is therefore sufficient here. But the question of the influence of magnetic field curvature upon the process of non-linear interaction of waves remains open. Indeed, in the presence of inhomogeneity there may be interaction not only of the longitudinal components E_z of the electric fields of waves, but also the transverse components E_x which, as we have seen (6.79), are largest for curvature-plasma waves. It is therefore necessary to consider the non-linear effects with allowance made for inhomogeneity of the medium, which is our prime concern in the following sections of this chapter.

6.3.3 Non-linear interaction in an inhomogeneous plasma

The quasi-linear approximation First of all we consider a quasilinear approximation taking into account the inverse effect of electromagnetic oscillations upon the evolution of the distribution functions of charged particles. The quasilinear equation for the case of a weakly inhomogeneous plasma (the conditions (6.39)) has the following form (Istomin, 1988):

$$\frac{dF^{\pm}}{dt} = \frac{e^2}{4}\sum_{\mathbf{k}} E_{\mu}^{0*}(\mathbf{k})E_{\lambda}^{0}(\mathbf{k})$$

$$\times\left\{\delta_{\mu\sigma}\left(1-\frac{\mathbf{kv}}{\omega}\right)+\frac{k_{\sigma}v_{\mu}}{\omega}-\frac{i}{2\omega}\delta_{\mu\sigma}v_{\alpha}\frac{\partial}{\partial r_{\alpha}}+\frac{i}{2\omega}v_{\mu}\frac{\partial}{\partial r_{\sigma}}\right\}$$

$$\times\frac{\partial}{\partial p_{\sigma}}\int_{-\infty}^{t} dt' \exp[i\omega(t-t')-i\mathbf{kR}^*]$$

$$\times \det{}^{-1}\left|\delta_{\alpha\beta}-\frac{\partial L_{\alpha}(\mathbf{r}+\mathbf{R}^*/2)}{2\,\partial r_{\beta}}\right|$$

$$\times\left[\delta_{\lambda\nu}\left(1-\frac{\mathbf{kv}'}{\omega}\right)+\frac{k_{\nu}v'_{\lambda}}{\omega}-\frac{i}{2\omega}\delta_{\lambda\nu}\frac{\partial}{\partial r_{\chi}}v'_{\chi}+\frac{i}{2\omega}\frac{\partial}{\partial r_{\nu}}v'_{\lambda}\right]\frac{\partial F^{\pm}}{\partial p'_{\nu}}+\text{c.c.} \tag{6.107}$$

The notation is here the same as in Section 6.1.1 (formulas (6.42)–(6.46)). The quantity $\mathbf{E}^0(\mathbf{k})$ is the amplitude of an electromagnetic wave with the local wave number $\mathbf{k}(\mathbf{r})$ determined by the wave phase $\psi_p(\mathbf{r})$

$$\mathbf{E}(\mathbf{r}) = \sum_{\mathbf{k}} \mathbf{E}^0(\mathbf{k})\exp[i\psi_p(\mathbf{r})], \qquad \mathbf{k} = \nabla\psi_p$$

The evolution of the distribution function described by equation (6.107) is due to the excitation of curvature-plasma oscillations. Since waves are emitted along the direction of particle motion with a small spread over angles $\Delta\theta \sim 1/\gamma$, the distribution function $F^{\pm}(\mathbf{r}, \mathbf{p}, t)$ also acquires a spread over the transverse momenta p_x. The quantity $p_x/p_{\parallel} \sim \gamma^{-1}$ is, however, not large, so that equation (6.107) can be integrated over p_x. As a result, taking into account the relation (6.49), we obtain the equation for the evolution of the longitudinal distribution function $F_{\parallel}^{\pm}(p_{\parallel})$:

$$\frac{dF_{\parallel}^{\pm}}{dt} = \frac{e^2}{4}\sum_{\mathbf{k}} E_{\mu}^{0*}(\mathbf{k})E_{\lambda}^{0}(\mathbf{k})\int_0^{\infty} d\tau\left(b_{\mu}+n_{\mu}\frac{v_{\parallel}}{2\rho_l}\tau\right)\frac{\partial}{\partial p_{\parallel}}$$

$$\times\left\{\exp\left[i(\omega-\mathbf{kb}v_{\parallel})\tau+i\mathbf{bk}\frac{v_{\parallel}^3}{24\rho_l^2}\tau^3\right]\right.$$

$$\left.\times\left(b_{\lambda}-n_{\lambda}\frac{v_{\parallel}}{2\rho_l}\tau\right)\frac{\partial F_{\parallel}^{\pm}}{\partial p_{\parallel}}\right\}+\text{c.c.} \tag{6.108}$$

From (6.108) it is seen that the relaxation of the longitudinal distribution function is due not only to the longitudinal components of the wave field (as in a homogeneous magnetic field), but also to the transverse components, the contribution of the transverse components being of the same order as the contribution of the longitudinal ones by virtue of (6.79). This is a result of non-locality of the interaction between resonance particles and waves, when at a given point the effect is influenced by the nearest neighbourhood of the point, where the transverse component of the electric field of the wave has a non-zero projection onto the direction of the magnetic field owing to its curvature. The right-hand side of (6.108) contains the Airy function (and its derivatives)

$$\mathrm{Ai}\left[2(\omega - k_z v_{\parallel})\frac{\rho_l^{2/3}}{k_z^{1/3} v_{\parallel}}\right]$$

as the dielectric permittivity tensor $\varepsilon_{\alpha\beta}$ (6.58), which determines the increment of unstable curvature-plasma waves. Instability exists only within the hydrodynamic limit, when the characteristic spread of the distribution function over the longitudinal velocities is not large, so that

$$\frac{1}{2\gamma^2} < \frac{\tilde{\omega}}{\omega} = |1 - n|$$

This means that after the stationary state is established, i.e. the right-hand side of (6.108) vanishes, the distribution function of particles must be cut off when $\gamma > \gamma_{\mathrm{cr}}$, where

$$\gamma_{\mathrm{cr}} \simeq \left(\frac{\gamma_{\min}^{3/2}\omega^2\rho_l}{\omega_p c}\right)^{1/5} \simeq \gamma_{\min}\left(\frac{\omega}{\omega_c}\right)^{1/5} A_p^{-1/5}$$

The quantity γ_{cr} for the characteristic frequencies of curvature radiation, $\omega \simeq \omega_c$ (3.31), is of the order of $\gamma_{\min}$. Thus, a quasi-linear relaxation should have led to a noticeable deceleration of a relativistic plasma flux and the corresponding stabilization at a high wave energy. However, as we have seen above, the non-linear interaction of waves yields instability saturation at a lower level. Therefore, the quasi-linear effects appear to be insignificant.

The three-wave interaction We shall now dwell on a three-wave interaction analogous to the one discussed in Section 6.3.2 in the quasi-homogeneous plasma approximation. Let there exist three waves with frequencies ω_1, ω_2, ω_3, and let their electric fields be, respectively,

$$\mathbf{E}_i^0(\mathbf{r}) \exp\{i\psi_{p_i}(\mathbf{r}) - i\omega_i t\}, \qquad i = 1, 2, 3 \tag{6.109}$$

The local wave numbers are equal to

$$\mathbf{k}_i = \nabla\psi_{p_i} = \mathbf{k}_i(\omega_i, \mathbf{r}) \tag{6.110}$$

Let us consider the case where $\omega_1 \simeq \omega_2 + \omega_3$. For the frequency $\omega_2 + \omega_3$, the non-linear current $\mathbf{j}$ describing the non-linear interaction of waves is given by

$$j_\alpha = \sigma^N_{\alpha\sigma\lambda} E^0_{\sigma_2} E^0_{\lambda_3} \exp[i(\psi_{p_2} + \psi_{p_3}) - i(\omega_2 + \omega_3)t] \tag{6.111}$$

Here $\sigma^N_{\alpha\sigma\lambda}$ is the non-linear plasma conductivity

$$\sigma^N_{\alpha\sigma\lambda} = e \int d\mathbf{p}\, v_\alpha[\hat{M}_\sigma(\omega_2 + \omega_3, \mathbf{k}_2 + \mathbf{k}_3)\hat{M}_\lambda(\omega_3, \mathbf{k}_3)$$
$$+ \hat{M}_\lambda(\omega_2 + \omega_3, \mathbf{k}_2 + \mathbf{k}_3)\hat{M}_\sigma(\omega_2, \mathbf{k}_2)]F^\pm(\mathbf{p}, \mathbf{r}) \tag{6.112}$$

where the operators $\hat{M}_\chi(\omega, \mathbf{k})$, are defined as follows:

$$\hat{M}_\chi(\omega, \mathbf{k})$$
$$= -\frac{e}{2}\int_{-\infty}^{t} dt' \exp[i\omega(t - t') - i\mathbf{k}\mathbf{R}^*] \det{}^{-1}\left|\delta_{\alpha\beta} - \frac{\partial L_\alpha(\mathbf{r} + \mathbf{R}^*/2)}{2\,\partial r_\beta}\right|$$
$$\times \left[\delta_{\nu\chi}\left(1 - \frac{\mathbf{k}\mathbf{v}'}{\omega}\right) + \frac{k_\nu v'_\chi}{\omega} + \frac{i}{2\omega}\frac{\partial}{\partial r_\nu} v'_\chi - \frac{i}{2\omega}\delta_{\nu\chi}\frac{\partial}{\partial r_\mu} v'_\mu\right]\frac{\partial}{\partial p'_\nu}\bigg|_{\mathbf{r}+\mathbf{R}^*/2} \tag{6.113}$$

The non-linear response at frequencies $\omega_1 - \omega_2$ and $\omega_1 - \omega_3$ can be written in a similar way.

For the distribution function of the form (6.48), ($\mathbf{p}_{\mathrm{dr}} \to 0$), $\hat{M}_\chi(\omega, \mathbf{k})$ is equal to

$$\hat{M}_\chi(\omega, \mathbf{k}) = -\frac{e}{2}\int_0^\infty d\tau \int_{-\infty}^{\infty} dp_\parallel \exp\left[i(\omega - \mathbf{k}\mathbf{b}v_\parallel)\tau + \frac{i}{24}\mathbf{k}\mathbf{b}\frac{v_\parallel^3}{\rho_l^2}\tau^3\right]$$
$$\times \left(b_\chi - n_\chi\frac{v_\parallel}{2\rho_l}\tau\right)\delta\left(\mathbf{p} - p_\parallel\mathbf{b} - \mathbf{n}p_\parallel\frac{v_\parallel}{2\rho_l}\tau\right)\frac{\partial}{\partial p_\parallel} \tag{6.114}$$

Substituting (6.114) into (6.112), we find

$$\sigma^N_{\alpha\sigma\lambda} = \frac{e^3}{4}\int_{-\infty}^{\infty} dp_{\parallel} v_{\parallel} \int_{-\infty}^{\infty}\int d\tau\, d\tau' \left[b_\alpha + n_\alpha \frac{v_\parallel}{2\rho_l}(\tau + \tau')\right]$$

$$\times \left[b_\sigma - n_\sigma \frac{v_\parallel}{2\rho_l}(\tau - \tau')\right] \exp[i(\omega_2 - \omega_3)\tau' - i(\mathbf{k}_2 - \mathbf{k}_3)\mathbf{R}^*_1] \frac{\partial}{\partial p_\parallel}$$

$$\times \left\{ \left[b_\lambda - n_\lambda \frac{v_\parallel}{2\rho_l}(\tau - \tau')\right][\exp(i\omega_3\tau - i\mathbf{k}_3\mathbf{R}^*) + \exp(i\omega_2\tau - i\mathbf{k}_2\mathbf{R}^*)] \right\} \frac{\partial F_\parallel}{\partial p_\parallel} \tag{6.115}$$

$$\mathbf{R}^*_1 = \mathbf{b}\left(v_\parallel \tau' - \frac{1}{24}\frac{v_\parallel^3}{\rho_l^2}\tau'^3 \right)$$

$$\mathbf{R}^* = \left(\mathbf{b} + \mathbf{n}\frac{v_\parallel}{2\rho_l}\tau' - \mathbf{b}\frac{v_\parallel^2}{8\rho_l^2}\tau'^2 \right) v_\parallel \tau - \mathbf{b}\frac{v_\parallel^3}{24\rho_l^2}\tau^3$$

6.3.4 The stationary spectrum of unstable modes

We shall consider a three-wave interaction of unstable curvature-plasma waves ($j = 5, 6$; formula (6.75)) among themselves and with plasma modes ($j = 2, 3$ (6.74)). The laws of conservation of momentum $\mathbf{k}_{5,6} = \mathbf{k}_2 + \mathbf{k}_3$ and of energy $\omega_{5,6} = \omega_2 + \omega_3$ are obeyed in the latter case only in a narrow cone of angles $k_x/k < \theta_\parallel$, $k_y/k < \theta_\perp$ (see (6.73) and (6.82)). Therefore in the entire cone of angles $(\theta_\parallel, \theta_\perp)$ (see Fig. 6.7), where the oscillation excitation is most effective, only the interaction of unstable curvature-plasma waves among themselves is effective.

The unstable modes have the dispersion law

$$\omega = k_z c + \tilde{\omega} \tag{6.116}$$

where the quantity $\tilde{\omega}$ is given by (6.71) and (6.75). They can be described as a single drift wave but with a complex amplitude rapidly changing in time:

$$\mathbf{E}_l = \mathbf{E}_l(\mathbf{r}, t)\exp[-i\omega_l t + i\psi_l(\mathbf{r})] \qquad \mathbf{k}_l = \nabla\psi_l, \qquad \omega_l = k_{zl}c$$

In view of the hydrodynamic character of instability, when $\mathrm{Re}\,\tilde{\omega} \simeq \mathrm{Im}\,\tilde{\omega}$, the excited waves are in fact monochromatic, and their electric field can be represented as the integral

$$\mathbf{E}_l = \int \mathbf{E}_l(\mathbf{k}_l, \mathbf{r}, t)\exp[-i\omega_l t + i\psi_l(\mathbf{r})]\, d\mathbf{k}_l \tag{6.117}$$

Individual harmonics can interact with one another in any order of non-linearity since the dispersion $\omega \simeq k_z c$ implies a strong connection between them. But in view of the fact that they exist only in a narrow cone of angles along the direction of the magnetic field, the three-wave interaction is the most effective because the phase volume occupied by these harmonics is small.

The calculation of the non-linear conductivity (6.115) is greatly simplified because for unstable modes we have $\tilde{\omega}/\omega \gg (c/\rho_l\omega)^{2/3}$ ($a \gg 1$) and the quantities $\mathbf{R}_1^*$ and $\mathbf{R}^*$ in the expression (6.115) can respectively be put equal to $\mathbf{R}_1^* = \mathbf{b}v_\parallel\tau'$ and $\mathbf{R}^* = \mathbf{b}v_\parallel\tau$ (as in a homogeneous medium). As a result we have

$$\sigma^N_{\alpha\sigma\lambda} = \frac{1}{2}\frac{e^3 n_e}{m_e^2}\omega_1\left\langle \frac{\text{sign}(e)}{v_\parallel\gamma^4}\left(3 + \frac{\omega}{\gamma^2}\frac{\partial}{\partial\tilde{\omega}}\right)\frac{1}{\tilde{\omega}_1^2}\right.$$
$$\times\left[b_\alpha b_\sigma b_\lambda I_1 + ib_\alpha(b_\sigma n_\lambda - n_\sigma b_\lambda)\frac{v_\parallel}{\rho_l}I_2 + in_\alpha b_\sigma b_\lambda\frac{v_\parallel}{\rho_l}I_3\right.$$
$$+ b_\alpha n_\sigma n_\lambda\frac{v_\parallel^2}{\rho_l^2}I_4 + n_\alpha(b_\sigma n_\lambda - b_\lambda n_\sigma)\frac{v_\parallel^2}{\rho_l^2}I_5$$
$$\left.\left. + in_\alpha n_\sigma n_\lambda\frac{v_\parallel^3}{\rho_l^3}I_6\right]\right\rangle \tag{6.118}$$

$$I_1 = \frac{1}{\tilde{\omega}_2} + \frac{1}{\tilde{\omega}_3}$$

$$I_2 = \frac{1}{\tilde{\omega}_1\tilde{\omega}_2} + \frac{1}{\tilde{\omega}_1\tilde{\omega}_3} - \frac{1}{2\tilde{\omega}_2^2} - \frac{1}{2\tilde{\omega}_3^2}$$

$$I_3 = \frac{1}{\tilde{\omega}_1\tilde{\omega}_2} + \frac{1}{\tilde{\omega}_1\tilde{\omega}_3} + \frac{1}{2\tilde{\omega}_2^2} + \frac{1}{2\tilde{\omega}_3^2}$$

$$I_4 = \frac{3}{2}\frac{1}{\tilde{\omega}_1^2\tilde{\omega}_2} + \frac{3}{2}\frac{1}{\tilde{\omega}_1^2\tilde{\omega}_3} - \frac{1}{\tilde{\omega}_1\tilde{\omega}_2^2} - \frac{1}{\tilde{\omega}_1\tilde{\omega}_3^2} + \frac{1}{2\tilde{\omega}_2^3} + \frac{1}{2\tilde{\omega}_3^3}$$

$$I_5 = \frac{1}{2\tilde{\omega}_2^3} + \frac{1}{2\tilde{\omega}_3^3} - \frac{3}{2}\frac{1}{\tilde{\omega}_1^2\tilde{\omega}_2} - \frac{3}{2}\frac{1}{\tilde{\omega}_1^2\tilde{\omega}_3}$$

$$I_6 = \frac{3}{4}\frac{1}{\tilde{\omega}_2^4} + \frac{3}{4}\frac{1}{\tilde{\omega}_3^4} - \frac{3}{4}\frac{1}{\tilde{\omega}_1^2\tilde{\omega}_2^2} - \frac{3}{4}\frac{1}{\tilde{\omega}_1^2\tilde{\omega}_3^2}$$
$$- \frac{1}{2}\frac{1}{\tilde{\omega}_1\tilde{\omega}_2^3} - \frac{1}{2}\frac{1}{\tilde{\omega}_1\tilde{\omega}_3^3} + 3\frac{1}{\tilde{\omega}_1^3\tilde{\omega}_2} + 3\frac{1}{\tilde{\omega}_1^3\tilde{\omega}_3}$$

In a homogeneous magnetic field only the longitudinal component of non-linear conductivity is non-zero. The expression (6.118) then becomes

$$\sigma_z^N = \frac{1}{2}\frac{e^3 n_e}{m_e^2}\frac{\omega_1}{v_\parallel \gamma^4}\left(3 + \frac{\omega}{\gamma^2}\frac{\partial}{\partial\tilde{\omega}}\right)\frac{1}{\tilde{\omega}_1\tilde{\omega}_2\tilde{\omega}_3}$$

which coincides with (6.99).

In the curvilinear field in (6.116) we have, in fact, a power series expansion of the quantity $c/\rho_l\tilde{\omega} \ll 1$. But since the transverse component of the electric field of unstable modes is $\rho_l\tilde{\omega}/c$ times as large as the longitudinal one (see (6.79)), all the terms in the expression (6.118) are of the same order of magnitude. Substituting (6.118) into the dispersion equation which takes into account the non-linear conductivity,

$$(n_1^2\delta_{\alpha\beta} - n_{1_\alpha}n_{1_\beta} - \varepsilon_{\alpha\beta})E_{1_\beta} = \frac{4\pi i}{\omega_1}\sigma^N_{\alpha\sigma\lambda}E_{2_\sigma}E_{3_\lambda}$$

we obtain

$$\begin{aligned}
&\left(\frac{4\tilde{\omega}_1^5\rho_l^2}{c^2\omega_p^2\omega_1}\langle\gamma^{-3}\rangle^{-1} - 1\right)E_{z_1} \\
&\quad = \frac{3}{2}i\frac{e}{m_e c}\langle\gamma^{-3}\rangle^{-1}\langle\gamma^{-4}\rangle E_{z_2}E_{z_3} \\
&\quad \times\left\{\tfrac{1}{2}\tilde{\omega}_1\left(\frac{\tilde{\omega}_3}{\tilde{\omega}_2^3} + \frac{\tilde{\omega}_2}{\tilde{\omega}_3^3}\right) + 2\left(\frac{1}{\tilde{\omega}_2} + \frac{1}{\tilde{\omega}_3} + \frac{1}{2}\frac{\tilde{\omega}_3}{\tilde{\omega}_2^2} + \frac{1}{2}\frac{\tilde{\omega}_2}{\tilde{\omega}_3^2}\right)\right. \\
&\quad - \frac{3}{2}\frac{1}{\tilde{\omega}_1}\left(\frac{\tilde{\omega}_3}{\tilde{\omega}_2} + \frac{\tilde{\omega}_2}{\tilde{\omega}_3}\right) + \left(\frac{4\tilde{\omega}_1^5\rho_l^2}{c^2\omega_p^2\omega_1}\langle\gamma^{-3}\rangle^{-1} - 3\right) \\
&\quad \left.\times\left[\frac{2}{\tilde{\omega}_1} + \frac{1}{2}\left(\frac{1}{\tilde{\omega}_2} + \frac{1}{\tilde{\omega}_3}\right)\right]\right\} \\
&\quad \times \exp[i(\omega_1 - \omega_2 - \omega_3)t - i(\psi_{p_1} - \psi_{p_2} - \psi_{p_3})] \qquad (6.119)
\end{aligned}$$

The quantity $\tilde{\omega}$ is proportional here to the derivative with respect to time which acts on the amplitudes of the wave packets: $\tilde{\omega} = i\,\partial/\partial t$.

In the left-hand side of (6.119) we neglect the turn of the wave vector at the expense of inhomogeneity ($d\theta/dt \simeq c/\rho_l$) since in the region $a \gg 1$

$$\frac{\rho_l}{c}|\tilde{\omega}| \gg \theta_\parallel^{-1}$$

Now going over to the spectral expansion (6.117) and solving equation (6.119), we obtain approximately

$$E_{z_{l_\omega}}(\theta) = \frac{m_e c}{e} \langle\gamma^{-4}\rangle^{-1} \langle\gamma^{-3}\rangle^{4/5} \omega_p^{-2/5} \left(\frac{c}{\rho_l}\right)^{3/5} \frac{\omega^{4/5}}{\omega_{\max}} \tag{6.120}$$

Here $\omega_{\max}$ is the maximal value of the frequency of the excited curvature-plasma mode (6.106). The wave amplitude $E_{l_\omega}(\theta)$ is constant inside the amplification cone and is equal to zero outside it.

The expansion (6.120) is close enough to the expression (6.105) obtained in the approximation of a homogeneous magnetic field. But formulas (6.120) and (6.105) have different spectral indices. An account of the influence of the magnetic field curvature upon the three-wave interaction thus leads to variation of the form of the curvature-plasma wave spectrum.

According to (6.79), the total spectral amplitude is

$$\begin{aligned} E_{l_\omega} = E_{x_{l_\omega}} &= \frac{\rho_l}{c} |\tilde{\omega}| E_{z_{l_\omega}}(\theta) \theta_{\parallel} \\ &= \frac{m_e c}{e} \langle\gamma^{-3}\rangle^{7/5} \langle\gamma^{-4}\rangle^{-1} \omega_p^{4/5} \left(\frac{c}{\rho_l}\right)^{-1/5} \omega^{2/5} \omega_{\max}^{-1} \end{aligned} \tag{6.121}$$

The saturation of instability up to the level determined by the state (6.121) proceeds within a short time of the order of $|\tilde{\omega}|^{-1}$. The corresponding length $c/|\tilde{\omega}|$ is small compared to the curvature radius, and therefore we may assume the stationary spectrum (6.121) to be reached, in fact, at each point of the space where there exists instability. The spectral energy density U_ω can be determined from the fact that the instability increment $|\tilde{\omega}|$ also specifies the time of wave coherence. Since $|\tilde{\omega}| \ll \omega_{\max}$, for the energy density U_ω we obtain the estimate

$$U_\omega \simeq \frac{1}{8\pi} E_{l_\omega}^2 |\tilde{\omega}| \simeq \frac{1}{8\pi} \left(\frac{m_e c}{e}\right)^2 \langle\gamma^{-3}\rangle^3 \langle\gamma^{-4}\rangle^{-2} \omega_p^2 \omega \omega_{\max}^{-2} \tag{6.122}$$

The total energy density in the entire frequency range then equals

$$U = \int U_\omega \, d\omega \simeq n_e m_e c^2 \langle\gamma\rangle \gamma_{\min}^{-2} \simeq \lambda^{-1} W_p \tag{6.123}$$

where $W_p = m_e c^2 n_e \langle\gamma\rangle$ is the energy density of particles.

Thus, we can see that the multiplicity factor $\lambda \sim 10^3$–10^5 determines the transformation coefficient for the energy of relativistic plasma into the energy of unstable curvature-plasma oscillations. It is a universal quantity depending only on the specificities of plasma generation in each individual pulsar. The transformation coefficient $\eta_{tr} = U/m_e c^2 n_e \langle\gamma\rangle$ is small and makes up about 10^{-3}–10^{-5}, which provides for the validity of the weak turbulence approximation (see Section 6.3).

6.3.5 Mode transformation

We now consider the question of the energy transformation from unstable curvature-plasma modes into normal waves $j = 1, 2$. The study of such a transformation is of importance since, as we shall see below, it is only these waves that can leave the pulsar magnetosphere. First of all we shall show that in a curved magnetic field the energy of unstable modes can be transformed into an extraordinary wave $j = 1$. As a result of a non-linear interaction of unstable waves $j = 5, 6$ among themselves, their frequency shift $\tilde{\omega}$ must tend to zero since the saturation is characterized by the condition $\partial/\partial t = 0$. This means that in the stationary state the normal modes $j = 5, 6$ are, in fact, indistinguishable from the mode $j = 4$ and, therefore, the energy density $U^{(4)}_{\omega}(\theta) \simeq U^{(4)}_{\omega}/\theta_{\parallel}$ of a normal wave $j = 4$ also reaches the values given by relation (6.122).

On the other hand, as shown in Section 6.2.2 for angles $\theta \sim \theta_{\perp}$ there occurs a linear transformation of the mode $j = 4$ into an ordinary wave $j = 1$. Since in a curved magnetic field the angle θ between the vectors **k** and **B** increases with characteristic velocity $d\theta/dt = c/\rho_l$, the energy of the normal wave $j = 4$ from the region $\theta < \theta_{\perp}$ will be permanently pumped into a normal wave $j = 1$ from angles $\theta > \theta_{\perp}$, as is seen in Fig. 6.8. Given this, the energy flow-off from the region $\theta < \theta_{\perp}$ will in no way affect the magnitude of the stationary spectrum (6.122) since the characteristic time of the variation of the angle $\tau_{\theta} \simeq \rho_l \theta_{\perp}/c$ greatly exceeds the oscillation stabilization time $\tau_{in} = |\tilde{\omega}|^{-1}$. Indeed, as can easily be verified, the condition $\tau_{\theta} \gg \tau_{in}$ coincides, according to (6.71) and (6.75), with the condition $a \gg 1$, which always holds in region I. Thus, we can see that owing to a permanent transformation of the mode $j = 4$, in a curved magnetic field there proceeds an effective generation of a transverse extraordinary electromagnetic wave.

Next, the energy transformation into an ordinary wave $j = 2$ $(\theta > \theta^*)$, can proceed at the expense of the decay of unstable modes $j = 5, 6$ into two plasma modes $j = 2$ and $j = 3$ $(\theta < \theta^*)$. This decay, as has already been mentioned, takes place at small angles $\theta < \theta_{\perp}$. But such a decay does not occur, because the low-frequency mode $j = 3$ is a mode of negative energy (Wilhelmson *et al.*, 1970). Indeed, if such a decay were realized, it would lead to increasing energy of a normal wave which has a negative energy. This energy increase would mean the fall of the amplitude to zero, which would result in a complete cessation of the three-wave interaction.

Nevertheless, an effective energy conversion into an ordinary wave $j = 2$ is still possible. The three-wave interaction causing the formation of a stationary spectrum of unstable modes has been considered above, in fact, within the framework of hydrodynamics, when the frequency shift $\tilde{\omega}$ exceeds greatly the kinetic velocity spread of particles:

$$\left|\frac{\tilde{\omega}}{\omega}\right| \gg \tfrac{1}{2}\gamma^{-2}$$

This condition holds well in the linear case. But an account of wave interaction leads to a decrease in the frequency shift; in particular, for a curvature-plasma wave the three-wave interaction appears to result in $\tilde{\omega} \to 0$. The stationary spectrum thus formed is the plasma density modulation, which drifts together with the plasma along the magnetic field lines. We should take into account the kinetic velocity spread of charged particles: when scattering on density inhomogeneities they radiate electromagnetic waves. The polarization of the inhomogeneities is such that waves are radiated with an electric field component lying in the plane of the magnetic field line. This is just the ordinary mode $j = 2$. This is, in fact, the scattering of a plasma wave on an inhomogeneity created by curvature-plasma oscillation: $\omega_2 = \omega_{5,6} + \omega_2'$. This process is most effective for waves propagating almost along the direction of the magnetic field, $\theta_2 \lesssim \theta^*$ (6.25).

Using expression (6.115) for non-linear conductivity under the condition $\tilde{\omega}_2 = \omega_p\langle\gamma^{-3}\rangle^{1/2} \gg \omega_2/2\gamma^2$ and $\tilde{\omega}_{5,6} = 0$, we obtain

$$\frac{\partial E_{z_2}}{\partial t} = \pi\frac{e}{m_e}\,\omega_p\langle\gamma^{-3}\rangle^{-1/2}|E_{z_l}|E'_{z_2}\frac{\rho_l^{2/3}}{\omega_l^{1/3}c^{5/3}}\left\langle\frac{1}{\gamma^4}\,[3\,\mathrm{Ai}(\xi) + 2\xi\,\mathrm{Ai}'(\xi)]\right\rangle$$

$$\xi = \left(\frac{\omega_l\rho_l}{c\gamma^3}\right)^{2/3} \qquad (6.124)$$

For a given wave E_l, equation (6.124) describes the plasma wave excitation. Knowing the spectrum of the curvature-plasma mode (6.120), we find from (6.124) the characteristic excitation increment for an ordinary wave:

$$\Gamma_2 \simeq \omega_p \left(\frac{c}{\omega_p \rho_l} \right)^{2/5} \gamma_{\min}^{-1/2} \tag{6.125}$$

The increment Γ_2 has the order $\tilde{\omega}$ for unstable waves $j = 5, 6$ with a characteristic frequency $\omega \simeq \omega_p \gamma_{\min}^{-3/2}$. This means that the excitation of an ordinary wave is as rapid as the linear growth of unstable modes $j = 5, 6$.

We now determine the energy density of the ordinary wave $j = 2$ in the range of angles $\theta \lesssim \theta^*$. This can be done easily by assuming that the increase in the amplitude of this wave will stop as soon as its inverse effect upon the curvature-plasma wave becomes significant. A calculation similar to (6.124) leads to the following relation for 'monochromatic' amplitudes:

$$(n^2 \delta_{\alpha\beta} - n_\alpha n_\beta - \varepsilon_{\alpha\beta}) E_{l_\beta} = -2\pi\omega_p \langle \gamma^{-3} \rangle^{-1/2} \frac{e}{m_e c} \frac{\rho_l^{4/3}}{\omega^{2/3} c^{4/3}} E_{z_2} E_{z_2}^*$$
$$\times \left\langle \frac{\text{sign } e}{\gamma^4} (3\,\text{Ai}' + 2\xi\,\text{Ai}'') \right\rangle \tag{6.126}$$

where $\xi = (\omega \rho_l / c\gamma^3)^{2/3}$. Now going over to the spectral amplitudes $E_\omega(\Omega_S) = E_\omega(\theta)/\theta_\perp$ and using (6.82), (6.120), we obtain

$$\mathscr{I} \cdot \mathscr{I}_1 \simeq \langle \gamma^{-3} \rangle^{11/5} \langle \gamma^{-4} \rangle^{-1} \left(\frac{\tilde{\omega}}{\omega} \right)^{-2} \left(\frac{m_e c}{e} \right)^2 \left(\frac{c}{\rho_l} \right)^{26/15} \omega_p^{2/5} \frac{\omega^{-2/15}}{\omega_{\max}} \tag{6.127}$$

where

$$\mathscr{I} = \int d\omega'\, d\theta'^2 \frac{\omega^2}{(\omega - \omega')^2} E_{2_{\omega'}}(\Omega_S') E_{2_{\omega-\omega'}}(\Omega_S - \Omega_S') \tag{6.128}$$

and

$$\mathscr{I}_1(\omega) = \left\langle \frac{\text{sign } e}{\gamma^3} [3\,\text{Ai}'(\xi) + 2\xi\,\text{Ai}''(\xi)] \right\rangle \tag{6.129}$$

First we shall consider the expression $\mathscr{I}_1$ in the angle brackets in (6.129). Clearly, the contributions of electrons and positrons will have opposite signs since the current quadratic in the amplitude is known to be proportional to the charge cubed. Remembering that within the range $\mathscr{E}_{\min}^{\pm} < \mathscr{E}^{\pm} < \mathscr{E}_{\max}^{\pm}$ both the components have a power-law spectrum with the index $\bar{\nu} \simeq 2$, we obtain for each of the components

$$\langle\ \rangle_{\pm} = \begin{cases} 3\,\mathrm{sign}(e)\,\mathrm{Ai}'(0)\langle\gamma^{-4}\rangle, & \omega \ll \dfrac{c}{\rho_l}(\gamma^{\pm}_{\min})^3 \\ c_1\gamma^{\pm}_{\min}\,\mathrm{sign}(e)\left(\dfrac{\omega\rho_l}{c}\right)^{-5/3}, & \dfrac{c}{\rho_l}(\gamma^{\pm}_{\min})^3 \ll \omega \ll \dfrac{c}{\rho_l}(\gamma^{\pm}_{\max})^3 \\ \sim 0, & \omega \gg \dfrac{c}{\rho_l}(\gamma^{\pm}_{\max})^3 \end{cases} \quad (6.130)$$

where

$$c_1 = \frac{1}{2}\int_0^{\infty} dx\, x^{3/2}(3\,\mathrm{Ai}' + 2x\,\mathrm{Ai}'') = \frac{1}{4\pi^{1/2}}\,3^{5/6}\Gamma(5/6) \simeq 0.40 > 0$$

Since $\mathrm{Ai}'(0) < 0$, the expression (6.130) is seen to have opposite signs for $\omega > (c/\rho_l)(\gamma^{\pm}_{\min})^3$ and $\omega < (c/\rho_l)(\gamma^{\pm}_{\min})^3$. Therefore, for instance, in our case, where $\gamma^{-}_{\min} < \gamma^{+}_{\min}$

$$\frac{c}{\rho_l}(\gamma^{-}_{\min})^3 < \omega < \frac{c}{\rho_l}(\gamma^{+}_{\min})^3 \quad (6.131)$$

the expression in the angle brackets will be equal to

$$\mathscr{I}_1(\omega) = \sum_{+,-}\langle\ \rangle_{\pm} = \begin{cases} c_1\gamma^{-}_{\min}(\omega\rho_l/c)^{-5/3}, & \omega < \omega_{\mathrm{br}} \\ 3\,\mathrm{Ai}'(0)\langle\gamma^{-4}\rangle_{+}, & \omega > \omega_{\mathrm{br}} \end{cases} \quad (6.132)$$

where

$$\omega_{\mathrm{br}} = \frac{c}{\rho_l}(\gamma^{-}_{\min})^3\left(\frac{\gamma^{+}_{\min}}{\gamma^{-}_{\min}}\right)^{12/5} \quad (6.133)$$

and the main contribution to (6.130) for $\omega < \omega_{\mathrm{br}}$ is made by the slow plasma component, while for $\omega > \omega_{\mathrm{br}}$ it is made by the fast component. Outside the frequency range (6.131), the expression (6.132) becomes negative, so that the stabilized state becomes impossible.

We now consider equation (6.127). We seek its solution in the form

$$E_{2_\omega}(\Omega_S) = E^A\omega^{\bar{\mu}}$$

The analysis of (6.127) shows that $\bar{\mu} = -2/15$ for $\omega > \omega_{\mathrm{br}}$ and $\bar{\mu} = 23/15$ for $\omega < \omega_{\mathrm{br}}$. Given this, the main contribution to the integral $\mathscr{I}$ determined by the relation (6.128) (in which integration over the angle θ' should be carried out up to the angle $\theta^*(\omega)$ (6.25)) will be made by the frequencies ω' concentrated near ω_{br}, where $\omega' \ll \omega$. Therefore, $E_{\omega-\omega'}(\Omega_S - \Omega'_S)$ in $\mathscr{I}$

could be taken outside the integral. We assume here for simplicity that $E_\omega(\Omega_S)$ does not depend on Ω_S in the range of angles $\theta \lesssim \theta^*$.

On the other hand, for $\omega' \ll \omega$ the integral $\mathscr{I}$ coincides with the spectral power of the square of the electric field amplitude, since

$$(E^2)_{\omega,\Omega_S} = \int d\omega'\, d\Omega'_S E_{\omega'}(\Omega'_S) E_{\omega-\omega'}(\Omega_S - \Omega'_S)$$

Therefore, the spectral energy density of an extraordinary wave, for which in the range of angles $\theta \lesssim \theta^*$ we have

$$U^{(2)} = \frac{1}{16\pi}\left[\frac{\partial}{\partial\omega}(\omega\varepsilon_{\alpha\beta})E_\alpha E_\beta^* + |B|^2\right] \simeq \frac{1}{8\pi}\,\omega\left\langle\frac{\omega_p^2}{\gamma^3}\right\rangle^{-1/2}|E_z|^2$$

can be estimated as

$$U_\omega^{(2)}(\Omega_S) = \frac{\omega_{\rm br}}{8\pi}\left\langle\frac{\omega_p^2}{\gamma^3}\right\rangle^{-1/2}\mathscr{I} \tag{6.134}$$

Using relations (6.127) and (6.134), we ultimately arrive at

$$U_\omega^{(2)}(\Omega_S) \simeq \left(\frac{\tilde{\omega}}{\omega}\right)^{-2}\langle\gamma^{-3}\rangle^{17/10}\langle\gamma^{-4}\rangle^{-1}\left(\frac{m_e c}{e}\right)^2$$
$$\times\left(\frac{c}{\rho_l}\right)^{26/15}\omega_p^{-3/5}\,\frac{\omega_{\rm br}}{\omega_{\rm max}}\,\omega^{-2/15}\mathscr{I}_1^{-1}(\omega) \tag{6.135}$$

where $\mathscr{I}_1(\omega)$ is given by (6.132). We see that the spectral energy density has a maximum for $\omega \sim \omega_{\rm br}$, and $U_\omega^{(2)}(\Omega_S) \propto \omega^{23/15}$ for $\omega < \omega_{\rm br}$, and $U_\omega^{(2)}(\Omega_S) \propto \omega^{-2/15}$ for $\omega > \omega_{\rm br}$. Finally, integrating (6.135) over $d\Omega_S = \pi\, d\theta^2$ and over the frequency, we obtain for the total transformation coefficient

$$\eta_{\rm tr}^{(2)} = \frac{U^{(2)}}{W_p} \simeq \gamma_c^{5/4}\left(\frac{\tilde{\omega}}{\omega}\right)^{-2}\left(\frac{c}{\rho_{l_{\rm min}}\omega_{p0}}\right)^{21/10}\left(\frac{\gamma_{\rm min}^+}{\gamma_{\rm min}^-}\right)^{152/25} \tag{6.136}$$

Here $U^{(2)}$ is the total radiation energy and W_p is the energy density of particles.

7
Comparison of theory with observational data

The theory we have constructed can be compared with observational results. But for the consideration of concrete processes, the theory in some cases should be detailed. In this chapter, we therefore make both a theoretical analysis, which allows us to obtain concrete calculational formulas, and a direct comparison of the theory with experiment. We analyse the structure of the active region (Section 7.1), the generation of electron–positron plasma and of high-frequency radiation (Section 7.2), the dynamics of a neutron star due to its current-induced deceleration (Section 7.3), the statistical distribution of pulsars (Section 7.4), the generation of radio emission (Sections 7.5 and 7.6) and, finally, non-stationary processes (Section 7.7).

7.1 The structure of the active region

7.1.1 The model of a partially filled magnetosphere

As shown in Chapters 3–5, the theory of plasma generation in the region of the double layer associates the pulsar rotation energy losses W_{tot} with the magnitude of the longitudinal electric current $j_{\parallel}$ flowing in the magnetosphere (see (4.215)). This current is specified by the 'compatibility relation' (4.174) and by the pulsar 'ignition' condition (5.66). Owing to this, we can reconstruct the structure of the active region in the polar cap and determine important parameters of a neutron star as the magnitude of the magnetic field B_0 from the observed quantities P and dP/dt.

It should be noted, however, that to be compared with observational data, the theory developed in Chapter 4 should be extended. The point is that, as mentioned in Section 5.3, a secondary plasma cannot be generated in the entire polar cap, for it cannot be generated in the circular region near the magnetic axes. Stationary plasma generation will take place only on those open field lines on which the drop of the potential $\Psi(f)$ in the region of a double layer satisfies the 'ignition' condition $\Psi(f) = \Psi_0(f)$, where according to (5.61),

$$\frac{e\Psi_0}{m_e c^2} = a_\Psi 10^7 \rho_{l_7}^{4/7} P^{-1/7} B_{12}^{-1/7} \cos^{1/7}\chi (1 - pi_0)^{1/7}, \qquad B_{12} \lesssim 5 \quad (7.1)$$

$$\frac{e\Psi_0}{m_e c^2} = a'_\Psi 10^7 \rho_{l_7}^{4/7} P^{-1/7} B_{12}^{1/7} \cos^{1/7}\chi (1 - pi_0)^{1/7}, \qquad B_{12} > 5 \quad (7.2)$$

Here $a_\Psi, a'_\Psi \simeq 1$ and f is a dimensionless dipole coordinate constant along a magnetic field line (see (4.47)). In polar regions of a neutron star it can be written in the form

$$f = \frac{f^*(\chi)\rho_\perp'^2}{R_0^2}$$

where $\rho'_\perp$ is the distance from the magnetic axis, R_0 is the polar radius, i.e. the distance from the last field line to the axis, and the function $f^*(\chi)$ is shown in Fig. 4.12. We shall see that for the majority of observed radio pulsars the magnetic field does not exceed 5×10^{12} G. We shall therefore mainly use expression (7.1).

Making use of the relation (5.65)

$$\rho_l = \tfrac{4}{3}R\left(\frac{\Omega R}{c}\right)^{-1/2} f^{-1/2}$$

valid for a dipole magnetic field, we obtain

$$\Psi_0 \propto f^{-2/7} \quad (7.3)$$

We can see that the potential $\Psi_0(f) \to \infty$ as $\rho'_\perp \to 0$ and therefore the condition $\Psi(f) = \Psi_0(f)$ cannot be fulfilled near the magnetic pole of a pulsar; more precisely for $\rho'_\perp < \rho_{\text{in}}$, where the quantity ρ_{in} is the internal radius of the ring in which a secondary electron–positron plasma can be generated. Thus, we see that there is no outflowing plasma on field lines with $f < f_1$ where, according to (7.2), $f_1 = f^*(\rho_{\text{in}}/R_0)^2$.

To find the internal radius ρ_{in} (and, accordingly, the quantity f_1), we can use the criterion $\rho'_\perp \gtrsim H$ (see Section 5.2.4), where we should put $\rho'_\perp = \rho_{\text{in}}$. Consequently, we have

$$H(f_1) = \rho_{\text{in}}$$

Making use of the expression (5.61) for $H(f)$ we are finally led to

$$\frac{f_1}{f^*} \simeq q_f^{7/9} \tag{7.4}$$

where

$$q_f = a_f B_{12}^{-8/7} P^{15/7} \cos^{1/7}\chi(1 - pi_0)^{1/7}, \qquad a_f \simeq 1 \tag{7.5}$$

We should note that (7.4) and (7.5) can also be derived from other considerations with the use of the restriction of the value of the maximal derivative $d\Psi/df$ (Beskin *et al.*, 1984).

Since $f_1 < f^*$, we can see that the quantity q_f must be less than some limiting value $\tilde{q} \simeq 1$:

$$q_f \lesssim \tilde{q} \tag{7.6}$$

If $q_f \to \tilde{q}$, the internal radius of the ring tends generally to the external radius, so that for $q_f > \tilde{q}$ stationary plasma generation becomes impossible. The relation $q_f < \tilde{q}$ (7.6) coincides with the condition $H < R_0$ introduced in Section 5.2.4, i.e. it is the 'ignition' condition for radio pulsars. The dashed line in Fig. 7.1 shows the dependence $\Psi_0(f)$ found from the condition (7.3), as well as the value f_1, satisfying relation (7.4).

It is noteworthy that the condition (7.6) certainly does not mean that for $q_f \simeq \tilde{q}$ plasma generation stops altogether, since even for $q_f \gtrsim \tilde{q}$ particles can be produced in separate 'islets' on the polar cap surface (Ruderman & Sutherland, 1975). For the determination of the quantity

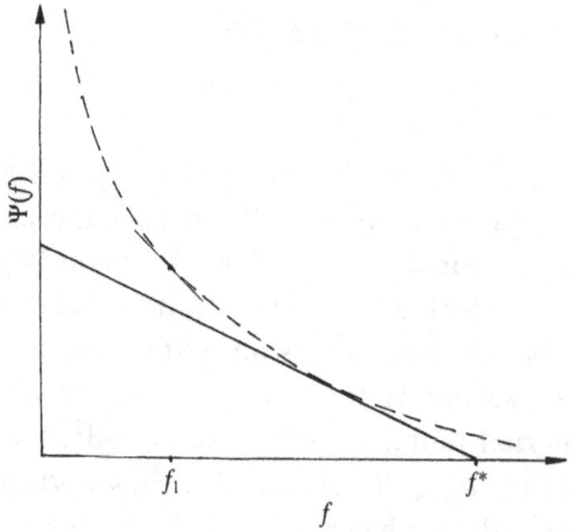

Fig. 7.1. The dependence $\Psi(f)$ for a real (formula (7.3), dashed line) and a model (solid line) potential Ψ.

$\tilde{q}$ we cannot use the expression (7.1) for the potential Ψ_0 since it was itself derived from the assumption that the double layer thickness H is much less than the polar cap radius R_0. At the same time we see that in the theory of plasma generation we can naturally distinguish between two cases: (1) $q_f \ll 1$, where the asymptotical expressions (7.4) and (7.5) hold; and (2) $q_f \gtrsim 1$, where plasma generation proceeds only at a relatively small part of the polar cap surface near its boundary. Since a more consistent theory can be constructed in the case $q_f < 1$ (more precisely, $q_f \ll 1$), this limiting case will be considered more often in what follows.

We shall first consider the case $q_f \ll 1$ when $f_1 \ll 1$ and plasma is produced on almost the entire polar cap. Note that in the region of open magnetosphere the picture of the current will be somewhat different from the case corresponding to a magnetosphere entirely filled with plasma. The point is that since in the region $f < f_1$ the plasma is absent, a current does not flow near the magnetic axis at all. Given this, along the magnetic surface an $f = f_1$ additional surface current $I_1 = (1/2c)B_0\Omega^2R^3g_1$ must flow, whose value can be found from the matching condition of the vacuum region $f < f_1$ and the region $f > f_1$, where a longitudinal current $i_0 = j_\parallel/j_c$ flows.

To establish the relationship between the current i_0 and the acceleration potential β_0, we consider for simplicity the axisymmetric case. The equations describing the magnetic field structure here have different forms in three regions: (1) In the region of a closed magnetosphere $i_0 = 0$, $\beta_0 = d\Psi/df = 0$ so that (4.49) is valid:

$$\Delta f[1 - \rho_\perp^2] - \frac{2}{\rho_\perp}\frac{\partial f}{\partial \rho_\perp} = 0 \qquad (f > f^*)$$

(2) In the region of an open magnetosphere, where a longitudinal electric current flows, we should employ (4.60):

$$\Delta f[1 - \rho_\perp^2(1 - \beta_0)^2] - \frac{2}{\rho_\perp}\frac{\partial f}{\partial \rho_\perp} + g\frac{dg}{df} = 0 \qquad (f_1 < f < f^*)$$

As in Section 4.2.2, we assume the dependence (4.58) of Ψ on f to be linear. (3) In the region of a polar magnetosphere, where plasma is not generated, the magnetic field has a dipole structure:

$$f = \frac{\rho_\perp^2}{(\rho_\perp^2 + z^2)^{3/2}}$$

The longitudinal current $i_\parallel$ is expressed, according to (4.45), in terms of the quantity g, $i_\parallel$ being equal to dg/df near the star's surface. Therefore, in the region $f_1 < f < f^*$,

$$g = g_1 + i_0(f - f_0)$$

As in Chapter 4, the matching between regions (1) and (2) will be on the light cylinder $\rho_\perp = 1$, $f = f^*$. The first equation implies that $\partial f/\partial\rho_\perp|_{\rho_\perp = 1} = 0$. The second equation will then yield the relation between the longitudinal current i_0 and the potential β_0:

$$i_0^2 = i_{\max}^2[1 - (1 - \beta_0)^2]\frac{f^*}{f^* - f_1} - \frac{i_0}{f^* - f_1}, \qquad i_{\max}^2 = -\frac{\Delta f}{f^*}\bigg|_{\rho_\perp = 1}$$

To find the surface current $I_1 \propto g_1$, we match regions (2) and (3) for $f = f_1$ and $\rho_\perp = (1 - \beta_0)^{-1}$:

$$i_0 g_1 = \frac{2}{\rho_\perp}\frac{\partial f}{\partial \rho_\perp}\bigg|_{\substack{f = f_1 \\ \rho_\perp = (1 - \beta_0)^{-1}}} = \frac{2}{\rho_\perp}\frac{\partial}{\partial\rho_\perp}\left[\frac{\rho_\perp^2}{(\rho_\perp^2 + z^2)^{3/2}}\right]$$

As a result we obtain that the dimensionless current i_0 in the region $f_1 < f < f^*$, the dimensionless accelerating potential β_0, and the parameter g_1 specifying the surface current I_1 on the internal boundary $f = f_1$ are related as (Beskin *et al.*, 1984)

$$i_0^2 = [1 - (1 - \beta_0)^2]\frac{f^*}{f^* - f_1} - 4\frac{f_1}{f^* - f_1}(1 - \beta_0)^{4/3}[(1 - \beta_0)^{2/3} - \tfrac{3}{2}f_1^{2/3}] \tag{7.7}$$

$$g_1 = 4\frac{f_1}{i_0}(1 - \beta_0)^{4/3}[(1 - \beta_0)^{2/3} - \tfrac{3}{2}f_1^{2/3}] \tag{7.8}$$

In this case, $q_f \ll 1$ and, therefore, $f_1 \ll 1$. Owing to this, (7.7) coincides with compatibility relations (4.86), and (7.8) yields

$$I_1 = \frac{B_0\Omega^2R^3}{2c}g_1 = 2\frac{B_0\Omega^2R^3}{c}\frac{f_1}{i_0} \tag{7.9}$$

i.e. the surface current I_1 flowing along the separatrix $f = f_1$ has the same direction as in the plasma generation region. In Fig. 7.1, the solid line shows the run of the potential in this model, which we consider in the sequel.

Thus, the picture of current flowing in the magnetosphere is as follows. In the region $f < f_1$, i.e. $\rho_\perp < \rho_{in}$ where, according to (7.2) and (7.4),

$$\frac{\rho_{in}}{R_0} = a_\rho P^{5/6} B_{12}^{-4/9} \cos^{-1/3}\chi, \qquad a_\rho \simeq 1 \tag{7.10}$$

there are no currents and plasma is not generated. Therefore, inside a plasma-filled magnetosphere, an empty region is formed. This fact reflects the peculiarity of plasma generation in our theory and corresponds to the 'hollow cone' model (Ruderman & Sutherland, 1975). Given this, a jet of surface current I_1 (7.9) flows, as follows from (7.8), along the boundary $f = f_1$; and then at $f > f_1$ there follows the region of permanent electron–positron plasma generation. The total current flowing in this region is determined by the compatibility relation (7.7) where, according to (7.1),

$$\beta_0 = \frac{\Psi_0}{\Psi_{max}} = a_\beta B_{12}^{-8/7} P^{15/7} \cos^{1/7}\chi(1 - pi_0)^{1/7}, \qquad a_\beta \simeq 1 \tag{7.11}$$

The reverse current flows as before along the separatrix $f = f^*$. Figure 7.2 shows the characteristic profiles of current (and, therefore, of densities) $G(f)$ of outflowing particles for pulsars with $q_f \ll 1$ and $q_f \gtrsim 1$.

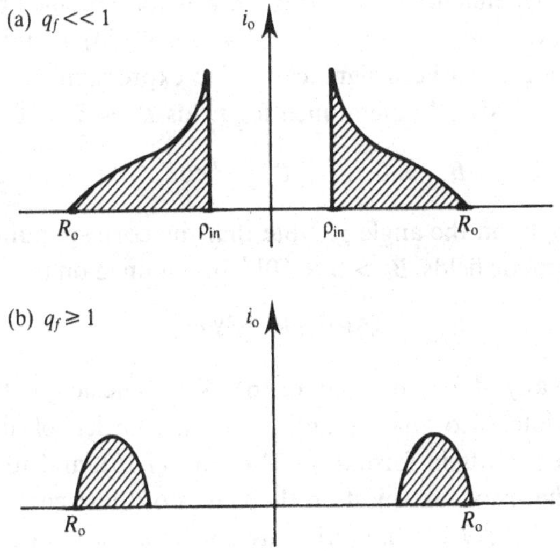

Fig. 7.2. Characteristic profiles of the density of secondary particles $G(f)$ for the case $q_f \ll 1$ and $q_f \gtrsim 1$.

7.1.2 Basic parameters

We shall determine the basic characteristics of pulsars by the observed quantities P and dP/dt. It is convenient to introduce a dimensionless parameter (Beskin *et al.*, 1984)

$$Q = 2P^{11/10}\dot{P}_{-15}^{-4/10} \tag{7.12}$$

where the period P is expressed in seconds and $\dot{P}_{-15} = 10^{15}\,dP/dt$ is the pulsar deceleration velocity. Employing (4.197) for the determination of the magnetic field B_0 in terms of the observed quantities P, $\dot{P}$ and the angle χ, and also using the asymptotics of the compatibility relation (4.174),

$$i_0 = i_{\max}(\chi)\left(\frac{2\beta_0}{\beta_{\max}(\chi)}\right)^{1/2} \tag{7.13}$$

which is valid just for $q_f \ll 1$ (i.e. for $i_0 \ll i_{\max}$, $\beta_0 \ll \beta_{\max}$), we obtain

$$\frac{\beta_0}{\beta_{\max}(\chi)} \simeq Q^2 \tag{7.14}$$

$$\frac{\rho_{\text{in}}}{R_0} \simeq Q^{7/9} \tag{7.15}$$

Note that the dependence on the angle χ, which is specified by an explicit form of the functions $\beta_{\max}(\chi)$, $i_{\max}(\chi)$ (4.172, 4.173) in formulas (7.14) and (7.15), appears to be insignificant. The expression for the magnetic field B_0 (valid, as already mentioned for fields $B_0 \lesssim 5 \times 10^{12}$ G),

$$B_{12} \simeq 0.5\dot{P}_{-15}^{7/10}P^{-1/10}\cos^{-1.1}\chi \tag{7.16}$$

depends strongly on the angle χ. Note that the corresponding expression for strong magnetic fields, $B_0 > 5 \times 10^{12}$ G, obtained on the basis of (7.2),

$$B_{12} \simeq 0.5\dot{P}_{-15}^{7/11}P^{-1/20}\cos^{-1.1}\chi$$

is not practically different from (7.16). We henceforth disregard this distinction. Note also that quantities of the order of unity are the numerical coefficients in formulas (7.14) and (7.15), and to preserve the accuracy of the theory in question the values of these coefficients a_Ψ, a_f, a_ρ, a_β in (7.1), (7.5), (7.10) and (7.11) are convenient to put equal to unity.

As is seen from (7.15), in pulsars with $Q < 1$ the internal radius ρ_{in} of outflowing plasma will be less than R_0 and $\beta_0 < \beta_{\max}(\chi)$. Consequently, in pulsars with $Q \ll 1$ the outflowing current occupies practically the whole surface of the polar cap. Thus, the condition $Q < 1$ coincides with the condition $q_f < 1$. For such pulsars we can use the compatibility relation (4.174), and more precisely, its asymptotics (7.13).

We should assume pulsars with $Q > 1$ to be located not far from the boundary of extinction, and therefore the cascade particle production is suppressed in them. This leads to underestimation of the mean value of longitudinal current i_0, which corresponds to the observed values $Q \gtrsim 1$. For such pulsars we should put $\rho_{in} = R_0$ and $\beta_0 = \beta_{max}(\chi)$. This allows us to derive the expression for the magnetic field of pulsars with $Q > 1$:

$$B_{12} \simeq 1.5P^{15/8} \cos^{-1} \chi, \qquad Q > 1 \tag{7.17}$$

Figure 7.3 shows the period distribution of 370 pulsars for which we know the deceleration velocity dP/dt (Lyne & Smith, 1990). The solid line corresponds to 183 pulsars with $Q < 1$, the dashed line to 187 pulsars with $Q > 1$. We see that the majority of pulsars with $Q > 1$ have periods $P > 0.7$ s, while the periods of pulsars with $Q < 1$ are for the most part less than 0.8 s. Note that in the analysis of observational data it was repeatedly pointed out (see, e.g., Malov & Sulejmanova, 1982) that the properties of pulsars with large and small periods differ significantly from each other. The observed distinction in the properties of radio pulsars is shown below to agree with the theory, since according to the theory it is just the quantity Q that is responsible for the details of plasma generation and of the structure of the active region.

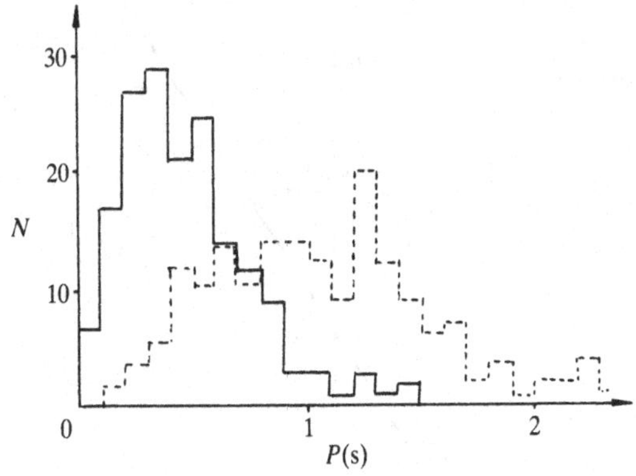

Fig. 7.3. Period distribution of pulsars with $Q > 1$ (dashed line) and $Q < 1$ (solid line).

7.2 Plasma generation

7.2.1 The line of extinction

As indicated above, in pulsars with $Q \gtrsim 1$ the electron–positron plasma generation should be suppressed since in such pulsars plasma is generated only in a thin ring with $\rho_{in} \simeq R_0$. In Figure 7.4 we can see pulsar distribution on the period–magnetic field diagram. The values were calculated using formulas (7.16) and (7.17) with $\cos\chi = \overline{\cos\chi} = 0.5$. We also see that all pulsars exhibiting extinction whose radio emission exhibited such irregularities as nullings, mode switching or drift of subpulses (see Section 1.3.3) actually lie near the extinction line and have $Q > 1$ (circles). All pulsars with $Q < 1$ are characterized by stable radio emission (points). This is an important argument in favour of the fact that the activity of radio pulsars is closely connected with electron–positron plasma generation in polar regions.

The criterion (5.66) defining the possibility of stationary plasma generation restricts the region of parameters for which radio pulsars exist. On the P–B_{12} diagram (Fig. 7.4), this region lies above the straight line (5.66):

$$PB_{12}^{-8/15} = 1.25b^{2/15}(K) \tag{7.18}$$

Here $b(K)$ is the coefficient in (5.59) describing the interaction of fast particles with the neutron star surface in the polar region.

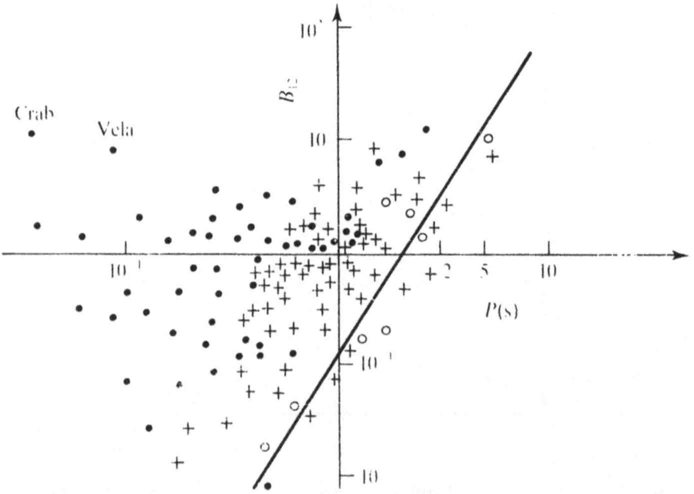

Fig. 7.4. Pulsar distribution on the period–magnetic field diagram.

As is seen from Fig. 7.4, this criterion agrees well with observational data. The comparison of (7.18) with the observed boundary of pulsar distribution implies that $b(K) \simeq 0.2$–0.4. From (5.60) it follows then that the mean coefficient of particle knock-down from the star surface, K (5.52), is approximately equal to 10–50. These values are in agreement with the K values obtained from the direct calculation for pulsars with mean magnetic fields $B_{12} \simeq 1$ (Bogovalov & Kotov, 1989) (see Fig. 5.3). For stronger magnetic fields $B_0 \simeq 10^{13}$ G ($B_{12} \simeq 10$), inherent in pulsars 0833−45 (Vela), 0531+21 (Crab), 1509−58, the K values as shown by Bogovalov & Kotov (1989), should be much smaller, i.e. of the order of 1–3. We will see below in Section 7.3.2 that these K values are also in agreement with observations.

7.2.2 'Relic' gamma-ray emission

Another observational test allowing a direct judgement of the processes proceeding in the vicinity of the magnetic poles of a neutron star is the discovery of the so-called 'relic' gamma-ray emission, i.e. curvature-radiation photons produced in the course of plasma generation near a star's surface. The spectrum of such photons is calculated on the basis of eqs. (5.6) and (5.7) for a rotating dipole magnetic field. For pulsars 0833−45 (Vela) and 0531+21 (Crab) the result is shown in Fig. 7.5. We can see sufficient agreement between computational data and observations carried out by satellites SAS-2 and COS-B for energies $\mathscr{E}_\gamma \lesssim 1$ GeV. It is noteworthy that in the calculation of flux of 'relic' gamma-ray quanta no arbitrary parameters were used, i.e. not only the form of the spectrum, but also a direct agreement between the number of observed and theoretically predicted photons within the range 1 MeV–1 GeV is fundamental.

It should be emphasized, however, that for energies $\mathscr{E}_\gamma > 1$ GeV the theory predicts an exponential fall in the spectrum of relic gamma-ray emission. This fall is due to the fact that high-energy gamma-ray quanta are effectively absorbed in the magnetic field of a pulsar, forming electron–positron pairs, which prevents them from going outside (Harding, Tademaru & Esposito, 1978). No such fall has yet been registered in experiments. However, within the energy range 10 GeV $< \mathscr{E}_\gamma < 10^3$ GeV there are as yet no observational data. But gamma-ray quanta of very high energy, 10^3–10^4 GeV, are systematically observed in some pulsars (see Section 1.4).

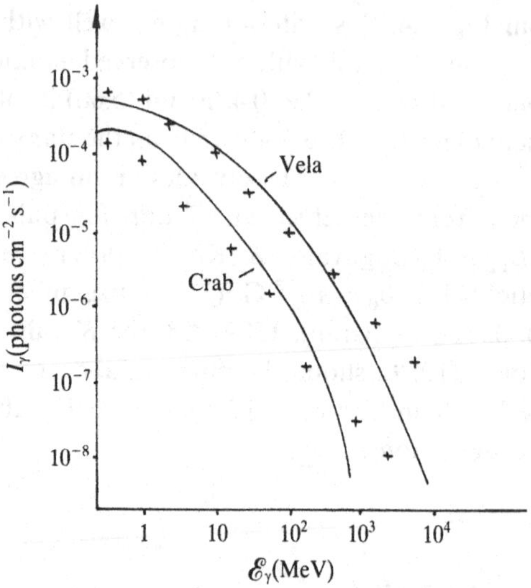

Fig. 7.5. The spectrum of 'relic' gamma-quanta for pulsars 0531+21 (Crab) and 0833−45 (Vela).

7.2.3 Polar cap heating. X-ray emission

The character of processes proceeding in polar regions of neutron stars can also be judged by X-ray emission observed from several radio pulsars (see Section 1.4). Indeed, as shown in Chapter 5, the cascade production and acceleration of particles in a double layer near the star surface must inevitably be accompanied by a reverse flux of energetic particles and gamma-ray quanta onto the surface. This flux is responsible for a heating of the neutron star's surface and for thermal X-ray emission from polar cap regions. Therefore, an observed flux of X-ray emission (provided that it is actually associated with the above-mentioned heating process) can also be an indication of the intensity of secondary-particle production in polar regions of a pulsar.

According to Section 5.2.3, the total energy flux towards the star is

$$W_H \simeq \frac{1}{1+K} i_0 \beta_0 \frac{\Omega^4 R^6 B_0^2}{c^3} \simeq \frac{1}{1+K} Q^2 W_{\text{tot}} \tag{7.19}$$

The greater part of this energy goes to the star's surface heating (Jones, 1986b). Given this, the temperature T and the size of the heated spot R_S will depend on the competition of two processes of energy transfer from the star's surface, namely, transfer by black-body radiation of the surface with total luminosity

$$L_X = \pi\sigma T^4 R_S^2 \tag{7.20}$$

and transfer by the thermal flux in the neutron star's volume

$$L_\kappa \simeq \kappa_\perp \nabla T \cdot R_S^2$$

Estimating ∇T as T/R_0, we find that for usual values of heat conductivity $\kappa_\perp \simeq 10^{12}$–$10^{18}$ erg cm^{-1} s^{-1} K^{-1} (see (2.19)), the heat conductivity can play a role in heat redistribution on the neutron star's surface. But for simplicity we will assume the size of the heated spot to coincide with the size of the polar cap R_0 (3.10) and the total dissipated energy W_H (7.19) to go to generation of thermal radiation from this spot.

As a result, the temperature of the heated spot must be

$$T \lesssim 6 \times 10^6 Q^{1/2} P^{-1/2} \dot{P}_{-15}^{1/4}(1 + K)^{-1/4} \tag{7.21}$$

Since, as shown in Section 2.4, the surface of the whole neutron star must have a temperature $T \lesssim 10^6$ K as a result of its cooling, we conclude that the polar caps may be strongly heated above the background of the rest of the star's surface. Given this, the radiation maximum of polar caps heated to the temperature $T \simeq 10^6$ K (7.21) lies just in the X-ray range $\mathscr{E}_X \simeq 0.3$ keV.

One of the important properties of such X-ray radiation must be its variability connected with neutron star rotation. Since, according to our theory, it is only heated polar regions that radiate, for the observed X-ray flux (Greenstein & Hartke, 1983) we have

$$L_X(t) = L_0(\cos^2 \chi + \sin^2 \chi \cos \Omega t) \tag{7.22}$$

In the derivation of this relation we have used the fact that the size of the polar cap is small and, therefore, the inclination angle χ of the axes in radio pulsars is close to the angle between the rotation axis and the direction to the observer β. As shown in Fig. 7.6, the profiles of observed X-ray radiation in radio pulsars 0540 − 693 and 1509 − 58 actually agrees rather well with the expected dependence (7.22). The observed frequency spectrum of the variable component is consistent with the hypothesis of its thermal origin (Seward & Harnden, 1982; Brinkmann & Ögelman, 1987; Ögelman & Buccheri, 1987). We can, therefore, assume this radiation actually to be due to the polar cap surface heating associated with electron–positron plasma generation. Note that the inclination angle χ of axes can be determined by the observed profile of X-ray radiation (Greenstein & Hartke, 1983). The influence of the effects of general

Table 7.1. *Pulsed X-ray radiation from radio pulsars*

Pulsar	P(s)	$\dot{P}_{-15}$	Q	η_m	$\log L_X^{obs}$ (erg s^{-1})	$\log L_X^{th}$ (erg s^{-1})	χ (deg)
0531+21	0.033	421	0.004	~100%	36.0	33.7	~90
0540−693	0.050	479	0.006	~20%	36.4	34.0	23–40
0656+14	0.385	55	0.141	~18%	32.0	32.2	14–20
1055−52	0.197	5.8	0.165	~40%	32.5	32.5	28–50
1509−58	0.150	1490	0.013	~60%	34.4	33.2	34–52
1951+32	0.039	5.9	0.028	~100%	34.2	33.8	90

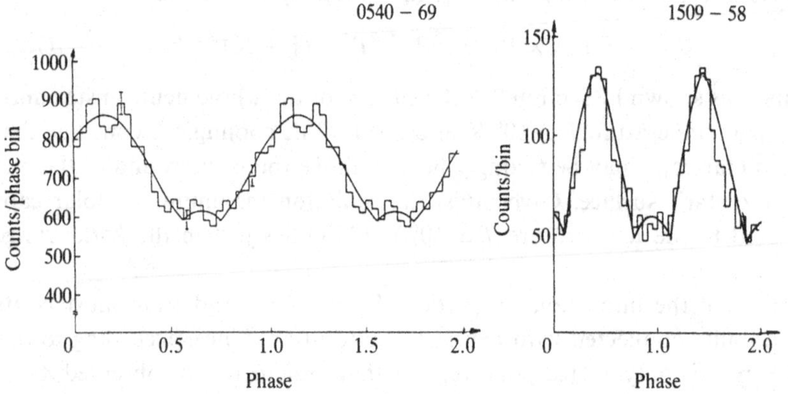

Fig. 7.6. Profiles of observed X-ray radiation for radio pulsars 0540−693 and 1509−58. The curves correspond to formula (7.22).

relativity and a strong magnetic field upon the profile of a pulsed X-ray radiation was discussed by Brinkmann (1980), Ftaclas *et al.* (1986) and Cheng & Shaham (1989).

In Table 7.1 we list the values of the observed X-ray flux of some pulsars taken from Table 1.12, the observed flux modulation depth $\eta_m = (L_X(\max) - L_X(\min))/L_X(\max)$, as well as theoretical estimates of the thermal flux determined by formula (7.19) and the inclination angle χ of the axes. We see that in most cases the observed flux L_X agrees with the value W_H (7.19).

In pulsar 0531+21 (Crab) the principal observed pulsed X-ray flux is not associated with thermal radiation since it has a much higher directivity than (7.22) and is a part of the total power-law non-thermal spectrum spread from the optical to the hard gamma-ray range (see Section 1.4). The thermal flux value for this pulsar does not, however, contradict observations, since the expected thermal flux makes up only 10^{-3} of the observed non-thermal flux. The X-ray emission of other sources (1509 − 58, 1055 − 52 and 1951 + 32), can be assumed to be due to thermal radiation of polar caps heated by a reverse particle flux.

7.2.4 Radiation from the light surface region

We shall briefly discuss radiation generated in the light surface region. Such radiation must be due to additional energy losses of particles which, as shown in Section 4.3, undergo a sharp acceleration up to the energies (4.111) near a light surface S_c:

$$\mathscr{E}_{\max} = \frac{eB_0\Omega^2R^3}{2\lambda c^2} i_0$$

We should emphasize from the very beginning that the relations presented in this section should be thought of as preliminary estimates, since an exact theory of the behaviour of electron–positron plasma outside the light surface has not yet been formulated.

To begin with, we determine the characteristic frequency of radiation generated in the region of the boundary layer. To this end, we employ relation (3.31) $\nu \simeq c(\mathscr{E}_{\max}/m_ec^2)^3/\rho_l$, and in our case

$$\rho_l \simeq \Delta = \frac{c}{\Omega\lambda} \tag{7.23}$$

is the characteristic curvature radius of a trajectory in a transition (see Section 4.3.1). As a result, we come to

$$\nu_{\rm ch} \simeq 10^{14}B_{12}^{-3}P^{-7}\lambda_4^{-2}i_0^3 \quad \text{Hz} \tag{7.24}$$

We see that for the majority of observed radio pulsars ($P \sim 1$ s, $B_0 \sim 10^{12}$ G, $\lambda \sim 10^4$) this frequency lies in the optical range. For moderately fast pulsars, $P = 0.1$–0.3 s, it shifts towards the X-ray range; while for very fast pulsars, such as 0531 + 21 (Crab) and 0833 − 45 (Vela), as well as for millisecond pulsars, the frequency $\nu_{\rm ch}$ reaches into the energy range 1–10 MeV.

We shall estimate the intensity of such high-frequency radiation. The power lost by one particle can be estimated using the well-known relation

(Landau & Lifshitz, 1983)

$$I_r = \frac{2}{3}\frac{e^4}{m_e^2 c^3}\gamma^2\left\{\left[\mathbf{E} + \frac{[\mathbf{vB}]}{c}\right]^2 - \frac{1}{c^2}(\mathbf{E}\cdot\mathbf{v})^2\right\}$$

Substituting into this expression the values for the electric and magnetic fields in the form (4.101), (4.102), as well as the particle velocities determined according to (4.113), (4.114), we obtain

$$I_r = \frac{2}{3}\frac{e^4}{m_e^2 c^3}\gamma^2 B_c^2\left[\frac{v_z^2}{c^2}\left(\frac{1}{x^2} + \kappa^2 - 1\right) - \left(1 - \frac{v_\phi}{cx} + \frac{v_{\rho_\perp}}{c}\kappa\right)^2\right]$$

where $x = \Omega\rho_\perp/c$. Since, owing to (4.95), (4.103) in the region of light surface $\kappa \simeq \kappa_0$, $[\kappa_0^2 - (1/x_0^2) - 1] = 0$ and $v_\phi/c = 1/x_0$, $\kappa_0 \simeq i_0$, we finally have

$$I_r \simeq \frac{2}{3}\frac{e^4}{m_e^2 c^3}\gamma^2 B_c^2 i_0^2 \tag{7.25}$$

Note that this value coincides with the expression for the power loss $I_r = \frac{2}{3}(e^2c/\rho_l^2)\gamma^4$ of a particle if its curvature radius ρ_l is defined by the expression (7.23). It should be emphasized that the curvature in the trajectory of a particle (7.23) is caused by its motion across the B_ϕ-component of the magnetic field, the ρ_l value (7.23) exactly coinciding with the cyclotron radius of particle motion in the field B_ϕ. Therefore, the energy losses of particles in the transition layer are of synchrotron character.

Clearly, each particle accelerated in the transition layer is capable of radiating only part of the energy $\mathscr{E}_{\max}$ gained. The portion of radiated energy can be estimated by

$$\delta_r = \frac{\mathscr{E}_r}{\mathscr{E}_{\max}} = \begin{cases} \tau_e/\tau_r, & \tau_e < \tau_r \\ 1, & \tau_e > \tau_r \end{cases}$$

Here $\tau_r = \mathscr{E}_{\max}/I_r$ is the time of synchrotron radiation and

$$\tau_e \simeq \frac{\Delta}{\langle v_{\rho_\perp}\rangle} \simeq \frac{1}{\Omega}(\lambda i_0)^{-1}$$

is the time of particle outflow from the transition layer region. As a result, we have

$$\delta_r = 10^{-9} B_{12}^3 P^{-7} \lambda_4^{-2} i_0^2 \tag{7.26}$$

It can readily be verified that the condition $\delta \ll 1$ holds for practically all pulsars. Since according to (4.111) the energy gained by the particles in the transition layer is close to the energy transferred by the electromagnetic field in the internal magnetosphere (and, therefore, is close to the total energy losses of a neutron star), we are finally led to

$$L_{R_L} \simeq \delta_r W_{tot} = 10^{23} B_{12}^5 P^{-11} \lambda_4^{-2} i_0^3 \quad \text{erg s}^{-1} \tag{7.27}$$

As a result, for ordinary radio pulsars whose frequency ν_{ch} (7.24) lies in the optical range, the luminosity for the characteristic distances $d \simeq 1$ kpc corresponds to stellar magnitudes $m \simeq 35$–40. Such radiation cannot be registered. Correspondingly, for moderate pulsars ($P \simeq 0.1$ s, $i_0 \simeq 0.1$, $B_0 \simeq 10^{12}$ G) whose frequency corresponds to the X-ray range, the radiation L_{R_L} amounts to 10^{30}–10^{32} erg s^{-1}, that is, it cannot be detected either. It is only for the very fast pulsars that their gamma-ray emission from the light surface can reach 1–10% of the total rotational energy losses and can, in principle, be observed.

For the region behind the light surface, $r > R_L$, a consistent theory has not yet been built. As mentioned in Section 4.3.1, there is reason to believe that here the waves are excited (4.118) with the same characteristic values of the magnetic field component B_ϕ as in the boundary layer. In this case, the energy losses of particles will again be determined by the relation (7.25), but the time of outflow is substantially larger. It can be written as

$$\tau_e \simeq \frac{R_L}{c} = \Omega^{-1}$$

Then for the portion of energy lost due to synchrotron radiation we obtain

$$\delta_r = 10^{-6} B_{12}^3 P^{-7} \lambda_4^{-1} i_0^3$$

This implies that even for moderate pulsars ($P \sim 0.1$ s) high-frequency radiation from the region $R_L < r < (2$–$3)R_L$ can make up a considerable part of the total energy losses of a neutron star, 10^{34}–10^{35} erg s^{-1}. Possibly, it is just this mechanism that provides the observed non-pulsed component of the X-ray flux in pulsars 0355+54, 1929+10, 1055−52, 1642−03 (Seward & Wang, 1988) and the constant component of the gamma-ray emission in pulsars 0531+21 and 0838−45 (Clear *et al.*, 1987). Table 7.2 presents the characteristic frequency and the intensity of synchrotron radiation expected according to the relations (7.24)–(7.27) for fast and moderate pulsars.

Table 7.2. *High-energy radiation from light surface region*

Pulsar	P (s)	$\dot{P}_{-15}$	$\mathscr{E}$ (MeV)	$\log L_{\rm RL}$ (erg s^{-1})	$\log L_{(2-3)\rm RL}$ (erg s^{-1})
0114+58	0.101	5.8	1.0	32	34
0531+21	0.033	421	300	37	37
0540−693	0.050	479	70	33	34
0740−28	0.167	16.8	0.03	31	32
0833−45	0.089	124	2	35	36
0906−49	0.107	15.2	0.3	33	35
1800−21	0.133	134	0.5	34	35
1821−24	0.003	0.002	40	34	35
1823−13	0.101	75.2	0.5	34	35
1830−08	0.085	9.2	0.3	33	35
1930+22	0.144	57.8	0.2	33	35
1937+214	0.002	0.0001	80	35	35
1951+32	0.039	5.9	4	35	36
1957+20	0.002	0.00001	0.1	32	34

7.3 Evolution: weakly radiating pulsars

7.3.1 The lifetime of radio pulsars

As has already been said, neutron stars are radio pulsars only if they rotate fast enough and can generate secondary electron–positron plasma. During this process they lose energy and decelerate. Pulsar evolution is thus connected with deceleration of the rotation.

We shall first consider the evolution of pulsars with $Q < 1$, since for these there exists a dynamic theory which allows us to take exact account of the rotation deceleration rate. Employing the asymptotic relations (7.13) and (7.16), we obtain for pulsars with $Q < 1$,

$$\dot{P}_{-15} = B_{12}^{10/7} P^{1/14} \cos^{2d} \chi \tag{7.28}$$

$$\dot{\chi} = \frac{\sin \chi_0}{P_0} B_{12}^{10/7} P^{1/14} \cos^{2d-1} \chi \tag{7.29}$$

where χ_0 and P_0 are the initial values of the angle χ and period P. In (7.28) and (7.29) we have presented the dependence of $\dot{P}$ and $\dot{\chi}$ on the angle χ in the form $\cos^{2d} \chi$, where $d(\chi)$ is determined from the concrete form of the functions $i_{\max}(\chi)$, $\beta_{\max}(\chi)$ given in Section 4.4.3 (see Fig. 4.13). According to (4.172), (4.173) and (4.215), we have

$$\cos^{d(\chi)} \chi = \frac{f^*(\chi)}{f^*(0)} \cos^{0.54} \chi \left[\frac{i_{\max}(\chi)}{i_{\max}(0)}\right]^{1/2} \beta_{\max}^{-1/4}(\chi) \tag{7.30}$$

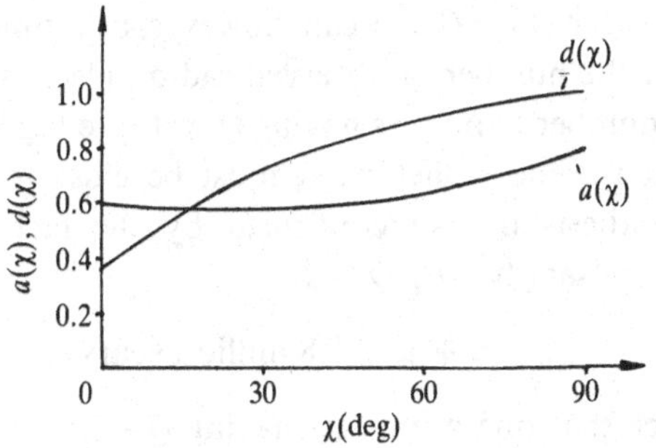

Fig. 7.7. Dependence of the functions $d(\chi)$ and $a(\chi)$ on the angle χ.

As shown in Fig. 7.7, the function $d(\chi)$ depends weakly on χ, so that we can assume with sufficient accuracy that

$$d(\chi) \simeq \bar{d} = 0.75, \qquad 2d \simeq 1.5 \tag{7.31}$$

Further on we shall need the quantity $\bar{a}$ defined by the condition $\beta_{\max}(\chi) = \cos^{\bar{a}}(\chi)$. As shown in Fig. 7.7, the function $a(\chi)$ also depends weakly on the angle χ, and we can put

$$a(\chi) = \bar{a} = 0.76 \tag{7.32}$$

We now find the characteristic lifetime of radio pulsars with $Q < 1$. To this end, we write (7.28) in a dimensionless form:

$$\frac{dP}{dt'} = \left[1 - \sin^2 \chi_0 \frac{P_0^2}{P^2}\right]^{\bar{d}} \tag{7.33}$$

where $t' = t/\tau_a$, and

$$\tau_a = 3 \frac{P_0}{\sin \chi_0} B_{12}^{-10/7} \text{ million years} \tag{7.34}$$

the quantity $P^{1/14}$ being neglected. The very form of eq. (7.33) shows that the value τ_a is just the characteristic lifetime of radio pulsars with $Q < 1$, that is, the time from the birth to the transition into the region $Q > 1$.

The τ_b lifetime within which particle generation processes in polar regions of a star stop completely cannot be determined for pulsars at the stage $Q > 1$ by the value of the period derivative, since the processes of particle production in such pulsars are suppressed and therefore the value $\dot{P}$ appears to be underestimated. This fact is illustrated in Fig. 1.8, showing a systematic exceeding of the dynamic age $\tau_d = P/\dot{P}$ compared with the kinematic age τ_{kin} (1.2).

The lifetime τ_b at the stage $Q > 1$ can, however, be estimated proceeding from the fact that the number of observed radio pulsars with $Q > 1$ is, in fact, close to the number of pulsars having $Q < 1$ (see Fig. 7.3). As a result, we conclude that the mean lifetime τ_b must be close to the value of τ_a (7.34). This hypothesis is also confirmed by the fact that the mean kinematic age of pulsars having $Q > 1$

$$\tau_{\text{tot}} = \tau_a + \tau_b \simeq 7.8 \text{ million years} \tag{7.35}$$

is just about twice that one of pulsars having $Q < 1$:

$$\langle \tau_a \rangle \simeq 3.5 \text{ million years} \tag{7.36}$$

Thus, the total active lifetime of radio pulsars

$$\tau_{\text{tot}} = \tau_a + \tau_b$$

comes on the average to 7–8 million years.

7.3.2 Evolution of individual pulsars

As we have seen, the mean lifetime of pulsars with $P \simeq 1$ s is several million years. However, there are cases when the pulsar period $P \ll 1$ s and at the same time the deceleration is rather large. In such cases the lifetime of a pulsar from its birth to the present moment is much less than its mean lifetime. For instance, the pulsar 0531+21 (Crab) is only 10^3 years old, and 0833−45 (Vela) is about 10^4 years. Equations (7.28) and (7.29) allow us to analyse the whole evolution of such a pulsar from its birth.

We shall calculate the evolution of pulsars 0531+21 and 0833−45. First, we determine the value of the magnetic field from the value of deceleration (see formula (7.16)). The presence of interpulses allows us to find today's value of the inclination angle χ. It was taken from the condition that the edge of the polar cap is at the present time observed at an angle of 90°. For pulsar 0531+21 (Crab) the following values of B_{12} and χ were obtained:

$$B_{12} = 37, \qquad \chi = 84^\circ \tag{7.37}$$

and for 0833−45 (Vela):

$$B_{12} = 15.6, \qquad \chi = 78^\circ \tag{7.38}$$

The variation of the rotation $P(t)$ and of the inclination angle $\chi(t)$ with time is determined according to formulas (7.28) and (7.29), which describe the evolution of rotation. The corresponding dependences are presented

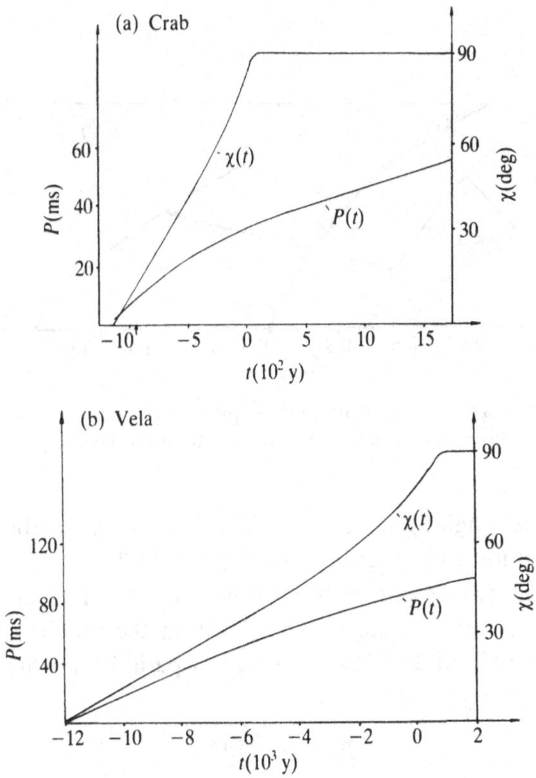

Fig. 7.8. Time-dependent variation of the period P and angle χ for pulsars 0531+21 (Crab) and 0833−45 (Vela).

in Fig. 7.8a,b. As is seen from Fig. 7.8a, the age of pulsar 0531+21 cannot generally exceed 1050 years. For the real age of 930 years (see Section 1.4) we obtain that at the moment of birth the Crab Nebula pulsar's period was $P_0 = 0.008$ s and the angle was $\chi_0 = 10°$. For pulsar 0833−45, Fig. 7.8b implies that its age is less than 12 000 years, which corresponds to observation of the Vela remnant (Lyne & Smith, 1990).

The evolution traces of individual pulsars with $Q < 1$ are conveniently shown on the P–$\sin\chi$ diagram (Fig. 7.9). Owing to the invariant $I_p = \Omega \sin\chi$ (4.199), pulsars in this diagram will move along the straight lines passing through the origin. Given this, motion is possible only to the extinction line $Q = 1$, shown in Fig. 7.9 for three characteristic values of the magnetic field B_0. For the neutron stars depicted in this diagram, above the extinction line the cascade particle production will be suppressed. Therefore, radio pulsars with large enough periods P cannot

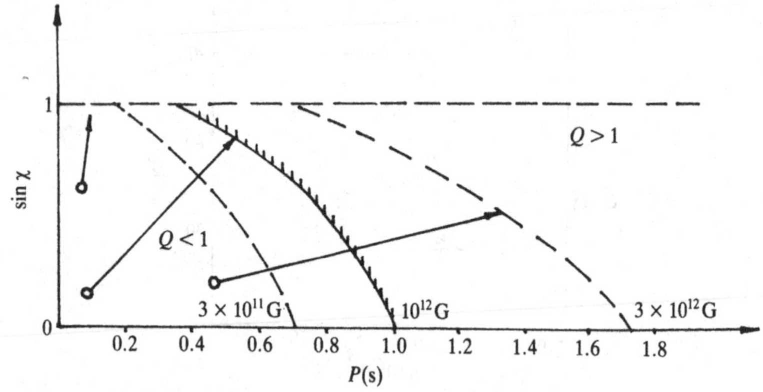

Fig. 7.9. Evolutionary tracks of individual pulsars in the P–$\sin\chi$ diagram. The extinction line is given for three values of the magnetic field B_0.

have inclination angles χ close to 90°. This has a considerable effect upon pulsar distribution over angles χ (see Section 7.4.3).

Thus, pulsars whose parameter Q was less than unity at the moment of birth go over into the region $Q > 1$ within the time $t \sim \tau_a$ (7.34). As follows from (7.12) and (7.16), the rotation period for which the pulsar reaches the value $Q = 1$ is

$$P_{Q=1} \simeq B_{12}^{8/15}\ \mathrm{s} \tag{7.39}$$

The dependence corresponds to the condition $Q = 1$ shown in Fig. 7.4. From the figure we can see that the relation (7.39) correctly estimates the maximal period of radio pulsars.

7.3.3 *The braking index*

Important information on the character of the evolution of radio pulsars is given by the braking index

$$n_b = \frac{\Omega\ddot{\Omega}}{\dot{\Omega}^2} = 2 - \frac{P\ddot{P}}{\dot{P}^2}$$

(eq. (1.8)) which is determined through the second derivative of the period $\ddot{P}$ (see Section 1.2.3). As shown in Section 4.5, the change of the rotation period P and of the inclination angle χ is not only determined by the regular deceleration (7.28, 7.29) but can also be caused by nutations induced by the deviation from the strict sphericity of the neutron star (see Section 4.5.3). It is therefore convenient to write the evolution equations in the general form

$$\dot{\Omega} = C_\Omega \Omega^{n_0} y(\chi)$$
$$\dot{\chi} = \frac{\dot{\Omega}}{\Omega} z(\chi) + \delta\chi\Omega_n \cos \Omega_n t \tag{7.40}$$

Here Ω_n and $\delta\chi$ are the frequency and amplitude of nutations. For angles χ not very close to 90° (so that $\tan \chi < (\Omega R/c)^{-1/2}$), when (7.28) and (7.29) hold, we have

$$n_0 = 1.93, \qquad y(\chi) = \cos^{-1.5} \chi, \qquad z(\chi) = -\tan \chi$$

For pulsars with $\chi \simeq 90°$, when their deceleration dynamics are determined by the asymmetric current i_A (4.201), we have

$$n_0 = 4, \qquad y(\chi) = i_A, \qquad z(\chi) = 0$$

Using the expression (7.40), we can easily derive the general formula for the braking index:

$$n_b = n_0 + z(\chi) \frac{dy/dx}{y(\chi)} + \delta\chi \frac{\Omega_n}{y(\chi)} \frac{\Omega}{\dot{\Omega}} \cos \Omega_n t \tag{7.41}$$

This consists of two parts: one of constant sign, determined by the regular part of the star's deceleration, and the other variable in time determined by nutations. Given this, for rapidly decelerated pulsars, for which

$$\frac{\dot{P}}{P} \gtrsim \frac{\delta\chi}{P_n}, \qquad P_n = \frac{2\pi}{\Omega_n} \tag{7.42}$$

the contribution of nutations must be insignificant, so that the braking index n_b will be determined by regular star deceleration processes. For these we should use the expression (7.41) without the last term:

$$n_b = n_0 + z(\chi) \frac{dy/dx}{y(\chi)}$$

For the cases $\chi \neq 90°$ and $\chi \simeq 90°$, this formula gives, respectively,

$$n_b \simeq 1.93 + 1.5 \tan^2 \chi, \qquad \chi < 90° \tag{7.43}$$

$$n_b \simeq 4 + \frac{\Omega}{i_A} \frac{di_A}{d\Omega}, \qquad \chi \simeq 90° \tag{7.44}$$

If we assume that the asymmetric current i_A is associated with the potential drop by a relation of the form (4.86), the current i_A will depend on the

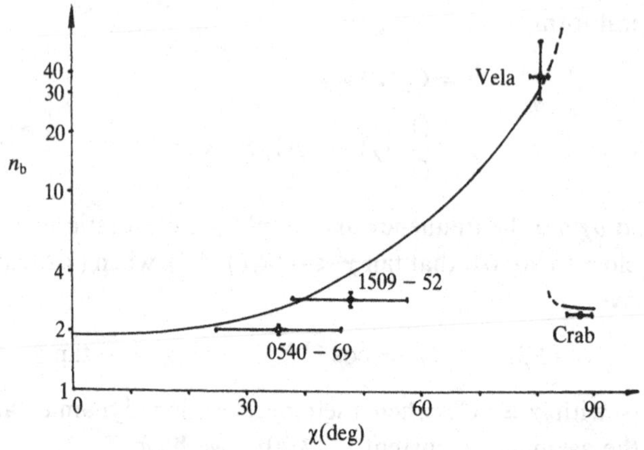

Fig. 7.10. Dependence of braking index n_b on the angle χ. Points indicate observed n_b values for four pulsars. The angles χ were determined from Table 7.1 and from the relations (7.37, 7.38).

frequency Ω in a power-law-like manner. Given this

$$n_b \simeq 2.6, \qquad \chi \simeq 90^\circ \tag{7.45}$$

In Fig. 7.10, the dependence of the braking index n_b on the angle χ (7.43–7.45) is compared with the observed values of n_b for four pulsars from Table 1.3, for which the values $\dot{P}/P$ are maximal: $\dot{P}/P \gtrsim 10^{-13}\ \mathrm{s}^{-1}$. From the figure we can see a close agreement between theory and experiment. It would, of course, be very desirable to have more pulsars for which the regular braking index is reliably established. For this purpose, attention should be paid to pulsars with high values of $\dot{P}$ and low values of the period P. The list of such pulsars is presented in Table 7.3.

For the rest of the pulsars from Table 1.3 which have $n_b \simeq \pm(10^2\text{–}10^3)$ there obviously holds the inverse of inequality (7.42):

$$\frac{P_n}{\delta\chi} \ll \frac{P}{\dot{P}} \tag{7.46}$$

For the characteristic values $\dot{P} \simeq 10^{-15}$, $P \simeq 1$ s (taking into account the fact that the amplitude of nutations $\delta\chi \lesssim 1$) we have from (7.41) the following estimate for the nutation period:

$$P_n \lesssim 10^{13}\ \mathrm{s}$$

Table 7.3. *Fast pulsars with small values of Q*

Pulsar	P(s)	$\dot{P}_{-15}$	Q	τ_d (10^6 y)	$\ddot{P}$ (10^{-27} s^{-1}) ($n_b = 3$)
0114+58	0.101	5.84	0.079	0.58	−0.34
0740−28	0.167	16.8	0.090	0.32	−1.69
0906−49	0.107	15.2	0.058	0.21	−2.16
1800−21	0.133	134	0.031	0.03	−135
1823−13	0.101	75.2	0.029	0.06	−26.0
1830−08	0.085	9.2	0.055	0.30	−0.95
1930+22	0.144	57.8	0.047	0.08	−22.6
1951+32	0.039	5.92	0.028	0.23	−0.90

It is possible that this characteristic P_n value corresponds to the star's asymmetry caused by the presence of the magnetic field (see Section 4.5.3).

As a result, in pulsars satisfying the condition (7.46) the n_b value must be determined by the sign-variable term in the general relation (7.41). Hence, in this case,

$$n_b = \delta\chi \frac{\Omega_n}{y(\chi)} \frac{\Omega}{\dot{\Omega}} \cos \Omega_n t \tag{7.47}$$

which is in qualitative agreement with the observed n_b values.

Finally, the determination of the second-order braking index $n_b^{(2)}$ (1.9) in pulsar 0531+21 in the Crab Nebula (see Section 1.2.3) also suggests an important conclusion about the character of neutron star evolution. Indeed, neglecting the nutation effect (as is the case with pulsar 0531+21), we obtain from (7.40)

$$n_b^{(2)} = n_b(2n_b - 1) + z(\chi)\frac{dn_b}{d\chi} \tag{7.48}$$

The parameter $n_b^{(2)}$ as a function of the angle χ, obtained from (7.43) and (7.48), is presented in Fig. 7.11.

Since pulsar 0531+21 has an angle $\chi \simeq 90°$ (see (7.37)), we have $z(\chi) \simeq 0$ and

$$n_b^{(2)} = n_b(2n_b - 1)$$

As is seen from Fig. 7.11, in this case, too, the theory is in agreement with experimental data since $n_b \simeq 2.5$ and $n_b^{(2)} = 10 \pm 1$.

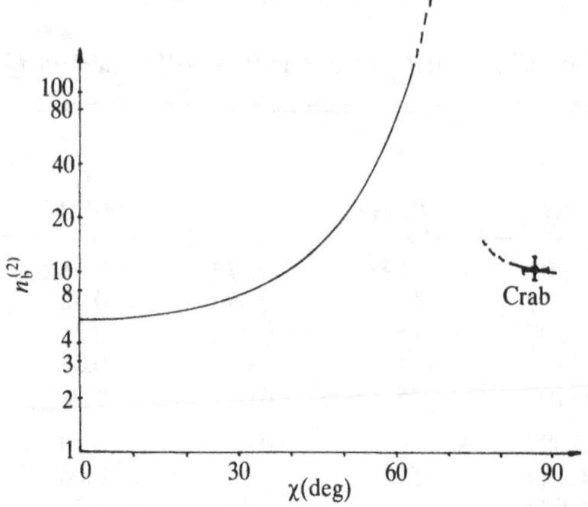

Fig. 7.11. Dependence of the second-order braking index $n_b^{(2)}$ on the angle χ.

7.3.4 *Weakly radiating pulsars*

As is seen from Fig. 7.9, if the initial period P_0 and the angle χ_0 are sufficiently small, within the active lifetime of a pulsar the quantities P and χ change substantially. If, in addition, the period $P_0 < 0.3$ s, then before reaching the extinction boundary, $\beta_0 = \beta_{max}(\chi)$, a radio pulsar goes over (within a finite time $t \sim \tau_a$), into the region of angles $\chi \simeq 90°$, where rotation deceleration is weakened appreciably. The extinction condition for pulsars with $\chi \simeq 90°$, with allowance made for (4.13), (5.66) and (7.1), can be written as $P = P_\perp$:

$$P_\perp = 0.3 B_{12}^{4/9} \text{ s} \tag{7.49}$$

Neutron stars with $\chi \simeq 90°$ can therefore be registered as radio pulsars only providing their period P is small enough: $P < P_\perp$.

The main specific feature of such pulsars is that their deceleration is determined by the asymmetric current i_A, i.e. by the current flowing down from the star in one part of the polar cap and returning onto the star in another part of it. As shown in Section 4.5.2, the asymmetric current makes the main contribution to the neutron star's deceleration only if the angle χ satisfies the condition

$$|\pi/2 - \chi| < (\Omega R/c)^{1/2} \tag{7.50}$$

Given this, the braking torque (4.192), (4.201) associated with the asymmetric current proves to be $(\Omega R/c)$ times less than the braking torque given by (4.215) for $\chi = 0°$.

We shall now establish the basic features of pulsars with $\chi \simeq 90°$. Note that below we shall show the direct connection of such pulsars with millisecond pulsars (see Section 1.5). In our estimations we always consider, along with normal pulsars with $P \lesssim P_{\perp}$, $B_0 \simeq 10^{12}$ G also millisecond pulsars with $P \simeq 1$–10 ms, $B_0 \simeq 10^9$ G.

Using (4.192) and (4.201) we obtain the estimate for the deceleration rate:

$$\dot{P} \simeq 0.3 \times 10^{-19} B_{12}^2 R_6^7 P^{-2} i_A$$
$$\dot{P} \simeq 0.3 \times 10^{-19} B_9^2 R_6^7 P_{-3}^{-2} i_A \tag{7.51}$$

where $R_6 = R \times 10^{-6}$ cm, $B_9 = B_0 \times 10^{-9}$ G^{-1}, $P_{-3} = P \times 10^3$ s^{-1}.

Thus, both for millisecond pulsars and for pulsars with $P \sim P_{\perp}$ we have $\dot{P} \sim 10^{-18}$–10^{-20}. Integrating (7.51) up to $P = P_{\perp}(B_0)$ we obtain for the lifetime of a radio pulsar with $\chi = 90°$,

$$\tau_a^{(\perp)} \simeq 10^4 B_{12}^{-2/3} R_6^{-7} i_A^{-1} \text{ million years}$$

We can see that the lifetime $\tau_a^{(\perp)}$ is extremely large, so that perhaps it exceeds the characteristic extinction time τ_B of a magnetic field (e.g., according to Flowers & Ruderman (1977), $\tau_B \simeq 20$ million years). In this case, the lifetime of pulsars at the stage $\chi \simeq 90°$ is determined by the value of τ_B.

We can easily estimate the radiation power of radio pulsars with $\chi \simeq 90°$ if we make use of the results of Chapter 5 (for more details see Sections 7.5 and 7.6). Indeed, as has already been shown, the intensity of radio emission L_r makes up a certain fraction of the particle energy flux W_p, i.e. $L_r = \alpha_T W_p$, where $\alpha_T \sim 10^{-4}$ (see (3.48)). On the other hand, for $i_A \simeq 1$ the energy flux transferred by particles coincides, in fact, with the total energy losses of a neutron star $W_{\text{tot}} = A_r \Omega \dot{\Omega}$. As a result we obtain

$$L_r^{(\perp)} = 3 \times 10^{22} \alpha_{-4} B_{12}^2 P^{-5} i_A \ \text{erg s}^{-1}, \qquad \alpha_{-4} = \alpha_T \times 10^4 \tag{7.52}$$

For pulsars with $P \simeq P_{\perp}$ we have

$$L_r^{(\perp)} \simeq 10^{25} \alpha_{-4} B_{12}^{-2/9} \ \text{erg s}^{-1} \tag{7.53}$$

For millisecond pulsars we have

$$L_r^{(\perp)} \simeq 10^{31} \alpha_{-4} B_9^2 P_{-3}^{-5} \ \text{erg s}^{-1} \tag{7.54}$$

Finally, the formulas for the magnetic field will be written as

$$B_{12} = P \dot{P}_{-19}^{1/2}, \qquad \dot{P}_{-19} = \dot{P} \times 10^{19}, \qquad B_9 = P_{-3} \dot{P}_{-19}^{1/2} \tag{7.55}$$

Thus, pulsars with angles $\chi \simeq 90°$ must have the following properties (Beskin *et al.*, 1984):

(a) fast ($P \lesssim 0.1$–0.3 s) or superfast ($P \lesssim 0.01$ s) rotation;
(b) an anomalously low value of $\dot{P} \simeq 10^{-18}$–10^{-20};
(c) a weak radio emission: $L_r \simeq 10^{24}$–10^{26} erg s^{-1} for pulsars with $P \lesssim P_\perp = 0.1$–0.3 s, and for millisecond radio pulsars $L_r \simeq 10^{29}$–10^{31} erg s^{-1};
(d) the presence of interpulse (which is connected with the possibility of observing both the magnetic poles);
(e) a reduced value of the magnetic field B_0.

The last circumstance is connected with the possibility of the magnetic field of a neutron star extinguished appreciably within the characteristic time $\tau_a^{(\perp)}$. Millisecond pulsars accelerated due to matter accretion cannot have magnetic fields $B_0 > 10^9$ G (see Section 2.1).

We see that with the present-day sensitivity of receiving devices (see Table 1.2) we can register only superfast 'weakly radiating' pulsars. Millisecond pulsars with period $P < 10$ ms, whose properties were discussed in Section 1.5 (see Table 1.13), obviously belong just to this group. Indeed, as is seen from Tables 1.6 and 1.13, five of the eight millisecond pulsars possess interpulses, their radio luminosities are 10^{29}–10^{31} erg s^{-1}, and the $\dot{P}$ values fall precisely into the interval 10^{-18}–10^{-20}.

It is, however, necessary to remember that according to the theory of evolution of millisecond pulsars they must undergo a binary-system stage when their deceleration dynamics may differ from those of a single neutron star (see Sections 1.5 and 2.1.3). Most millisecond pulsars are actually members of binary systems. For this reason we cannot, in principle, exclude the possibility that their transition into the range of angles $\chi \simeq 90°$ could be influenced by the external moment of forces determined by accreting matter. But a neutron star becomes a radio pulsar only after accretion has ceased. For this reason, irrespective of the transition process, the fact that millisecond pulsars are in the state $\chi \simeq 90°$ undoubtedly confirms the specific feature and stability of this state in agreement with the theory.

The observation concerns weakly radiating pulsars with period $P \lesssim 0.1$–0.3 s is possible only if the sensitivity of current receiving devices is increased by 1–2 orders of magnitude. The discovery of such pulsars would be an important confirmation of the theory.

7.4 Pulsar statistics

7.4.1 The kinetic equation

In a statistical analysis of pulsars it is convenient to consider their distribution function over the magnetic field B_0, the period P, the inclination angle of axes χ and the time t: $N(B_0, P, \chi, t)$. The change of the distribution function is described by the kinetic equation

$$\frac{\partial N}{\partial t} + \frac{\partial}{\partial P}\left(N\frac{dP}{dt}\right) + \frac{\partial}{\partial \chi}\left(N\frac{d\chi}{dt}\right) = U - V_d \tag{7.56}$$

Here dP/dt and $d\chi/dt$ are determined according to (7.28) and (7.29), i.e. they are known functions B_0, P and χ. On the right-hand side of eq. (7.56) the source U describes the production and V_d the disappearance of pulsars.

The characteristic variation time of the source U, determined by the mean lifetime of stars, is much larger than the lifetime of pulsars (7.35). Therefore, the distribution function of pulsars may be thought of as quasi-stationary, i.e. time-independent. Below, we assume for simplicity that the source U is independent of B_0 and P and is equiprobable with respect to the initial angles:

$$U = \frac{2}{\pi} U_P(P) U_B(B_0) \tag{7.57}$$

The extinction function V_d should be non-zero only in the pulsar extinction region, i.e. in the region of parameters corresponding to the condition $Q > 1$. It is therefore convenient first to consider pulsars with $Q < 1$, for which we can put $V_d = 0$. In this case, the 'extinction' will be the intersection of the boundary $\beta_0 = \beta_{\max}(\chi)$ corresponding to the value $Q = 1$.

Substituting into the kinetic equation (7.56) the asymptotic expressions (7.28) and (7.29), which are valid for pulsars with $Q < 1$, taking into account stationarity of the distribution function ($\partial N/\partial t = 0$) and using the integral of motion (4.199), we finally arrive at

$$N_{Q<1}(P, B_0, \chi) = \frac{2}{\pi} \frac{U_B(B)}{B_{12}^{10/7} P \cos^{2\bar{d}-1}\chi} \int_0^P \frac{P' U_P(P')\, dP'}{[1 - \sin^2\chi P'^2/P^2]^{1/2}} \Theta[\beta_m - \beta_0] \tag{7.58}$$

The step function $\Theta[\beta_{\max}(\chi) - \beta_0(P, \chi, B_0)]$ just singles out pulsars with $Q \leqslant 1$.

The distribution function $N_{Q>1}$ for pulsars with $Q > 1$ differs from the corresponding distribution function $N_{Q<1}$ for pulsars with $Q < 1$ since for 'old' pulsars the extinction function V_d becomes important. On the other hand, as has already been mentioned, pulsars with $Q > 1$ are near the extinction boundary $Q = 1$. For such pulsars it therefore suffices to consider the distribution function $N_{Q>1}$ of only two variables, for instance, of the angle χ and the magnetic field B_0. The period P will be determined from the relation

$$P = P_c(B_0, \chi) \simeq B_{12}^{8/15}(\cos \chi)^{\frac{7}{15}(\bar{a} - 1/7)} \tag{7.59}$$

which corresponds to the condition $Q(B_0, P, \chi) = 1$. The quantity $\bar{a} = 0.76$ (see (7.32)). The kinetic equation (7.56) for pulsars with $Q > 1$ can be approximated in the form

$$N_{Q<1}\dot{P} + \tilde{U} - \tilde{V} = 0 \tag{7.60}$$

where the term $N_{Q<1}\dot{P}$ describes the supply of pulsars from the region $Q < 1$, the quantity $\tilde{V}$ corresponds to their extinction, and the creation function $\tilde{U}$ is related to the quantity U, which enters in the kinetic equation (7.56) as

$$\tilde{U} = \int_{P_c}^{\infty} U_{Q>1}\, dP \tag{7.61}$$

where integration is over the region $Q > 1$.

It should be emphasized at once that, as shown in Section 7.1, the rotational dynamics of pulsars with $Q > 1$ is characterized by noticeable irregularities. Therefore, the solution of eq. (7.60) for the function $N_{Q>1}(\chi, B_0)$ cannot be obtained with the same accuracy as for the function $N_{Q<1}$. Nevertheless, we can give a sufficiently reliable estimate of $N_{Q>1}$ from the following simple considerations.

As shown in Section 7.3.1, the lifetimes of pulsars at stages $Q > 1$ and $Q < 1$ approximately coincide. The extinction function can therefore be approximately written as

$$\tilde{V} = \frac{N_{Q>1}}{\tau_a} \tag{7.62}$$

where τ_a is the characteristic lifetime of pulsars at the stage $Q < 1$. In addition, it is natural to assume that the real part of pulsar production in the region $Q > 1$ is smaller than that in the region $Q < 1$, so that $U_{Q<1} > U_{Q>1}$. This allows us to neglect the function $\tilde{U}$ as compared to $\tilde{V}$ in (7.60). As a result, from (7.60) and (7.62) it follows that the distribution function $N_{Q>1}$ of pulsars with $Q > 1$ can be approximately represented as

$$N_{Q>1}(B_0, \chi) = N_{Q<1}[P_c(B_0, \chi), B_0, \chi]\dot{P}\tau_a \tag{7.63}$$

where $N_{Q<1}$ is the distribution function of pulsars with $Q < 1$ for $P = P_c$, the expression for P_c is determined by the relation (7.59), and $\dot{P} = \dot{P}(P_c, B_0, \chi)$ is given by (7.28) and τ_a by (7.34).

We should emphasize that strictly speaking the distribution functions (7.58) and (7.63) should not coincide with the expected distribution functions of observed pulsars. The point is that, as has been said in Section 1.1, radio emission can be registered only when the flux density in a pulse I_ν exceeds some quantity $I_{\min}(P, \dot{P}, DM, d, \ldots)$, characterizing a given radiation receiver (DM is the dispersion measure of a pulsar and d is the distance to it). For example, a pulsar cannot be observed if the Earth does not enter its directivity pattern. Therefore, the distribution function of observed pulsars, $N^{\text{obs}}(P, B_0, \chi)$ must be connected with the stationary functions (7.58) and (7.63) by a relation of the form (1.4):

$$N^{\text{obs}} = NA \tag{7.64}$$

where $A(P, DM, \ldots,)$ is the observability function. Further on, for this coefficient we use a simple expression

$$A = \begin{cases} \pi \sin \chi \dfrac{W_r}{360^\circ}, & \chi > W_r \\[2ex] \pi \sin W_r \dfrac{W_r}{360^\circ}, & \chi < W_r \end{cases} \tag{7.65}$$

(W_r, in degrees, is the window-width of observed radiation) which reflects only the geometric properties of the directivity pattern. More exact expressions for the quantity A which take into account the specificity and sensitivity of different receivers and therefore permit the analysis of the statistical effects connected with different distances to pulsars, are presented in Lyne *et al.* (1985) and Dewey *et al.* (1985) (see Section 1.1 and Table 1.2).

Thus, for a consistent comparison of statistical observational material with theory we should know both the distribution function of pulsars $N(P, B_0, \chi)$ and the observability function A. Knowing these quantities, we can determine any statistical characteristics of pulsars as their mean with respect to the distribution function $N^{\text{obs}}(P, B_0, \chi)$. In particular, such an approach permits reconstruction of the source functions $U_P(P)$ and $U_B(B_0)$ (7.57).

7.4.2 The source function

To begin with, we note that the distribution function of pulsars over the initial periods can be estimated from observations in an independent way. The point is that when we know the kinematic age τ_{kin} of a pulsar and its deceleration rate $\dot{P}$, we can single out those pulsars whose initial period P_0 is not significantly different from the observed value (i.e. pulsars with $\tau_{\text{kin}} \ll \tau_d$). Their distribution in periods is shown in Fig. 7.12. We see that this distribution is close to equiprobable and in any case does not have a noticeable maximum for small periods. It is therefore natural to put

$$U_P(P) = U_P = \text{const} \tag{7.66}$$

The distribution function $U_P(P)$ thus chosen shows that it is only a small part of neutron stars that become radio pulsars with very small periods $P \lesssim 0.1$ s, like the young pulsars 0531+21 (Crab) and 0833−45 (Vela). The conclusion about equal distribution of pulsars over initial periods is in agreement with the results of the paper of Vivekanand & Narayan (1981) in which the necessity of 'injection', i.e. the appearance of pulsars with periods $P \gtrsim 0.5$ s, was shown on the basis of an independent study.

The form of the source function $U_B(B_0)$ is established on the basis of the analysis of observed distribution pulsars over P and $\dot{P}$. As shown below, a rather good interpolation for it is the function

$$U_B(B_0) = \frac{\Gamma(\bar{\gamma} + \bar{\delta} + 1)}{\Gamma(\bar{\gamma} + 1)\Gamma(\bar{\delta})} \frac{1}{B_k} \left(\frac{B_0}{B_k}\right)^{\bar{\gamma}} \left(1 + \frac{B_0}{B_k}\right)^{-1-\bar{\gamma}-\bar{\delta}} \tag{7.67}$$

with certain values of $\bar{\gamma}$, $\bar{\delta}$ and B_k; these constants will be found below. Employing relations (7.58), (7.66) and (7.67), as well as the fact that the

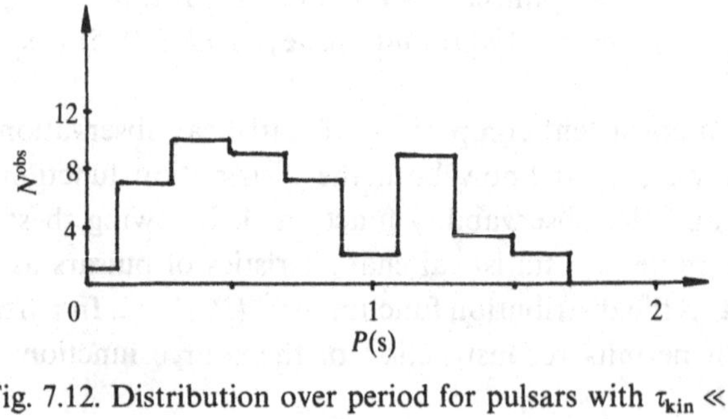

Fig. 7.12. Distribution over period for pulsars with $\tau_{\text{kin}} \ll \tau_d$.

width of the radio window W_r for pulsars with $Q < 1$ can be written as $W_r \propto P^{\bar{v}}$, where (see Section 7.6 below)

$$\bar{v} = -0.1 \pm 0.1 \tag{7.68}$$

we finally obtain for the distribution function (7.64) $N^{\text{obs}}_{Q<1}$

$$N^{\text{obs}}_{Q<1}(P, B_0, \chi) = k_N N_f P^{\bar{v}+1} \frac{1}{B_k}\left(\frac{B_0}{B_k}\right)^{\bar{\gamma}-\frac{10}{7}}\left(1+\frac{B_0}{B_k}\right)^{-1-\bar{\gamma}-\bar{\delta}} \times F_d(\chi)\Theta(\beta_{\max}-\beta_0) \tag{7.69}$$

where

$$F_d(\chi) = 2\frac{1-\cos\chi}{\sin\chi}\cos^{1-2\bar{a}}\chi, \qquad \chi > W_r \tag{7.70}$$

N_f is the total number of pulsars and k_N is the normalization factor of the order of unity. Similarly, for pulsars with $Q > 1$ we have

$$N^{\text{obs}}_{Q>1}(B_0, \chi) = k'_N N_f \frac{1}{B_k}\left(\frac{B_0}{B_k}\right)^{\bar{\gamma}-\frac{10}{7}+\frac{8}{15}(\bar{v}+2)}\left(1+\frac{B_0}{B_k}\right)^{-1-\bar{\gamma}-\bar{\delta}}\tilde{F}(\chi) \tag{7.71}$$

$$\tilde{F}(\chi) = 2\frac{1-\cos\chi}{\sin^2\chi}(\cos\chi)^{1+\frac{7}{15}(\bar{v}+2)(\bar{a}-\frac{1}{7})} \tag{7.72}$$

and according to (7.32) and (7.68), $\bar{a} = 0.76$, $\bar{v} = -0.1$.

We now find the values of the parameters $\bar{\gamma}$, $\bar{\delta}$ and B_k which enter (7.67) and (7.69). To this end we should compare the predictions of the theory with the distribution of observed pulsars over the value of the magnetic field B_0, which for each pulsar is given either by (7.16) or by (7.17). For simplicity we put $\cos\chi \simeq \overline{\cos\chi} = 0.5$; use of the exact distribution of pulsars over the angle χ does not lead to a noticeable difference.

Integrating the distribution function (7.69) over P, χ and the distribution function (7.71) over χ we obtain that both for pulsars with $Q > 1$ and for those with $Q < 1$

$$N^{\text{obs}}(B_0) \propto \left(\frac{B_0}{B_k}\right)^{\frac{8}{15}(\bar{v}+2)-\frac{10}{7}+\bar{\gamma}}\left(1+\frac{B_0}{B_k}\right)^{-1-\bar{\delta}-\bar{\gamma}} \tag{7.73}$$

We compare this distribution with the observed distribution of pulsars with $Q > 1$ and $Q < 1$. We note first that, as is seen from Fig. 7.13, the distribution of observed pulsars for large magnetic fields B_0 has a

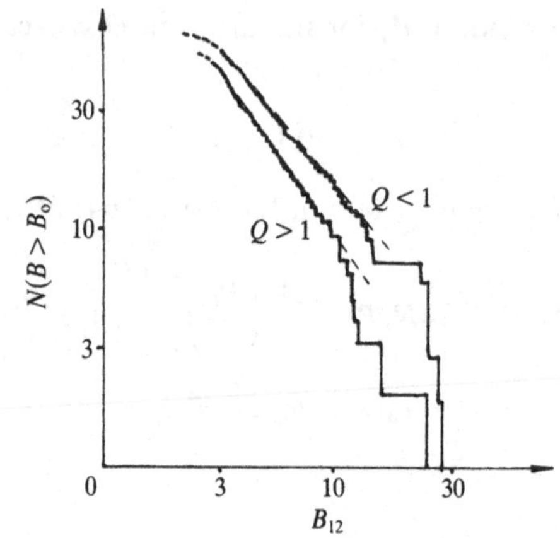

Fig. 7.13. Distribution of observed pulsars over the magnetic field for high B_0 values.

power-like form $N(B_0 > B_1) \propto B_1^{-\bar{s}}$, where

$$\bar{s}_{Q<1} = 1.4 \pm 0.3 \tag{7.74}$$

$$\bar{s}_{Q>1} = 1.9 \pm 0.4 \tag{7.75}$$

Consequently, the source function $U_B(B_0)$ chosen in power-like form (7.67) is in agreement with observations. Since, according to (7.73), the theoretical estimate of the quantity $\bar{s}$ is

$$\bar{s}_{Q>1} = \bar{s}_{Q<1} = -\tfrac{8}{15}(\bar{v} + 2) + \tfrac{10}{7} + \bar{\delta}$$

we are finally led to

$$\bar{\delta} = 0.7 \pm 0.2 \tag{7.76}$$

Similar estimations give the following values of the quantities $\bar{\gamma}$ and B_k:

$$\bar{\gamma} = 2 \pm 1, \qquad B_k = (1 \pm 0.5) \times 10^{12}\ \mathrm{G} \tag{7.77}$$

As shown in Fig. 7.14, the distribution function (7.73) with values of the parameters (7.76) and (7.77) is in agreement with observations for pulsars both with $Q > 1$ and with $Q < 1$.

The last fact seems to be important. The point is that, as is seen already from the kinetic equation (7.56), we carry out the analysis of statistical distribution of pulsars on the assumption that the magnetic field of each individual pulsar does not change substantially within the pulsar lifetime.

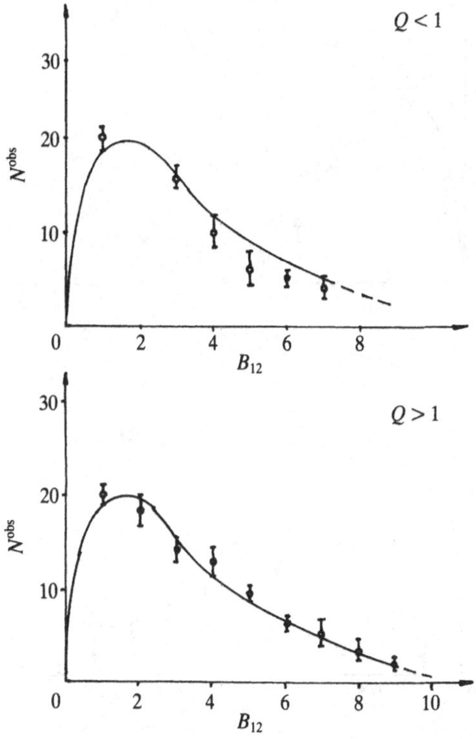

Fig. 7.14. The distribution functions of pulsars over the magnetic field $N^{\text{obs}}(B_0)$ (formula (7.73)). Circles indicate observed values. Vertical bars signify $\pm$SEM.

Good agreement between the theory and experiment both for 'young' ($Q < 1$) and 'old' ($Q > 1$) pulsars confirms this assumption.

The validity of the distribution functions (7.69) and (7.71) can readily be verified on an example of the distribution of observed pulsars over the value of the period P. We obtain

$$N^{\text{obs}}_{Q<1}(P) = \int_0^{\pi/2} d\chi \int_{B_b(P,\chi)}^{\infty} dB_0 N^{\text{obs}}_{Q<1}(P, B_0, \chi) \qquad (7.78)$$

$$N^{\text{obs}}_{Q>1}(P) = \int_0^{\pi/2} d\chi \, N^{\text{obs}}_{Q>1}(B_0, \chi) \frac{dB_b}{dP} \qquad (7.79)$$

where $B_b(P, \chi)$ is determined from (7.59). As is seen from Fig. 7.15, we have here, too, satisfactory agreement between the theory and observations.

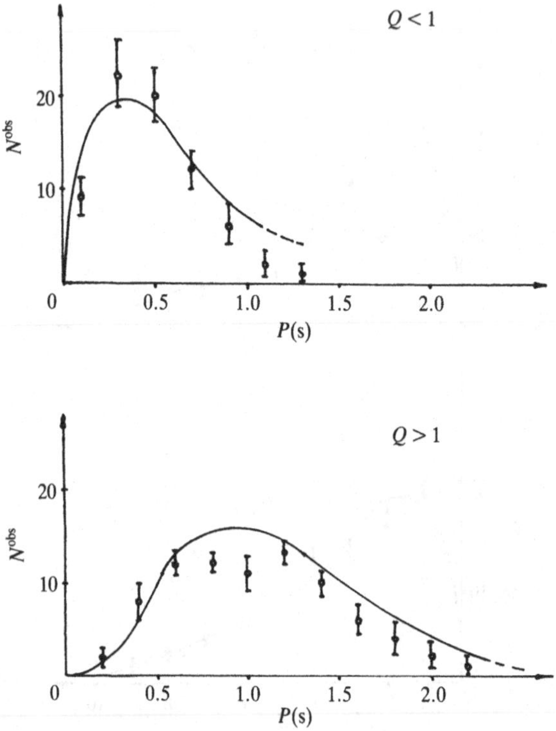

Fig. 7.15. The distribution functions of pulsars over the period $N^{obs}(P)$ (formulas (7.78), (7.79)). Circles correspond to the observed distribution (see Fig. 7.3). Vertical bars signify $\pm$SEM.

Thus, the theory of neutron star evolution allows us to determine with rather good accuracy the distribution function of observed radio pulsars. We should stress that the expressions for the distribution function are distinct for pulsars with $Q > 1$ and with $Q < 1$.

7.4.3 The inclination angle of axes χ

We now turn to the distribution of pulsars over the inclination angle χ. We first consider pulsars with $Q < 1$. Here, for a detailed comparison with observations we should consider pulsars with different values of the period P. However, since the data are insufficient for such a detailed comparison, we divide all pulsars with $Q < 1$ into two groups – those with small ($P < P_1 = 0.5$ s) and those with large ($P > P_1$) periods. The value $P_1 = 0.5$ s is chosen from the condition that the number of pulsars with known angles χ should be close in these two groups. It is of importance that the dependence $N(\chi)$ should be different for small and large periods. So, according to (7.78), for $P < P_1$

$$N_1^{\text{obs}}(\chi) = \int_0^{P_1} dP \int_0^{\infty} dB_0\, N_{Q<1}^{\text{obs}}(P, B_0, \chi) \propto F_d(\chi) \qquad (7.80)$$

where $F_d(\chi)$ is given by (7.70). Correspondingly, for pulsars with $P > P_1$, according to (7.71)

$$N_2^{\text{obs}}(\chi) = \int_{P_1}^{\infty} dP \int_0^{\infty} dB_0\, N_{Q<1}^{\text{obs}}(P, B_0\, \chi) \propto F_d(\chi) \cos^{0.6}\chi \qquad (7.81)$$

The mean value of χ as a function of the period P is determined by the relation

$$\bar{\chi}(P) = \frac{\int\int d\chi\, dB_0\, \chi N_{Q<1}^{\text{obs}}}{\int\int d\chi\, dB_0\, N_{Q<1}^{\text{obs}}} \qquad (7.82)$$

Figure 7.16 gives the comparison of the theoretical predictions with the observed distribution of pulsars over χ (we have chosen only pulsars with $Q < 1$ from Table 1.5). We can see that the distributions of pulsars with small and large periods over the angles χ are essentially different. There is a deficit of pulsars with $\chi \simeq 90°$ for large $P > 0.5$ s values. This is exactly as it should be, since, as mentioned in Section 7.3.2 (Fig. 7.9), the majority of pulsars with $P > 0.5$ s would be outside the limits of the extinction line for high χ values. In such pulsars, the cascade particle production is therefore possible only for sufficiently small angles χ.

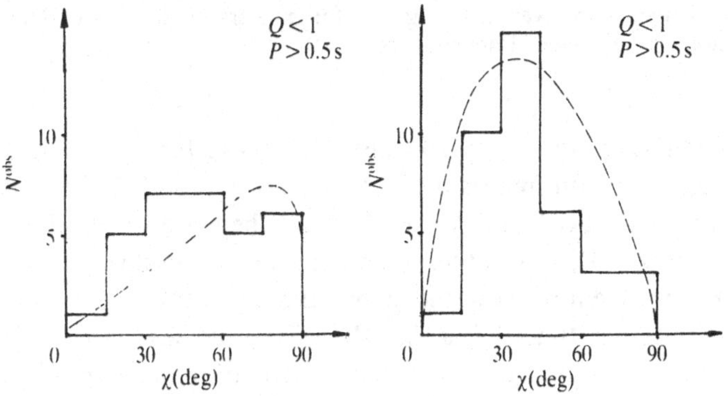

Fig. 7.16. Observed distribution of pulsars with $Q < 1$ over the angle χ. The curves correspond to expected dependences (7.80), (7.81).

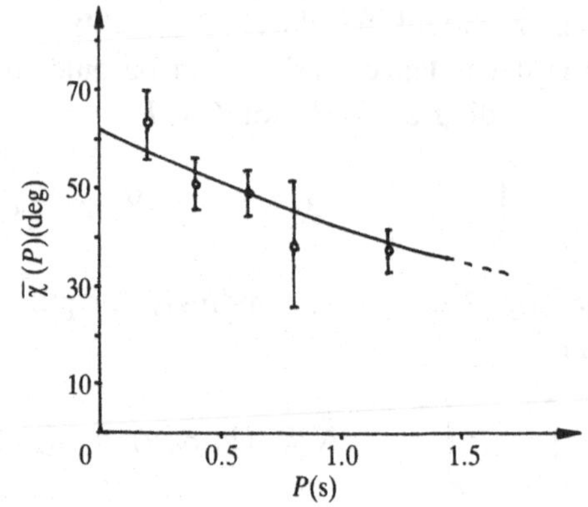

Fig. 7.17. Dependence of the mean $\bar{\chi}(P)$ value (formula (7.82)) on the period P. Circles correspond to the observed distribution.

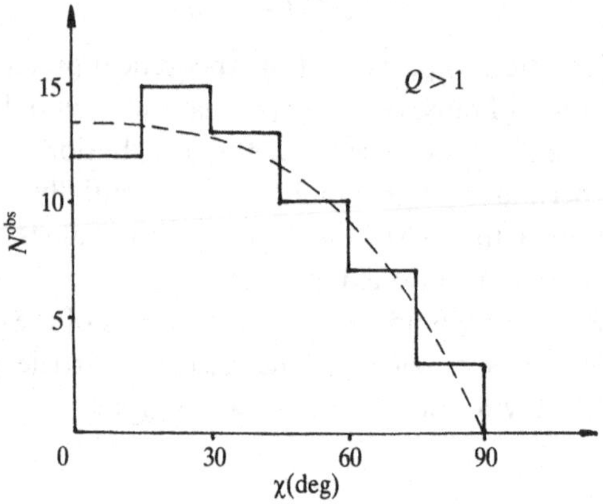

Fig. 7.18. Distribution over the angle χ for pulsars with $Q > 1$. The curve corresponds to the expected dependence (7.72).

For the same reason, the mean value $\bar{\chi}$ (7.82) as a function of the period P must decrease with increasing P. As shown in Fig. 7.17, the observed dependence $\bar{\chi}(P)$, according to Table 1.5 (where we again chose only pulsars with $Q < 1$), is in rather good agreement with theory.

A good agreement between the theory and observations occurs also for pulsars with $Q > 1$. Indeed, Fig. 7.18 shows that the observed distribution over χ for such pulsars agrees with their distribution function $\tilde{F}(\chi)$ (7.71), (7.72). We can see that there exists here a considerable deficit of pulsars for angles χ close to 90°.

7.4.4 Interpulse pulsars

As shown in Section 1.2.11, an interpulse in a radio pulsar is observed in two cases. First, in the case when radio emission comes to the observer from both magnetic poles of a neutron star. The inclination angle χ of the axes of such pulsars is close to 90°, and the observability condition is written in the form (1.27)

$$|\chi - \pi/2| < W_r \tag{7.83}$$

Second, the interpulse radiation can be imitated for angles $\chi \simeq 0°$ when the condition (1.28)

$$\chi < W_r \tag{7.84}$$

holds. In this case, the appearance of two radiation pulses is connected with a double intersection by the line of sight of the directivity pattern, which has the form of a hollow cone (see Fig. 1.24).

When comparing the observed distribution of interpulse pulsars with theory, we should bear in mind that singling out pulsars with $\chi \simeq 90°$ requires a detailed analysis, including an account of sensitivity of receiving devices. The point is that as is seen from (4.179) and (4.201) (see Section 7.3.4), the total energy losses of pulsars with $\chi \simeq 90°$ must on average be less than the characteristic losses of radio pulsars. The radio luminosities of such pulsars must, accordingly, also be less than the characteristic radio luminosities of the majority of sources. Another factor reducing the observed flux from interpulse pulsars with $\chi \simeq 90°$ may be a purely geometric effect (see (7.65)), due to which the observation of an interpulse becomes possible only for a lateral crossing of the directivity pattern.

As is seen from the analysis of observational data, the mean radio luminosity of interpulse pulsars with $P > 0.1$ s,

$$\langle L_r \rangle_{\text{int}} \simeq 10^{27}\ \text{erg s}^{-1} \tag{7.85}$$

is lower by approximately an order of magnitude than the mean luminosity of non-interpulse pulsars with $Q < 1$ (for $P > 0.1$ s):

$$\langle L_r \rangle_{Q<1} \simeq 10^{28}\ \text{erg s}^{-1} \tag{7.86}$$

Singling out pulsars with $P > 0.1$ s only is connected with the fact that radio luminosity of pulsars increases sharply with decreasing P, so that fast pulsars appear to be singled out. However, their number is insufficient to establish statistically reliable relations of the type (7.69), (7.71) in the region of low P values. This fact can also be illustrated by an example of the mean distance to observed interpulse pulsars with $P > 0.1$ s, which is

$$\langle d \rangle_{\text{int}} = 1.0 \pm 0.2 \text{ kpc} \tag{7.87}$$

as distinct from the mean distance to non-interpulse pulsars with $Q < 1$ (for $P > 0.1$ s)

$$\langle d \rangle_{Q<1} = 2.5 \pm 0.5 \text{ kpc} \tag{7.88}$$

As might be expected, interpulse pulsars are not only weakly radiating (7.85), but are observed mostly at small distances, and the condition $\langle L_r \rangle_{\text{int}}/\langle L_r \rangle_{Q<1} \simeq \langle d \rangle^2_{\text{int}}/\langle d \rangle^2_{Q<1}$ holds rather well.

Taking into account that practically all interpulse radio pulsars have $Q < 1$, we obtain from (7.69) that the expected distributions of interpulse pulsars can, subject to the period P, be written in the form $N_{\text{int}} = N_0^{\text{int}} + N_{90}^{\text{int}}$:

$$N_0^{\text{int}}(P) = \int dB_0 \int_0^{W_r} d\chi \, N_{Q<1}^{\text{obs}}(P, B_0, \chi) \tag{7.89}$$

$$N_{90}^{\text{int}}(P) = \frac{\langle d \rangle_{Q<1}^{-2}}{\langle d \rangle_{\text{int}}^{-2}} \int dB_0 \int_{\pi/2 - W_r}^{\pi/2} d\chi \, N_{Q<1}^{\text{obs}}(P, B_0, \chi)\left(1 - \frac{\pi/2 - \chi}{W_r}\right) \tag{7.90}$$

The additional factor $\langle d \rangle_{Q<1}^{-2}/\langle d \rangle_{\text{int}}^{-2}$ takes into account the reduced possibility of observing interpulse pulsars with $\chi \simeq 90°$. A more detailed study of this question should, of course, be carried out on the basis of the general relation (7.64), which makes allowance also for sensitivity of different receivers. In what follows we shall not distinguish between interpulse pulsars with $\chi \simeq 90°$ and $\chi \simeq 0°$ since the quantity χ is not always well known for them.

Figure 7.19 shows the relative number of interpulse pulsars N_{int}/N as a function of the period P, determined from observations on pulsars with $Q < 1$, as well as the expected distribution implied by (7.89) and (7.90). The theory thus not only estimates correctly the total number of pulsars with interpulses, but also explains well the dependence of their relative number N_{int}/N on the period P. Note that the absence of pulsars with interpulses for high P values is first of all due to the fact that such pulsars reach the extinction boundary $Q = 1$ for angles χ much smaller than 90° (see Fig. 7.9). Moreover, the opening of the directivity pattern (and, therefore, the probability of discovering a radio pulsar) decreases with increasing P. For small P, in contrast, the overwhelming majority of observed pulsars must have an interpulse, as is the case, for instance, with

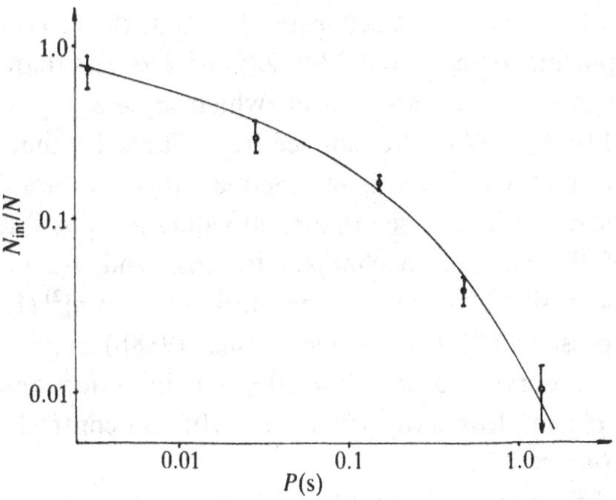

Fig. 7.19. The relative number of interpulse pulsars, N_{int}/N, as a function of the period P.

millisecond pulsars. This is also connected, in particular, with the fact that pulsars with initial rotation periods $P_0 \ll 1$ s go rather rapidly into the region of angles $\chi = 90°$, exactly where their effective accumulation must take place, as is seen from (7.34).

7.4.5 Comparison with the theory of magnetodipole losses

As shown in Chapter 4, the presence of a rather dense plasma in the magnetosphere of a neutron star leads inevitably to a complete suppression of magnetodipole losses – the dynamics of star deceleration is determined by the longitudinal current only. The results presented in this chapter (including those which refer to the properties of pulsar radio emission, Section 7.6) are a convincing confirmation of this theory. In particular, the statistical analysis of observed pulsar distribution, which we have carried out above on the basis of the current-induced deceleration of a neutron star, has shown good agreement between theory and observations. But in the analysis of observational data reported earlier in the literature, the slowing down was usually hypothesized to be due to magnetodipole vacuum losses. It therefore seems reasonable to emphasize the points where the theory of magnetodipole losses does not suggest an explanation of observations and also to formulate briefly certain distinctions in the conclusions following from these two models of pulsar deceleration.

To begin with, we note that according to (7.43) the braking index can, in the case of plasma losses $n_b = 1.93 + 2\bar{d}\tan^2\chi$, be less than 3, as distinct from the magnetodipole mechanism which $n_b \geqslant 3$: $n_b = 3 + 2\cot^2\chi$ (Davis & Goldstein, 1970). We can see from Table 1.3 that in three out of four cases when the index n_b is well defined, the n_b values lie within the interval $2 < n_b < 3$. We also see that such values are possible only in the framework of the current mechanism for loss and are impossible for magnetodipole losses. The second-order braking index $n_b^{(2)}$ (1.9), measured recently for pulsar 0531+21 (Lyne *et al.*, 1988b) suggests the same conclusion. Its observed value $n_b^{(2)} = 10 \pm 1$ is in perfect agreement with the theory of plasma losses ($n_b^{(2)}$(theory) = 10) and contradicts magnetodipole losses ($n_b^{(2)} > 15$).

Next, in the case of current-induced deceleration, the quantity $I_p = \Omega \sin\chi$ (4.199) is invariant. In the case of magnetodipole losses, the quantity (Davis & Goldstein, 1970)

$$I_{\rm md} = \Omega\cos\chi \tag{7.91}$$

is invariant. Hence, irrespective of the deceleration mechanism, pulsars which change period considerably during the course of their evolution must accordingly change the inclination angle of axes considerably.

One such pulsar is 0531+21 (Crab) whose dynamic age of $\tau_d = 2400$ years appreciably exceeds its real age of $\tau = 935$ years. This means that during its lifetime the rotational frequency Ω has fallen significantly. Therefore, the closeness to 90° of the inclination angle of the axes of this pulsar is undoubtedly evidence of the fact that this angle has actually increased with time. This agrees with the theory of plasma losses (4.199) (see Fig. 7.8), and contradicts the magnetodipole loss theory.

Note that the observed decrease of the mean inclination angle of axes with increasing pulsar period P is usually regarded as an important argument in favour of the magnetodipole mechanism of neutron star deceleration (Lyne & Manchester, 1988; Malov, 1990). But as shown in Section 7.4.3, such a dependence is also realized in the framework of the current mechanism (showing good agreement between theory and experiment) in spite of the fact that for each individual pulsar the angle χ increases with time. Therefore, the observed dependence $\bar{\chi}(P)$ only reflects the fact that the conditions for the existence of radio pulsars are connected with particle generation processes in polar regions of a neutron star, which allow neutron stars with a considerable period P to be active radio pulsars only for comparatively low values of the angle χ.

Finally, one further distinction in the conclusions of the theory of plasma and magnetodipole losses is that, as shown in Section 7.4.2, within the model of plasma losses there is no need to introduce an additional assumption about the magnetic field evolution. This is evidently connected with the relatively small active lifetime of a pulsar, $\tau \simeq (4\text{–}8) \times 10^6$ years (7.35), during which the magnetic field does not change substantially. In the magnetodipole mechanism, on the other hand, satisfactory agreement between theory and observations can be obtained only with the assumption that the magnetic field of a neutron star extinguishes noticeably (with the characteristic time $\tau_B = (3\text{–}10) \times 10^6$ years) during the lifetime of a pulsar (Phinney & Blandford, 1981; Chevalier & Emmering, 1986; Narayan, 1987; Stollman, 1987). Note, in conclusion, that a direct experimental determination of the quantity $d\chi/dt$ for an individual pulsar would be an important direct test. Indeed, according to (4.199) and (7.91), the values of $d\chi/dt$ have different signs for current and magnetodipole losses. Such measurements have not yet been carried out.

7.5 Generation of radio emission. The directivity pattern

7.5.1 Basic parameters

We now proceed to the discussion of the properties of observed radio emission of pulsars. To this end, we should first of all determine the basic characteristics of the directivity pattern of radio emission.

As shown in Chapter 4, in internal regions of the magnetosphere the magnetic field can be regarded as dipole. In this case, the curvature radius of the magnetic field line $\rho_l(\mathbf{r})$ and its inclination angle to the magnetic axis $\alpha(\mathbf{r})$ near the axis can be written as

$$\rho_l(\mathbf{r}) = \frac{4}{3}\frac{r^2}{\rho'_\perp} \tag{7.92}$$

$$\alpha(\mathbf{r}) = \frac{3}{2}\frac{\rho'_\perp}{r} \tag{7.93}$$

where $\rho'_\perp$ is the distance from the magnetic dipole axis and r is the distance from the star's centre.

We henceforth use the dipole coordinates f and l, where l is the coordinate along the magnetic field line and the dimensionless coordinate f (see the definitions (4.47) and (4.149)) is constant along the field line. Clearly, for field lines going out of the polar cap, i.e. near the magnetic dipole axis, we can put $l = r$. As a result, the relations (7.92) and (7.93) can be written as

$$\rho_l(\mathbf{r}) = \rho_{l_0} f^{-1/2}\left(\frac{l}{R}\right)^{1/2} \tag{7.94}$$

$$\alpha(\mathbf{r}) = \frac{3}{2}\left(\frac{\Omega R}{c}\right)^{1/2} f^{1/2}\left(\frac{l}{R}\right)^{1/2} \tag{7.95}$$

where

$$\rho_{l_0} \simeq 0.75 \times 10^8 R_6^{1/2} P^{1/2} \quad \text{cm}$$

and R is the star's radius. The particle density also changes along the field line. Employing (4.13) and (4.19), we obtain

$$\omega_p^2 = 2\lambda\omega_B\Omega = 10^{24}\lambda_4 B_{12} P^{-1} G(f)\left(\frac{l}{R}\right)^{-3} \quad \text{s}^{-2} \tag{7.96}$$

where the factor $G(f)$ determines the profile of plasma density generated in the magnetosphere (see Fig. 7.2). Here and below,

$$R_6 = R \times 10^{-6}\ \text{cm}^{-1}, \qquad \lambda_4 = \lambda \times 10^{-4}, \qquad B_{12} = B_0 \times 10^{-12}\ \text{G}^{-1}$$

Thus, in the pulsar magnetosphere there exist two independent parameters – the wave frequency $\nu = \omega/2\pi$ and the distance l to the star. All the other parameters are expressed via these using the relations (7.94)–(7.96). In particular, the parameter a (6.68) introduced in Section 6.3.2 has the form

$$a = 1.5 \times 10^9 \frac{\lambda_4 B_{12} R_6^{2/3}}{\nu_{\text{GHz}}^{2/3} P^{1/3} \gamma_{100}^3} G(f) f^{-1/3}\left(\frac{l}{R}\right)^{-7/3} \tag{7.97}$$

where $\nu_{\text{GHz}} = \nu \times 10^{-9}\ \text{Hz}^{-1}$ and $\gamma_{100} = \gamma_c \times 10^{-2}$.

In Fig. 7.20, on the ν–l plane there are three sectors corresponding to three different regions considered in Section 6.3.2 (see Fig. 6.3). The coordinates of the 'knot' point are

$$\nu^* \simeq \frac{c^{3/4}\Omega^{1/2}\gamma_c^{17/4}}{\lambda^{1/4} R^{3/4} \omega_{B_0}^{1/4}} \simeq 0.6 \frac{\gamma_{100}^{17/4} G^{-1/4}(f) f^{3/4}}{\lambda_4^{1/4} B_{12}^{1/4} R_6^{3/4} P^{1/2}} \quad \text{MHz}$$

$$l^* = \frac{\lambda^{1/2}\omega_{B_0}^{1/2} R^{3/2}}{\gamma_c^{5/2} c^{1/2}} \simeq 3 \times 10^4 R \frac{\lambda_4^{1/2} B_{12}^{1/2} R_6^{1/2} f^{-1/2}}{G^{1/2}(f)\gamma_{100}^{5/2}}$$

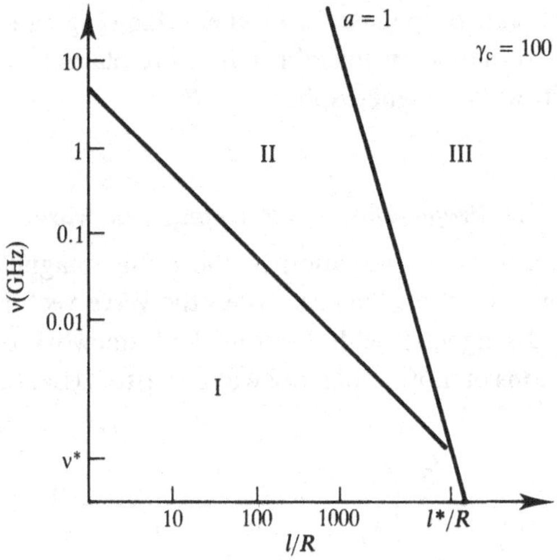

Fig. 7.20. Three regions of parameters distinguished in the pulsar magnetosphere. Maser amplification of curvature-plasma waves is realized in region I. In region III, only two transverse waves can propagate.

We can see that for the characteristic values $\gamma_c \simeq 100$, the frequency ν^* lies much lower than the observed frequency range, and l^* corresponds to distances comparable with the light cylinder radius $R_L = c/\Omega$.

Next, as can readily be obtained from (7.97), the level $a = 1$ will, depending on the frequency, be at a distance

$$l_a \simeq 6 \times 10^3 R \frac{\lambda_4^{3/7} B_{12}^{3/7} R_6^{3/7} G^{3/7}(f)}{\nu_{\mathrm{GHz}}^{2/7} P^{1/7} \gamma_{100}^{9/7}} f^{-2/7}$$

The boundary between regions I and II, according to (6.88), can be written as

$$\nu_{\mathrm{I-II}}(l) = 3.5 \frac{\lambda_4^{1/4} B_{12}^{1/4} \gamma_{100}^{7/4} G^{1/4}(f) f^{1/4}}{R_6^{1/4} P^{1/2}} \left(\frac{l}{R}\right)^{-1} \quad \mathrm{GHz} \tag{7.98}$$

or, which is the same,

$$l(\nu) = 3.5R \frac{\lambda_4^{1/4} B_{12}^{1/4} \gamma_{100}^{7/4} G^{1/4}(f) f^{1/4}}{R_6^{1/4} P^{1/2}} \nu_{\mathrm{GHz}}^{-1} \tag{7.99}$$

Thus, we see that the external regions of pulsar magnetosphere $r > l_a$ will correspond to region III owing to a strong dependence of the particle density on the distance l. As has already been mentioned, only two transverse modes can propagate here, where damping can be neglected. Region I, in which there exist unstable curvature-plasma waves, occupies the internal part of the magnetosphere $r < l(v)$.

7.5.2 Propagation of electromagnetic waves

We now consider wave propagation in the pulsar magnetosphere. For simplicity we dwell only on the case when the wave vector **k** lies in the plane of a curved magnetic field. Then in the framework of geometrical optics, the equations of motion of a ray will be written (Landau & Lifshitz, 1983) as

$$\frac{d\rho'_\perp}{dl} = \frac{\partial}{\partial k_\perp}\left(\frac{k}{n_j}\right), \qquad \frac{dk_\perp}{dl} = -\frac{\partial}{\partial \rho'_\perp}\left(\frac{k}{n_j}\right) \tag{7.100}$$

where the index $\perp$ corresponds to components perpendicular to the dipole axis, for example, $\theta_\perp = k_\perp/k$. Moreover, we assume, for simplicity, the plasma density n_e to be independent on the transverse coordinate $\rho'_\perp$, i.e. $G(f) \equiv 1$ (see (7.96)).

The expressions for the coefficients $n_j(k, \mathbf{r})$ can, in fact, be borrowed from the theory of the homogeneous magnetic field, which was presented in Section 6.2. Indeed, as we have already said, an account of magnetic field inhomogeneity leads only to splitting of the Alfven wave $j = 4$ into three curvature-plasma modes whose dispersion properties do not change substantially, as shown in Figs. 6.1 and 6.4. Therefore, to determine n_j, we can employ relations (6.16) and (6.24a) in which, however, it is necessary to put $\theta = \theta_\perp - \alpha(\mathbf{r})$. The latter change is connected with the fact that (6.24)–(6.27) involve the angle between the vectors **k** and **B**.

As a result, in region III, where there exist only two transverse waves with $n_j = 1$, we have

$$\frac{d\rho'_\perp}{dl} = \theta_\perp, \qquad \frac{d\theta_\perp}{dl} = 0$$

That is, both the waves propagate rectilinearly along the wave vector **k**. In regions I and II, it is only an extraordinary wave $j = 1$ that propagates along a straight line, as shown in Fig. 7.21. Clearly, for this wave the angle $\Theta^{(1)} = \theta_\perp(\infty)$ will practically coincide with the value $\theta_\perp(l_r, f)$ taken at the radiation point $\mathbf{r}_r$. Since, as shown in Section 6.4, the energy conversion into an extraordinary wave is possible only for small angles $\theta \ll a^{-3/20}\theta^*$, we obtain that for an extraordinary wave the directivity pattern will be

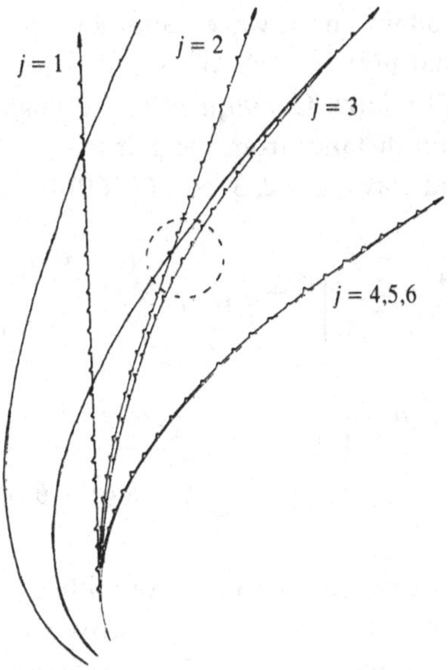

Fig. 7.21. Trajectories of normal mode propagation in a curved magnetic field. The circle indicates the 'tearing-off region'.

determined by the opening of magnetic field lines at the point of its radiation. Consequently,

$$\Theta^{(1)} = \frac{3}{2}\left(\frac{\Omega R}{c}\right)^{1/2} f_r^{1/2}\left(\frac{l_r}{R}\right)^{1/2} \tag{7.101}$$

For Alfven modes with $n_j = 1/\cos\theta$ (and, therefore, also for curvature-plasma modes $j = 4, 5, 6$ in region I), equation (7.100), after the change to the variables f, l, becomes

$$\frac{df}{dl} = 0 \tag{7.102}$$

$$\frac{d\theta_\perp}{dl} = \frac{3}{2}\frac{\alpha(\mathbf{r}) - \theta_\perp}{l} \tag{7.103}$$

Equation (7.102) shows that curvature-plasma waves propagate along magnetic field lines. This has already been pointed out in Section 6.2.2 (see Fig. 6.4). Equation (7.103) can easily be integrated. Making allowance for (7.94) and (7.95), we obtain (Barnard & Arons, 1986)

$$\theta_\perp(l) = \alpha(l_0, f)\left[\frac{1}{4}\left(\frac{l}{l_0}\right)^{-3/2} + \frac{3}{4}\left(\frac{l}{l_0}\right)^{1/2}\right]$$

Here l_0 is the radius on which the wave vector **k** is tangent to the magnetic field vector **B**, so that $\alpha(\mathbf{r}) = \theta_\perp(\mathbf{r})$. We see that for $l \gg l_0$ the angle $\theta_\perp$ tends in magnitude to $\frac{3}{4}\alpha(\mathbf{r})$. Consequently, the angle between **k** and **B** increases as $\frac{1}{4}\alpha(\mathbf{r})$ with distance from the star.

Finally, for normal waves $j = 2, 3$, eqs. (7.100) become

$$\frac{d\rho'_\perp}{dl} = \theta_\perp + \frac{\alpha - \theta_\perp}{2}\left[1 \pm \frac{(\alpha - \theta_\perp)^2}{\left(\dfrac{16}{\omega^2}\left\langle\dfrac{\omega_p^2}{\gamma^3}\right\rangle + (\alpha - \theta_\perp)^4\right)^{1/2}}\right] \tag{7.104}$$

$$\frac{d\theta_\perp}{dl} = \frac{3}{4}\frac{\alpha - \theta_\perp}{l}\left[1 \pm \frac{(\alpha - \theta_\perp)^2}{\left(\dfrac{16}{\omega^2}\left\langle\dfrac{\omega_p^2}{\gamma^3}\right\rangle + (\alpha - \theta_\perp)^4\right)^{1/2}}\right] \tag{7.105}$$

where the plus sign corresponds to a wave with a negative energy and the minus sign to an ordinary wave $j = 2$. Given this, in the region of angles $\theta < \theta^*$ both normal waves $j = 2, 3$ propagate equally, since eqs. (7.104) and (7.105) for $\theta < \theta^*$ are, for both the modes,

$$\frac{d\rho'_\perp}{dl} = \frac{\alpha + \theta_\perp}{2} \tag{7.106}$$

$$\frac{d\theta_\perp}{dl} = \frac{3}{4}\frac{\alpha - \theta_\perp}{l} \tag{7.107}$$

In the region of angles $\theta \gg \theta^*$, the ordinary wave $j = 2$ propagates, as an extraordinary wave rectilinearly, whereas the wave with negative energy (the same as for curvature-plasma modes) propagates along the magnetic field lines. The branching region, where $\theta \simeq \theta^*$ for these two modes, and where, accordingly, their trajectories start to diverge, is also depicted in Fig. 7.21.

The solutions of eqs. (7.106), (7.107) for angles $\theta < \theta^*$ can be written in an explicit form:

$$f(l) = f_0\left(\frac{l}{l_0}\right)^{-3}\left[1.12\left(\frac{l}{l_0}\right)^{1.29} - 0.12\left(\frac{l}{l_0}\right)^{-0.29}\right]^2 \tag{7.108}$$

$$\theta_\perp(l) = 0.82\alpha_0\left(\frac{l}{l_0}\right)^{0.29} + 0.18\alpha_0\left(\frac{l}{l_0}\right)^{-1.29} \tag{7.109}$$

where $f_0 = f(l_0)$ and $\alpha_0 = \alpha(l_0)$. Here again l_0 is the distance at which

$\theta_\perp(l) = \alpha(l)$. Using eq. (7.95) for $\alpha(r)$, we obtain that, for $l \gg l_0$,

$$\theta = \theta_\perp - \alpha = -0.3\alpha_0 \left(\frac{l}{l_0}\right)^{0.29} \tag{7.110}$$

and, therefore, the angle between the vectors **k** and **B** increases gradually with increasing l, as for curvature-plasma waves. Finally, combining (6.25) and (7.110), we obtain the following expression for the angle $\Theta^{(2)} = \theta_\perp^{(2)}(\infty)$:

$$\Theta^{(2)} = \left(\frac{\Omega R}{c}\right)^{0.36} \left[\frac{1}{\omega^2}\left\langle\frac{\omega_{po}^2}{\gamma^3}\right\rangle\right]^{0.070} f_r^{0.36} \left(\frac{l_r}{R}\right)^{0.15} \tag{7.111}$$

where $\omega_{po} = (4\pi n_{eo} e^2/m_e)^{1/2}$ is the plasma frequency value near the star's surface. We see that the quantity $\Theta^{(2)}$ depends not only on the level of generation l_r, f_r, but also on the frequency ν. This is connected with the fact that the 'tearing off level' l_t defined by the condition $\theta = \theta^*$ and equal, according to (6.25) and (7.110), to

$$l_t = 2R\left(\frac{\Omega R}{c}\right)^{-0.48}\left[\frac{\omega_{po}^2}{\omega^2}\langle\gamma^{-3}\rangle\right]^{0.24} f_r^{-0.48}\left(\frac{l_r}{R}\right)^{-0.20}$$

$$\simeq 40RP^{0.24}\nu_{\mathrm{GHz}}^{-0.48}\gamma_{100}^{-0.72}B_{12}^{0.24}\lambda_4^{0.24}f_r^{-0.48}\left(\frac{l_r}{R}\right)^{-0.20}$$

is seen to depend on the frequency of the ordinary wave.

The character of normal wave propagation in internal regions of pulsar magnetosphere is thus essentially different. In particular, an extraordinary wave propagates practically along a straight line, as shown in Fig. 7.21, whereas an ordinary wave deviates from the dipole axis for $\theta < \theta^*$, and it is only at heights $l > l_t$ that the divergence stops. Consequently, the opening of the directivity pattern, determined by the mode $j = 2$, will be much larger than that determined by an extraordinary wave. We shall use this fact in our further analysis of observations. Unstable curvature-plasma waves propagate strictly along magnetic field lines. Note that this conclusion holds rigorously only in the approximation $G(f) \equiv 1$, which disregards the influence of transverse plasma density gradients.

7.5.3 Amplification of curvature-plasma waves

We now turn to estimation of the total optical depth passed by unstable waves when they propagate in internal regions of pulsar magnetosphere.

Making use of the asymptotics (6.78) and (6.79), we obtain

$$\tau_{j=5,6} = 2\frac{\omega}{c}\int \text{Im}\, n_{5,6}\, dl \simeq -890 s_{5,6} J_{5,6} \frac{\nu_{\text{GHz}}^{1/5} R_6^{1/5} \lambda_4^{1/5} B_{12}^{1/5} f^{1/5}}{P^{2/5}\gamma_{100}^{3/5}} \tag{7.113}$$

where for two unstable normal modes $s_5 = \sin(2\pi/5) \simeq 0.95$; $s_6 = \sin(4\pi/5) \simeq 0.59$ and

$$J_{5,6} = \frac{1}{5}\int_1^\infty dx\, x^{-4/5} q_{5,6}(x) \tag{7.114}$$

Here $x = l/R$ and $q_{5,6} = \text{Im}\, \xi_{5,6} s_{5,6}^{-1}(2a/\pi)^{-1/5}$. Clearly, $q_{5,6}(x) \simeq 1$ if $\xi_{5,6}$ is given by the asymptotic expression (6.75) and $q_{5,6} < 1$ for all other asymptotics. For this reason, the upper integration limit in (7.114) can, in fact, be replaced by $x_1 = l(\nu)/R$. Hence

$$J_{5,6} = \left[\frac{l(\omega)}{R}\right]^{1/5} - 1$$

Figure 7.22 shows the value of the optical depth, τ_5, determined using (7.113) and (7.114), as a function of the frequency ν. We see that for characteristic pulsar parameters ($\gamma_c = 100$–500, $P \simeq 1$ s, $B_0 \simeq 10^{12}$ G) the optical depth modulo is several hundred, which corresponds to an amplification by a factor $e^{-\tau} \gtrsim 10^{100}$. It is clear that such an amplification is unrealistic, since the wave energy can by no means exceed the energy of the outflowing plasma. In practice, as shown in Section 6.4, the

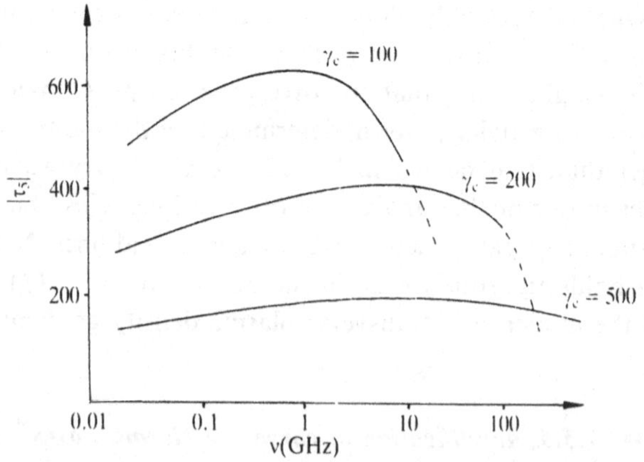

Fig. 7.22. The value of the optical depth τ_5 as a function of frequency ν for $P = 1$ s, $B_0 = 10^{12}$ G.

amplification is stopped earlier by the non-linear effects. Given this, the distance Δr at which such non-linear interaction becomes substantial, can be estimated as

$$\Delta r \simeq \frac{\Lambda R}{|\tau_{5,6}|}$$

where $\Lambda \sim 10$–30 is a logarithmic factor. We see that the value Δr is, in fact, less than or of the order of the star's radius.

It is also of importance that within the framework of geometrical optics, unstable normal waves $j = 5, 6$ cannot freely leave the star magnetosphere. Indeed, as shown in Fig. 7.20, to large distances from the star there corresponds region III in which only two transverse waves $j = 1$ and $j = 2$ can propagate freely. This is because, owing to (6.73) and (6.78), normal modes $j = 4, 5, 6$ leave the amplification cone (6.73) and begin damping rapidly as the angle θ between the vectors $\mathbf{k}$ and $\mathbf{B}$ increases.

Thus, the picture of physical processes leading to generation of intense radio emission of pulsars is of the following nature. The plasma produced in the polar regions of a neutron star and flowing along open field lines is unstable under curvature-plasma wave excitation in it. The instability increment (6.78) proves to be so large that already at a distance of several radii from the star surface, where the formation of a secondary electron–positron plasma ceases, the growth of perturbations is stopped by non-linear processes. The non-linear wave interaction considered in Section 6.4 leads, as we have seen, not only to saturation of curvature-plasma oscillations, but also to their effective transformation into two orthogonal modes – ordinary ($j = 2$) and extraordinary ($j = 1$) – which are capable of leaving the magnetosphere of a neutron star. The transformation and propagation of transverse modes in the magnetosphere determines the basic properties and the directivity pattern of observed pulsar radio emission.

7.5.4 *Intensity of radio emission*

We now proceed to the determination of the spectral radiation density $I_\omega(\Theta)$ in the element of a solid angle $d\Omega_S = \pi\Theta\, d\Theta$; that is, to the determination of the directivity pattern of pulsar radio emission. We shall, as before, consider only the case of a dipole magnetic field and, besides, assume an axial symmetry of the directivity pattern.

We have seen that in a curved magnetic field the wave vectors $\mathbf{k}$ of two normal waves $j = 1, 2$, which are capable of leaving the pulsar magnetosphere, will when propagating deviate from the direction of the magnetic field line. As is seen from (7.103) and (7.105), in the region of small angles $\theta = \alpha - \theta_\perp$ we can put for both the modes

$$\frac{d\theta}{dt} = \frac{c}{\rho_l} \tag{7.115}$$

The transport velocity of electromagnetic oscillations (7.115) out of the interaction region is in fact precisely what determines the efficiency of transformation of curvature-plasma waves into two orthogonal modes which leave the pulsar magnetosphere.

We now make use of the fact that the most efficient transformation is in the region of small angles θ. In this case we may assume radiation from each volume element dV to go only to an angle $\theta = 0$ to which, as shown above, there corresponds quite a definite angle Θ between the dipole axis and the propagation direction of the wave leaving the magnetosphere. The angle Θ, as is seen from (7.101) and (7.111), depends here on the coordinate of the radiation point.

Writing the expression for the volume element as

$$dV = \pi \frac{\Omega l}{c} l^2 \, dl \, df \tag{7.116}$$

we obtain for the spectral energy radiated from the volume element dV within the time dt into the element of angle $d\Theta$:

$$dI_\omega^{(j)}(\Theta) \, d\Theta \, dt = U_\omega^{(j)}(\theta) \frac{\Omega l}{c} l^2 \, d\theta \, df \, dl \tag{7.117}$$

Here $U_\omega^{(j)}(\theta)$ is the spectral density of the established oscillations, integrated over 'transverse' angles $\theta_\perp$, and the index $j = 1, 2$ corresponds to two orthogonal modes capable of leaving the pulsar magnetosphere. In the derivation of (7.117) we have used the fact that $dU_\omega^{(j)} = U_\omega^{(j)}(\theta) \, d\theta$. Integrating (7.117) over all the volume elements from which radiation goes into the given angle Θ, and using (7.116), we have

$$dI_\omega^{(j)}(\Theta) = c \int df \frac{\Omega l^3(f)}{c} U_\omega^{(j)}(\theta) \frac{1}{\rho_l} \frac{dl}{d\Theta} \tag{7.118}$$

where the derivatives $dl/d\Theta$ should be determined by (7.101) and (7.111). Formula (7.118) just determines the structure of the directivity pattern of radio emission.

For our further purposes it will be convenient to rewrite (7.118) in a somewhat different form. To this end we introduce the quantity $\eta_{\rm tr}^{(j)} = U^{(j)}/U_p$, where $U_p = n_e(l)\, m_e c^2 \langle\gamma\rangle = U_0 G(f)(l/R)^3$ is the energy density of outflowing plasma. Then (7.118) can be written as

$$dI_\omega^{(j)}(\Theta) = \frac{\pi c U_0 R^2 \Omega R}{c} \int df\, \eta_{\rm tr}^{(j)} \frac{1}{\rho_l} \frac{dl}{d\Theta} \frac{U_\omega^{(j)}(\theta)}{U^{(j)}} G(f)$$

On the other hand, the quantity $\pi c U_0 R^2(\Omega R/c)$ is simply the flux of energy W_p transferred by particles within the light cylinder (see (4.213)). As a result, we have

$$dI_\omega^{(j)}(\Theta) = W_p \int df\, \eta_{\rm tr}^{(j)} \frac{dl}{\rho_l\, d\Theta} \frac{U_\omega^{(j)}(\theta)}{U^{(j)}} G(f) \tag{7.119}$$

Now we are in a position to show how we can, using (7.119), obtain simple expressions for the basic characteristics of observed radio emission. We consider, for example, the energy spectrum of pulsars. Recall that in the analysis of the spectra one customarily uses the so-called energy in a pulse (1.19):

$$\tilde{E}_\omega = \int_0^P I_\omega(t)\, dt \tag{7.120}$$

i.e. the total energy received at a given frequency ω per pulsar period P. For not very small angles χ, when the 'reduced' pulse width $W_r/\sin\chi \ll 1$, there must correspond to (7.120) the value

$$\tilde{E}_\omega^{(j)} = \frac{2}{\Omega \sin\chi} \int_{|\chi-\beta|}^{\infty} d\Theta \frac{\Theta}{[\Theta^2 - (\chi-\beta)^2]^{1/2}} I_\omega^{(j)}(\theta) \tag{7.121}$$

where β is the angle between the rotation axis of a neutron star and the direction to the observer. In what follows, we consider, for simplicity, the case of the central passage through the directivity pattern $\beta = \chi$, for which (7.121) becomes

$$\tilde{E}_\omega^{(j)} = \frac{2}{\Omega \sin\chi} \int_0^{\infty} d\Theta\, I_\omega^{(j)}(\Theta) \tag{7.122}$$

Now employing (7.119), we obtain

$$\tilde{E}_\omega^{(j)} = \frac{W_p}{\Omega \sin\chi} \int_R^{l_m(\omega)} dl \int_0^{f^*} df\, \eta_{\rm tr}^{(j)}(l, f) \frac{G(f)}{\rho_l} \frac{U_\omega^{(j)}(\theta)}{U^{(j)}} \tag{7.123}$$

where $l_m(\omega)$ corresponds to the maximal frequency $\omega_{\max}(l)$ (formulas (6.106) and (7.98)).

Similarly, for the coefficient of plasma energy transformation into radio emission energy, $\alpha_T = L_r/W_p$, where in the case of central passage $\beta = \chi$,

$$L_r^{(j)} = \int d\omega \int d\Theta\, \Theta I_\omega^{(j)}(\Theta)$$

we have

$$\alpha_T^{(j)} = \int d\omega \int_R^{l_m(\omega)} dl \int_0^{f^*} df \frac{G(f)\eta_{\rm tr}^{(j)}}{\rho_l} \frac{U_\omega^{(j)}(\theta)}{U^{(j)}} \Theta^{(j)}(l, f) \qquad (7.124)$$

Here the quantity $\Theta^{(j)}(l, f)$ is determined by (7.101) or (7.111).

We now proceed to the calculation of the integrals (7.123) and (7.124). For an extraordinary mode, as we have seen (see Section 6.3), $\eta_{\rm tr}^{(1)} = 1/\lambda$, and the expression for $U_\omega^{(1)}(\theta)$ can be written in the form

$$U_\omega^{(1)}(\theta) = \frac{U_\omega^{(1)}}{\theta_\parallel}$$

Therefore, from (6.122),

$$\frac{U_\omega^{(1)}(\theta)}{U^{(1)}} = \frac{\omega}{\omega_{\max}^2} \frac{1}{\theta_\parallel} \qquad (7.125)$$

Substituting these values into (7.122), we obtain

$$\tilde{E}_\omega^{(1)} = \frac{W_p}{\Omega \sin\chi} \lambda^{-1} \omega \int_R^{l_m(\omega)} dl \int_0^{f^*} \frac{df\, G(f)}{\rho_l(l, f)\omega_{\max}^2(l, f) a^{3/20}(\omega, l, f)\theta^*(\omega, l, f)} \qquad (7.126)$$

where the dependences $\rho_l(l, f)$, $\omega_{\max}(l, f)$, $a(\omega, l, f)$ and $\theta^*(\omega, l, f)$ are given, respectively, by (7.94), (7.98), (7.97) and (6.25). Further calculations will be carried out in a rather rough approximation which permits estimations of the basic dependences. In this approximation, integration over df is trivial and uses the fact that for sufficiently smooth functions $G(f)$ falling as $f \to 0$, the integral over df is determined by the upper integration limit f^*, so that $\int df\, G(f) \cdots \simeq 1$. Taking into account the explicit form of the functions entering (7.126), we obtain

$$\tilde{E}_\omega^{(1)} \simeq \frac{W_p}{\Omega \sin\chi} \lambda^{-1} \frac{R\omega^{8/5}}{\rho_{l_0}^{1/5}\omega_{po}^{9/5}c^{4/5}} \int_1^{l_m(\omega)/R} dx\, x^{13/5}\gamma_c^{-23/10} \qquad (7.127)$$

where $x = l/R$, and the index '0' henceforth means that the corresponding quantity is taken on the star's surface $x = 1$.

We thus see that a concrete form of the spectral density $\tilde{E}_\omega^{(1)}$ depends on the variation of the value of the Lorentz factor γ_c with the height l. Assuming the Lorentz factor γ_c to depend weakly on l, we obtain, integrating in (7.127) with allowance made for the expression (7.98) for $l_m(\omega)$,

$$\tilde{E}_\omega^{(1)} = W_p \lambda^{-1} \gamma_c^4 \Omega \omega^{-2} \tag{7.128}$$

Given this, the characteristic opening of the directivity pattern is equal to

$$\Theta_{\max}^{(1)} \simeq \tfrac{3}{2} f^{*1/2} \left(\frac{\Omega R}{c}\right)^{1/2} \left(\frac{\omega_{\max}^{(0)}}{\omega}\right)^{1/2} \tag{7.129}$$

The transformation coefficient $\alpha_{\mathrm{T}}^{(1)}$ (7.129) then becomes

$$\alpha_{\mathrm{T}}^{(1)} = \lambda^{-1} \left(\frac{\Omega R}{c}\right)^{1/2} \Omega \gamma_c^4 \frac{\omega_{\max}^{1/2}}{\omega_{\min}^{3/2}} \tag{7.130}$$

Numerical estimates of all the quantities are given in the section that follows.

For the ordinary mode $j = 2$, we can, as shown in Section 6.3.5, assume

$$U_\omega^{(2)}(\theta) = U_\omega^{(2)}(\Omega_S)\theta^* \tag{7.131}$$

Here the angle θ^* (formula (6.25)) corresponds to the characteristic scale of angles between **k** and **B**, within which the non-linear energy transformation to the mode $j = 2$ is the most efficient. As a result, owing to (6.135), we have

$$\begin{aligned} \frac{U_\omega^{(2)}(\theta)}{U^{(2)}} &= \gamma_c^{3/4} \omega_p^{-1/2} \omega_{\mathrm{br}}^{-23/15} \omega^{31/30}, \qquad \omega < \omega_{\mathrm{br}} \\ \frac{U_\omega^{(2)}(\theta)}{U^{(2)}} &= \gamma_c^{3/4} \omega_p^{-1/2} \omega_{\mathrm{br}}^{2/15} \omega^{-19/30}, \qquad \omega > \omega_{\mathrm{br}} \end{aligned} \tag{7.132}$$

where the break frequency ω_{br} in the spectrum is determined by (6.133). Finally,

$$\eta_{\mathrm{tr}}^{(2)} = \gamma_c^{5/4} \left(\frac{\tilde{\omega}}{\omega}\right)^{-2} \left(\frac{\gamma^+}{\gamma_c}\right)^{152/25} \left(\frac{c}{\omega_p \rho_l}\right)^{21/10} \tag{7.133}$$

As a result, carrying out integration analogous to (7.126), we obtain for the frequencies $\omega < \omega_{br}$

$$\tilde{E}^{(2)}_{\omega} \simeq \frac{W_p}{\Omega \sin \chi} \frac{Rc^{17/30}}{\rho_{l_0}^{47/30}} \omega_{po}^{-13/5} \omega^{31/30} \int_1^{l_m(\omega)/R} dx\, x^{187/60} \left(\frac{\gamma^+}{\gamma_c}\right)^{12/5} \gamma_c^{17/5} \tag{7.134}$$

and, accordingly, for $\omega > \omega_{br}$

$$\tilde{E}^{(2)}_{\omega} \simeq \frac{W_p}{\Omega \sin \chi} \frac{Rc^{17/30}}{\rho_{l_0}^{47/30}} \omega_{po}^{-13/5} \omega^{-19/30} \int_1^{l_m(\omega)/R} dx\, x^{187/60} \left(\frac{\gamma^+}{\gamma_c}\right)^{12/5} \gamma_c^{17/5} \omega_{br}^{5/3} \tag{7.135}$$

We see that the form of the spectrum of the ordinary wave $j = 2$ depends not only on the character of variation of the slow plasma component but also on that of the fast component. It is therefore reasonable to consider here different cases which can occur depending on the character of variation of γ_c and γ^+ as functions of l.

If the mean values $\gamma_c = \langle \gamma^{-3} \rangle^{-1/3}$ and $\gamma^+ = \langle \gamma^{-4} \rangle_+^{-1/4}$ decrease rather rapidly with increasing distance from the star surface (so that the integrand functions in (7.134) and (7.135) fall faster than x^{-1}), these expressions will be determined by the lower integration limit $l = R$. From here on, we call this case 'internal'. Then

$$\tilde{E}^{(2)}_{\omega \text{in}} \simeq \frac{W_p}{\Omega \sin \chi} \frac{Rc^{0.6}}{\rho_{l_0}^{1.6}} \gamma_c^3 \omega_{po}^{-2.6} \omega \tag{7.136}$$

for $\omega < \omega_{br}$ and

$$\tilde{E}^{(2)}_{\omega \text{in}} \simeq \frac{W_p}{\Omega \sin \chi} \frac{Rc^{0.6}}{\rho_{l_0}^{1.6}} \gamma_c^3 \omega_{po}^{-2.6} \omega_{br}^{1.7} \omega^{-0.7} \tag{7.137}$$

for $\omega > \omega_{br}$. Here and below we approximate exponents to two digits.

We see that near the frequency ω_{br} (6.133) in the spectrum determined by the normal mode $j = 2$, there must be a break $\Delta\bar{\alpha}$ whose value for the internal case is

$$\Delta\bar{\alpha}_{\text{in}} = 5/3 \tag{7.138}$$

Given this, the characteristic angular dimension of the directivity pattern is equal to

$$\Theta^{(2)}_{\text{in}} = f^{*0.36} \left(\frac{\Omega R}{c}\right)^{0.36} \left[\frac{1}{\omega^2} \left\langle \frac{\omega_{po}^2}{\gamma^3} \right\rangle\right]^{0.07} \left(\frac{l_r}{R}\right)^{0.15} \tag{7.139}$$

and the transformation coefficient is

$$\alpha_{T_{\rm in}}^{(2)} \simeq \gamma_c^{10} \frac{c^{2.5}}{R^{2.5}} \left(\frac{\Omega R}{c}\right)^{2.1} \omega_{po}^{-2.5} \tag{7.140}$$

If the quantities γ^+ and γ_c depend weakly on l, then the integrals (7.134) and (7.135) will be determined by the upper limit $l_m(\omega)$. In particular, in the limiting 'external' case, when γ^+ and γ_c do not depend on x at all, we obtain

$$\tilde{E}_{\omega_{\rm out}}^{(2)} \simeq \frac{W_p}{\Omega \sin \chi} \gamma_c^{10} \frac{Rc^{3.0}}{\rho_{l_0}^{4.0}} \omega_{po}^{-1.0} \omega^{-3.0} \tag{7.141}$$

for $\omega < \omega_{\rm br}$ and

$$\tilde{E}_{\omega_{\rm out}}^{(2)} \simeq \frac{W_p}{\Omega \sin \chi} \gamma_c^{10} \frac{Rc^{3.0}}{\rho_{l_0}^{4.0}} \omega_{po}^{-1} \omega_{\rm br}^{0.8} \omega^{-3.8} \tag{7.142}$$

for $\omega > \omega_{\rm br}$. In this case, as we see, the break in the spectrum $\Delta\bar{\alpha}$ is equal to

$$\Delta\bar{\alpha}_{\rm out} = \tfrac{5}{6} \tag{7.143}$$

The distinction between the values (7.138) and (7.143) is simply due to the fact that the additional factor $\omega_{\rm br}^{5/3} \propto \rho_l^{5/3}$ in the integrand in (7.134) depends, according to (7.98), on the coordinate l, and the upper limit $l_m(\omega)$ on the frequency ω. According to (7.98) and (7.111), the directivity pattern is now given by

$$\Theta_{\rm out}^{(2)} = f^{0.36} \left(\frac{\Omega R}{c}\right)^{0.36} \left[\frac{1}{\omega^2} \left\langle \frac{\omega_{po}^2}{\gamma^3} \right\rangle\right]^{0.07} \left(\frac{\omega}{\omega_{\rm max}^{(0)}}\right)^{-0.15} \tag{7.144}$$

Finally, the transformation coefficient for the internal case is

$$\alpha_{T_{\rm out}}^{(2)} = \gamma_c^{10} \frac{c^{2.5}}{R^{2.5}} \left(\frac{\Omega R}{c}\right)^{2.1} \omega_{po}^{-0.85} (\omega_{\rm max}^{(0)})^{0.15} \omega_{\rm min}^{-1.8} \tag{7.145}$$

It should be emphasized that the expressions for the opening of the directivity pattern of radiation (7.129), (7.139) and (7.144) do not depend on the approximations used in the calculation of the integrals (7.134) and (7.135) and are, therefore, more exact.

7.6 Radio emission. Comparison with observations

7.6.1 Linear polarization. Two orthogonal modes

We now turn to comparing the predictions of the theory of radio emission generation with the observed properties of radio pulsars. First of all, we consider polarization characteristics of radiation.

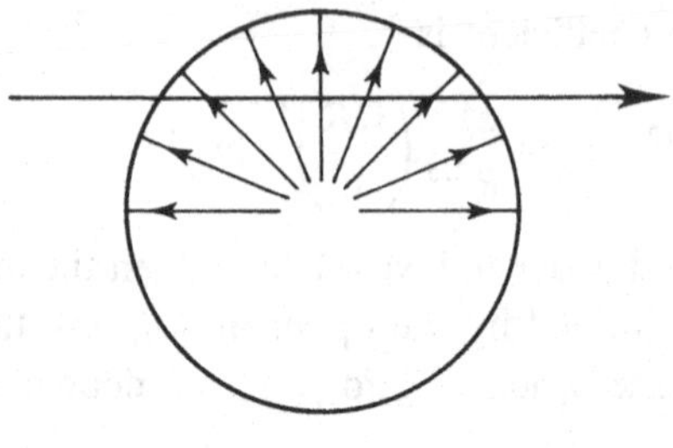

Fig. 7.23. Magnetic field orientation reversal (in the picture plane) relative to the line of sight (the so-called vector model of radio emission (Radhakrishnan & Cocke, 1969)).

As shown in Chapter 6, the energy of unstable curvature-plasma modes is transformed into two transverse waves which leaves the pulsar magnetosphere. These waves are linearly polarized. Their polarization is determined by the structure of the magnetic field in the region of wave generation and propagation. In particular, as is seen from (6.27), the plane of polarization of an extraordinary wave is always orthogonal, while that of an ordinary wave is parallel to the projection of the magnetic field onto the picture plane.

Since, as is seen from Fig. 7.23, the star's rotation reverses the magnetic field orientation relative to the line of sight, the observed position angle ϕ_p (i.e. the angle between the direction of the electric field of the wave and a given direction lying in the picture plane) in each of the two orthogonal modes must change along the mean profile of the pulsar. Introducing, as usual, the longitude $\phi = 2\pi t/P$ which characterizes the position of the signal in the mean profile, we obtain

$$\tan(\phi_p^{(j)} - \phi_{po}^{(j)}) = \frac{\sin\chi \sin\phi}{\sin\beta\cos\chi - \cos\beta\sin\chi\cos\phi}, \qquad \phi_{po}^{(1)} - \phi_{po}^{(2)} = \frac{\pi}{2} \tag{7.146}$$

where β and χ are the angles between the rotation axis and, respectively, the line of sight and the magnetic dipole axis. Formula (7.146) describes the position angle ϕ_p as a function of the longitude ϕ.

Thus, the theory predicts the existence of two radiation modes with mutually orthogonal linear polarizations, which change their position angle along the mean profile according to formula (7.146). As we have seen in Section 1.2.6, it is just these orthogonal modes that are actually observed. For some pulsars (0525+21, 0833−45, 2021+51, 2045−16) the actual run of the position angle agrees so well with (7.146) that this relation is used for a direct estimation of the angle χ (see Section 1.2.8).

7.6.2 The radio window width

The relations (7.129), (7.139) and (7.144) are, in fact, the opening of the directivity pattern of radio emission $W_r^{(j)}$ since (cf. (1.16), (7.121))

$$W_r^{(j)} = \frac{2}{\sin\chi}[(\Theta^{(j)})^2 - (\beta - \chi)^2]^{1/2}$$

We see that the decrease of the pattern's width W_r due to non-central passage ($\beta \neq \chi$) is on average compensated by the fact $1/\sin\chi > 1$. It is therefore natural to limit ourselves in the statistical analysis of pulsars, as in Section 7.5.4, to the case of central passage through the directivity pattern, assuming

$$W_r^{(j)} = 2\Theta^{(j)} \tag{7.147}$$

For individual pulsars, the factor $1/\sin\chi$ can generally be substantial (see Section 1.2.8). As a result, for small enough frequencies $\nu \lesssim 1$ GHz, when the asymptotics (7.98) and (7.99) hold, we obtain

$$W_r^{(1)} \simeq 3.6° P^{-3/4} \nu_{\rm GHz}^{-1/2} \lambda_4^{1/8} B_{12}^{1/8} \gamma_{100}^{7/8} f_r^{5/8} \tag{7.148}$$

$$W_{r_{\rm in}}^{(2)} \simeq 7.8° P^{-0.43} \nu_{\rm GHz}^{-0.14} \lambda_4^{0.07} B_{12}^{0.07} \gamma_{100}^{-0.11} \left(\frac{l}{3R}\right)^{0.15} f_r^{0.36} \tag{7.149}$$

$$W_{r_{\rm out}}^{(2)} \simeq 10° P^{-0.5} \nu_{\rm GHz}^{-0.29} \lambda_4^{0.1} B_{12}^{0.1} \gamma_{100}^{-0.05} f_r^{0.4} \tag{7.150}$$

When the frequencies $\nu \gtrsim 1$–2 GHz, that is, when the level of radio emission generation is close to the star's surface, we have

$$W_r^{(1)} \simeq 2.8° P^{-1/2} f_r^{1/2} \tag{7.151}$$

and the window width $W_{r_{\rm out}}^{(2)}$ does not practically differ from $W_{r_{\rm in}}^{(2)}$ (7.148). We see that the window width should generally be determined by an ordinary mode $j = 2$. But for suppressed ordinary modes the pattern width W_r is determined by an extraordinary wave $j = 1$.

We shall now make a more detailed analysis of some quantitative relations following from the theory. First of all, the values for W_r agree rather well with the characteristic width of the mean profiles of pulsars (Backer, 1976; Rankin, 1983a). Moreover, the relations (7.148)–(7.151) allow us also to explain the observed power-like dependence of the window width on the frequency ν and on the pulsar period P. Indeed, according to (7.148)–(7.151), the quantity W_r for each pulsar must depend on the frequency ν in a power-like manner: $W_r \propto \nu^{-\bar{\beta}}$, and in a wide frequency range

$$\bar{\beta}^{(1)} = 0.5 \qquad (7.152)$$

$$\bar{\beta}^{(2)}_{\text{in}} = 0.14 \qquad (7.153)$$

$$\bar{\beta}^{(2)}_{\text{out}} = 0.29 \qquad (7.154)$$

These values are in close agreement with observations. Thus, Fig. 7.24 illustrates the $\bar{\beta}$-distribution of pulsars (taken from Rankin, 1983b; Slee *et al.*, 1987). We can see that pulsars are concentrated near the $\bar{\beta}$ values determined by (7.152)–(7.154). In addition, according to Kuzmin *et al.* (1986), the averaged $\bar{\beta}$ values are close to (7.152)–(7.154), as is seen from Table 1.4. The pulsars with the pattern width determined by the extraordinary mode $j = 1$ are obviously 0919+06 ($\bar{\beta} = 0.6 \pm 0.07$) and 1842+14 ($\bar{\beta} = 0.57$) (Slee *et al.*, 1987).

The dependence of the window width W_r on the period P will be determined not only by the power-like factors $P^{-0.43}$ and $P^{-0.5}$ in (7.148)–(7.151) but also by the fact that the value of the parameter f_r of the field line responsible for the opening of the directivity pattern is period-dependent. In pulsars with $Q < 1$, the generation of secondary plasma proceeds practically over the entire polar cap surface, as shown in Section 7.1.1. However, near the internal boundary of a hollow cone, $\rho'_\perp = \rho_{\text{in}}$ (i.e. for $f = f_1$) an intense jet of surface current I_1 (7.9) flows

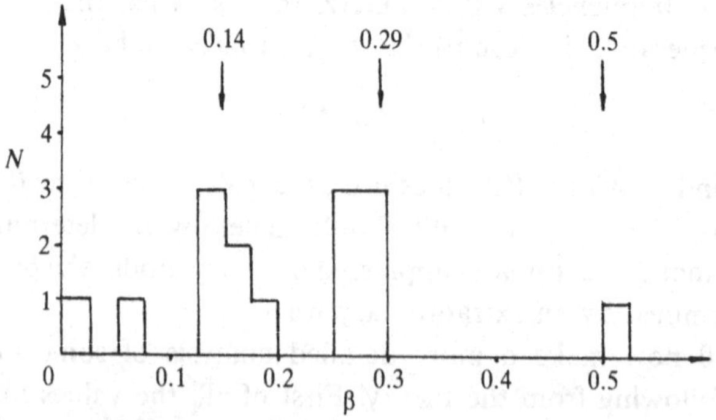

Fig. 7.24. Pulsar distribution over the quantity $\bar{\beta}$ (Rankin, 1983b). Arrows indicate the expected values (7.152)–(7.154).

(see Fig. 7.2). Given this, the value of f_1 is, according to (7.15), in one-to-one correspondence with the parameter Q specified according to (7.12) as

$$f_1 = Q^{14/9} \tag{7.155}$$

On the other hand, the intensity of radio emission $I_\nu = \eta_{tr} W_p$ increases, naturally, with increasing energy flux density of radiating particles. For pulsars with $Q < 1$, we should therefore assume $f_r \simeq f_1$. For pulsars with $Q > 1$, particles are generated in a ring with $f_r \simeq 1$, as shown in Fig. 7.2.

As a result, for pulsars with $Q > 1$ we might expect the dependence $W_r \propto P^{\bar{\upsilon}}$, where $\bar{\upsilon}_{in} = -0.43$ or $\bar{\upsilon}_{out} = -0.5$, while for pulsars with $Q < 1$ we find $\bar{\upsilon} = -0.07$ due to the change of $f_1(Q)$ according to (7.155). The analysis of observations has shown that for pulsars with $Q > 1$ the quantity $\bar{\upsilon}$ calculated using the least-squares method is

$$\langle \bar{\upsilon} \rangle_{Q>1} = -0.48 \pm 0.07 \tag{7.156}$$

For pulsars with $Q < 1$, in contrast (cf. Malov & Sulejmanova, 1982),

$$\langle \bar{\upsilon} \rangle_{Q<1} = -0.12 \pm 0.14 \tag{7.157}$$

We see that the observed dependence of the radio window width on the period is also in satisfactory agreement with the theory. Note that the value $\bar{\upsilon} = -0.12 \pm 0.14$ (7.157) has already been used above in Section 7.4 (see (7.68)).

7.6.3 *The structure of the mean profile*

The structure of the mean profile is one of the most important characteristics of pulsar radio emission. According to the theory developed above, the mean profile must contain at least two components corresponding to two orthogonal modes of radiation. As has already been mentioned, for each of the modes the directivity pattern must follow the complicated density profile $G(f)$ shown in Fig. 7.2. This must yield a considerable variety of forms of the mean profiles of radio pulsars.

We shall discuss the specifics of the structure of the mean profiles which follow from the theory (see Fig. 7.25). To begin with, we have seen that in pulsars with $Q < 1$ the largest current flows near the internal boundary of the plasma outflow cone. In the integral radiation profile of such pulsars there must therefore exist an intense single central component. As shown in Fig. 7.25a, it can be associated both with ordinary mode radiation by

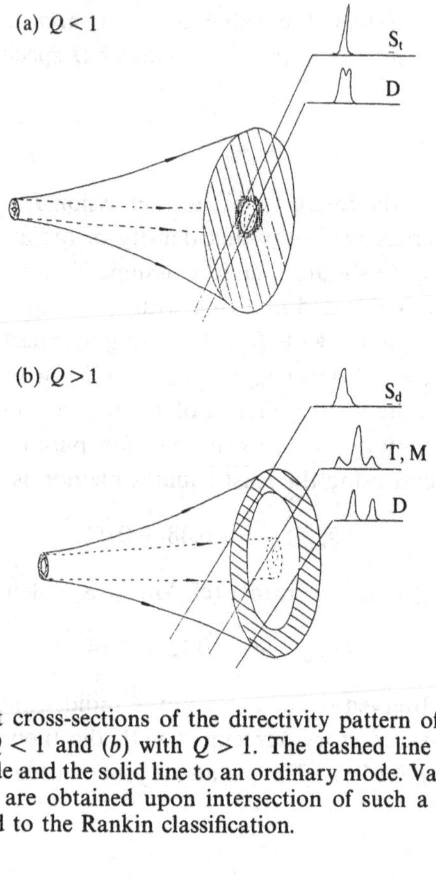

Fig. 7.25. Different cross-sections of the directivity pattern of radio emission of pulsars (*a*) with $Q < 1$ and (*b*) with $Q > 1$. The dashed line corresponds to an extraordinary mode and the solid line to an ordinary mode. Various mean profiles are shown, which are obtained upon intersection of such a directivity pattern. Letters correspond to the Rankin classification.

plasma flowing near the internal boundary of the hollow cone and merely with an extraordinary mode $j = 1$.

As to pulsars with $Q > 1$, as shown in Fig. 7.25b, the plasma outflow diagram has the form of a rather thin ring. Depending on the mutual orientation of the directivity pattern and the line of sight, both single and double profiles are possible here, the current amplification near the internal boundary of plasma outflow being insignificant in such pulsars (see Section 7.1.1). Consequently, in pulsars with $Q > 1$ the radiation must be of smooth peripheral character. The central component can be associated only with an extraordinary mode whose directivity pattern has an opening (7.148) much smaller than the observed radio window width (7.149), (7.150). Thus, the theory naturally explains the existence of two components in pulsar radio emission. Given this, the core component should be most clearly pronounced in pulsars with $Q < 1$. As a result, as

Table 7.4. *Comparison of the Rankin (1983a, 1986, 1990) classification with the values of parameter Q*

	Rankin classification				
	M	T	D	S_d	S_t
All	14	43	17	21	50
$Q < 1$	1	23	0	2	48
$Q > 1$	13	20	17	19	2

shown in Fig. 7.25, the theory suggests the following classification of possible mean profiles of pulsar radio emission:

I. Single-hump profile associated with the core component.
II. Single-hump profiles formed due to side passage through the peripheral component.
III. Two-hump profiles due to central passage through the peripheral component.
IV. Complex profiles existing in the case when the radiation components have similar intensities.

This picture completely corresponds to the detailed phenomenological classification of mean profiles given in Section 1.2.7. Indeed, pulsars of class I are naturally referred to as pulsars of class S_t. As shown in Table 7.4, practically all S_t pulsars actually have the parameter $Q < 1$. Pulsars of classes II and III with mean profiles determined by the peripheral component must, on the contrary, belong to classes S_d and D. The parameter Q of such pulsars must exceed unity. As we can see from Table 7.4, it is exactly this picture that is observed in reality.

Finally, pulsars of class IV can naturally be associated with pulsars of classes M and T. They are intermediate in the sense that the intensities of their central and peripheral components are of the same order of magnitude. As might be expected, these are pulsars with $Q \simeq 1$. Moreover, in some cases it is also possible to single out the central component obviously associated with just the extraordinary mode. Such a component is observed, in particular, in pulsars 1541+09, 1737+13, 1821+05, 1944+17 and 1952+29 (Hankins & Rickett, 1986) (all of them have parameter $Q > 1$ and belong to classes M and T). Furthermore, in pulsars

0834+06 (Class D) (Gil, 1987) and 1604−00 (class T) (Rankin, 1988), one of the orthogonal modes is observed only in the central coffer-dam and the other in the region of the main double-hump profile. Pulsars 0919+06 and 1842+14 (which have the values $\bar{\beta} \simeq 0.5$ and are therefore naturally associated with the extraordinary mode $j = 1$) belong to class S_t as follows from the theory.

7.6.4 The frequency range of observed radio emission

As shown in Section 6.2, an amplification of unstable curvature-plasma waves is only possible for frequencies $\omega < \omega_{max}$, where ω_{max} is given by (7.98). Thus, the value

$$\nu_{max} \simeq 3.5P^{-1/2}\gamma_{100}^{7/4} \quad \text{GHz} \tag{7.158}$$

is just the upper boundary of frequencies generated in pulsar magnetosphere. The lowest frequency ν_{min} can be determined by the propagation conditions. For example, the refractive index n_2 of an ordinary wave becomes equal to zero for $\nu < \nu_{min}$ (see (6.24a)), and hence for $\nu < \nu_{min}$ (and for angles $\theta < \theta^*$) such a wave can propagate only towards the star. In this case

$$\nu_{min} \simeq 120P^{-1/2}\gamma_{100}^{-3/2}\lambda_4^{1/2}B_{12}^{1/2} \quad \text{MHz} \tag{7.159}$$

We shall compare these values with those of high-frequency and low-frequency cutoffs observed in the spectra of many pulsars (see Section 1.2.4):

$$\nu_{max}^{obs} \simeq 3P^{-(0.62 \pm 0.19)} \quad \text{GHz} \tag{7.160}$$

$$\nu_{min}^{obs} \simeq 100P^{-(0.38 \pm 0.09)} \quad \text{MHz} \tag{7.161}$$

We can see good agreement between the theory and observations. Comparing the theoretical dependences (7.158), (7.159) with the observed ones (7.160), (7.161) we can also see a correct theoretical cutoff frequency dependence on the pulsar period P.

Note that the nature of low-frequency cutoff could also be associated with other processes. For example, as is seen from (6.130), the low-frequency cutoff can be due to cessation of energy conversion into an ordinary wave. In this case,

$$\nu_{min} = 80P^{-1/2}\gamma_{100}^{3} \quad \text{MHz}$$

which estimate is also sufficiently close to the observed value (7.159).

7.6.5 The radio emission spectrum

As shown in Section 7.5.4, the theory predicts a power-law spectrum of pulsar radio emission, so that $\tilde{E}_\nu \propto \nu^{-\bar{\alpha}}$. For an extraordinary wave $j = 1$ the spectral index $\bar{\alpha}$ must be close to 2 (see (7.128)), so that

$$\tilde{E}_\nu^{(1)} \propto \nu^{-2} \tag{7.162}$$

Insignificant variations of the spectral index $\bar{\alpha}$ may occur only due to a weak height dependence of the Lorentz factor γ_c. The spectrum of the ordinary wave $j = 2$ can be described by a whole set of possible $\bar{\alpha}$ values and is characterized, in addition, by the presence of a break at the frequency (6.133)

$$\nu_{\rm br} \simeq 3P^{-1/2}\gamma_{100}^3\left(\frac{\gamma_{300}^+}{\gamma_{100}}\right)^{12/5} \quad \text{GHz} \tag{7.163}$$

As shown in Section 7.5.4, the parameter $\bar{\alpha}$, as is the value of the break $\Delta\bar{\alpha}$, is determined by the magnetosphere region making the basic contribution to pulsar radio emission. So, if radiation is generated mainly at small distances from the star, the internal case holds, where the value of the break $\Delta\bar{\alpha}_{\rm in} = 5/3$ and the spectrum has the form

$$\tilde{E}_{\nu_{\rm in}}^{(2)} \propto \begin{cases} \nu^{1.0}, & \nu < \nu_{\rm br} \\ \nu^{-0.7}, & \nu > \nu_{\rm br} \end{cases} \tag{7.164}$$

If radiation is formed at distances $r = l_m(\nu)$ (7.99), the external case holds when the spectral index can take on the values (7.142), (7.143):

$$\tilde{E}_{\nu_{\rm out}}^{(2)} \propto \begin{cases} \nu^{-3.0}, & \nu < \nu_{\rm br} \\ \nu^{-3.8}, & \nu > \nu_{\rm br} \end{cases} \tag{7.165}$$

It is clear that in reality the intermediate case is also possible, and so the limits of the spectral index are $-1 \leqslant \bar{\alpha} \leqslant 3$. At the same time, as has already been emphasized, the value of the break in the spectrum is independent of the spectral index $\bar{\alpha}$ and is equal to $\Delta\bar{\alpha}_{\rm out} = 5/6$.

It is noteworthy that there is a certain agreement between the theoretical and experimental data presented in Section 1.2.9. First of all, the observed spectrum of pulsar radio emission is indeed of a power-law form. Given this, the absolute values of the spectral index $\bar{\alpha}$ are on the whole in reasonable agreement with observations. Hence, the spectral index of an extraordinary wave $j = 2$ (7.162) coincides with the mean value $\langle\bar{\alpha}\rangle = 2.0 \pm 0.1$ (1.21) obtained by Malov & Malofeev (1981) for 43 pulsars in the high-frequency part of the spectrum. In many cases, a break is observed in the radio emission spectrum (see Fig. 1.19). The analysis made by Kuzmin *et al.* (1986) for 21 pulsars exhibiting a break in the spectrum has shown that $\langle\bar{\alpha}\rangle = 1.7 \pm 0.3$ (1.22) for frequencies $\nu < \nu_{br}$ and $\langle\bar{\alpha}\rangle = 3.1 \pm 1.1$ (1.23) for $\nu > \nu_{br}$. In addition, as shown in Fig. 7.26 for 12 pulsars with $Q > 1$, the value $\langle\bar{\alpha}\rangle = 1.7 \pm 0.3$ for $\nu < \nu_{br}$ and $\langle\bar{\alpha}\rangle = 2.9 \pm 0.4$ above the break frequency. These values can be compared with the limiting values $\bar{\alpha} = 3.0$ ($\nu < \nu_{br}$) and $\bar{\alpha} = 3.8$ ($\nu > \nu_{br}$) which must, according to (7.165), be observed in the spectrum determined by the ordinary wave provided that radio emission is formed at large distances from the star (the 'external' case). It is only for the 'internal' version (7.164) that the theory leads to the spectrum with a maximum in the region of the frequency ν_{br}, which is not the case in reality.

Finally, for pulsars showing a break in the spectrum as depicted in Fig. 7.27, the ratio $\nu_{br}^{obs}/\nu_{br}^{th}$, where ν_{br}^{th} is given by (7.163), is close to unity.

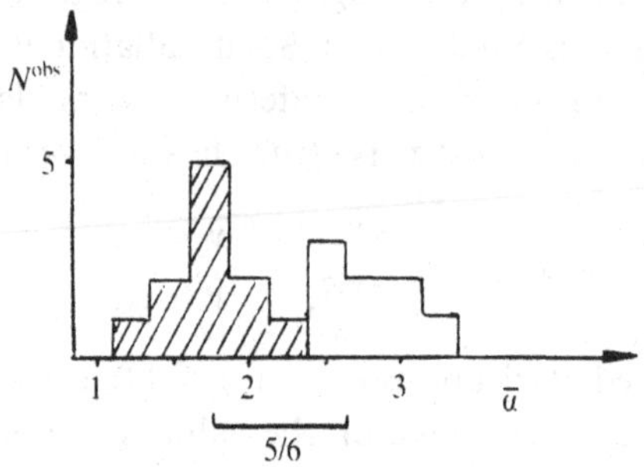

Fig. 7.26. Distribution of 'old' ($Q > 1$) pulsars over the spectral index $\bar{\alpha}$. Shaded area is the distribution for frequencies $\nu < \nu_{br}$.

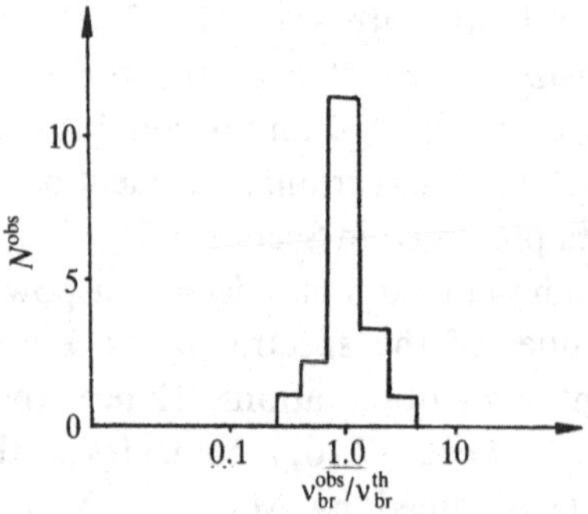

Fig. 7.27. Pulsar distribution over the value $\nu_{br}^{obs}/\nu_{br}^{th}$.

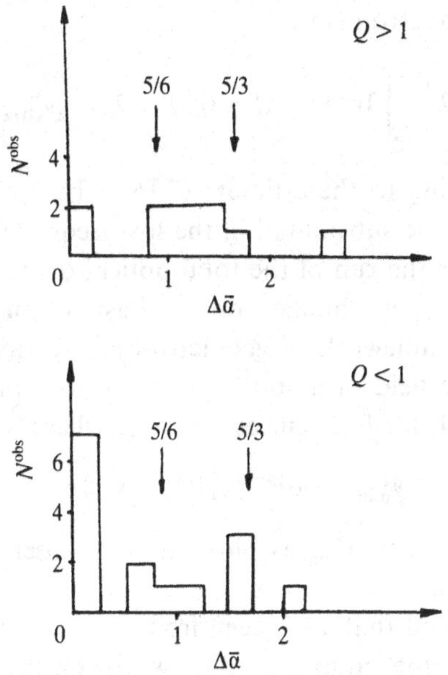

Fig. 7.28. Pulsar distribution over the break in the spectrum $\Delta\bar{\alpha}$. Arrows indicate the expected values (7.138) and (7.143).

The difference in the spectral indices $\Delta\bar{\alpha}$, as is seen from Fig. 7.28, is also concentrated in the range of values specified by the theoretical formulas (7.138) and (7.143).

7.6.6 Cyclotron absorption and circular polarization

Another cause of a low-frequency cutoff can be an absorption of electromagnetic waves in the region of cyclotron resonance $\omega_B/\gamma = \omega - k_z v_\parallel$ (6.29). According to (7.95) and (7.96), the cyclotron resonance condition is fulfilled at heights $r \simeq r_c$, where

$$r_c = 2 \times 10^3 R \lambda_4^{1/3} B_{12}^{1/3} \nu_{\rm GHz}^{-1/3} \gamma_{100}^{-1/3} \qquad (7.166)$$

i.e. in a distant region of the magnetosphere, perhaps even in the region of the light surface (in fast pulsars with $P < 0.1$ the region of cyclotron resonance can generally be outside the light surface). Using (6.38), we obtain the estimate for the total optical depth associated with cyclotron absorption, for two orthogonal modes $j = 1, 2$ propagating at large distances from the neutron star,

$$\tau_{1,2} = 2\frac{\omega}{c}\int \mathrm{Im}\, n_{1,2}\, dl \simeq 0.3P^{-1}\lambda_4 B_{12}\nu_{\mathrm{GHz}}^{-1/3}\gamma_{100}^{1/3} \qquad (7.167)$$

We see that according to the estimate (7.167) the cyclotron absorption can indeed prove to be substantial in the low-frequency range.

Figure 7.29 shows the run of the total optical depth $\tau_{1,2}$ as a function of the wave frequency ν, obtained on the basis of an exact calculation carried out in the framework of geometro-optical wave propagation in the dipole magnetic field of a star. Cyclotron absorption is seen to be rather large, $\tau_{1,2} > 1$, for frequencies $\nu < \nu_{\min}^c$, where

$$\nu_{\min}^c \simeq 30P^{-3}\lambda_4^3 B_{12}^3 \quad \mathrm{MHz} \qquad (7.168)$$

Note that the frequency $\nu_{\min}^c$ is close to the observed values of the low-frequency cutoff.

It should be stressed that, as is seen from Fig. 7.29, the frequency $\nu_{\min}^c$ depends strongly on the value λ, in other words on the density of plasma flowing in the pulsar magnetosphere. Therefore, a detailed comparison of observation of cyclotron absorption with the theory would make it possible to determine this important parameter. From Fig. 7.29 and formula (7.168) we can conclude that λ is of the order of 10^4–10^5, which agrees with the results of the theory of particle generation near the neutron star's surface (see Sections 3.2.2, 5.3).

The analysis of cyclotron interaction between normal waves and plasma

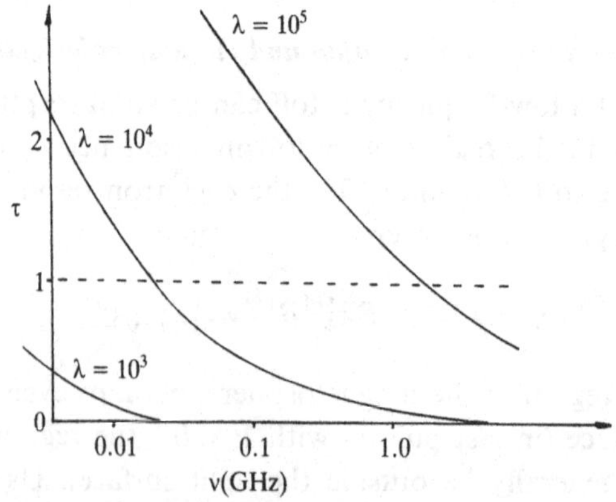

Fig. 7.29. The total optical depth of cyclotron absorption as a function of the frequency ν for various values of λ ($P = 1$ s, $B_0 = 10^{12}$ G).

at heights $r \simeq r_c$ (7.166) also predicts the possibility of circular polarization observed in some cases in two orthogonal modes (Stinebring *et al.* 1984a,b). Indeed, according to Barnard (1986), the region of limiting polarization of radio emission (for pulsars with $P > 0.06$ s) is in the region of cyclotron resonance. In other words, the polarization characteristics of radio emission going to infinity can be determined by normal wave polarization in the region $r \simeq r_c$. And in Chapter 6 it was shown (see (6.36)) that already in the region of cyclotron resonance the polarization of normal waves is not strictly linear. The degree of circular polarization here can reach 10–30%, which agrees with observations (see Section 1.2.6).

Consider, in particular, the question of the appearance of circular polarization for an extraordinary mode $j = 1$, which we associated in Section 7.6.3 with the 'core' component of pulsar radio emission. According to (6.37), the polarization of normal waves becomes circular:

$$\left(\frac{E_x}{E_y}\right)_{1,2} = \mp i$$

The polarization vector of the extraordinary mode $j = 1$ rotates counter-clockwise and that of the ordinary mode in the opposite direction. The refractive indices for left- and right-polarized waves are not equal, according to (6.38):

$$n_{1,2} = 1 \pm \theta^2 \left\langle \frac{\omega_p^2}{4\omega(\omega_B \mp \gamma\tilde{\omega})} \right\rangle \tag{7.169}$$

one of them having resonance on positrons, the other on electrons. This difference can cause the Faraday effect, i.e. the rotation of the polarization vector on leaving the resonance region. But we are interested in another effect, namely, the appearance of a circularly polarized component of radiation due to the non-uniform profile of radiation intensity. Indeed, the group velocities of two modes are also different:

$$\mathbf{v}_{g_1} - \mathbf{v}_{g_2} = c \left\langle \frac{\omega_p^2 \omega_B}{\omega(\omega_B^2 - \gamma^2 \tilde{\omega}^2)} \right\rangle \mathbf{b} \tag{7.170}$$

($\mathbf{b}$ is a unit vector along the magnetic field direction, $\frac{1}{2}\theta^2 = 1 - \cos\theta = 1 - \mathbf{kb}/k$). For this reason, left- and right-polarized waves come to us from different regions of magnetosphere, where the radiation intensities are distinct. Therefore, after undergoing resonance the circular component of radiation does not vanish completely. With the intensity of a circularly polarized wave denoted by I_ν^c, we have

$$I_\nu^c(x) = I_\nu(x_1) - I_\nu(x_2) = \frac{dI_\nu}{dx}(x_1 - x_2)$$

where x is the coordinate along the line of sight in the picture plane. Introducing the variable $\phi = x/l$, i.e. the longitude on which the intensities I_ν and I_ν^c depend, and taking into account $d(x_1 - x_2) = (v_{g_1} - v_{g_2})_x \, dl/c$, we obtain

$$I_\nu^c(\phi) = \frac{dI_\nu}{d\phi} \int \frac{dl}{l} \left\langle \frac{\omega_p^2(l)\omega_B(l) b_x(l, \phi)}{\omega[\omega_B^2(l) - \gamma^2 \tilde{\omega}^2(l)]} \right\rangle \tag{7.171}$$

Clearly, the x-component of the unit vector depends not only on the longitude ϕ, but also on the distance l from the star. The point is that an extraordinary wave propagates along a straight line, and during the time necessary for it to reach the resonance region the star turns through a finite angle. After averaging in (7.171) over the distribution functions of positrons and electrons and integration over the distance l from (7.171), we come to

$$I_\nu^c(\phi) = \frac{4\lambda}{P\nu} \left| \cos\chi \ln \frac{\gamma_{\min}^+}{\gamma_{\min}^-} \right| \frac{dI_\nu}{d\phi} \tag{7.172}$$

Figure 7.30 presents the intensity profiles for several typical 'core' components of individual pulsars (see also Fig. 1.15). We can see that the behaviour of the circular polarization component corresponds qualitatively to the expression (7.172). In particular, the sign reversal of $I_\nu^c(\phi)$ correlates very well with the maximal of the total intensity I_ν.

The value of λ can be found directly from the relation between the total intensity I_ν and the intensity of a circularly polarized wave I_ν^c. To obtain an approximate estimate of λ, in (7.172) we replace the derivative $dI_\nu/d\phi$ by $2I_\nu^{\max}/W_r$:

$$\lambda_4 \ln(\lambda_4 W_{r_{10}}) \simeq 20 P \nu_{\text{GHz}} W_{r_{10}} \left(\frac{I_\nu^c}{I_\nu} \right)_{\%} \cos^{-1}\chi \tag{7.173}$$

Here $W_{r_{10}}$ is the width of the radiation profile in degrees normalized to 10°, and I_ν^c is the maximal value of the intensity of the circularly polarized component. For pulsars 2002+31 and 1933+16 we have $\lambda_4 \simeq 10$, which corresponds to $\lambda \simeq 10^5$, i.e. it agrees with the theory presented in Section 5.3. Note that for a frequency of 430 MHz, for which the intensity is given, the value $\lambda \simeq 10^5$ corresponds to a small cyclotron absorption (as is seen

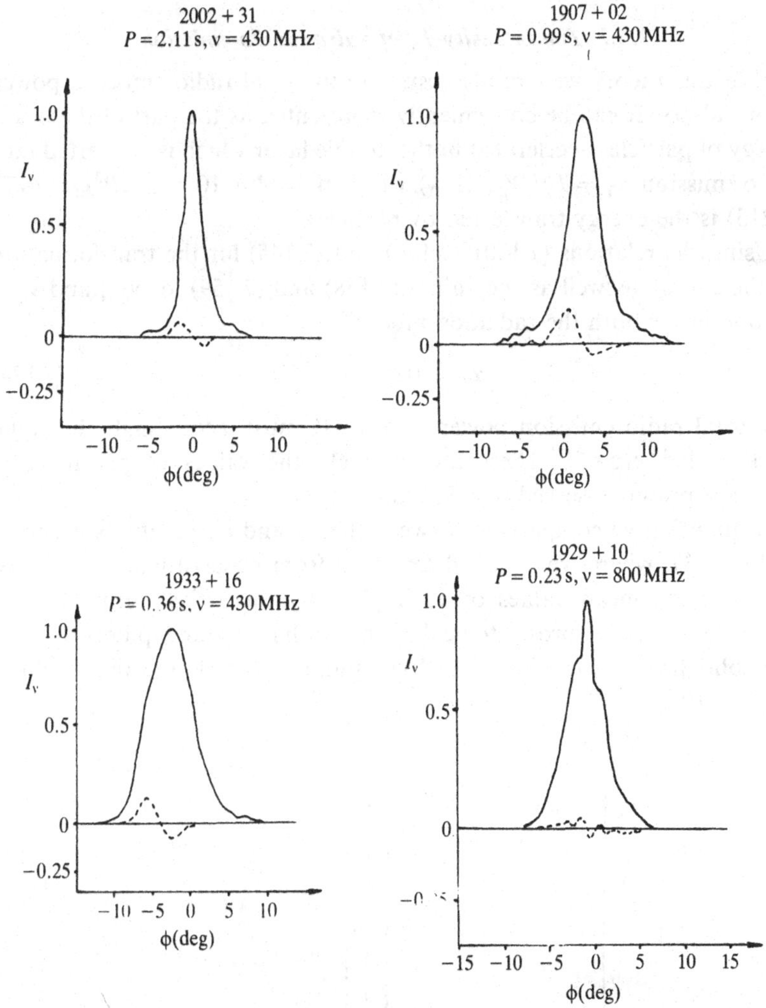

Fig. 7.30. Examples of S_t type pulsars. Circular polarization reversal takes place at maximum intensity. T type pulsar 1929 + 10 exhibits a similar effect.

from Fig. 7.29) but for frequencies $\nu < 100$ MHz this absorption already becomes significant.

The dependence $I_\nu^v \propto dI_\nu/d\phi$ is also exhibited by pulsars having a triple intensity profile (class T). A considerable part of the intensity for these pulsars is represented by the 'core' component. A corresponding example is presented in Fig. 7.30.

7.6.7 *The intensity L_r of pulsar radio emission*

Within this theory we can also estimate the total radio emission power L_r of pulsars. It can be conveniently represented as the part of the beam energy of particles accelerated in the double layer which is converted into radio emission, $\alpha_T = L_r / W_p$ (7.124), where $W_p = 4 \times 10^{31} \dot{P}_{-15} P^3 i_0$ erg s^{-1} (4.213) is the energy transferred by particles.

Using the relations (7.130), (7.140) and (7.145) for the transformation coefficient α_T, as well as the values (7.158) and (7.159) for ν_{max} and ν_{min}, we obtain for both the radiation modes,

$$\alpha_T = 10^{-3}\text{–}10^{-5} \qquad (7.174)$$

The total radio emission power $L_r = \alpha_T W_p$ must, accordingly, be equal to 10^{26}–10^{28} erg s^{-1}. These are precisely the values of pulsar radio emission power observed (see Section 1.4).

A quantitative comparison between theory and observations is shown in Fig. 7.31, where we see α_T determined from observations. The arrow indicates the mean values of α_T implied by theoretical formulas. The distribution over α_T proves to be the same both for 'young' pulsars ($Q < 1$, the solid line in Fig. 7.31) and 'old' pulsars ($Q > 1$, the dashed line), $\langle \alpha_T \rangle \simeq 10^{-4}$.

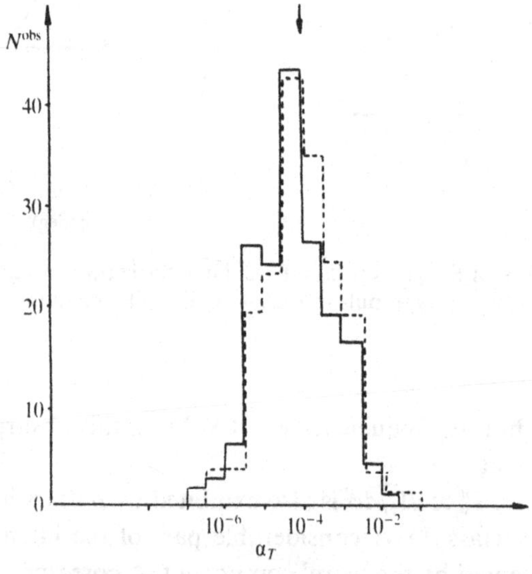

Fig. 7.31. Pulsar distribution over the transformation coefficient α_T for pulsars with $Q < 1$ (solid line) and $Q > 1$ (dashed line).

Thus, both for pulsars with $Q > 1$ and for those with $Q < 1$ there exists reasonable agreement between theory and observations. This confirms once again that the principal mechanism of coherent radio emission generation is the same in all the observed pulsars.

7.7 Non-stationary structure of radio pulses

The theory presented above has made it possible to determine the coherent mechanism of generation of observed radio emission. The basic predictions of the theory agree with observations, as we have seen. This refers, however, only to stationary radio emission characteristics averaged over many pulses. Observations show (see Section 1.3), in addition, the presence of a complicated space–time radio signal structure caused by various non-stationary processes. The origin of non-stationary phenomena should be sought in the instabilities developed in the pulsar magnetosphere. Such instabilities have not been considered in the theory of 'ideal' pulsars formulated above. It therefore seems reasonable to analyse briefly the nature of the instabilities capable of affecting the observed characteristics of pulsar radio emission.

7.7.1 Qualitative considerations

The instabilities responsible for non-stationary structure of a radio signal can be due, firstly, to the instability of the plasma generation mechanism itself and, secondly, to non-linear processes accompanying the propagation of electromagnetic radiation flux in the magnetosphere.

The instability of the plasma generation mechanism can be of a quasi-stationary nature. The essence of this instability is the division of the entire electron–positron plasma generation region into quasi-stationary subregions in the polar cap plane, which act independently or are very weakly interrelated (Fig. 7.32). These subregions move across the magnetic field due both to the general plasma drift in the pulsar magnetosphere and to the action of their own electric polarization. Such an instability is characterized by the timescales $\tau_S \simeq P r_S / R_0 \lesssim 0.1P$, where P is the pulsar period and R_0 is the polar cap radius. The spatial dimension of subregions is $r_S \gtrsim H$, where H is the double-layer height (3.21, 5.61), and their velocities are of the order of corotation velocities ΩR_0 (the scales and velocities are given in projections onto the polar cap plane).

Non-stationary processes can also be associated with transverse ionization motion in the region of plasma generation. The point is that

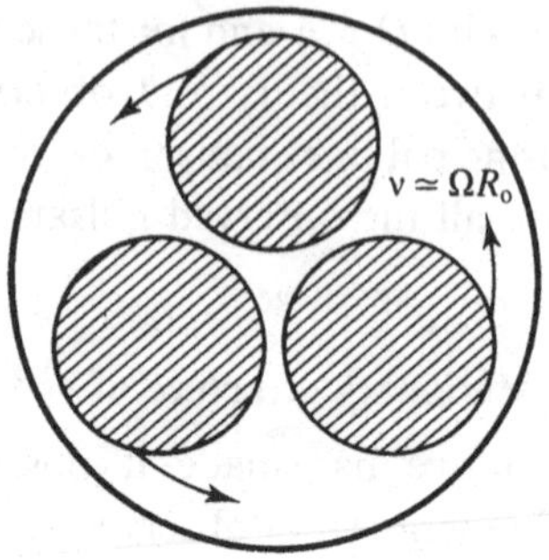

Fig. 7.32. Plasma generation region in the polar cap. The size of the region r_S is of the order of the height H of the generation layer.

due to a rectilinear motion of curvature radiation quanta, the region where electron–positron pairs are generated by these quanta always shifts towards the magnetic axis (Fig. 3.3) (Ruderman & Sutherland, 1975; Cheng & Ruderman, 1977). Therefore, transverse generation waves can appear which move across the polar cap towards the axis at a velocity $v_\perp \simeq 10^6$ cm s^{-1}. Associated with these waves are radio emission oscillations with characteristic times $t \simeq r_S/v_\perp \simeq 10^{-2}$–$10^{-3}$ s. The instability is of the drift type. Such ionization oscillations must occur near the external polar cap boundary owing to plasma generation instability in the transition layer region between the closed and open parts of the magnetosphere. Some characteristic processes in this region were considered by Jones (1981).

There may also exist longitudinal generation waves propagating along the magnetic field at a velocity close to that of light. These waves are excited owing to instability of primary electron–positron plasma generation in the double layer near the neutron star's surface. The frequency of these waves is $F_l \gtrsim 10^6$–10^7 s^{-1}, and the wavelength is $\lambda_l \gtrsim 10^4$ cm. The theory of excitation of longitudinal generation waves is developed in Section 7.7.2 below.

A non-stationary structure of radio pulsars can also be induced by a non-linear interaction between electromagnetic radiation and plasma in the course of propagation of both curvature-plasma and electromagnetic waves in the magnetosphere. In particular, there may occur modulation-type instabilities leading to excitation of modulation frequency $F_m \simeq 10^4$–10^5 s^{-1}. An example of modulation instability is considered in Section 7.7.3. Finally, in Section 7.7.4 we give an example of a possible mechanism of the appearance of low-frequency relaxation oscillations

$F_r \simeq 10^{-1}$–$10^{-2}\,s^{-1}$ accompanied by mode switching, and under conditions close to extinction, by pulse nulling.

Thus, the instabilities mentioned here overlap the observed time correlation range of the radio emission structure (see Section 1.3), although they are certainly far from exhausting all possible mechanisms of radio signal non-stationarity. Therefore, to fully comprehend the real nature of non-stationary processes in pulsar magnetosphere, it is necessary to determine from observations not only the time scales but also the spatial scales of inhomogeneities, as well as the magnitude and direction of the velocities of their motion.

7.7.2 Excitation of longitudinal generation waves

In Chapter 5 we considered stationary generation of primary electron–positron plasma in the double layer near the neutron star's surface. Under certain conditions this process may prove to be unstable, which leads to the appearance of ionization pulsations propagating along the magnetic field at a velocity close to the velocity of light.

Indeed, let us imagine that the electron density n^- or the positron density n^+ in the double layer has increased as compared to the stationary value. This will lead to the appearance of the electric potential $\delta\Psi$ and, therefore, to a variation in the particle energy $\delta\gamma^{\pm}$. The latter induces variation in the particle multiplication coefficients K_1, K_m and K. As a result, the flux stationarity condition $K_m(K + K_1) = 1$ (5.53) is violated, and the number of generated particles will therefore change. Under certain conditions, the fluctuations of n^+ and n^- can increase with time. We shall describe this process in more detail.

We shall consider, for definiteness, the case where the electric current transferred by fast particles in the polar region of the magnetosphere is created by positrons ($\rho_c > 0$). Given this, positrons produced in the double layer move away from the star, while electrons move towards the star. Since particles multiply in a narrow vicinity of the star's surface and in the upper boundary of the double layer, the equations for n^+ and n^- can be written as

$$\begin{aligned}\frac{\partial n^+}{\partial t} + v_\parallel \frac{\partial n^+}{\partial h} &= v_\parallel (K_1 + K) n^-(t, h, \rho'_\perp + \delta)\,\delta(h)\\ \frac{\partial n^-}{\partial t} - v_\parallel \frac{\partial n^-}{\partial h} &= v_\parallel K_m n^+(t, h, \rho'_\perp + \delta)\,\delta(h - H)\end{aligned} \tag{7.175}$$

Here $v_{\parallel}$ is the particle velocity close to the velocity of light $v_{\parallel} = c(1 - 1/2\gamma^2)$ (γ is the Lorentz factor of the particles, $\gamma \gg 1$), h is the coordinate along the magnetic field line, and K_1, K_m, K are the multiplication coefficients depending on the value of the potential $\Psi(t, \mathbf{r})$. In (7.175) allowance is made for the fact that due to the magnetic field curvature the gamma-quanta generate secondary particles at a point displaced along the radius inwards (Fig. 3.3) relative to the magnetic field line along which the particles are accelerated. δ is the magnitude of the displacement of the gamma-quanta which pass, before pair creation, the distance $H/4$ (see Section 5.2.1):

$$\delta = \frac{1}{16}\frac{H^2}{\rho_l(\mathbf{r})} \tag{7.176}$$

Here H is the double-layer thickness (5.61), and $\rho_l(\mathbf{r})$ is the curvature radius of the magnetic field line (7.94). Since

$$n^+(t, h) = n_0^+\left(t - \frac{h}{v_{\parallel}}\right), \qquad n^-(t, h) = n_0^-\left(t + \frac{h}{v_{\parallel}}\right)$$

it follows from (7.175) that

$$n_0^+(t, \rho'_\perp) = K_m\left(t - \frac{H}{c}, \rho'_\perp + \delta\right)[K_1(t_1, \rho'_\perp) + K(t_1, \rho'_\perp)] \times n_0^+\left(t - \frac{2H}{c}, \rho'_\perp + 2\delta\right) \tag{7.177}$$

Relation (7.177) coincides with the generation condition (5.53) in the stationary and homogeneous cases. We shall linearize eq. (7.177) assuming the deviations from the stationary state to be small, $n_0^+(t, \rho'_\perp) = n_0^+ + \delta n_0^+(t, \rho'_\perp)$, $K(\Psi) = K(\Psi_0) + (\partial K/\partial \Psi)\, \delta\Psi$, $\delta n/n$, $\delta\Psi/\Psi_0 \ll 1$:

$$\delta n_0^+(t, \rho'_\perp) = (K_1 + K)\frac{\partial K_m}{\partial \Psi}\,\delta\Psi\left(t - \frac{H}{c}, \rho'_\perp + \delta\right)n_0^+ + K_m\frac{\partial(K_1 + K)}{\partial \Psi}\,\delta\Psi(t, \rho'_\perp)n_0^+ + \delta n_0^+\left(t - \frac{2H}{c}, \rho'_\perp + 2\delta\right) \tag{7.178}$$

We have taken into account that by virtue of stationarity

$$K_m(K_1 + K) = 1$$

The potential perturbations $\delta\Psi$ are related to the density perturbations $\delta n^{\pm}$ as

$$\frac{1}{c^2}\frac{\partial^2 \delta\Psi}{\partial t^2} - \Delta\,\delta\Psi = 4\pi e(\delta n^+ - \delta n^-) \tag{7.179}$$

Since on the right-hand side of (7.179) the dependence δn on t and h enters the combinations $t \pm h/v_{\parallel}$, the solution of (7.179) in the quasi-classical approximation, i.e. when

$$\delta n_0 = \delta n_A\left(\frac{R_0}{\rho'_{\perp}}\right)^{1/2} \exp\left\{-i\omega t + i\int k_{\perp}\,d\rho'_{\perp}\right\}$$

is

$$\delta\Psi = \left(\frac{R_0}{\rho'_{\perp}}\right)^{1/2} \frac{4\pi e}{\omega^2/\gamma^2 c^2 + k_{\perp}^2}(\delta n_A^+ - \delta n_A^-)\,e^{-i\omega t + i\int k_{\perp}\,d\rho'_{\perp}} \tag{7.180}$$

Substituting (7.180) into (7.178) and taking into account that

$$\delta n_0^- = K_m\,\delta n_0^+ - K_m^2\frac{\partial(K_1+K)}{\partial\Psi}n_0^+\,\delta\Psi$$

we derive the dispersion equation

$$\left[H^2\left(\frac{\omega^2}{\gamma^2c^2}+k_{\perp}^2\right)(1-pi_0) - \frac{2iK_m^2}{1+K_m}\Psi\frac{\partial(K_1+K)}{\partial\Psi}\right]X^2$$

$$+\,2(K+K_1)\Psi\frac{\partial K_m}{\partial\Psi}i_0 pX$$

$$-\left[H^2\left(\frac{\omega}{\gamma^2c^2}+k_{\perp}^2\right)(1-pi_0) - \frac{2K_m}{1+K_m}i_0\Psi\frac{\partial(K_1+K)}{\partial\Psi}\right] = 0 \tag{7.181}$$

$$X = \exp\left\{i\frac{\omega H}{c} + ik_{\perp}\delta\right\}, \qquad p = \frac{1-K_m}{1+K_m}$$

From (7.181) it follows that for the frequencies $\omega < \gamma c/H$ and for the long-wave perturbations $k_{\perp}H < 1$ there exists one unstable root:

$$\omega + \frac{c}{H}k_{\perp}\delta = 2\pi\frac{c}{H}(l+\tfrac{1}{2}) + i\frac{c}{H}\ln\left|\frac{\partial K_m^{-1}/\partial\Psi}{\partial(K_1+K)/\partial\Psi}\right| \tag{7.182}$$

where l is an integer, $l = 0, 1, 2, \ldots$. For the short-wave perturbations

$k_\perp H > 1$, the instability increment falls:

$$\omega + \frac{c}{H} k_\perp \delta = 2\pi \frac{c}{H} (l + \tfrac{1}{2}) + i \frac{2c}{H} \frac{i_0 p}{(1 - i_0 p)} \frac{\Psi_0 |\partial(K_1 + K)/\partial\Psi|}{H^2(\omega^2/\gamma^2 c^2 + k_\perp^2)} \tag{7.183}$$

Thus, the stationary generation of plasma in a double layer is unstable: ionization oscillations are excited which then propagate along the magnetic field at a velocity close to c. The minimal oscillation frequency $F_{l_{\min}} = c/2H$ is about $10^6\ \mathrm{s}^{-1}$. The layer here oscillates as a whole (since $k_\perp H < 1$ and, therefore, the transverse wavelength $\lambda_\perp = 2\pi/k_\perp > 2\pi H \simeq R_0$, where R_0 is the dimension of the polar cap). Smaller oscillations $k_\perp H > 1$ are also excited, but their increment falls proportionally to $1/(k_\perp H)^2$. The wavelength along the magnetic field is

$$\lambda_\parallel < 2H \simeq 200\ \mathrm{m}$$

The instability considered here may cause a noise micropulse structure of pulsar radio signals (see Section 1.3).

7.7.3 Modulation instability of curvature oscillations

It has been shown above that curvature electromagnetic oscillations excited in the polar region of a pulsar magnetosphere are capable, when transforming into transverse modes, of providing the observed radio emission. The intensity of curvature modes is rather high here. Their non-linear interaction with electron–positron plasma can therefore lead to wave self-modulation effects.

We shall consider this process in more detail. Since, as already mentioned (Section 6.2.2), instability occurs in those regions of the magnetosphere where the cyclotron frequency is large, $\omega_B/\gamma \gg \omega, \omega_p/\gamma^{3/2}$, particle motion is only possible along the magnetic field. Besides, the frequency shift of curvature-plasma waves, $\tilde{\omega} = \omega - k_z v_\parallel$, is large as compared to the frequency spread associated with the energy dispersion of particles ($\tilde{\omega} \gg \omega/\gamma^2$). The non-linear self-action can therefore be considered in the framework of hydrodynamics.

A high-frequency wave with amplitude $E_z = E^{(1)}$ induces corresponding oscillations of longitudinal velocity, momentum and density of plasma (see (6.97)):

$$\delta v_\parallel^{(1)} = \frac{ieE^{(1)}}{\tilde{\omega} m_e \gamma^3}, \qquad \delta p_\parallel^{(1)} = \frac{ieE^{(1)}}{\tilde{\omega}}, \qquad \delta n^{(1)} = \frac{i n_0 k_z e}{\tilde{\omega}^2 m_e \gamma^3} E^{(1)} \tag{7.184}$$

Non-linear beats of high-frequency oscillations induce low-frequency density, velocity and potential modulations δn, $\delta v_\parallel$ and $\delta\Psi$. The continuity equations for the density and momentum together with the Poisson equation make it possible to determine these quantities:

$$\begin{gathered}\frac{\partial \delta n}{\partial t} + v_\parallel \frac{\partial \delta n}{\partial z} + n_0 \frac{\partial \delta v_\parallel}{\partial z} = -\frac{1}{2}\frac{\partial}{\partial z}(\delta n^{(1)} \cdot \delta v_\parallel^{(1)*}) \\ \frac{\partial \delta p_\parallel}{\partial t} + v_\parallel \frac{\partial \delta p_\parallel}{\partial z} = e\frac{\partial \delta\Psi}{\partial z} - \frac{1}{2}\delta v_\parallel^{(1)} \cdot \frac{\partial \delta p_\parallel^{(1)*}}{\partial z} \\ \frac{\partial^2 \delta\Psi}{\partial z^2} = -4\pi e(\delta n^+ - \delta n^-) \\ \delta v_\parallel = \frac{\delta p_\parallel}{m_e \gamma^3} - \frac{3}{4}\frac{v_\parallel}{m_e^2 c^2 \gamma^4}\delta p_\parallel^{(1)} \cdot \delta p_\parallel^{(1)*}\end{gathered} \tag{7.185}$$

Introducing the dimensionless quantities

$$x = \frac{\delta n}{n_0}, \qquad y = \frac{\delta p_\parallel}{m_e v_\parallel \gamma^3} \tag{7.186}$$

and making use of (7.184), we can write eqs. (7.185) as

$$\begin{gathered}\left(\frac{\partial}{\partial t} + v_\parallel \frac{\partial}{\partial z}\right)x + v_\parallel \frac{\partial y}{\partial z} = \frac{e^2}{m_e^2 \gamma^4 \tilde{\omega}^2 c}\left(\frac{3}{4} - \frac{1}{2}\frac{k_z c}{\tilde{\omega}\gamma^2}\right)\frac{\partial}{\partial z}|E^{(1)}|^2 \\ v_\parallel \frac{\partial}{\partial z}\left(\frac{\partial}{\partial t} + v_\parallel \frac{\partial}{\partial z}\right)y \pm \Omega_p^2(x^+ - x^-) = -\frac{1}{4}\frac{e^2}{m_e^2 \gamma^6 \tilde{\omega}^2}\frac{\partial^2}{\partial z^2}|E^{(1)}|^2 \\ \Omega_p^2 = \frac{4\pi n_0 e^2}{m_e \gamma^3}\end{gathered} \tag{7.187}$$

The ponderomotive action of a high-frequency field induces variations in plasma density and velocity. In the stationary case, for sufficiently long-wave perturbations ($\mathscr{L} > c/\Omega_p$, and quasi-neutrality holds, $x^+ = x^-$) the solution of (7.187) is

$$\begin{gathered}y_0 = -\frac{1}{4}\frac{e^2}{m_e^2 \gamma^6 \tilde{\omega}^2 c^2}|E^{(1)}|^2 \\ x_0 = \frac{e^2}{m_e^2 \gamma^4 \tilde{\omega}^2 c^2}\left[\frac{3}{4} + \frac{1}{4}\frac{1}{\gamma^2} - \frac{1}{2}\frac{\omega}{\tilde{\omega}\gamma^2}\right]|E^{(1)}|^2\end{gathered} \tag{7.188}$$

Here and below we understand γ as the mean value of the Lorentz factor of the slower plasma component.

The change in the mean characteristics of an electron–positron flux leads to a change in the propagation conditions for the curvature wave $E^{(1)}$. Since the dispersion equation for unstable modes has the form (3.37),

$$\omega - k_z v_{\|} = \Gamma_L = \omega^{1/5}\omega_p^{2/5}\left(\frac{c}{\rho_l}\right)^{2/5}\gamma^{-3/5}$$

the equation for the field $E^{(1)}$, with allowance made for the density and velocity modulations, is

$$i\left(\frac{\partial}{\partial t} + v_{\|}\frac{\partial}{\partial z}\right)E^{(1)} - \Gamma_L E^{(1)} = \tfrac{1}{5}\Gamma_L x E^{(1)} - iv_{\|}\, y\frac{\partial E^{(1)}}{\partial z} \tag{7.189}$$

We shall investigate the system of equations (7.187)–(7.189) for stability. We introduce small deviations from the equilibrium values:

$$x = x_0 + \delta x, \qquad y = y_0 + \delta y, \qquad a_m = a_0 + \delta a_m, \qquad \phi = \phi_0 + \delta\phi$$

$$E^{(1)} = a_m e^{i\phi}$$

The frequency and the wave vector of perturbations are respectively equal to $\mathbf{\Omega}_m$ and κ_m. We obtain

$$\delta\phi = 0$$

$$[\Omega_m - c\kappa_m(1 + y_0)]\,\delta a_m = \tfrac{1}{5}\Gamma_L a_0\,\delta x$$

$$-(\Omega_m - c\kappa_m)\,\delta x + c\kappa_m\,\delta y = \frac{e^2}{m_e^2\gamma^4\Gamma_L^2 c}\left(\frac{3}{2} - \frac{\omega}{\Gamma_L\gamma^2}\right)\kappa_m a_0\,\delta a_m \tag{7.190}$$

$$(\Omega_m - c\kappa_m)\,\delta y = \frac{1}{2}\frac{e^2}{m_e^2\gamma^6\Gamma_L^2 c}\kappa_m a_0\,\delta a_m$$

The dispersion equation corresponding to the system (7.190) has the form

$$\frac{5}{\Gamma_L}[\Omega_m - c\kappa_m(1 + y_0)] = \frac{e^2 a_0^2\kappa_m}{m_e^2\gamma^4\Gamma_L^2 c(\Omega_m - c\kappa_m)} \times\left[\left(\frac{\omega}{\Gamma_L\gamma^2} - \frac{3}{2}\right) + \frac{\kappa_m c}{2\gamma^2(\Omega_m - c\kappa_m)}\right] \tag{7.191}$$

Assuming that $\Omega_m - c\kappa_m > c\kappa_m/\gamma^2,\ c\kappa_m y_0$, we find that eq. (7.191) has an unstable root

$$\Omega_m - c\kappa_m = \pm i\frac{a_0 e}{m_e\gamma^2 c}\left(\frac{3\kappa_m c}{10\Gamma_L}\right)^{1/2} \tag{7.192}$$

under the condition $\omega/\gamma^2\Gamma_L < 3/2$, which corresponds exactly to the excitation range of curvature-plasma oscillations (6.88). For shorter waves $\kappa_m c > a_0^2 e^2/m_e^2 c^2 \Gamma_L$ the increment of modulation instability increases,

$$\Omega_m - c\kappa_m = i\frac{\sqrt{3}}{2}\left(\frac{a_0 e}{m_e c\gamma^3}\right)^{2/3}\left(\frac{\kappa_m^2 c^2}{10\Gamma_L}\right)^{1/3} \tag{7.193}$$

and reaches its maximum for the wave numbers $c\kappa_m = \omega_p/\gamma^{3/2}$, when quasilinearity is violated

$$(\mathrm{Im}\,\Omega_m)_{\max} \simeq 0.4\left(\frac{ea_0}{m_e c}\right)^{2/3}\omega_p^{8/15}\omega^{-1/15}\left(\frac{c}{\rho_l}\right)^{-2/15}\gamma^{-14/15}$$

Substituting the characteristic value a_0 from (7.97), we estimate the instability increment:

$$(\mathrm{Im}\,\Omega_m)_{\max} \simeq 0.4\omega^{-1/15}\omega_p^{13/15}\left(\frac{c}{\rho_l}\right)^{1/5}\gamma^{-13/30}$$

Numerically this yields

$$\mathrm{Im}\,\Omega_m \simeq 8\times 10^4 \nu_{\mathrm{GHz}}^{-1/15}(B_{12}\lambda_4)^{13/30}P^{-8/15}\gamma_{100}^{-13/30}\left(\frac{r}{R\times 10^3}\right)^{-7/5}\ \mathrm{s}^{-1} \tag{7.194}$$

Such an instability results in the characteristic times $t = 10^{-4}$–10^{-5} s in the variation of intensity of a received signal.

7.7.4 *The appearance of relaxation oscillations*

We shall give here an interesting example showing that the details of interaction between fast particles and the neutron star surface can under certain conditions excite slow relaxation oscillations accompanied by a change in the properties of the plasma flux generated and, therefore, of the pulsar radio emission flux.

According to the calculations made by Bogovalov and Kotov (1988), the dependence of the multiplication coefficient K on the neutron star's surface temperature for a given energy γ of particles incident onto the surface has the form presented in Fig. 7.33. A sharp maximum or minimum in the T-dependence corresponds to a temperature T_0 at which the cross-section of the capture of a neutron by a ^{56}Fe atomic nucleus has its minimal value. Given this, for energetic primary particles ($\gamma \simeq 10^7$) the cascade goes deep from the star's surface, and the coefficient K is maximal

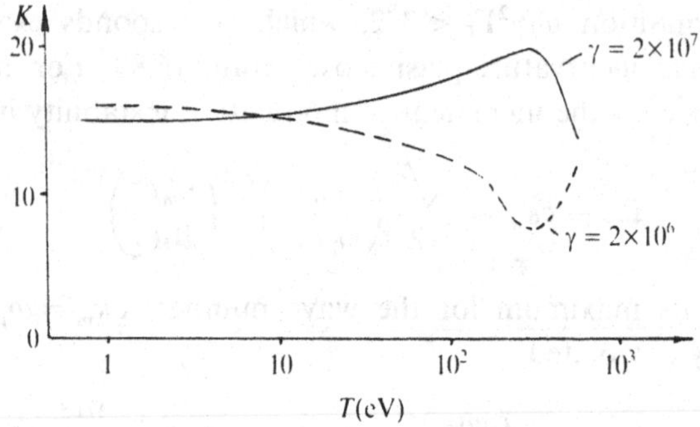

Fig. 7.33. Temperature dependence of energetic gamma-quanta (capable of pair production) knock-out of the star's surface. Two curves correspond to different energies of fast particles (10^7 MeV and 10^6 MeV). From Bogovalov & Kotov (1988).

for this temperature; for less energetic particles ($\gamma \simeq 10^6$) the K value is minimal. The value $T_0 \simeq 2 \times 10^6$ K.

It is, however, important that in the case of a pulsar the energy of particles incident onto the star's surface is proportional to the potential difference in the double layer, $\gamma = e\Psi_0/m_e c^2$ (5.59), and this value is in turn proportional to the K-dependent factor $b^{-2/7}$ (5.61). The quantity γ is therefore K-dependent as well. Taking account of this fact, the total T-dependence of K has the qualitative form shown in Fig. 7.34. A break in this dependence is seen to occur at $T^* \simeq 10^6$ K.

The star's surface temperature in the polar region is determined by the balance between the energy of primary plasma generated in the double layer and expended to surface heating and the energy radiated by the heated surface. The former is determined by the quantity

$$W_H = \frac{1}{1+K}\int j_{\parallel}\Psi_0\, ds = \frac{B_0\Omega^2 m_e c R^3}{2e(1+K)}\cos\chi \int_0^{f^*} i(f)\varphi_0(f)\, df \quad (7.195)$$

The latter energy can be approximated by black-body radiation:

$$W_T = \sigma T^4 \pi R^3 \frac{\Omega}{c} f^* \quad (7.196)$$

The stationary temperature T is thus

$$T = 4\times10^6 (B_{12}P^{-1}\cos\chi)^{1/4}(1+K)^{-1/4}\left[\int_0^1 \varphi_7(f')i_0(f')\, df'\right]^{1/4} \text{K} \quad (7.197)$$

$$\varphi_7 = 10^{-7}\frac{e\Psi_0}{m_e c^2}, \qquad i_0 = \frac{j_{\parallel}}{j_c}, \qquad f' = \frac{f}{f^*}$$

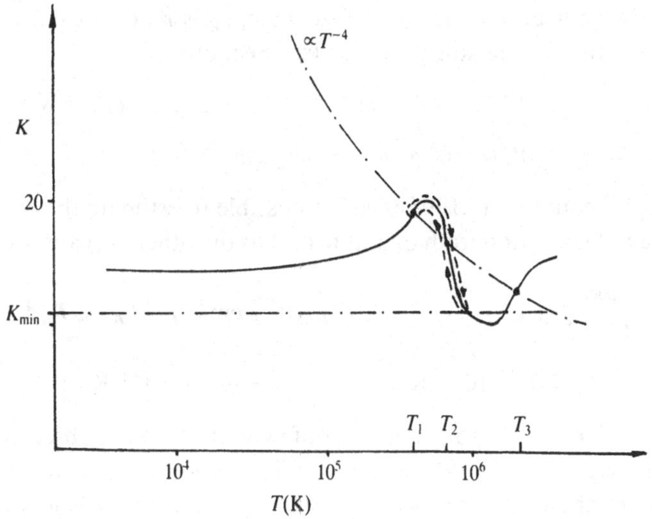

Fig. 7.34. Three stationary values of neutron star surface temperature T_1, T_2, T_3. K_{min} is the value of the knock-out coefficient below which e^+,e^- plasma production is impossible. The dashed line indicates a possible temperature run leading to pulsar nulling.

For definite parameters of a neutron star, the stationary value determined by (7.197) can be close to T^*, in which case three roots $T_1 < T_2 < T_3$ can exist. We can readily make sure that the state corresponding to the mean root T_2 is unstable.

Thus, two stationary solutions, T_1 and T_3, are possible, between which there will be transitions provided that the temperature fluctuations δT are large enough, $\delta T \gtrsim (T_2 - T_1) = \Delta T$. To describe the non-stationary transition process, it is necessary to solve the problem of heat propagation in the iron core of a neutron star in the case that a heat flux is incident onto its surface. The solution of the heat conductivity equation

$$\frac{\partial T_i}{\partial t} - \frac{\kappa_\parallel}{c_V}\frac{\partial^2 T_i}{\partial h^2} = 0 \tag{7.198}$$

with the boundary condition

$$\kappa_\parallel \left.\frac{\partial T_i}{\partial h}\right|_{h=0} = \frac{1}{1+K} j_\parallel \Psi - \sigma T^4, \qquad T = T_i(h=0)$$

determines the star's surface temperature in the polar region:

$$T - T_S = (\pi\kappa_\parallel c_V)^{-1/2}\int_0^t d\tau (t-\tau)^{-1/2}\left[\frac{j_\parallel \Psi}{1+K(\tau)} - \sigma T^4(\tau)\right] \tag{7.199}$$

Here $\kappa_{\parallel}$ is the heat conductivity of the core, c_V is its heat capacity, T_S is the temperature of the star's interior (see Section 2.5):

$$\kappa_{\parallel} = 1.7 \times 10^{15}(\rho/10^6 \text{ g cm}^{-3})^{4/3} T_6^{-1} \quad \text{erg s}^{-1}\text{ cm}^{-1}\text{ K}^{-1}$$

$$c_V = 5 \times 10^{12}(\rho/10^6 \text{ g cm}^{-3}) \quad \text{erg cm}^{-3}\text{ K}^{-1}$$

The integral equation (7.199) makes it possible to estimate the characteristic time t of transition from one state (T_1) to the other (T_3) and vice versa

$$t \simeq \frac{\pi \kappa c_V}{16\sigma^2 T^2}(\Delta T)^{-4} = 5 \times 10^3(\rho/10^6 \text{ g cm}^{-3})^{7/3} T_6^{-3}\, \Delta T_5^{-4} \quad \text{s}$$

$$T_6 = T \times 10^{-6}\text{ K}^{-1}, \qquad \Delta T_5 = \Delta T \times 10^{-5}\text{ K}^{-1} \tag{7.200}$$

The time (7.200) corresponds in order of magnitude to the observed mode switching times for pulsars (see Section 1.3.3). Moreover, for pulsars which are close to the excitation boundary ($Q > 1$), not only mode switching but also nulling is possible, because the 'breakdown' potential cannot exceed its maximal value Ψ_{max} (3.13). We shall take into account that, since $\Psi_{\text{max}} \propto b^{-2/7}(K)$, there exists a value K_{min} for which $\Psi(K_{\text{min}}) = \Psi_{\text{max}}$, and therefore for $K < K_{\text{min}}$ the generation of plasma and current in the polar region stops. If $K(T_3) < K_{\text{min}}$, then plasma is not generated in the state with $T = T_3$, and therefore there is no radio emission. In this case, the transition to the region $T > T_2$ is associated with pulsar switch-off. If the values of T_1 and T_2 are close (see Fig. 7.34), the state $T = T_2$ cannot exist for long, and the surface temperature will gradually increase until the pulsar is switched off. This is how the relaxation cycle is formed, which is also illustrated in Fig. 7.34. The time of going back to the state $T = T_2$ is characterized by the same time of core surface cooling and is given by the expression (7.200).

Conclusion

The theory presented above provides insight into the basic physical processes proceeding in the magnetosphere of a neutron star. Within this theory we can describe the principal details of the dynamics and evolution of pulsars, clarify the nature of their activity, establish the origin of the electron–positron plasma, and describe the coherent generation mechanism of observed radio emission. The main predictions of the theory are in agreement with observations.

There are, of course, many questions requiring further theoretical investigations. For example, it is necessary to carry out a reliable calculation of free electron energy on the neutron star's surface (Section 2.5), to investigate the wave excitation processes in the boundary layer near the light surface and in an extensive region behind it (Chapter 4), to study the transition layer structure on the boundary between the open and closed regions of the magnetosphere (Chapter 4), and so on. It is noteworthy that only first steps have been made in the study of the nature of non-stationary processes in the pulsar magnetosphere (Section 7.7). Nevertheless, the physical picture of the basic processes in the pulsar magnetosphere seems on the whole to be clear.

Appendix

Table I

Table I includes the most thoroughly investigated pulsars. We present the basic parameters of their radio spectra (the spectral index $\bar{\alpha}_1$, frequencies of the low-frequency ν_{min} and high-frequency ν_{max} cut-offs and, wherever possible, the break frequency ν_{br} and the spectral index above the break $\bar{\alpha}_2$). In addition, we present the exponent $\bar{\beta}$ in the relation $W_r \propto \nu^{\bar{\beta}}$ (see formula (1.8)) and the time during which the pulsar is switched off. The letters in the second column correspond to Rankin's classification.

Table I

Pulsar		P (s)	$\dot{P}_{-15}$	ν_{min} (MHz)	$\bar{\alpha}_1$	ν_{br} (GHz)	$\bar{\alpha}_2$	ν_{max} (GHz)	$\bar{\beta}$	Nullings (%)
0031−07	S_d	0.943	0.41	70	1.6	5	3	–	0.30	37.7
0138+59		1.223	0.39	90	–	–	–	–	0.17	–
0301+19	D	1.388	1.26	102	–	–	–	–	0.15	~10
0329+54	T	0.714	2.05	180	2.0	8	4.4	>10	0.07	<0.25
0355+54	S_t	0.156	4.39	195	0.9	5	1.5	>24	–	–
0525+21	D	3.745	40.1	80	2.3	–	–	2.5	0.16	25
0531+21	S_t	0.033	421	–	3.2	–	–	–	–	–
0628−28	S_d	1.244	7.11	–	1.6	4	2.9	–	0.17	<2
0740−28	S_t	0.167	16.8	160	1.8	–	–	–	–	<25
0809+74	S_d	1.292	0.17	50	1.6	4.7	2.5	–	0.25	1.4
0823+26	S_d	0.531	1.72	50	1.2	3	2	2.5	–	<5
0834+06	D	1.274	6.80	64	2	2.4	5.2	–	0.0	7.1
0919+06	S_t	0.431	13.7	85	2.2	–	–	–	0.60	<0.05
0950+08	S_d	0.253	0.23	75	2.2	5.0	–	5	0.14	–
1133+16	D	1.188	3.73	60	1.4	1.4	2.5	>10	0.30	15
1237+25	M	1.382	0.96	80	1.8	4.9	3.4	–	0.11	6
1508+55	T	0.740	5.03	75	2.3	2.7	4.3	–	–	–
1604−00	T	0.442	0.31	90	1.7	–	–	–	–	<1
1642−03	S_t	0.388	1.78	90	3.2	–	–	–	0.14	–
1706−16	S_t	0.653	6.38	100	1.6	–	–	>24	0.40	–
1749−28	S_t	0.563	8.15	150	2.6	–	–	–	0.31	<0.75
1857−26	M	0.612	0.16	–	1.7	2.7	2.5	–	–	10
1919+21	M	1.337	1.35	55	–	1.5	–	1.4	0.09	<0.25
1929+10	S_t	0.226	1.16	130	1.4	4.8	2.4	>8	0.19	<1
1933+16	S_t	0.359	6.00	250	2	5	5.2	–	0.14	<0.25
2016+28	S_d	0.558	0.15	140	2.2	2.5	–	–	0.15	–
2020+28	T	0.343	1.89	160	–	–	–	–	0.11	<3
2021+51	S_d	0.529	3.05	330	1.6	4.7	–	11	–	<5
2111+46	T	1.015	0.72	–	1.5	3	3	–	–	12
2154+40	D	1.525	3.41	100	1.3	1.7	2.7	–	–	7.5
2319+60	T	2.256	7.04	–	0.9	–	–	–	–	25

Table II

Table II contains 403 pulsars for which the quantity $\dot{P}$ is known. For these pulsars we present physical parameters of neutron stars determined within the theory developed in the book. These parameters are: the period P (s), the deceleration velocity $\dot{P}$ (in 10^{-15} s s^{-1}), the parameter Q (7.12) (for pulsars with $Q > 1$ we here put $Q = 1$), the magnetic field B_0 (7.16), (7.17) in 10^{12} G, the total energy losses $A_r\Omega\dot{\Omega}$ (on the logarithmic scale), as well as the transformation coefficient $\alpha_T = L_r/W_p$. The letters in the third column again correspond to the Rankin classification. The astronomical characteristics of radio pulsars, as the precise coordinates, distances, DM, etc., can be found in the book by Lyne & Smith (1990).

Table II

Pulsar		P (s)	$\dot{P}_{-15}$	Q	B_{12}	log W_{tot}	α_T
1 0011+47		1.241	5.61×10^{-1}	1.000	3.748	31.07	1.02×10^{-4}
2 0031−07	S_d	0.943	4.08×10^{-1}	1.000	2.239	31.29	1.32×10^{-5}
3 0037+56		1.118	2.88	1.000	3.082	31.91	4.51×10^{-5}
4 0045+33		1.217	2.35	1.000	3.613	31.72	2.15×10^{-5}
5 0052+51		2.115	9.54	1.000	10.183	31.60	1.80×10^{-5}
6 0053+47		0.472	3.57	0.526	1.807	33.13	2.91×10^{-7}
7 0059+65		1.679	5.95	1.000	6.606	31.70	3.30×10^{-4}
8 0105+68		1.071	4.90×10^{-2}	1.000	2.843	30.20	9.06×10^{-4}
9 0105+65	S_t	1.284	13.1	0.941	4.271	32.39	1.35×10^{-6}
10 0114+58		0.101	5.84	0.079	2.755	35.35	8.24×10^{-7}
11 0136+57		0.272	10.7	0.185	4.006	34.33	2.50×10^{-5}
12 0138+59		1.223	3.90×10^{-1}	1.000	3.646	30.93	3.30×10^{-4}
13 0144+59		0.196	2.56×10^{-1}	0.574	0.299	33.13	1.04×10^{-6}
14 0148−06	D	1.465	4.45×10^{-1}	1.000	5.115	30.75	1.04×10^{-3}
15 0149−16		0.833	1.30	1.000	1.775	31.95	1.14×10^{-6}
16 0153+39		1.811	1.80×10^{-1}	1.000	7.613	30.08	1.37×10^{-3}
17 0154+61	S_t	2.352	1.89×10^{-2}	0.630	26.843	32.76	1.65×10^{-6}
18 0203−40		0.631	1.20	1.000	1.054	32.28	1.04×10^{-7}
19 0226+70		1.467	3.12	1.000	5.129	31.59	2.01×10^{-5}
20 0254−53		0.448	3.10×10^{-2}	1.000	0.555	31.14	6.03×10^{-6}
21 0301+19	D	1.388	1.26	1.000	4.623	31.27	2.02×10^{-5}
22 0320+39	S_d	3.032	7.10×10^{-1}	1.000	20.007	30.01	5.91×10^{-4}
23 0329+54	T	0.714	2.05	1.000	1.329	32.35	7.54×10^{-5}
24 0331+45		0.269	7.34×10^{-3}	1.000	0.213	31.18	1.67×10^{-5}
25 0339+53		1.935	13.4	1.000	8.619	31.87	7.09×10^{-6}
26 0353+52		0.197	4.77×10^{-1}	0.450	0.461	33.40	2.22×10^{-6}
27 0355+54	S_t	0.156	4.39	0.143	2.208	34.66	1.03×10^{-5}
28 0402+61		0.595	5.57	0.568	2.439	33.02	1.14×10^{-5}
29 0403−76		0.545	1.54	0.863	0.996	32.58	1.37×10^{-6}
30 0410+69		0.391	7.65×10^{-2}	1.000	0.430	31.71	7.43×10^{-7}
31 0447−12		0.438	1.03×10^{-1}	1.000	0.532	31.69	6.60×10^{-5}
32 0450+55	T	0.341	2.36	0.434	1.375	33.37	1.17×10^{-6}
33 0450−18	T	0.549	5.75	0.514	2.504	33.14	4.32×10^{-5}
34 0458+46		0.638	5.58	0.613	2.434	32.93	2.09×10^{-6}
35 0523+11	M	0.354	7.10×10^{-2}	1.000	0.357	31.80	1.33×10^{-4}
36 0525+21	D	3.745	40.1	1.000	29.728	31.48	2.60×10^{-4}
37 0531+21	T	0.033	421.0	0.004	37.000	38.67	1.40×10^{-5}
38 0538−75		1.246	5.70×10^{-1}	1.000	3.776	31.07	1.12×10^{-4}
39 0540+23	S_t	0.246	15.4	0.143	5.195	34.61	1.06×10^{-5}
40 0540−693		0.050	479.0	0.006	62.396	38.18	—
41 0559−05		0.396	1.31	0.648	0.904	32.92	1.15×10^{-5}
42 0559−57		2.261	2.78	1.000	11.541	30.98	3.00×10^{-6}

Table II (*cont.*)

Pulsar		P (s)	$\dot{P}_{-15}$	Q	B_{12}	log W_{tot}	α_T
43 0609+37		0.298	5.89×10^{-2}	1.000	0.258	31.95	2.19×10^{-7}
44 0611+22	S_t	0.335	59.6	0.117	13.191	34.80	7.09×10^{-6}
45 0621−04	M	1.039	8.46×10^{-1}	1.000	2.686	31.48	6.46×10^{-5}
46 0626+24	S_t	0.477	1.99	0.673	1.200	32.86	1.07×10^{-5}
47 0628−28	S_d	1.244	7.11	1.000	3.765	32.17	8.14×10^{-5}
48 0643+80		1.214	3.81	1.000	3.596	31.93	1.90×10^{-5}
49 0655+64	S_d	0.196	6.77×10^{-4}	1.000	0.118	30.55	6.67×10^{-5}
50 0656+14	T	0.385	55.0	0.141	12.384	34.58	1.22×10^{-7}
51 0727−18		0.510	18.9	0.294	5.781	33.75	2.67×10^{-6}
52 0736−40	T	0.375	1.61	0.562	1.047	33.08	3.92×10^{-4}
53 0740−28	S_t	0.167	16.8	0.090	5.629	35.16	3.15×10^{-5}
54 0743−53		0.215	2.73	0.247	1.558	34.04	4.02×10^{-5}
55 0751+32	D	1.442	1.07	1.000	4.966	31.15	6.85×10^{-5}
56 0756−15		0.682	1.62	1.000	1.220	32.31	3.31×10^{-6}
57 0808−47		0.547	3.09	0.656	1.622	32.88	3.38×10^{-4}
58 0809+74	S_d	1.292	1.68×10^{-1}	1.000	4.042	30.49	2.73×10^{-5}
59 0818−13	S_d	1.238	2.11	1.000	3.731	31.65	1.18×10^{-4}
60 0818−41	D	0.545	2.70×10^{-2}	1.000	0.801	30.82	6.39×10^{-3}
61 0820+02	S_d	0.865	1.05×10^{-1}	1.000	1.905	30.81	8.67×10^{-5}
62 0823+26	S_d	0.531	1.72	0.802	1.078	32.66	3.01×10^{-6}
63 0826−34	M	1.849	1.00	1.000	7.915	30.80	4.56×10^{-4}
64 0833−45	T	0.089	124.0	0.020	15.600	36.85	1.81×10^{-5}
65 0834+06	D	1.274	6.80	1.000	3.937	32.12	1.78×10^{-6}
66 0835−41	S_t	0.752	3.55	0.881	1.759	32.52	5.21×10^{-5}
67 0839−53		0.721	1.65	1.000	1.354	32.24	5.68×10^{-5}
68 0840−48		0.644	9.50	0.501	3.530	33.15	5.59×10^{-6}
69 0841+80		1.602	4.42×10^{-1}	1.000	6.049	30.63	4.86×10^{-6}
70 0844−35		1.116	1.57	1.000	3.071	31.65	2.16×10^{-5}
71 0853−33		1.267	6.50	1.000	3.896	32.10	3.04×10^{-6}
72 0855−61		0.962	1.68	1.000	2.325	31.88	1.32×10^{-5}
73 0901−63		0.660	1.07×10^{-1}	1.000	1.147	31.17	2.08×10^{-5}
74 0903−42		0.965	1.89	1.000	2.338	31.92	1.80×10^{-5}
75 0904−74		0.550	4.63×10^{-1}	1.000	0.815	32.04	1.46×10^{-5}
76 0905−51		0.254	1.83	0.348	1.168	33.65	1.98×10^{-5}
77 0906−17		0.402	6.71×10^{-1}	0.861	0.565	32.61	1.03×10^{-7}
78 0906−49		0.107	15.2	0.058	5.367	35.69	—
79 0909−71		1.363	3.33×10^{-1}	1.000	4.468	30.72	8.11×10^{-5}
80 0918+63		1.568	3.61	1.000	5.811	31.57	2.61×10^{-6}
81 0919+06	S_t	0.431	13.7	0.278	4.654	33.83	4.23×10^{-6}
82 0922−52		0.746	35.5	0.347	8.819	33.53	4.51×10^{-6}
83 0923−58		0.739	4.83	0.764	2.184	32.68	6.67×10^{-6}
84 0932−52		1.445	4.65	1.000	4.985	31.79	1.59×10^{-5}
85 0940−55	S_t	0.664	27.2	0.366	6.486	33.49	2.90×10^{-5}

Table II (*cont.*)

Pulsar		P (s)	$\dot{P}_{-15}$	Q	B_{12}	log W_{tot}	α_T
86 0940+16	M	1.087	9.00×10^{-1}	1.000	2.923	31.45	3.74×10^{-5}
87 0941−56		0.808	39.6	0.363	9.482	33.48	4.81×10^{-6}
88 0942−13	S_t	0.570	4.60×10^{-2}	1.000	0.871	31.00	3.04×10^{-6}
89 0943+10	S_d	1.098	3.53	1.000	2.979	32.03	2.30×10^{-6}
90 0950−38		1.374	5.80×10^{-1}	1.000	4.536	30.95	4.66×10^{-4}
91 0950+08	S_d	0.253	2.29×10^{-1}	0.795	0.273	32.75	1.65×10^{-6}
92 0953−52		0.862	3.52	1.000	1.892	32.34	1.61×10^{-5}
93 0957−47	D	0.670	8.20×10^{-2}	1.000	1.180	31.04	2.64×10^{-4}
94 0959−54	S_t	1.437	51.7	0.615	11.103	32.84	1.99×10^{-5}
95 1001−47		0.307	22.1	0.158	6.616	34.48	1.02×10^{-6}
96 1010−23		2.518	1.40	1.000	14.123	30.54	1.50×10^{-5}
97 1014−53		0.770	1.93	1.000	1.531	32.23	6.19×10^{-7}
98 1015−56		0.503	3.13	0.595	1.643	32.99	1.47×10^{-4}
99 1016−16		1.805	1.75	1.000	7.565	31.07	1.53×10^{-4}
100 1030−58		0.464	3.00	0.554	1.602	33.08	4.40×10^{-4}
101 1039−19	D	1.386	1.10	1.000	4.610	31.22	2.83×10^{-5}
102 1039−55		1.171	6.73	1.000	3.361	32.22	1.64×10^{-4}
103 1044−57		0.369	1.14	0.634	0.823	32.96	8.28×10^{-5}
104 1054−62		0.422	3.57	0.465	1.817	33.28	8.05×10^{-4}
105 1055−52	T	0.197	5.83	0.165	2.661	34.48	6.73×10^{-6}
106 1056−78		1.347	1.33	1.000	4.370	31.34	1.92×10^{-5}
107 1056−57		1.185	4.29	1.000	3.437	32.01	2.55×10^{-5}
108 1105−59		1.516	3.40×10^{-1}	1.000	5.454	30.59	3.62×10^{-4}
109 1110−65		0.334	8.24×10^{-1}	0.647	0.659	32.94	1.96×10^{-4}
110 1110−69		0.820	2.84	1.000	1.723	32.31	3.77×10^{-5}
111 1112+50	T	1.656	2.49	1.000	6.437	31.34	2.23×10^{-6}
112 1114−41		0.943	7.94	0.819	3.055	32.58	1.37×10^{-6}
113 1118−79		2.281	3.67	1.000	11.733	31.09	5.47×10^{-6}
114 1119−54		0.536	2.77	0.670	1.504	32.85	1.05×10^{-4}
115 1133+16	D	1.188	3.73	1.000	3.453	31.95	2.14×10^{-6}
116 1133−55		0.365	8.23	0.284	3.285	33.83	1.21×10^{-5}
117 1143−60		0.273	1.79	0.380	1.146	33.54	1.60×10^{-5}
118 1154−62	S_t	0.400	3.93	0.422	1.949	33.39	7.74×10^{-4}
119 1159−58		0.453	2.13	0.619	1.262	32.96	2.92×10^{-5}
120 1221−63	T	0.216	4.95	0.195	2.363	34.29	1.11×10^{-5}
121 1222−63		0.420	9.46×10^{-1}	0.787	0.718	32.71	3.38×10^{-4}
122 1236−68		1.302	11.9	0.993	3.990	32.33	5.16×10^{-6}
123 1237+25	M	1.382	9.59×10^{-1}	1.000	4.586	31.16	9.95×10^{-6}
124 1237−41		0.512	1.74	0.767	1.088	32.71	5.31×10^{-7}
125 1240−64	S_t	0.388	4.50	0.387	2.146	33.49	5.45×10^{-4}
126 1254−10		0.617	3.62×10^{-1}	1.000	1.011	31.79	1.07×10^{-5}
127 1256−67		0.663	1.21	1.000	1.157	32.22	5.13×10^{-6}

Table II (*cont.*)

Pulsar		P (s)	$\dot{P}_{-15}$	Q	B_{12}	log W_{tot}	α_T
128 1302−64		0.572	4.03	0.620	1.949	32.93	1.12×10^{-3}
129 1309−53		0.728	1.48×10^{-1}	1.000	1.379	31.18	9.42×10^{-4}
130 1309−12		0.447	1.51×10^{-1}	1.000	0.552	31.83	9.77×10^{-6}
131 1309−55		0.849	5.71	0.832	2.438	32.57	4.05×10^{-5}
132 1317−53		0.280	9.26	0.202	3.615	34.23	1.32×10^{-5}
133 1322−66		0.543	5.31	0.524	2.370	33.12	3.04×10^{-4}
134 1322+83		0.670	5.65×10^{-1}	1.000	1.180	31.87	—
135 1323−63		0.793	3.10	0.986	1.595	32.39	7.19×10^{-4}
136 1323−58		0.478	3.21	0.557	1.677	33.07	3.70×10^{-3}
137 1323−62		0.530	18.9	0.307	5.770	33.70	3.72×10^{-4}
138 1325−43		0.533	3.01	0.644	1.594	32.90	3.56×10^{-6}
139 1325−49		1.479	6.10×10^{-1}	1.000	5.208	30.88	8.18×10^{-4}
140 1336−64		0.379	5.05	0.360	2.329	33.57	3.37×10^{-6}
141 1352−51		0.644	2.81	0.815	1.505	32.62	1.46×10^{-5}
142 1356−60		0.127	6.34	0.099	2.885	35.09	1.73×10^{-3}
143 1358−63		0.843	16.9	0.535	5.213	33.05	1.23×10^{-5}
144 1417−54		0.936	2.37×10^{-1}	1.000	2.208	31.06	4.14×10^{-4}
145 1424−55		0.570	2.09	0.802	1.231	32.65	2.38×10^{-5}
146 1426−66		0.785	2.77	1.000	1.588	32.36	4.57×10^{-5}
147 1436−63		0.460	1.12	0.814	0.804	32.66	2.93×10^{-5}
148 1449−64	S_t	0.179	2.75	0.201	1.580	34.28	3.55×10^{-5}
149 1451−68	T	0.263	9.88×10^{-2}	1.000	0.204	32.33	4.82×10^{-6}
150 1454−51		1.748	5.29	1.000	7.124	31.60	1.56×10^{-6}
151 1503−51		0.841	6.37	0.788	2.633	32.63	1.13×10^{-6}
152 1503−66		0.356	1.16	0.605	0.834	33.01	2.21×10^{-5}
153 1504−43		0.287	1.60	0.420	1.056	33.43	2.19×10^{-6}
154 1507−44		0.944	6.10×10^{-1}	1.000	2.244	31.46	2.28×10^{-4}
155 1508+55	T	0.740	5.03	0.753	2.247	32.69	3.56×10^{-6}
156 1509−58		0.150	1490.0	0.013	130.707	37.24	1.05×10^{-6}
157 1510−48		0.455	9.25×10^{-1}	0.868	0.704	32.59	1.95×10^{-6}
158 1523−55		1.049	11.3	0.799	3.890	32.59	2.58×10^{-4}
159 1524−39		2.418	19.1	1.000	13.089	31.73	9.27×10^{-6}
160 1530+27	D	1.125	8.03×10^{-1}	1.000	3.118	31.35	1.98×10^{-6}
161 1530−53		1.369	1.43	1.000	4.505	31.35	2.77×10^{-5}
162 1540−06	S_d	0.709	8.83×10^{-1}	1.000	1.312	31.99	2.48×10^{-6}
163 1541−52		0.179	6.10×10^{-2}	0.923	0.110	32.63	3.32×10^{-6}
164 1541+09	T	0.748	4.30×10^{-1}	1.000	1.450	31.61	6.25×10^{-4}
165 1550−54		1.081	15.7	0.724	4.890	32.69	8.39×10^{-5}
166 1552−31		0.518	5.00×10^{-2}	1.000	0.728	31.16	1.52×10^{-4}
167 1552−23		0.533	7.00×10^{-1}	1.000	0.768	32.26	4.30×10^{-6}
168 1555−55		0.957	20.5	0.569	5.930	32.97	4.99×10^{-5}
169 1556−44	S_t	0.257	1.02	0.445	0.775	33.38	1.87×10^{-5}
170 1556−57		0.194	2.12	0.244	1.312	34.06	8.14×10^{-5}

Table II (*cont.*)

Pulsar			P (s)	$\dot{P}_{-15}$	Q	B_{12}	log W_{tot}	α_T
171	1557−50	S_t	0.193	5.06	0.171	2.413	34.45	—
172	1558−50	T	0.864	69.6	0.312	14.024	33.63	4.33×10^{-7}
173	1600−27		0.778	2.92	0.988	1.531	32.39	4.63×10^{-6}
174	1600−49		0.327	1.01	0.583	0.761	33.06	4.97×10^{-5}
175	1601−52	D	0.658	2.56×10^{-1}	1.000	1.141	31.55	4.31×10^{-5}
176	1604−00	T	0.422	3.06×10^{-1}	1.000	0.496	32.21	7.91×10^{-7}
177	1607−13		1.018	2.30×10^{-1}	1.000	2.585	30.94	1.70×10^{-4}
178	1609−47		0.382	6.32×10^{-1}	0.834	0.544	32.65	3.99×10^{-5}
179	1612+07	S_d	1.207	2.36	1.000	3.557	31.73	1.38×10^{-6}
180	1612−29		2.478	2.50	1.000	13.705	30.82	1.45×10^{-5}
181	1620−42		0.365	1.02	0.655	0.762	32.92	5.30×10^{-4}
182	1620−09		1.276	3.00	1.000	3.948	31.76	2.62×10^{-5}
183	1620−26		0.011	8.20×10^{-4}	0.240	0.006	34.39	—
184	1630−59		0.529	1.37	0.875	0.919	32.57	1.89×10^{-5}
185	1633+24	T	0.490	1.19×10^{-1}	1.000	0.656	31.60	2.31×10^{-5}
186	1641−45	S_t	0.455	20.1	0.253	6.070	33.93	1.03×10^{-3}
187	1641−68		1.786	1.70	1.000	7.417	31.07	1.56×10^{-4}
188	1642−03	S_t	0.388	1.78	0.561	1.121	33.08	1.22×10^{-5}
189	1647−528		0.890	2.07	1.000	2.009	32.07	7.25×10^{-5}
190	1647−52		0.635	1.81	0.957	1.107	32.45	5.98×10^{-5}
191	1648−17		0.973	3.04	1.000	2.375	32.12	1.07×10^{-6}
192	1657−13		0.641	6.21×10^{-1}	1.000	1.086	31.97	2.12×10^{-5}
193	1659−60		0.306	9.10×10^{-1}	0.565	0.709	33.10	3.26×10^{-5}
194	1700−32	T	1.212	7.00×10^{-1}	1.000	3.585	31.19	8.78×10^{-4}
195	1700−18		0.804	1.73	1.000	1.661	32.12	3.76×10^{-6}
196	1701−75		1.191	1.88	1.000	3.470	31.65	1.74×10^{-6}
197	1702−19	T	0.299	4.14	0.300	2.051	33.79	5.41×10^{-7}
198	1706−16	S_t	0.653	6.38	0.596	2.670	32.96	1.65×10^{-6}
199	1707−53		0.899	15.5	0.594	4.891	32.93	5.77×10^{-6}
200	1709−15		0.869	1.11	1.000	1.921	31.83	7.07×10^{-5}
201	1717−16		1.565	5.82	1.000	5.790	31.78	2.44×10^{-5}
202	1717−29		0.620	8.00×10^{-1}	1.000	1.020	32.13	1.39×10^{-5}
203	1718−02		0.478	8.70×10^{-2}	1.000	0.626	31.50	4.54×10^{-4}
204	1718−32		0.477	7.00×10^{-1}	1.000	0.624	32.41	4.66×10^{-5}
205	1719−37		0.236	10.8	0.158	4.061	34.51	9.72×10^{-6}
206	1726−00		0.386	1.12	0.671	0.811	32.89	2.85×10^{-6}
207	1727−47	T	0.830	164.0	0.212	25.604	34.06	1.14×10^{-4}
208	1729−41		0.628	12.8	0.432	4.355	33.31	2.59×10^{-5}
209	1732−02		0.839	4.20×10^{-1}	1.000	1.799	31.45	1.28×10^{-4}
210	1735−32		0.768	8.00×10^{-1}	1.000	1.524	31.85	—
211	1736−31		0.529	18.4	0.310	5.663	33.69	—
212	1736−29	S_t	0.322	7.86	0.252	3.201	33.97	—
213	1737+13	M	0.803	1.45	1.000	1.657	32.05	5.38×10^{-5}

Table II (*cont.*)

Pulsar		P (s)	$\dot{P}_{-15}$	Q	B_{12}	log W_{tot}	α_T
214 1737−30		0.606	466.0	0.099	54.028	34.92	—
215 1737−39		0.512	1.81	0.755	1.119	32.73	9.16×10^{-5}
216 1738−08	M	2.043	2.27	1.000	9.543	31.03	5.93×10^{-4}
217 1740−03		0.445	3.17	0.517	1.668	33.16	4.72×10^{-7}
218 1740−13		0.405	4.81×10^{-1}	0.992	0.448	32.46	2.79×10^{-5}
219 1742−30	T	0.367	10.7	0.257	3.946	33.94	1.00×10^{-5}
220 1745−12		0.394	1.21	0.665	0.855	32.90	1.09×10^{-5}
221 1745−56		1.332	2.12	1.000	4.279	31.55	5.31×10^{-6}
222 1747−46	S_t	0.742	1.29	1.000	1.429	32.10	4.88×10^{-6}
223 1749−28	S_t	0.563	8.15	0.459	3.193	33.26	9.21×10^{-5}
224 1750−24		0.528	14.1	0.344	4.701	33.58	—
225 1753+52		2.391	1.57	1.000	12.817	30.66	1.09×10^{-4}
226 1753−24		0.670	2.84×10^{-1}	1.000	1.180	31.58	—
227 1754−24		0.234	13.0	0.145	4.626	34.61	4.15×10^{-5}
228 1756−22		0.461	10.8	0.329	3.927	33.64	2.69×10^{-5}
229 1758−03		0.921	3.32	1.000	2.143	32.23	1.48×10^{-4}
230 1800−21	S_t	0.133	134.0	0.031	24.359	36.36	—
231 1802+03		0.218	1.00	0.374	0.771	33.58	5.84×10^{-6}
232 1804−20		0.918	17.1	0.585	5.234	32.94	—
233 1804−27		0.828	12.2	0.597	4.154	32.93	2.20×10^{-4}
234 1804−08		0.164	2.90×10^{-2}	1.000	0.084	32.42	1.02×10^{-4}
235 1806−53		0.261	3.83×10^{-1}	0.670	0.390	32.93	4.49×10^{-6}
236 1806−21		0.702	3.82	0.793	1.858	32.64	—
237 1809−173		1.205	19.1	0.755	0.579	32.64	—
239 1809−175		0.538	1.50	0.860	0.979	32.58	—
239 1810+02		0.794	3.60	0.930	1.771	32.46	2.66×10^{-5}
240 1811+40		0.931	2.55	1.000	2.186	32.10	7.22×10^{-6}
241 1813−17		0.782	8.00	0.664	3.100	32.82	—
242 1813−36		0.387	2.05	0.528	1.238	33.15	1.56×10^{-5}
243 1815−14		0.291	2.03	0.388	1.247	33.52	—
244 1817−13		0.921	4.50	1.000	2.143	32.36	—
245 1818−04	S_t	0.598	6.34	0.543	2.670	33.07	3.95×10^{-5}
246 1819−22		1.874	6.00×10^{-1}	1.000	8.117	30.56	2.74×10^{-3}
247 1819−14		0.215	4.50×10^{-1}	0.507	0.441	33.26	—
248 1820−31		0.284	2.92	0.326	1.611	33.71	1.57×10^{-6}
249 1821−19		0.189	5.24	0.165	2.475	34.49	3.57×10^{-4}
250 1821−11		0.436	3.58	0.482	1.818	33.24	—
251 1821−24		0.003	1.55×10^{-3}	0.045	0.010	36.36	—
252 1821+05	T	0.753	2.25×10^{-1}	1.000	1.469	31.32	9.25×10^{-5}
253 1822−09	T	0.769	52.3	0.308	11.549	33.66	5.77×10^{-7}
254 1822−14		0.279	22.7	0.141	6.773	34.62	—
255 1822−00		0.779	8.89×10^{-1}	1.000	1.565	31.87	8.78×10^{-5}

Table II (*cont.*)

Pulsar		P (s)	$\dot{P}_{-15}$	Q	B_{12}	log W_{tot}	α_T
256 1823−11		2.093	4.89	1.000	9.986	31.33	—
257 1823−13		0.101	75.2	0.029	16.482	36.46	—
258 1824−09		0.246	1.48	0.366	1.008	33.60	—
259 1826−17		0.307	5.59	0.274	2.528	33.89	7.35×10^{-4}
260 1828−60		1.889	2.70×10^{-1}	1.000	8.239	30.20	1.14×10^{-4}
261 1828−10		0.405	60.0	0.144	13.128	34.56	—
262 1829−08		0.647	63.9	0.235	13.402	33.97	—
263 1830−08		0.085	9.16	0.055	3.808	35.77	—
264 1831−03		0.687	41.5	0.298	9.878	33.71	3.91×10^{-4}
265 1831−04	M	0.290	1.97×10^{-1}	0.981	0.244	32.51	1.65×10^{-4}
266 1831−00		0.521	1.50×10^{-2}	1.000	0.736	30.63	3.04×10^{-3}
267 1832−06		0.305	40.0	0.124	10.025	34.75	—
268 1834−10		0.563	11.8	0.396	4.137	33.42	1.05×10^{-3}
269 1834−04		0.354	1.79	0.506	1.131	33.21	—
270 1834−06		1.906	7.90×10^{-1}	1.000	8.379	30.66	—
271 1838−04		0.186	6.39	0.150	2.846	34.60	—
272 1839−04	D	1.840	5.10×10^{-1}	1.000	7.843	30.51	—
273 1839+09	S_t	0.381	1.09	0.668	0.796	32.89	2.71×10^{-6}
274 1839+56		1.653	1.70	1.000	6.415	31.18	1.96×10^{-4}
275 1841−05		0.256	9.70	0.180	3.751	34.36	—
276 1841−04		0.991	3.92	1.000	2.458	32.21	—
277 1842−04		0.161	20.0	0.081	6.372	35.28	—
278 1842+14	S_t	0.375	1.87	0.529	1.163	33.15	2.09×10^{-6}
279 1842−02		0.508	15.1	0.321	4.942	33.66	—
280 1844−04	S_t	0.598	51.9	0.234	11.632	33.99	1.16×10^{-4}
281 1845−19		4.308	23.3	1.000	38.655	31.06	5.17×10^{-6}
282 1845−01	T	0.659	5.20	0.654	2.313	32.86	1.51×10^{-4}
283 1846−06		1.451	45.7	0.653	10.180	32.77	3.11×10^{-5}
284 1848+13		0.346	1.49	0.531	0.996	33.16	3.82×10^{-6}
285 1848+04		0.284	1.60×10^{-3}	1.000	0.236	30.44	1.13×10^{-3}
286 1848+12		1.205	11.5	0.924	3.911	32.42	1.07×10^{-5}
287 1849+00		2.180	100.0	0.747	17.256	32.58	—
288 1851−79		1.279	1.86	1.000	3.966	31.55	1.12×10^{-5}
289 1851−14		1.147	4.17	1.000	3.233	32.04	3.00×10^{-5}
290 1855+02		0.416	40.3	0.174	9.923	34.35	—
291 1855+09		0.005	1.70×10^{-5}	0.514	0.001	33.73	8.11×10^{-8}
292 1857−26	M	0.612	1.60×10^{-1}	1.000	0.996	31.44	4.11×10^{-4}
293 1859+01		0.288	2.36	0.361	1.387	33.59	1.15×10^{-5}
294 1859+03	S_t	0.655	7.49	0.561	2.987	33.03	2.06×10^{-3}
295 1859+07		0.644	1.58	1.000	1.095	32.37	—
296 1900+05	S_t	0.747	12.9	0.522	4.341	33.09	4.70×10^{-5}
297 1900+01	S_t	0.729	4.03	0.809	1.925	32.62	2.16×10^{-4}

Table II (*cont.*)

Pulsar			P (s)	$\dot{P}_{-15}$	Q	B_{12}	$\log W_{tot}$	α_T
298	1900−06		0.432	3.40	0.487	1.754	33.23	9.08×10^{-5}
299	1902−01		0.643	3.06	0.787	1.598	32.66	2.72×10^{-4}
300	1903+07		0.648	4.94	0.655	2.233	32.86	—
301	1904+06		0.267	2.14	0.345	1.300	33.65	—
302	1905+39	M	1.236	5.30×10^{-1}	1.000	3.719	31.05	8.70×10^{-5}
303	1906+09	D	0.830	9.80×10^{-2}	1.000	1.763	30.83	1.92×10^{-3}
304	1907+00	T	1.017	5.51	1.000	2.580	32.32	3.79×10^{-6}
305	1907+02	S_t	0.990	2.76	1.000	2.453	32.05	6.37×10^{-5}
306	1907+10	S_t	0.284	2.64	0.340	1.501	33.66	3.34×10^{-5}
307	1907+03	T	2.330	4.53	1.000	12.210	31.15	7.65×10^{-4}
308	1907−03	S_t	0.505	2.19	0.689	1.279	32.83	1.23×10^{-4}
309	1907+12		1.442	8.25	1.000	4.966	32.04	9.09×10^{-5}
310	1910+20	M	2.233	10.2	1.000	11.275	31.56	2.07×10^{-5}
311	1911+13	T	0.521	8.05×10^{-1}	1.000	0.736	32.36	4.29×10^{-5}
312	1911−04	S_t	0.826	4.07	0.924	1.926	32.46	3.78×10^{-5}
313	1911+11		0.601	6.56×10^{-1}	1.000	0.962	32.08	5.86×10^{-6}
314	1913+167	T	1.616	4.08×10^{-1}	1.000	6.148	30.59	1.33×10^{-4}
315	1913+10	S_t	0.404	15.2	0.248	5.022	33.96	1.33×10^{-4}
316	1913+16	T	0.059	8.64×10^{-3}	0.595	0.030	33.22	2.11×10^{-5}
317	1914+09	T	0.270	2.52	0.327	1.456	33.71	2.82×10^{-6}
318	1914+13	S_t	0.282	3.62	0.297	1.873	33.81	1.43×10^{-4}
319	1915+13	S_t	0.195	7.20	0.150	3.087	34.59	5.71×10^{-6}
320	1916+14	T	1.181	211.0	0.282	30.009	33.71	2.13×10^{-7}
321	1917+00	T	1.272	7.68	1.000	3.925	32.17	2.27×10^{-5}
322	1918+19	T	0.821	8.95×10^{-1}	1.000	1.727	31.81	1.35×10^{-3}
323	1918+26		0.786	3.51×10^{-2}	1.000	1.592	30.46	2.04×10^{-4}
324	1919+14	T	0.618	5.61	0.591	2.447	32.98	1.54×10^{-5}
325	1919+21	M	1.337	1.35	1.000	4.310	31.35	4.13×10^{-5}
326	1920+21	T	1.078	8.19	0.937	3.101	32.42	2.88×10^{-4}
327	1922+20	S_t	0.238	2.09	0.307	1.286	33.79	2.84×10^{-5}
328	1923+04	S_d	1.074	2.46	1.000	2.858	31.90	1.74×10^{-5}
329	1924+16	S_t	0.580	18.0	0.346	5.551	33.56	2.73×10^{-5}
330	1924+14	D	1.325	2.23×10^{-1}	1.000	4.237	30.58	3.36×10^{-3}
331	1925+22		1.431	7.71×10^{-1}	1.000	4.895	31.02	1.04×10^{-3}
332	1927+13	S_t	0.760	3.66	0.880	1.796	32.52	1.99×10^{-5}
333	1929+10	S_t	0.226	1.16	0.367	0.854	33.60	5.58×10^{-8}
334	1929+20	S_t	0.268	4.18	0.265	2.076	33.94	2.98×10^{-4}
335	1930+22	S_t	0.144	57.8	0.047	13.468	35.89	5.37×10^{-4}
336	1933+16	S_t	0.359	6.00	0.316	2.635	33.71	4.02×10^{-4}
337	1933+15		0.967	4.04	1.000	2.348	32.25	1.41×10^{-5}
338	1935+25		0.201	1.56	0.287	1.057	33.88	4.78×10^{-6}
339	1937−26		0.403	9.60×10^{-1}	0.748	0.726	32.77	2.11×10^{-6}
340	1937+214		0.002	1.05×10^{-4}	0.064	0.002	36.04	2.22×10^{-4}

Table II (*cont.*)

Pulsar			P (s)	$\dot{P}_{-15}$	Q	B_{12}	log W_{tot}	α_T
341	1940−12	S_d	0.972	1.66	1.000	2.370	31.86	2.05×10^{-6}
342	1941−17		0.841	9.80×10^{-1}	1.000	1.807	31.82	1.15×10^{-5}
343	1942−00	D	1.046	5.36×10^{-1}	1.000	2.720	31.27	3.22×10^{-4}
344	1943−29		0.959	1.70	1.000	2.311	31.88	3.18×10^{-6}
345	1944+22		1.334	8.89×10^{-1}	1.000	4.291	31.17	1.27×10^{-4}
346	1944+17	T	0.441	2.40×10^{-2}	1.000	0.539	31.05	9.14×10^{-5}
347	1946−25		0.958	3.40	1.000	2.307	32.19	2.57×10^{-7}
348	1946+35	S_t	0.717	7.05	0.635	2.850	32.88	4.47×10^{-4}
349	1951+32		0.039	5.92	0.028	2.917	36.60	—
350	1952+29	T	0.427	2.01×10^{-3}	1.000	0.507	30.01	5.97×10^{-5}
351	1953+50	S_t	0.519	1.37	0.857	0.920	32.59	2.63×10^{-6}
352	1953+29	S_t	0.006	3.00×10^{-5}	0.475	0.001	33.74	8.37×10^{-6}
353	1957+20		0.002	1.20×10^{-5}	0.157	0.001	35.10	—
354	2000+32		0.697	105.0	0.209	18.904	34.09	—
355	2000+40		0.905	1.74	1.000	2.073	31.97	8.27×10^{-4}
356	2002+31	S_t	2.111	74.6	0.811	14.079	32.50	4.27×10^{-5}
357	2003−08	T	0.581	4.00×10^{-2}	1.000	0.903	30.91	3.79×10^{-5}
358	2012+38		0.230	8.86	0.166	3.540	34.46	5.20×10^{-5}
359	2016+28	S_d	0.558	1.49×10^{-1}	1.000	0.837	31.53	2.98×10^{-5}
360	2020+28	T	0.343	1.89	0.478	1.177	33.27	8.69×10^{-6}
361	2021+51	S_d	0.529	3.05	0.635	1.610	32.91	3.07×10^{-6}
362	2022+50		0.373	2.51	0.468	1.429	33.28	—
363	2027+37		1.217	12.3	0.910	4.098	32.43	—
364	2028+22	M	0.630	8.84×10^{-1}	1.000	1.051	32.15	9.76×10^{-6}
365	2034+19		2.074	2.04	1.000	9.816	30.96	4.56×10^{-5}
366	2035+36		0.619	4.54	0.644	2.110	32.88	3.22×10^{-5}
367	2036+53		1.425	1.06	1.000	4.857	31.16	5.55×10^{-4}
368	2043−04	S_d	1.547	1.48	1.000	5.665	31.20	2.32×10^{-5}
369	2044+15	D	1.138	1.85×10^{-1}	1.000	3.186	30.70	1.23×10^{-4}
370	2045+56		0.477	11.1	0.338	3.996	33.61	9.99×10^{-6}
371	2045−16	T	1.942	11.0	1.000	8.846	31.76	5.55×10^{-6}
372	2048−72		0.341	1.96×10^{-1}	1.000	0.333	32.29	4.02×10^{-6}
373	2053+36	S_t	0.221	3.65×10^{-1}	0.569	0.380	33.13	4.77×10^{-5}
374	2053+21		0.815	1.34	1.000	1.704	31.99	1.76×10^{-5}
375	2106+44		0.415	8.60×10^{-2}	1.000	0.481	31.68	1.17×10^{-3}
376	2110+27	S_d	1.203	2.62	1.000	3.535	31.78	3.17×10^{-5}
377	2111+46	T	1.015	7.20×10^{-1}	1.000	2.571	31.44	1.10×10^{-2}
378	2113+14	S_t	0.440	2.90×10^{-1}	1.000	0.536	32.13	1.93×10^{-5}
379	2122+13		0.694	7.69×10^{-1}	1.000	1.260	31.96	8.06×10^{-6}
380	2123−67		0.326	2.26×10^{-1}	1.000	0.306	32.41	2.91×10^{-6}
381	2148+63	S_d	0.380	1.68×10^{-1}	1.000	0.407	32.09	1.21×10^{-4}
382	2148+52		0.332	10.1	0.236	3.809	34.04	2.64×10^{-5}

Table II (*cont.*)

Pulsar		P (s)	$\dot{P}_{-15}$	Q	B_{12}	log W_{tot}	α_T
383 2151−56		1.374	4.23	1.000	4.536	31.81	2.27×10^{-7}
384 2152−31		1.030	1.23	1.000	2.642	31.65	1.40×10^{-6}
385 2154+40	D	1.525	3.41	1.000	5.515	31.58	1.93×10^{-4}
386 2210+29	M	1.005	4.95×10^{-1}	1.000	2.523	31.29	1.41×10^{-4}
387 2217+47	S_t	0.538	2.76	0.674	1.500	32.85	5.28×10^{-6}
388 2224+65	T	0.682	9.67	0.530	3.564	33.08	4.85×10^{-6}
389 2227+61		0.443	2.25	0.590	1.312	33.01	9.61×10^{-5}
390 2241+69		1.664	4.82	1.000	6.495	31.62	1.41×10^{-5}
391 2255+58	S_t	0.368	5.75	0.331	2.555	33.66	1.31×10^{-4}
392 2303+30	S_d	1.576	2.90	1.000	5.866	31.47	3.45×10^{-5}
393 2303+46		1.067	5.68×10^{-1}	1.000	2.823	31.27	8.28×10^{-5}
394 2306+55	D	0.475	2.02×10^{-1}	1.000	0.619	31.88	3.26×10^{-5}
395 2310+42	S_d	0.349	1.16×10^{-1}	1.000	0.347	32.04	2.35×10^{-6}
396 2315+21	S_d	1.445	1.05	1.000	4.985	31.14	4.64×10^{-6}
397 2319+60	T	2.256	7.04	1.000	11.493	31.39	1.20×10^{-3}
398 2321−61		2.347	2.60	1.000	12.378	30.90	3.20×10^{-6}
399 2323+63	D	1.436	2.89	1.000	4.927	31.59	1.25×10^{-5}
400 2324+60		0.234	3.09×10^{-1}	0.647	0.338	32.98	6.36×10^{-5}
401 2327−20	T	1.664	4.63	1.000	6.495	31.60	3.51×10^{-7}
402 2334+61		0.495	192.0	0.113	29.339	34.80	2.37×10^{-6}
403 2351+61		0.945	16.2	0.617	5.032	32.88	1.60×10^{-5}

References

Ables, J. G. & Manchester, R. N. (1976). Hydrogen-line absorption observations of distant pulsars. *Astronomy and Astrophysics*, **50**, 177–84.

Abramowitz, M. & Stegun, I. A. (1964). *Handbook of mathematical functions.* New York: Dover Publications.

Alcock, C., Farhi, E. & Olinto, A. (1986). Strange stars. *Astrophysical Journal*, **310**, 261–72.

Alpar, M. A., Cheng, A. F., Ruderman, M. A. & Shaham, J. (1982). A new class of radio pulsars. *Nature*, **300**, 728–30.

Alpar, M. A., Anderson, P. W., Pines, D. & Shaham, J. (1984a). Vortex creep and the internal temperature of neutron stars. I. General theory. *Astrophysical Journal*, **276**, 325–34.

Alpar, M. A., Anderson, P. W., Pines, D. & Shaham, J. (1984b). Vortex creep and the internal temperature of neutron stars. II. Vela pulsar. *Astrophysical Journal*, **278**, 791–805.

Alpar, M. A., Nandkumar, R. & Pines, D. (1985). Vortex creep and the internal temperature of neutron stars: the Crab pulsar and PSR 0525+21. *Astrophysical Journal*, **288**, 191–5.

Arnett, W. D. & Bowers, R. L. (1977). A microscopic interpretation of neutron star structure. *Astrophysical Journal, Supplement Series*, **33**, 415–36.

Atteia, J.-L., Barat, C., Chernenko, A., Dolidze, V., Dyachkov, A., Jourdain, E., Khavenson, N., Kozlenkov, A., Kucherova, R., Mitrofanov, I., Moskaleva, L., Niel, M., Pozanenko, A., Scheglov, O., Surkov, Yu., Vedrenne, G. & Vilchinskaya, A. (1991). Rapid variability of the strong cosmic gamma-ray burst GB 881024. *Planetary and Space Science*, **39**, 23–37.

Backer, D. C. (1976). Pulsar average wave forms and hollow-cone beam models. *Astrophysical Journal*, **209**, 895–907.

Backer, D. C. & Hellings, R. W. (1986). Pulsar timing and general relativity. *Annual Review of Astronomy and Astrophysics*, **24**, 537–76.

Backer, D. C. & Rankin, J. M. (1980). Statistical summaries of polarized pulsar radiation. *Astrophysical Journal, Supplement Series*, **42**, 143–74.

Backer, D. C. & Sramek, R. A. (1982). Apparent proper motions of the galactic center compact radio sources and PSR 1929+10. *Astrophysical Journal*, **260**, 512–9.

Bailes, M., Manchester, R. N., Kesteven, M. J., Norris, R. P. & Reynolds, J. E. (1990). The parallax and proper motion of PSR 1451−68. *Nature*, **343**, 240–1.

Barnard, J. J. (1986). Probing the magnetic field of radio pulsars: a reexamination of polarization position angle swings. *Astrophysical Journal*, **303**, 280–91.

Barnard, J. J. & Arons, J. (1986). Wave propagation in pulsar magnetospheres: refraction of rays in the open flux zone. *Astrophysical Journal*, **302**, 138–62.

Bartel, N., Sieber, W. & Wolszczan, A. (1980). Pulse to pulse intensity modulation from radio pulsars with particular reference to frequency dependence. *Astronomy and Astrophysics*, **90**, 58–64.

Bartel, N., Morris, D., Sieber, W. & Hankins, T. H. (1982). The mode-switching phenomenon in pulsars. *Astrophysical Journal*, **258**, 776–89.

Baym, G. & Pethick, C. (1979). Physics of neutron stars. *Annual Review of Astronomy and Astrophysics*, **17**, 415–44.

Berestetskii, V. B., Lifshitz, E. M. & Pitaevskii, L. P. (1971). *Relativistic quantum theory*. New York: Pergamon Press.

Bernstein, I. B. & Friedland, L. (1983). Geometrical optics of nonstationary and nonhomogeneous plasma. In *Basic plasma physics*, ed. R. Z. Sagdeev & M. Rosenbluth, pp. 367–418. Amsterdam: North-Holland.

Beskin, G. M., Neizvestnyi, S. I., Pimonov, A. A., Plakhotnichenko, V. L. & Shvartsman, V. F. (1983). A fine-resolution optical light curve of the Crab Nebula pulsar. *Soviet Astronomy Letters*, **9**, 148–51.

Beskin, V. S. (1982a). On the pair creation in a strong magnetic field. *Astrofizika*, **18**, 439–49.

Beskin, V. S. (1982b). Dynamical screening of the acceleration region in the magnetosphere of a pulsar. *Soviet Astronomy*, **26**, 443–6.

Beskin, V. S., Gurevich, A. V. & Istomin, Ya. N. (1983). Electrodynamics of pulsar magnetospheres. *Soviet Physics – JETP*, **58**, 235–53.

Beskin, V. S., Gurevich, A. V. & Istomin, Ya. N. (1984). Spin-down of pulsars by the current: comparison of theory with observations. *Astrophysics and Space Science*, **102**, 301–26.

Beskin, V. S., Gurevich, A. V. & Istomin, Ya. N. (1987a). On the theory of pulsar radio emission. *Radiophysics and Quantum Electronics*, **30**, 115–37.

Beskin, V. S., Gurevich, A. V. & Istomin, Ya. N. (1987b). Permittivity of a weakly inhomogeneous plasma. *Soviet Physics – JETP*, **64**, 715–26.

Beskin, V. S., Gurevich, A. V. & Istomin, Ya. N. (1988). Theory of the radio emission of pulsars. *Astrophysics and Space Science*, **146**, 205–81.

Bhat, P. N., Ramana Murthy, P. V., Sreekantan, P. V. & Vishwanath, P. R. (1986). A very high energy γ-ray burst from the Crab pulsar. *Nature*, **319**, 127–8.

Bhat, P. N., Gupta, S. K., Ramana Murthy, P. V., Sreekantan, B. V., Tonwar, S. C. & Vishwanath, P. R. (1987). Very high energy gamma rays from the Vela pulsar. *Astronomy and Astrophysics*, **178**, 242–6.

Biggs, J. D., McCulloch, P. M., Hamilton, P. A., Manchester, R. N. & Lyne, A. G. (1985). A study of PSR 0826 – 34 – a remarkable pulsar. *Monthly Notices of the Royal Astronomical Society*, **215**, 281–94.

Bisnovatyi-Kogan, G. S. & Komberg, B. V. (1974). Pulsars and close binary systems. *Soviet Astronomy*, **18**, 217–21.

Bisnovatyi-Kogan, G. S. & Komberg, B. V. (1976). Possible evolution of a binary-system radio pulsar as an old object with a weak magnetic field. *Soviet Astronomy Letters*, **2**, 130–2.

Blandford, R. D., Applegate, J. H. & Hernquist, L. (1983). Thermal origin of neutron star magnetic fields. *Monthly Notices of the Royal Astronomical Society*, **204**, 1025–48.

Bogovalov, S. V. & Kotov, Yu. D. (1988). Development of electromagnetic cascades in a neutron star crust. In *Physics of neutron stars, pulsars and bursters*, ed. D. G. Yakovlev, pp. 91–106. Leningrad: FTI Publisher (In Russian).

Bogovalov, S. V. & Kotov, Yu. D. (1989). Electromagnetic cascades in a pulsar magnetosphere, and the processes on the neutron star surface. *Soviet Astronomy Letters*, **15**, 185–9.

Boriakoff, V. (1983). On the radio pulse emission mechanism of PSR 1133+16: simultaneous dual-frequency high time resolution observations. *Astrophysical Journal*, **272**, 687–701.

Boriakoff, V., Buccheri, R. & Fauci, F. (1983). Discovery of a 6.1 ms binary pulsar PSR 1953+29. *Nature*, **304**, 417–9.

Boynton, P. E., Groth, E. J., Hutchinson, D. P., Nanos, G. R., Pardridge, R. B. & Wilkinson, D. T. (1972). Optical timing of the Crab pulsar NP 0532. *Nature*, **175**, 217–41.

Brinkmann, W. (1980). Thermal radiation from highly magnetized stars. *Astronomy and Astrophysics*, **82**, 352–61.

Brinkmann, W. & Ögelman, H. (1987). Soft X-ray observations of the radio pulsar PSR 1055−52. *Astronomy and Astrophysics*, **182**, 71–4.

Bruck, Yu. M. & Ustimenko, B. Yu. (1977). Some features of the pulsed radiation from the pulsar 1919+21 at 16.7, 20 and 25 MHz. *Astrophysics and Space Science*, **49**, 349–66.

Canuto, V. (1974). Equation of state at ultrahigh densities. Part I. *Annual Review of Astronomy and Astrophysics*, **12**, 167–214.

Canuto, V. (1975). Equation of state at ultrahigh densities. Part II. *Annual Review of Astronomy and Astrophysics*, **13**, 335–80.

Cheng, A. F. (1984). X-ray emission from radio pulsar winds. *Astrophysical Journal*, **275**, 790–801.

Cheng, A. F. (1989). Radio pulsar spin-down tests of pulsar models. *Astrophysical Journal*, **337**, 803–13.

Cheng, A. F. & Ruderman, M. A. (1977). Bunching mechanism for coherent curvature radiation in pulsar magnetospheres. *Astrophysical Journal*, **212**, 800–6.

Cheng, K. S. & Shaham, J. (1989). Pulse sharpness and asymmetry in millisecond pulsars. *Astrophysical Journal*, **339**, 279–90.

Cheng, K. S., Alpar, M. A., Pines, D. & Shaham, J. (1988). Spontaneous superfluid unpinning and the inhomogeneous distribution of vortex lines in neutron stars. *Astrophysical Journal*, **330**, 835–46.

Chevalier, R. A. & Emmering, R. T. (1986). Are pulsars born as slow rotators? *Astrophysical Journal*, **304**, 140–53.

Chugai, N. N. (1984). Pulsar space velocities and neutrino chirality. *Soviet Astronomy Letters*, **10**, 87–8.

Clear, J., Bennett, K., Buccheri, R., Grenier, I. A., Hermsen, W., Mayer-Hasselwander, H. A. & Sacco, B. (1987). A detailed analysis of the high energy gamma-ray emission from the Crab pulsar and nebula. *Astronomy and Astrophysics*, **174**, 85–94.

Clifton, T. R. & Lyne, A. G. (1986). A high frequency survey for short period pulsars. *Nature*, **320**, 43–5.

Clifton, T. R., Frail, D. A., Kulkarni, S. R. & Weisberg, J. M. (1988). Neutral hydrogen absorption observations toward high-dispersion measure pulsars. *Astrophysical Journal*, **333**, 332–40.

Cocke, W. J., Disney, M. J. & Taylor, D. J. (1969). Discovery of optical signals from pulsar NP0532. *Nature*, **221**, 525–7.

Cohen, J. M. & Rosenblum, A. (1972). Pulsar magnetosphere. *Astrophysics and Space Science*, **16**, 130–6.

Coppi, B. & Pegoraro, F. (1979). Magnetic equation for a rotating neutron star. *Annals of Physics*, **119**, 97–116.

Cordes, J. M. (1976a). Correlation analyses of microstructure and noise-like intensity fluctuations from pulsars 2016+28. *Astrophysical Journal*, **208**, 944–54.

Cordes, J. M. (1976b). Pulsar radiation as polarized short noise. *Astrophysical Journal*, **210**, 780–91.

Cordes, J. M. (1979). Coherent radio emission from pulsars. *Space Science Review*, **24**, 567–600.

Cordes, J. M. (1986). Space velocities of radio pulsars from interstellar medium. *Astrophysical Journal*, **311**, 183–96.

Cordes, J. M. & Downs, G. S. (1985). JPL pulsar timing observations. III Pulsar rotation fluctuations. *Astrophysical Journal, Supplement Series*, **59**, 343–82.

Cordes, J. M. & Hankins, T. N. (1979). Frequency structure of micropulses from pulsar PSR 0950+08. *Astrophysical Journal*, **233**, 981–6.

Cordes, J. M., Downs, G. S. & Krause-Polstorff, J. (1988). JPL pulsar timing observations. V. Macro- and microjumps in the Vela pulsar 0833−45. *Astrophysical Journal*, **330**, 847–69.

Cordova, F. A., Hjellming, R. M., Mason, K. O. & Middleditch, J. (1989). Soft X-ray emission from the radio pulsar PSR 0656+14. *Astrophysical Journal*, **345**, 451–63.

Damashek, M., Backus, P. R., Taylor, J. H. & Burkhardt, R. K. (1982). Northern hemisphere pulsar survey: a third pulsar in a binary system. *Astrophysical Journal*, **253**, L57–60.

D'Amigo, N., Manchester, R., Durdin, J. M., Stokes, G. H., Stinebring, D. R., Taylor, J. H. & Brissenden, R. J. V. (1988). A survey for millisecond pulsars at Molonglo. *Monthly Notices of the Royal Astronomical Society*, **234**, 437–44.

Daugherty, J. K. & Harding, A. K. (1982). Electrodynamic cascades in pulsars. *Astrophysical Journal*, **252**, 337–47.

Daugherty, J. K. & Harding, A. K. (1983). Pair production in superstrong magnetic fields. *Astrophysical Journal*, **273**, 761–73.

Davis, L. & Goldstein, M. (1970). Magnetic-dipole alignment in pulsars. *Astrophysical Journal*, **159**, L81–5.

Deich, W. T. S., Cordes, J. M., Hankins, T. H. & Rankin, J. M. (1986). Null transition times, quantized drift modes, and no memory across nulls for PSR 1944+17. *Astrophysical Journal*, **300**, 540–50.

Deutsch, A. J. (1955). The electromagnetic field on an idealized star in rigid rotation in vacuo. *Annales d'Astrophysique*, **18**, 1–10.

Dewey, R. J., Taylor, J. H., Weisberg, J. M. & Stokes, G. H. (1985). A search for low-luminosity pulsars. *Astrophysical Journal*, **294**, L25–9.

Dewey, R. J., Taylor, J. H., Maguire, C. M. & Stokes, G. H. (1988). Period derivatives and improved parameters for 66 pulsars. *Astrophysical Journal*, **332**, 762–9.

Downs, G. S. (1981). JPL pulsar timing observations: I. The Vela pulsar. *Astrophysical Journal*, **249**, 687–97.

Downs, G. S. (1982). JPL pulsar timing observations: spinups in PSR 0525+21. *Astrophysical Journal*, **257**, L67–70.
Durdin, J. M., Large, M. I., Little, A. G., Manchester, R. N., Lyne, A. G. & Taylor, J. H. (1979). An unusual pulsar – PSR 0826−34. *Monthly Notices of the Royal Astronomical Society*, **186**, 39p–41p.
Endean, V. G. (1974). 'Lorentz force-free' pulsar rotating fields. *Astrophysical Journal*, **187**, 359–60.
Endean, V. G. (1976). Self-consistent equilibria in the pulsar magnetosphere. *Monthly Notices of the Royal Astronomical Society*, **174**, 125–35.
Endean, V. G. (1983). The pulsar oblique rotator: numerical solution of an illustrative problem. *Monthly Notices of the Royal Astronomical Society*, **204**, 1067–79.
Ergma, E. V. (1983). Thermonuclear flashes in neutron-star envelopes. In *Astrophysics and space physics reviews*, vol. 2, ed. R. A. Syunyaev, pp. 163–88. Amsterdam: Harwood Academic Publishers GmbH and OPA.
Ferguson, D. C. & Seiradakis, J. H. (1978). A detailed, high time resolution study of high frequency radio emission from PSR 1133+16. *Astronomy and Astrophysics*, **64**, 27–42.
Flowers, E. & Itoh, N. (1976). Transport properties of dense matter. I. *Astrophysical Journal*, **206**, 218–42.
Flowers, E. & Itoh, N. (1979). Transport properties of dense matter. II. *Astrophysical Journal*, **230**, 847–58.
Flowers, E. & Ruderman, M. A. (1977). Evolution of pulsar magnetic fields. *Astrophysical Journal*, **215**, 302–10.
Flowers, E., Lee, J.-F., Ruderman, M. A., Sutherland, P., Hillebrandt, W. & Muller, E. (1977). Variation calculation of ground-state energy of iron atoms and condensed matter in strong magnetic fields. *Astrophysical Journal*, **215**, 291–301.
Friedman, J. L., Ipser, J. R. & Parker, L. (1986). Rapidly rotating neutron star models. *Astrophysical Journal*, **304**, 115–39.
Friedman, J. L., Ipser, J. R. & Parker, L. (1989). Implications of a half-milli-second pulsar. *Physical Review Letters*, **62**, 3015–9.
Friman, B. L. & Maxwell, O. V. (1979). Neutrino emissivities of neutron stars. *Astrophysical Journal*, **232**, 541–57.
Fruchter, A. S., Stinebring, D. R. & Taylor, J. H. (1988). A millisecond pulsar in an eclipsing binary. *Nature*, **333**, 237–9.
Ftaclas, C., Kearney, M. W. & Pechenick, K. (1986). Hot spots on neutron stars. II. The observer's sky. *Astrophysical Journal*, **300**, 203–8.
Fujimoto, M. Y. & Taam, R. E. (1986). The mass and radius of a neutron star in X-ray burster MXB 1636−536. *Astrophysical Journal*, **305**, 246–50.
Fujimoto, M. Y., Sztajno, M., Lewin, W. H. G. & van Paradijs, J. (1987). On the theory of type I X-ray bursts: the energetics of bursts and the nuclear fuel reservoir in the envelope. *Astrophysical Journal*, **319**, 902–15.
Gedalin, M. E. & Machabeli, G. Z. (1983). Oblique wave propagation in the relativistic electron–positron plasma. *Astrofizika*, **19**, 153–60.
Ghosh, P. & Lamb, F. K. (1979a). Accretion by rotating neutron stars. II. Radial and vertical structure of the transition zone in disk acceleration. *Astrophysical Journal*, **232**, 259–76.
Ghosh, P. & Lamb, F. K. (1979b). Accretion by rotating magnetic neutron stars. III. Accretion torques and period changes in pulsating X-ray sources. *Astrophysical Journal*, **234**, 296–316.

Giacconi, R., Gorenstein, P., Gurski, H. & Waters, J. R. (1967). An X-ray survey of the Cygnus region. *Astrophysical Journal*, **148**, L119–28.

Gil, J. A. (1987). Orthogonal polarization modes in PSR 0834+06. *Astrophysical Journal*, **314**, 629–33.

Ginzburg, V. L. (1969). Superfluidity and superconductivity in the universe. *Soviet Physics Uspekhi*, **12**, 241–51.

Ginzburg, V. L. (1970). *The propagation of electromagnetic waves in plasmas.* Oxford: Pergamon Press.

Ginzburg, V. L. & Usov, V. V. (1972). Concerning the atmosphere of magnetic neutron stars (pulsars). *JETP Letters*, **15**, 196–8.

Ginzburg, V. L., Zheleznyakov, V. V. & Zaitsev, V. V. (1969). Coherent radio-emission mechanisms and magnetic pulsar models. *Soviet Physics Uspekhi*, **12**, 378–98.

Glen, G. & Sutherland, P. (1980). On the cooling of neutron stars. *Astrophysical Journal*, **239**, 671–84.

Godfray, B. B., Shanahan, W. R. & Thode, L. E. (1975). Linear theory of a cold relativistic beam propagating along an external magnetic field. *Physics of Fluids*, **18**, 346–55.

Gold, T. (1968). Rotating neutron stars and the origin of the pulsating radio sources. *Nature*, **218**, 731–2.

Goldreich, P. (1970). Neutron star crusts and alignment of magnetic axes in pulsars. *Astrophysical Journal*, **160**, L11–15.

Goldreich, P. & Julian, W. H. (1969). Pulsar electrodynamics. *Astrophysical Journal*, **157**, 869–80.

Good, M. L. & Ng, K. K. (1985). Electromagnetic torques secular alignment, and spin down of neutron stars. *Astrophysical Journal*, **299**, 706–22.

Good, R. H. & Muller, E. W. (1956). Field emission. *Handbuch der Physik*, **21**, 1–78.

Gott, J. R., Gunn, J. E. & Ostriker, J. P. (1970). Runaway stars and the pulsars near the Crab Nebula. *Astrophysical Journal*, **160**, 91–5.

Graser, V. & Schönfelder, V. (1983). Search for pulsed gamma-ray emission at MeV energies from 24 radio pulsars. *Astrophysical Journal*, **273**, 681–7.

Greenstein, G. & Hartke, G. I. (1983). Pulselike character of blackbody radiation from neutron stars. *Astrophysical Journal*, **271**, 283–93.

Grigoryan, M. Sh. & Sahakyan, G. S. (1979). Pionization effect and its astrophysical aspects. *Physics of Elementary Particles and Atomic Nuclei*, **10**, 1075–113 (in Russian).

Grindlay, J., Gursky, H., Schnopper, H., Parsignault, D. R., Heise, J., Brinkman, A. C. & Schrijver, J. (1976). Discovery of intense X-ray bursts from the globular cluster NGC6624. *Astrophysical Journal*, **205**, L127–30.

Groth, E. J. (1975). Timing of the Crab Pulsar. I. Arrival times. II. Method of analysis. III. The slowing down and the nature of the random process. *Astrophysical Journal, Supplement Series*, **29**, 431–66.

Gurevich, A. V. & Istomin, Ya. N. (1985). Production of an electron–positron plasma in a pulsar magnetosphere. *Soviet Physics – JETP*, **62**, 1–11.

Gurevich, A. V., Krylov, A. L. & Tsedilina, E. E. (1976). Electric fields in the earth's magnetosphere and ionosphere. *Space Science Review*, **19**, 59–160.

Guseinov, O. Kh. & Yusifov, I. M. (1984). Spatial distribution of pulsars. *Soviet Astronomy*, **28**, 415–23.

Gwinn, C. R., Taylor, J. H., Weisberg, J. M. & Rawley, L. A. (1986). Measurement of pulsar parallaxes by VLBI. *Astronomical Journal*, **91**, 338–42.

Hankins, T. H. (1971). Short-time scale structure in two pulsars. *Astrophysical Journal*, **177**, L11–15.
Hankins, T. H. & Boriakoff, V. (1978). Submicrosecond time resolution observations of PSR 0950+08. *Nature*, **276**, 45–7.
Hankins, T. H. & Boriakoff, V. (1981). Microstructure in the pulsar 0950+08 interpulse at radio wavelengths. *Astrophysical Journal*, **249**, 238–40.
Hankins, T. H. & Fowler, L. A. (1986). Frequency dependence of the main pulse to interpulse separation for seven pulsars. *Astrophysical Journal*, **304**, 256–64.
Hankins, T. H. & Rickett, B. J. (1986). Frequency dependence of pulsar properties. *Astrophysical Journal*, **311**, 684–93.
Hardee, P. E. & Rose, W. K. (1976). A mechanism for the production of pulsar radio radiation. *Astrophysical Journal*, **210**, 533–8.
Harding, A. K. & Preece, R. (1987). Quantized synchrotron radiation in strong magnetic fields. *Astrophysical Journal*, **319**, 939–50.
Harding, A. K., Tademaru, E. & Esposito, L. W. (1978). A curvature-radiation-pair-production model for γ-ray pulsars. *Astrophysical Journal*, **225**, 226–36.
Harrison, E. R. & Tademaru, E. (1975). Acceleration of pulsars by asymmetric radiation. *Astrophysical Journal*, **201**, 447–61.
Heiles, C., Kulkarni, S. R., Stevens, M. A., Backer, D. C., Davis, M. M. & Goss, W. M. (1983). Distance to the 1.5 millisecond pulsar and other 4C 21.53 objects. *Astrophysical Journal*, **273**, L71–4.
Heintzmann, H. (1981). Pulsar slow-down epochs. *Nature*, **292**, 811–14.
Helfand, D. J. (1984). X-ray synchrotron nebulae and the origin of neutron stars. *Advances in Space Research*, **3**, 10–12.
Helfand, D. J., Chanan, G. A. & Novick, R. (1980). Thermal X-ray emission from neutron stars. *Nature*, **283**, 337–43.
Henriksen, R. N. & Norton, J. A. (1975). Oblique rotating pulsar magnetosphere with wave zones. *Astrophysical Journal*, **201**, 719–28.
Herring, C. & Nichols, M. H. (1949). Thermionic emission. *Review of Modern Physics*, **21**, 185–270.
Hewish, A., Bell, S. J., Pilkington, J. D., Scott, P. F. & Collins, R. A. (1968). Observation of a rapidly pulsating radio source. *Nature*, **217**, 709–13.
Hinata, S. & Jackson, E. A. (1974). On the axisymmetric pulsar atmosphere. *Astrophysical Journal*, **192**, 703–11.
Il'in, V. G., Ilyasov, Yu. P., Kuzmin, A. D., Pushkin, S. B., Palij, G. N., Shabanova, T. V. & Shitov, Yu. P. (1984). Scale of pulsar time. *Doklady Akademy Nauk SSSR. Mathematics & Physics*, **275**, 835–8 (in Russian).
Ilyasov, Yu. P., Kuzmin, A. D., Shabanova, T. V. & Shitov, Yu. P. (1989). Scale of pulsar time. In *Pulsars*, Proceedings of the Lebedev Physics Institute of the Academy of Sciences of the USSR, vol. 199, ed. A. D. Kuzmin, 149–59 (in Russian).
Imshennik, V. S. & Nadezhin, D. K. (1983). The terminal phases of stellar evolution and the supernova phenomenon. In *Astrophysics and space physics reviews*, vol. 2, ed. R. A. Syunyaev, pp. 75–161. Amsterdam: Harwood Academic Publishers GmbH and OPA.
Ingraham, R. L. (1973). Algorithm for solving the nonlinear pulsar equation. *Astrophysical Journal*, **186**, 625–9.
Istomin, Ya. N. (1988). Nonlinear interactions of waves in an inhomogeneous plasma. *Soviet Physics – JETP*, **68**, 1380–5.

Izvekova, V. A., Kuzmin, A. D., Malofeev, V. M. & Shitov, Yu. P. (1981). Radio spectra of pulsars. I. Observations of flux densities at meter wavelengths and analysis of the spectra. *Astrophysics and Space Science*, **78**, 45–72.

Izvekova, V. A., Malofeev, V. M. & Shitov, Yu. P. (1989a). Pulsar mean pulse profiles at 102.5 MHz. *Soviet Astronomy*, **33**, 175–82.

Izvekova, V. A., Kuzmin, A. D., Malofeev, V. M. & Shitov, Yu. P. (1989b). Frequency dependence of mean profiles of pulsar radio emission. In *Pulsars*, Proceedings of the Lebedev Physics Institute of the Academy of Sciences of the USSR, vol. 199, ed. A. D. Kuzmin, 13–41 (in Russian).

Jones, P. B. (1978). Particle acceleration at the magnetic poles of a neutron star. *Monthly Notices of the Royal Astronomical Society*, **178**, 807–23.

Jones, P. B. (1981). A model of the normal and null states of pulsars. *Monthly Notices of the Royal Astronomical Society*, **197**, 1103–24.

Jones, P. B. (1985a). Density-functional calculations of the cohesive energy of condensed matter in very strong magnetic fields. *Physical Review Letters*, **55**, 1338–40.

Jones, P. B. (1985b). Density functional calculations of the ground-state energies of atoms and infinite linear molecules in very strong magnetic fields. *Monthly Notices of the Royal Astronomical Society*, **216**, 503–10.

Jones, P. B. (1986a). Properties of condensed matter in very strong magnetic fields. *Monthly Notices of the Royal Astronomical Society*, **218**, 477–87.

Jones, P. B. (1986b). Pair creation in radio pulsars: the return mode changes and normal-null transitions. *Monthly Notices of the Royal Astronomical Society*, **222**, 577–91.

Joss, P. C. & Rappaport, S. A. (1984). Neutron stars in interacting binary systems. *Annual Review of Astronomy and Astrophysics*, **22**, 537–92.

Julian, W. H. (1973). Pulsar electrodynamics II. *Astrophysical Journal*, **183**, 967–71.

Kadomtsev, B. B. & Kudryavtsev, V. S. (1971). Molecules in an ultrastrong magnetic field. *JETP Letters*, **13**, 9–12.

Kaminker, A. D., Pavlov, G. G., Shibanov, Yu. A., Kurt, V. G., Smirnov, A. S., Shamolin, V. M., Kopaeva, I. F. & Sheffer, E. K. (1989). Spectral evolution of a burst from MXB 1728 – 34 and constraints of burster parameters. *Astronomy and Astrophysics*, **220**, 117–27.

Kardashev, N. S., Kuzmin, A. D., Nikolaev, N. Ya., Novikov, A. Yu., Popov, M. V., Smirnova, T. V., Soglasnov, V. A., Shabanova, T. V., Shinskij, M. D. & Shitov, Yu. P. (1987). Observations of pulsars with 10 μsec time resolution at the frequency of 102.5 MHz. *Soviet Astronomy*, **22**, 1024–30.

Kennel, C. F. & Coroniti, F. V. (1984a). Confinement of the Crab pulsar's wind by its supernova remnant. *Astrophysical Journal*, **283**, 694–709.

Kennel, C. F. & Coroniti, F. V. (1984b). Magneto-hydrodynamic model of Crab nebula radiation. *Astrophysical Journal*, **283**, 710–30.

Kirzhnits, D. A. (1972). External states of matter (ultrahigh pressures and temperatures). *Soviet Physics Uspekhi*, **14**, 512–23.

Kirzhnits, D. A. & Nepomnyaschii, Yu. A. (1971). Coherent crystallization of quantum liquid. *Soviet Physics – JETP*, **32**, 1191–8.

Klebesadel, R. W., Strong, I. B. & Olson, R. A. (1973). Observations of gamma-ray burst of cosmic origin. *Astrophysical Journal*, **182**, L85–6.

Knight, F. K. (1982). Observations and interpretations of the pulsed emission from the Crab pulsar. *Astrophysical Journal*, **260**, 538–52.

Kulkarni, S. R. (1986). Optical identification of binary pulsars: implications for magnetic field decay in neutron stars. *Astrophysical Journal*, **306**, L85–9.

Kulkarni, S. R., Clifton, T. C., Backer, D. C., Foster, R. S., Fruchter, A. S. & Taylor, J. H. (1988). A fast pulsar in radio nebula CTB 80. *Nature*, **331**, 50–3.

Kuzmin, A. D. (1986). Pulsar radio pulses: the frequency dependent dispersion measure and the extra arrival-time delay. *Soviet Astronomy Letters*, **12**, 325–8.

Kuzmin, A. D., Dagkesamanskaya, I. M. & Pugachev, V. D. (1984). The evolving orientation of the magnetic axes of pulsars. *Soviet Astronomy Letters*, **10**, 357–9.

Kuzmin, A. D., Malofeev, V. M., Izvekova, V. A., Sieber, W. & Wielebinski, R. (1986). A comparison of high-frequency and low-frequency characteristics of pulsars. *Astronomy and Astrophysics*, **161**, 183–94.

Kuzmin, O. A. (1985). Periodic microstructure in the pulses of PSR 1133+16 from synchronous observations at three meter wavelengths. *Soviet Astronomy*, **29**, 133–6.

Lamb, F. K. (1989). Accretion by magnetic neutron stars. In *Timing neutron stars*, ed. H. Ögelman & E. P. J. van den Heuvel, pp. 649–722. Dordrecht: Kluwer.

Landau, L. D. & Lifshitz, E. M. (1960a). *Mechanics*. Oxford: Pergamon Press.

Landau, L. D. & Lifshitz, E. M. (1960b). *Electrodynamics of continuous media*. Oxford: Pergamon Press.

Landau, L. D. & Lifshitz, E. M. (1965). *Quantum mechanics*. Oxford: Pergamon Press.

Landau, L. D. & Lifshitz, E. M. (1983). *The classical theory of fields*. Oxford: Pergamon Press.

Lang, K. R. (1974). *Astrophysical formulae*. Berlin: Springer-Verlag.

Lapidus, I. I., Syunyaev, R. A. & Titarchuk, L. G. (1986). Comptonization and X-ray burster spectra. *Soviet Astronomy Letters*, **12**, 383–6.

Lebedev, V. S., Neizvestnyi, S. I., Pimonov, A. A. & Plachotnichenko, V. L. (1983). Search for optical emission from radio pulsars on the six-meter telescope. *Soviet Astronomy*, **27**, 429–31.

Liang, E. P. (1986). Gamma burst annihilation lines and neutron star structure. *Astrophysical Journal*, **304**, 682–7.

Lindblom, L. (1986). Estimates of the maximum angular velocity of rotating neutron stars. *Astrophysical Journal*, **303**, 146–53.

Lipunov, V. M. (1991). *Astrophysics of neutron stars*. Berlin: Springer-Verlag.

Lohsen, E. (1975). Third speed-up of the Crab pulsar. *Nature*, **258**, 688–9.

Lominadze, D. G. & Patarya, A. D. (1982). Some nonlinear mechanisms of pulsar emission. *Physica Scripta*, **T2: 1**, 206–14.

Lominadze, D. G., Mikhaylovsky, A. B. & Sagdeev, R. Z. (1979). Langmuir turbulence of a relativistic plasma in a strong magnetic field. *Soviet Physics – JETP*, **50**, 927–32.

London, R. A., Taam, R. E. & Howard, W. M. (1984). The spectra of X-ray bursting neutron stars. *Astrophysical Journal*, **287**, L27–30.

Lyne, A. G. (1987). A massive glitch in an old pulsar. *Nature*, **326**, 569–70.

Lyne, A. G. & Manchester, R. N. (1988). The shape of pulsar radio beams. *Monthly Notices of the Royal Astronomical Society*, **234**, 477–508.

Lyne, A. G. & Pritchard, R. S. (1987). A glitch in the Crab pulsar. *Monthly Notices of the Royal Astronomical Society*, **229**, 223–5.

Lyne, A. G. & Smith, F. G. (1982). Interstellar scintillation and pulsar velocities. *Nature*, **298**, 825–7.

Lyne, A. G. & Smith, F. G. (1990). *Pulsar astronomy*. Cambridge: Cambridge University Press.

Lyne, A. G., Manchester, R. N. & Taylor, J. H. (1985). The galactic population of pulsars. *Monthly Notices of the Royal Astronomical Society*, **213**, 613–40.

Lyne, A. G., Brinklow, A., Middleditch, J., Kulkarni, S. R., Backer, D. C. & Clifton, T. R. (1987). The discovery of a millisecond pulsar in the globular cluster M28. *Nature*, **328**, 339–40.

Lyne, A. G., Biggs, J. D., Brinklow, A., Ashworth, M. & McKenna, J. (1988a). Discovery of a binary millisecond pulsar in the globular cluster M4. *Nature*, **332**, 45–7.

Lyne, A. G., Pritchard, R. S. & Smith, F. G. (1988b). Crab pulsar timing 1982–87. *Monthly Notices of the Royal Astronomical Society*, **233**, 667–76.

Malofeev, V. M. (1989). Method of energy measurement of pulsar radio emission. In *Pulsars*, Proceedings of the Lebedev Physics Institute of the Academy of Sciences of the USSR, vol. 199, ed. A. D. Kuzmin, 125–46 (in Russian).

Malov, I. F. (1990). On the angles between magnetic field and rotation axis in pulsars. *Soviet Astronomy*, **34**, 189–97.

Malov, I. F. & Malofeev, V. M. (1981). Radio spectra of pulsars. II. The interpretation. *Astrophysics and Space Science*, **78**, 73–83.

Malov, I. F. & Sulejmanova, S. A. (1982). Two types of pulsars? *Astrofizika*, **18**, 107–18.

Manchester, R. (1971). Observations of pulsar polarization at 410 and 1665 MHz. *Astrophysical Journal, Supplement Series*, **23**, 283–322.

Manchester, R. N. & Peterson, B. A. (1989). A braking index for PSR 0540 – 69. *Astrophysical Journal*, **342**, L23–6.

Manchester, R. N. & Taylor, J. H. (1977). *Pulsars*. San Francisco: W. H. Freeman.

Manchester, R. N., Taylor, J. H. & Huguenin, G. R. (1975). Observations of pulsar radio emission. *Astrophysical Journal*, **196**, 83–102.

Manchester, R. N., Lyne, A. G., Goss, W. M., Smith, F. G., Disney, M. J., Hartley, K. F., Jones, D. H. P., Danziger, I. J., Murdin, P. G., Peterson, D. A. & Wallace, P. T. (1978a). Optical observations of southern pulsars. *Monthly Notices of the Royal Astronomical Society*, **184**, 159–70.

Manchester, R. N., Lyne, A. G. & Taylor, J. H. (1978b). The second Molonglo pulsar survey – discovery of 155 pulsars. *Monthly Notices of the Royal Astronomical Society*, **185**, 409–21.

Manchester, R. N., Newton, L. M., Goss, W. H. & Hamilton, P. A. (1978c). Detection of a large period discontinuity in the longer period pulsar PSR 1641 – 45. *Monthly Notices of the Royal Astronomical Society*, **184**, 35p–37p.

Manchester, R. N., Wallace, P. T., Peterson, B. A. & Elliott, K. H. (1980). The optical pulse profile of the Vela pulsar. *Monthly Notices of the Royal Astronomical Society*, **190**, 9p–13p.

Manchester, R. N., Tuohy, I. R. & D'Amigo, N. (1982). Discovery of radio pulsations from the X-ray pulsar in the supernova remnant G 32.4–1.2. *Astrophysical Journal*, **262**, L31–4.

Manchester, R. N., D'Amigo, N. & Tuohy, I. R. (1985a). A search for short period pulsars. *Monthly Notices of the Royal Astronomical Society*, **212**, 975–86.

Manchester, R. N., Durdin, J. M. & Newton, L. M. (1985b). A second measurement of a pulsar braking index. *Nature*, **313**, 374–5.
Margon, B., Ford, H. C., Katz, J. I., Kwitter, K. B., Ulrich, R. K., Stone, R. P. S. & Klemova, A. (1979). The bizarre spectrum of SS 433. *Astrophysical Journal*, **230**, L41–6.
Maxwell, O. V. (1979). Neutron star cooling. *Astrophysical Journal*, **231**, 201–10.
Mazets, E. P. (1988). Gamma-ray bursts: current status. In *The physics of compact objects*, ed. N. E. White & L. G. Filipov, pp. 669–77. Oxford: Pergamon Press.
McCulloch, P. M., Hamilton, P. A., Ables, J. G. & Hunt, A. G. (1983). A radio pulsar in the Large Magellanic Cloud. *Nature*, **303**, 307–8.
McCulloch, P. M., Hamilton, P. A., Manchester, R. N. & Ables, J. G. (1978). Polarization characteristics of southern pulsars. *Monthly Notices of the Royal Astronomical Society*, **183**, 645–76.
McKenna, J. & Lyne, A. G. (1990). PSR 1737−30 and period discontinuities in young pulsars. *Nature*, **343**, 349–50.
Mestel, L. (1971). Pulsar magnetosphere. *Nature Physical Science*, **233**, 149–52.
Mestel, L. (1973). Force-free pulsar magnetosphere. *Astrophysics and Space Science*, **24**, 289–97.
Mestel, L. & Wang, Y.-M. (1979). The axisymmetric pulsar magnetosphere. *Monthly Notices of the Royal Astronomical Society*, **188**, 799–812.
Michel, F. C. (1973a). Rotating magnetospheres: an exact 3-D solution. *Astrophysical Journal*, **180**, L133–7.
Michel, F. C. (1973b). Rotating magnetosphere: a simple relativistic model. *Astrophysical Journal*, **180**, 207–25.
Migdal, A. B. (1972). Vacuum stability and maximum fields. *Soviet Physics – JETP*, **34**, 1184–98.
Müller, E. (1984). Variation calculation of iron and helium atoms and molecular chains in superstrong magnetic fields. *Astronomy and Astrophysics*, **130**, 415–18.
Murakami, T., Fujii, M., Hayashida, K., Itoh, M., Nishimura, J., Yamagami, T., Conner, J. P., Evans, W. D., Fenimore, E. E., Klebesadel, R. W., Yoshida, A., Kondo, J. & Kawai, N. (1988). Evidence for cyclotron absorption from spectral features in gamma-ray bursts seen with Ginga. *Nature*, **335**, 234–5.
Nagase, F. (1989). Accretion-powered X-ray pulsars. *Publications of the Astronomical Society, Japan*, **41**, 1–79.
Narayan, R. (1987). The birthrate and initial spin period of single radio pulsars. *Astrophysical Journal*, **319**, 162–79.
Neuhauser, D., Koonin, E. & Langanke, K. (1987). Structure of matter in strong magnetic fields. *Physical Review A*, **36**, 4163–75.
Nikishov, A. I. & Ritus, V. N. (1986). Asymptotic representations for some functions and integrals connected with the Airy function. In *Group theoretical methods in physics*, ed. M. A. Markov, pp. 126–47. Moscow: Nauka.
Nomoto, K. & Tsuruta, S. (1986). Confronting X-ray observations of young supernova remnants with neutron star cooling models. *Astrophysical Journal*, **305**, L19–22.
Ögelman, H. & Buccheri, R. (1987). Soft X-ray imaging observations of the 39 millisecond pulsar PSR 1951+32. *Astronomy and Astrophysics*, **186**, L17–19.

Okamoto, I. (1974). Force-free pulsar magnetosphere. I. The steady axisymmetric theory for the charge-separated plasma. *Monthly Notices of the Royal Astronomical Society*, **167**, 457–74.

Oppenheimer, J. R. & Volkoff, G. M. (1939). On massive neutron cores. *Physical Review*, **55**, 374–81.

Palmer, R. G. (1975). Neutron star cores. *Astrophysics and Space Science*, **34**, 209–22.

Pandharipande, V. R., Pines, D. & Smith, R. A. (1976). Neutron star structure: theory, observations and speculation. *Astrophysical Journal*, **208**, 550–66.

Perry, T. E. & Lyne, A. G. (1985). Unpulsed radio emission from pulsars. *Monthly Notices of the Royal Astronomical Society*, **212**, 489–96.

Phinney, E. S. & Blandford, R. D. (1981). Analysis of the pulsar P–$\dot{P}$ distribution. *Monthly Notices of the Royal Astronomical Society*, **194**, 137–48.

Popov, M. V. (1986). Correlation of intensity fluctuations of individual pulses of PSR 0329+54 at different longitudes of the average profile. *Soviet Astronomy*, **30**, 577–87.

Popov, M. V., Smirnova, T. V. & Soglasnov, V. A. (1987). Microstructure of pulsars PSR 0809+74, 0950+08, and 1133+16 in the 67–102 MHz range. *Soviet Astronomy*, **31**, 529–37.

Pravdo, S. H., Becker, R. H., Boldt, E. A., Holt, S. S., Rothschild, R. E., Serlemitsos, P. T. & Swank, J. H. (1976). X-ray observations of the Vela pulsar: statistics and spectrum. *Astrophysical Journal*, **208**, L67–9.

Radhakrishnan, V. & Cocke, D. J. (1969). Magnetic poles and the polarization structure of pulsar radiation. *Astrophysical Letters*, **3**, 225–9.

Raich, M. E. & Yakovlev, D. G. (1982). Thermal and electrical conductivities of crystals in neutron stars and degenerate dwarfs. *Astrophysics and Space Science*, **87**, 193–204.

Rankin, J. M. (1983a). Toward an empirical theory of pulsar emission. I. Morphological taxonomy. *Astrophysical Journal*, **274**, 333–58.

Rankin, J. M. (1983b). Toward an empirical theory of pulsar emission. II. On the spectral behaviour of component width. *Astrophysical Journal*, **274**, 359–68.

Rankin, J. M. (1986). Toward an empirical theory of pulsar emission. III. Mode changing, drifting subpulses, and pulse nulling. *Astrophysical Journal*, **301**, 901–22.

Rankin, J. M. (1988). On the polarization–modal construction of triplicity in the profile of pulsar 1604−00. *Astrophysical Journal*, **325**, 314–19.

Rankin, J. M. (1990). Toward an empirical theory of pulsar emission. IV. Geometry of the core emission region. *Astrophysical Journal*, **352**, 247–57.

Rankin, J. M., Wolszczan, A. & Stinebring, D. R. (1988). Mode changing and quasi-periodic modulation in pulsar 1737+13, a bright, five-component pulsar. *Astrophysical Journal*, **324**, 1048–55.

Rankin, J. M., Stinebring, D. R. & Weisberg, J. M. (1989). Arecibo 21-cm polarimetry of 64 pulsars: a guide to classification. *Astrophysical Journal*, **346**, 869–97.

Rawlay, L. A., Taylor, J. H. & Davis, M. M. (1988). Fundamental astronomy and millisecond pulsars. *Astrophysical Journal*, **326**, 947–53.

Ray, A. & Datta, B. (1984). Rotating neutron star structure: implications of the millisecond pulsar PSR 1937+214. *Astrophysical Journal*, **282**, 542–9.

Richardson, M. B., Van Horn, H. M., Ratcliff, K. F. & Malone, R. C. (1982). Neutron star evolutionary sequences. *Astrophysical Journal*, **255**, 624–53.

Rickett, B. J. (1975). Amplitude-modulated noise: an empirical model for the radio radiation received from pulsars. *Astrophysical Journal*, **197**, 185–91.

Rickett, B. J., Hankins, T. H. & Cordes, J. M. (1975). The radio spectrum of micropulses from pulsar PSR 0950+08. *Astrophysical Journal*, **201**, 425–30.

Ritchings, R. T. (1976). Pulsar single pulse intensity measurements and pulse nulling. *Monthly Notices of the Royal Astronomical Society*, **176**, 249–63.

Ruderman, M. A. (1971). Matter in superstrong magnetic fields: the surface of a neutron star. *Physical Review Letters*, **27**, 1306–8.

Ruderman, M. A. & Sutherland, P. G. (1975). Theory of pulsars: polar gaps, sparks and coherent microwave radiation. *Astrophysical Journal*, **196**, 51–72.

Salvati, M. (1973). On self-consistent model for the pulsar magnetosphere. *Astronomy and Astrophysics*, **27**, 413–15.

Sato, K. (1979). Nuclear composition in the inner crust of neutron stars. *Progress of Theoretical Physics*, **62**, 957–68.

Sawyer, R. F. (1972). Condensed π-phase in neutron-star matter. *Physical Review Letters*, **29**, 382–5.

Scalapino, D. J. (1972). π-Condensate in dense nuclear matter. *Physical Review Letters*, **29**, 386–8.

Schaaf, M. E. (1987). *Thermische Strahlung von Neutronensternen*, MPI Report 203. Garching: Max-Planck-Institut für Physik und Astrophysik (in German).

Scharlemann, E. T. & Wagoner, R. V. (1973). Aligned rotating magnetosphere. I. General analysis. *Astrophysical Journal*, **182**, 951–60.

Schreier, E., Levinson, R., Gursky, H., Kellog, E., Tananbaum, H. & Giacconi, R. (1972). Evidence for binary nature of Centaurus X-3 from UHURU X-ray observations. *Astrophysical Journal*, **172**, L79–80.

Sedrakyan, D. M. & Movsisyan, A. G. (1986). Magnetic moments of neutron stars with different equations of state. *Astrofizika*, **24**, 163–6.

Seward, F. D. & Harnden, F. R. (1982). A new, fast X-ray pulsar in the supernova remnant MSH15 − 52. *Astrophysical Journal*, **256**, L45–7.

Seward, F. D. & Wang, Z.-R. (1988). Pulsars, X-ray synchrotron nebulae, and guest stars. *Astrophysical Journal*, **332**, 199–205.

Seward, F. D., Harnden, F. R. & Helfand, D. J. (1984). Discovery of a 50 millisecond pulsar in the Large Magellanic Cloud. *Astrophysical Journal*, **287**, L19–22.

Shafranov, V. D. (1967). Electromagnetic waves in plasma. In *Review of plasma physics*, vol. 3, ed. M. A. Leontovich, pp. 1–47. New York: Consultants Bureau.

Shapiro, S. L. & Teukolsky, S. A. (1983). *Black holes, white dwarfs, and neutron stars. The physics of compact objects.* New York: Wiley.

Shapiro, S. L., Teukolsky, S. A. & Wasserman, I. (1983). Implications of the millisecond pulsars from neutron star model. *Astrophysical Journal*, **272**, 702–7.

Shaposhnikov, V. E. (1981). Curvature radiation on longitudinal waves in magnetosphere of a neutron star. *Astrofizika*, **17**, 749–64.

Shibazaki, N. & Lamb, F. K. (1989). Neutron star evolution with internal heating. *Astrophysical Journal*, **346**, 808–22.

Shitov, Yu. P. & Malofeev, V. M. (1985). A superdispersion delay of the meter-wave pulses of PSR 0809 + 74. *Soviet Astronomy Letters*, **11**, 39–42.

Shitov, Yu. P., Kuzmin, A. D., Kutuzov, S. M., Ilyasov, Yu. P., Alekseev, Yu. I. & Alekseev, I. A. (1980). New pulsars discovered of 3-m wavelength. *Soviet Astronomy Letters*, **6**, 85–6.
Shitov, Yu. P., Malofeev, V. M. & Izvekova, V. A. (1988). Superdispersion delay of low-frequency pulsar pulses. *Soviet Astronomy Letters*, **14**, 181–5.
Shklovsky, I. S. (1971). Pulsars and type II supernovae. *Astrophysical Letters*, **8**, 101–3.
Shklovsky, I. S. (1978). *Stars—their birth, life, and death.* San Francisco: W. H. Freeman.
Sieber, W. (1973). Pulsar spectra: a summary. *Astronomy and Astrophysics*, **28**, 237–52.
Sieber, W., Reinecke, R. & Wielebinski, R. (1975). Observations of pulsars at high frequencies. *Astronomy and Astrophysics*, **38**, 169–82.
Slee, O. B., Alurkar, S. K. & Bobra, A. D. (1986). Flux densities, spectra and variability of pulsars at metre wavelengths. *Australian Journal of Physics*, **39**, 103–14.
Slee, O. B., Bobra, A. D. & Alurkar, S. K. (1987). Spectral behaviour of pulsar width in pulsars. *Australian Journal of Physics*, **40**, 557–86.
Smirnova, T. V. & Shabanova, T. V. (1988). Between-pulse radio emission of pulsars. *Soviet Astronomy*, **32**, 61–7.
Smirnova, T. V., Soglasnov, V. A., Popov, M. V. & Novikov, A. Yu. (1986). Dual-frequency correlation of pulsar micropulses. *Soviet Astronomy*, **30**, 51–6.
Smith, F. G. (1977). *Pulsars.* Cambridge: Cambridge University Press.
Soglasnov, V. A., Smirnova, T. V., Popov, M. V. & Kuzmin, A. D. (1981). A statistical analysis of the fine-scale structure of PSR 0809 + 74. *Soviet Astronomy*, **25**, 442–5.
Stinebring, D. R., Cordes, J. M., Rankin, J. M., Weisberg, J. M. & Boriakoff, V. (1984a). Pulsar polarization fluctuations. I. 1404 MHz statistical summaries. *Astrophysical Journal, Supplement Series*, **55**, 247–77.
Stinebring, D. R., Cordes, J. M., Weisberg, J. M., Rankin, J. M. & Boriakoff, V. (1984b). Pulsar polarization fluctuations. II. 800 MHz statistical summaries. *Astrophysical Journal, Supplement Series*, **55**, 278–88.
Stokes, G. H., Taylor, J. H., Weisberg, J. M. & Dewey, R. J. (1985). A survey for short-period pulsars. *Nature*, **317**, 787–8.
Stokes, G. H., Segelstein, D. J., Taylor, J. H. & Dewey, R. J. (1986). Results of two surveys for fast pulsars. *Astrophysical Journal*, **311**, 694–700.
Stollman, G. M. (1987). Pulsar statistics. *Astronomy and Astrophysics*, **178**, 143–52.
Strickman, M. S., Kurfess, J. D. & Johnson, W. N. (1982). A transient 77 keV emission feature from the Crab pulsar. *Astrophysical Journal*, **253**, 123–6.
Sturrock, P. A. (1971). A model of pulsars. *Astrophysical Journal*, **164**, 529–56.
Sulejmanova, S. A. (1989). Linear polarization of mean impulses at frequencies 102.5, 60, and 40 MHz. In *Pulsars*, Proceedings of the Lebedev Physics Institute of the Academy of Sciences of the USSR, vol. 199, ed. A. D. Kuzmin, 42–67 (in Russian).
Sulejmanova, S. A. & Izvekova, V. A. (1984). Two-mode emission for PSR 0943 + 10. *Soviet Astronomy*, **28**, 32–5.
Sulejmanova, S. A., Volodin, Yu. V. & Shitov, Yu. P. (1988). Polarization of the mean profiles of pulsar pulses at 102.5 MHz. *Soviet Astronomy*, **32**, 177–85.
Suvorov, E. V. & Chugunov, Yu. V. (1975). The electromagnetic waves in a relativistic plasma with a strong magnetic field. *Astrofizika*, **11**, 305–18.

Syrovatskii, S. I. (1981). Pinch sheets and reconnection in astrophysics. *Annual Review of Astronomy and Astrophysics*, **19**, 163–230.

Sztajno, M., Fujimoto, M. Y., van Paradijs, J., Vacca, W. D., Lewin, W. H. G., Penninx, W. & Trümper, J. (1986). Constraints on the mass-radius relation of the neutron star in 4U 1746-37/NGC 6441. *Monthly Notices of the Royal Astronomical Society*, **226**, 39–56.

Taam, R. E. & van den Heuvel, E. P. J. (1986). Magnetic field decay and the origin of neutron star binaries. *Astrophysical Journal*, **305**, 235–45.

Tademaru, E. (1973). On the energy spectrum of relativistic electrons in the Crab nebula. *Astrophysical Journal*, **183**, 625–35.

Taylor, J. H. & Stinebring, D. R. (1986). Recent progress in the understanding of pulsars. *Annual Review of Astronomy and Astrophysics*, **24**, 285–328.

Taylor, J. H. & Weisberg, J. M. (1982). A new test of general relativity: gravitational radiation and the binary pulsar PSR 1913+16. *Astrophysical Journal*, **253**, 908–20.

Taylor, J. H. & Weisberg, J. M. (1989). Further experimental tests of relativistic gravity using the binary pulsar PSR 1913+16. *Astrophysical Journal*, **345**, 434–50.

Townes, C. H., Lacy, J. H., Geballe, T. R. & Hollenbach, D. J. (1983). The centre of the Galaxy. *Nature*, **301**, 661–6.

Trümper, J., Pietsch, W., Reppin, C., Voges, W., Staubert, R. & Kendziorra, E. (1978). Evidence for strong cyclotron line emission in the hard X-ray spectrum of Hercules X-1. *Astrophysical Journal*, **219**, 105–10.

Tsuruta, S. (1979). Thermal properties and detectability of neutron stars. I. Cooling and heating of neutron stars. *Physical Reports*, **56**, 237–78.

Tsytovich, V. N. (1971). *Nonlinear effects in plasma*. New York: Plenum Press.

Tuohy, I. R., Garmire, G. P., Manchester, R. N. & Dopita, M. A. (1983). The central X-ray source in RCW 103: evidence for blackbody emission. *Astrophysical Journal*, **268**, 778–81.

Urpin, V. A., Levshakov, S. A. & Yakovlev, D. G. (1986). Generation of neutron star magnetic fields by thermomagnetic effects. *Monthly Notices of the Royal Astronomical Society*, **219**, 703–18.

van den Heuvel, E. P. J. (1984). Models for the formation of binary and millisecond radio pulsars. *Journal of Astrophysics and Astronomy*, **5**, 209–33.

van den Heuvel, E. P. J. & Taam, R. E. (1984). Two types of binary radio pulsars with different evolutionary histories. *Nature*, **309**, 235–7.

van Paradijs, J. (1978). Average properties of X-ray burst sources. *Nature*, **274**, 650–3.

van Riper, K. A. (1988). Magnetic neutron star atmosphere. *Astrophysical Journal*, **329**, 339–75.

van Riper, K. A. & Lamb, D. Q. (1981). Neutron star evolution and results from the Einstein X-ray observatory. *Astrophysical Journal*, **244**, L13–17.

Vivekanand, M. & Narayan, R. (1981). A new look at pulsar statistics – birthrate and evidence for injection. *Journal of Astrophysics and Astronomy*, **2**, 315–37.

Vivekanand, M., Narayan, R. & Radhakrishnan, V. (1982). On selection effects in pulsar searches. *Journal of Astrophysics and Astronomy*, **3**, 237–47.

Weeks, T. C. (1988). Very high energy gamma-ray astronomy. *Physics Reports*, **160**, 1–121.

Wheaton, W. A., Doty, J. P., Primini, F. A., Cooke, B. A., Dobson, C. A., Goldman, A., Hecht, M., Hoffman, J. A., Howe, S. K., Scheepmaker, A., Tsiang, E. Y., Lewin, W. H. G., Matteson, J. L., Gruber, D. E., Baity, W. A., Rothschild, R., Knight, F. K., Nolan, P. & Peterson, L. E. (1979). An absorption feature in the spectrum of the pulsed hard X-ray flux from 4U0115−63. *Nature*, **282**, 240–3.

White, R. S., Sweeney, W., Tumer, T. & Zych, A. (1985). Low-energy and medium-energy gamma rays from PSR 0531+21. *Astrophysical Journal*, **229**, L23–8.

Wilhelmson, H., Stenflo, L. & Engelman, F. (1970). Explosive instabilities in the well-defined phase description. *Journal of Mathematical Physics*, **11**, 1738–42.

Witten, E. (1984). Cosmic separation of phases. *Physical Review D*, **30**, 272–85.

Wolszczan, A., Kulkarni, S. R., Middleditch, J., Backer, D. C., Fruchter, A. S. & Dewey, R. J. (1989). A 110-ms pulsar, with negative period derivative, in globular cluster M15. *Nature*, **337**, 531–3.

Wright, G. A. E. (1981). The geometry of pulsar polar-cap drift. *Monthly Notices of the Royal Astronomical Society*, **196**, 153–7.

Wright, G. A. E. & Fowler, L. A. (1981). Mode-changing and quantized subpulse drift rates in pulsar PSR 2319+60. *Astronomy and Astrophysics*, **101**, 356–61.

Wright, G. A. E. & Loh, E. D. (1987). The companion star of a binary millisecond pulsar. *Nature*, **324**, 127–8.

Yakovlev, D. G. (1984). Transport properties of the degenerate electron gas of neutron stars along the quantizing magnetic field. *Astrophysics and Space Science*, **98**, 37–60.

Yakovlev, D. G. & Urpin, V. A. (1980). Thermal and electric conductivity in white dwarfs and neutron stars. *Soviet Astronomy*, **24**, 303–10.

Yungelson, L. R. & Masevich, A. G. (1983). Evolution in close binary systems. In *Astrophysics and space physics reviews*, vol. 2, ed. R. A. Syunyaev, pp. 29–74. Amsterdam: Harwood Academic Publishers GmbH and OPA.

Zeldovich, Ya. B. & Novikov, I. D. (1971). *Stars and relativity*. Chicago: The University of Chicago Press.

Zhuravlev, V. I. & Popov, M. V. (1990). Deterministic chaos in microstructure of radio pulses of 0809+74. *Soviet Astronomy*, **34**, 377–82.

Additional literature

We present here the main theoretical works that we have not included in the list of References. We have tried to cover all the basic ideas developed in connection with the problem of radio pulsars. Here we give references in chronological order.

1. Reviews

In addition to the monographs and reviews cited in the text, the following review papers and monographs are devoted to radio pulsars.

Kennel, C. F., Fujimura, F. S. & Pellat, R. (1979). Pulsar magnetosphere. *Space Science Reviews*, **24**, 407–36.
Arons, J. (1979). Some problems of pulsar physics. *Space Science Reviews*, **24**, 437–510.
Michel, F. C. (1981). Theory of pulsar magnetosphere. *Reviews of Modern Physics*, **54**, 1–66.
Srinivasan, G. (1989). Pulsars: their origin and evolution. *Astronomy and Astrophysics Review*, **1**, 209–60.

2. Magnetosphere of a neutron star

2.1 The vacuum model

The model of vacuum magnetic rotator in application to neutron stars has been considered in the following works.

Pacini, F. (1967). Energy losses of a rotating neutron star. *Nature*, **221**, 567–8.
Ostriker, J. P. & Gunn, J. E. (1969). On the nature of pulsars. I. Theory. *Astrophysical Journal*, **157**, 1395–417.
Jackson, E. A. (1976). A new pulsar atmospheric model. I. Aligned magnetic and rotation axes. *Astrophysical Journal*, **206**, 831–41.
Jackson, E. A. (1978). Theory of the pulsar atmosphere. II. Arbitrary magnetic and rotation axes. *Astrophysical Journal*, **222**, 675–88.
Martin, R. F. (1985). On the existence of an exterior toroidal region in the nonaligned pulsar magnetosphere. *Astrophysical Journal*, **288**, 665–71.

These works have mainly been considered with the evolution of neutron star rotation and the motion of test particles in a given field of a rotating magnetic dipole.

2.2 The model of magnetosphere completely filled with plasma

Besides the papers cited in the book, the model of a magnetosphere completely filled with plasma has also been considered in the following papers.

Tademaru, E. (1974). Space-charge flows and electric fields in the magnetosphere of rotating magnetic stars. *Astrophysics and Space Science*, **30**, 179–86.

Kuo-Petravic, L. G., Petravic, M. & Roberts, K. V. (1975). Numerical studies of the axisymmetric pulsar magnetosphere. *Astrophysical Journal*, **202**, 762–72.

Okamoto, I. (1975). Force-free pulsar magnetosphere. II. The steady, axisymmetric theory of a normal plasma. *Monthly Notices of the Royal Astronomical Society*, **170**, 81–93.

Cohen, J. M., Kegels, L. S. & Rosenblum, A. (1975). Magnetospheres and pulsars with net charge. *Astrophysical Journal*, **201**, 783–91.

Mestel, L., Wright, G. A. E. & Westhold, K. C. (1976). Plasma-wave interaction in non-aligned pulsar model. *Monthly Notices of the Royal Astronomical Society*, **175**, 257–78.

Buckley, R. (1977). Pulsar magnetosphere with arbitrary geometry in the f–f approximation. *Monthly Notices of the Royal Astronomical Society*, **180**, 125–40.

Okamoto, I. (1978). Relativistic centrifugal winds. *Monthly Notices of the Royal Astronomical Society*, **185**, 69–107.

Jackson, E. A. (1981). Pulsars: polar pumps, interpolar currents, and induced Landau radiation. *Astrophysical Journal*, **251**, 665–73.

Mestel, L. & Wang, Y.-M. (1982). The non-aligned pulsar magnetosphere: an illustrative model for small obliquity. *Monthly Notices of the Royal Astronomical Society*, **198**, 405–27.

Asseo, E., Beaufils, D. & Pellat, R. (1984). Pulsar magnetospheres. *Monthly Notices of the Royal Astronomical Society*, **209**, 285–306.

Cheng, A. F. (1985). Interstellar grains and current flow in pulsar magnetospheres. *Astrophysical Journal*, **299**, 917–24.

Bogovalov, S. V. (1989). Plasma ejection by pulsars. *Soviet Astronomy Letters*, **15**, 469–74.

Lyubarskii, Yu. E. (1990). Equilibrium of the return current sheet and structure of the pulsar magnetosphere. *Soviet Astronomy Letters*, **16**, 16–20.

2.3 Other directions

2.3.1 The model of magnetosphere partially filled with plasma

The model suggested that plasma occupies a finite volume near a neutron star. The case of an axisymmetric magnetosphere has been considered for which the shape of the plasma-filled region has been determined.

Pillipp, W. (1974). On the electrodynamics equilibrium of charge region around rotating neutron star. *Astrophysical Journal*, **190**, 391–401.

Rylov, Yu. A. (1977). On the electron cap shape of a rotating neutron star with a strong magnetic field. *Astrophysics and Space Science*, **51**, 59–76.

Rylov, Yu. A. (1979). Acceleration of electrons in an internal zone of the pulsar electron cap. *Astrophysics and Space Science*, **66**, 401–28.

Jackson, E. A. (1979). Minimum energy state of the constrained pulsar atmosphere. *Astrophysical Journal*, **227**, 266–74.

Michel, F. C. (1979). Vacuum gaps in pulsar magnetospheres. *Astrophysical Journal*, **227**, 579–89.
Rylov, Yu. A. (1984). Influence of electron–positron pair production on a pulsar magnetospheric structure. *Astrophysics and Space Science*, **107**, 381–401.
Krause-Polstorff, J. & Michel, F. C. (1985). Electrosphere of an aligned magnetized neutron star. *Monthly Notices of the Royal Astronomical Society*, **213**, 43p–49p.
Krause-Polstorff, J. & Michel, F. C. (1985). Pulsar space charging. *Astronomy and Astrophysics*, **144**, 72–80.

2.3.2 The Ψ*-formalism

The Ψ*-formalism allows, in the MHD approximation, the mass of particles occupying the magnetosphere to be taken into account. This makes it possible, in principle, to extend the equations of the theory of the magnetosphere completely filled with plasma out to the limits of the light surface. The most detailed presentation of the Ψ*-formalism can be found in the following.

Scharlemann, E. T. (1974). Aligned rotating magnetospheres. II. Inclusion of inertial forces. *Astrophysical Journal*, **193**, 217–23.
Henriksen, R. N. & Norton, J. A. (1975). Does a steady axisymmetric pulsar magnetosphere accelerate charges? *Astrophysical Journal*, **201**, 431–9.
Endean, V. G. (1976). Self-consistent equilibria in the pulsar magnetosphere. *Monthly Notices of the Royal Astronomical Society*, **174**, 125–35.
Ardavan, H. (1976). Magnetospheric shock discontinuities in pulsars. Analysis of the inertial effects at the light cylinder. *Astrophysical Journal*, **203**, 226–32.
Ardavan, H. (1976). The pulsar equation including the inertial term: its first integrals and its alfvenic singularity. *Astrophysical Journal*, **204**, 889–95.
Ardavan, H. (1976) Incompatibility of the continuous steady-state models of pulsar magnetospheres with relativistic magnetohydrodynamics. *Monthly Notices of the Royal Astronomical Society*, **175**, 645–51.
Ardavan, H. (1976). Quasi-steady pulsar magnetospheres. *Monthly Notices of the Royal Astronomical Society*, **177**, 661–72.
Wang, Y.-M. (1978). On the role of finite inertia and resistivity in axisymmetric pulsar magnetospheres. *Monthly Notices of the Royal Astronomical Society*, **182**, 157–77.
Wright, G. A. E. (1978). The properties of charge-separated pulsar magnetosphere. *Monthly Notices of the Royal Astronomical Society*, **182**, 735–49.
Buckley, R. (1978). Pulsar magnetosphere with parallel electric fields. I. *Monthly Notices of the Royal Astronomical Society*, **183**, 771–8.
Mestel, L., Phillips, P. & Wang, Y.-M. (1979). The axisymmetric pulsar magnetosphere. I. *Monthly Notices of the Royal Astronomical Society*, **188**, 385–415.
Schmalz, R., Ruder, H. & Herold, H. (1979). On the self-consistent description of axisymmetric pulsar magnetospheres. *Monthly Notices of the Royal Astronomical Society*, **189**, 709–22.
Burman, R. R. (1980). Flow dynamics in pulsar magnetosphere models with particle inertia. *Australian Journal of Physics*, **33**, 771–87.
Buckley, R. (1981). Some exact axisymmetric force-free magnetosphere – I. *Monthly Notices of the Royal Astronomical Society*, **196**, 1021–49.

We should also mention recent papers developing the model of the magnetosphere containing no region of particle acceleration in polar regions of a neutron star.

Mestel, L., Robertson, J. A., Wang, Y.-M. & Westfold, K. C. (1985). The axisymmetric pulsar magnetosphere. *Monthly Notices of the Royal Astronomical Society*, **217**, 443–84.

Fitzpatrick, R. & Mestel, L. (1988). Pulsar electrodynamics – I. *Monthly Notices of the Royal Astronomical Society*, **232**, 277–302.

Fitzpatrick, R. & Mestel, L. (1988). Pulsar electrodynamics – II. *Monthly Notices of the Royal Astronomical Society*, **232**, 303–21.

Shibata, S. (1988). Appearance of the trans-field flow in the pulsar magnetosphere. *Monthly Notices of the Royal Astronomical Society*, **233**, 405–22.

2.3.3 The Michel disc model

This model suggests the existence of a MHD disc in the vicinity of a neutron star, in which there occurs closure of electric currents circulating in the magnetosphere. The axisymmetric case is considered in the following.

Michel, F. C. & Dessler, A. J. (1981). Pulsar disk systems. *Astrophysical Journal*, **251**, 654–64.

Michel, F. C. (1983). Radio pulsar disk electrodynamics. *Astrophysical Journal*, **266**, 188–200.

2.4 Region behind the light surface

Interaction between the electromagnetic field and plasma outside the light surface has been considered in the following.

Gunn, J. & Ostriker, J. P. (1969). Acceleration of high-energy cosmic ray by pulsars. *Physical Review Letters*, **22**, 728–31.

Michel, F. C. (1969). Relativistic stellar-wind torques. *Astrophysical Journal*, **158**, 727–38.

Max, C. & Perkins, F. (1971). Strong electromagnetic waves in overdense plasmas. *Physical Review Letters*, **27**, 1342–5.

Gunn, J. & Ostriker, J. P. (1971). On the motion and radiation of charged particles in strong electromagnetic waves: I Motion in plane and spherical waves. *Astrophysical Journal*, **165**, 523–41.

Ferrari, A. & Trussoni, E. (1973). Magnetic fields around highly magnetized objects. *Astrophysics and Space Science*, **24**, 3–15.

Michel, F. C. (1974). Rotating magnetosphere: far-field solutions. *Astrophysical Journal*, **187**, 585–8.

Asseo, E., Kennel, F. C. & Pellat, R. (1975). Flux limit of cosmic ray acceleration by strong spherical pulsar waves. *Astronomy and Astrophysics*, **44**, 31–40.

Kennel, C. F. & Pellat, R. (1976). Relativistic nonlinear plasma waves in a magnetic field. *Journal of Plasma Physics*, **15**, 335–55.

Asseo, E., Kennel, C. F. & Pellat, R. (1976). Synchro-Compton radiation damping of relativistically strong linearly polarized plasma waves. *Astronomy and Astrophysics*, **65**, 401–8.

Kaburaki, O. (1980). Determination of the electromagnetic field produced by a magnetic oblique rotator. II. Exact vacuum solution. *Astrophysics and Space Science*, **67**, 3–18.
Kaburaki, O. (1981). Determination of the electromagnetic field produced by a magnetic oblique rotator III. *Astrophysics and Space Science*, **74**, 333–56.
Kaburaki, O. (1982). Determination of the electromagnetic field produced by a magnetic oblique rotator. *Astrophysics and Space Science*, **82**, 441–56.
Asseo, E. & Beaufils, D. (1983). Role of the pressure anisotropy in the relativistic pulsar wind. *Astrophysics and Space Science*, **89**, 133–41.
Michel, F. C. (1984). Relativistic charge-separated winds. *Astrophysical Journal*, **284**, 384–8.
Asseo, E., Llobet, X. & Pellat, R. (1984). Spherical propagation of large amplitude pulsar waves. *Astronomy and Astrophysics*, **139**, 417–25.
da Costa, A. A. & Kahn, F. D. (1985). Pulsar electrodynamics: the back reaction of the motion of charged particles. *Monthly Notices of the Royal Astronomical Society*, **215**, 701–11.

3. Acceleration and generation of plasma in pulsar magnetosphere

3.1 General questions

General questions of quantum electrodynamical processes in a strong magnetic field in application to pulsar magnetosphere have been discussed in the following.

Baier, V. N. & Katkov, V. M. (1967). Processes involved in the motion of high energy particles in a magnetic field. *Soviet Physics – JETP*, **26**, 854–60.
Tsai, W. & Erber, T. (1974). Photon pair creation in intense magnetic field. *Physical Review D*, **10**, 492–9.
Shabad, A. E. (1975). Photon dispersion in a strong magnetic field. *Annals of Physics*, **90**, 166–95.
Daugherty, J. K. & Lerche, I. (1975). On pair production in intense electromagnetic fields occurring in astrophysical situations. *Astrophysics and Space Science*, **38**, 437–45.
Daugherty, J. K. & Lerche, I. (1976). Theory of pair production in strong electric and magnetic fields and its application pulsars. *Physical Review D*, **14**, 340–55.
Gnedin, Yu. N., Pavlov, G. G. & Shibanov, Yu. A. (1978). The effect of vacuum birefringence in a magnetic field on the polarization and beaming of X-ray pulsars. *Soviet Astronomy Letters*, **4**, 117–19.
Pavlov, G. G. & Shibanov, Yu. A. (1979). Influence of vacuum polarization by a magnetic field on the propagation of electromagnetic waves in a plasma. *Soviet Physics – JETP*, **49**, 741–9.
Daugherty, J. K. & Bussard, K. W. (1980). Pair annihilation in superstrong magnetic fields. *Astrophysical Journal*, **238**, 296–310.
Herold, H., Ruder, H. & Wunner, G. (1981). Relativistic effects on the photon propagation in a strong magnetized plasma. *Plasma Physics*, **23**, 775–92.
Shabad, A. E. & Usov, V. V. (1982). Effects of capturing of hard photons by strong magnetic field and the role it may play in pulsars. *Nature*, **295**, 215–18.
Shabad, A. E. & Usov, V. V. (1984). Propagation of γ-radiation in strong magnetic fields of pulsars. *Astrophysics and Space Science*, **102**, 327–58.
Herold, H., Ruder, H. & Wunner, G. (1985). Can γ-quanta really be captured by pulsar magnetic fields. *Physical Review Letters*, **54**, 1452–5.

Harding, A. K. (1986). One-photon pair annihilation in magnetized relativistic plasmas. *Astrophysical Journal*, **300**, 167–77.

3.2 The theory of double layer (internal gap)

3.2.1 The Ruderman–Sutherland scheme

The Ruderman–Sutherland scheme suggests a hampered particle outflow from the star surface. Besides the papers cited in the book, the scheme has been discussed in the following.

Tademaru, E. (1974). Space-charge flows and electric fields in the magnetospheres of rotating magnetic stars. *Astrophysics and Space Science*, **30**, 179–86.
Holloway, N. J. (1975). Particle acceleration at pulsar magnetic poles. *Monthly Notices of the Royal Astronomical Society*, **171**, 619–35.
Cheng, A. F. & Ruderman, M. A. (1977). Pair production discharges above pulsar polar caps. *Astrophysical Journal*, **214**, 598–606.
Cheng, A. F. & Ruderman, M. A. (1977). A Crab pulsar model: X-ray, optical and radio emission. *Astrophysical Journal*, **216**, 865–72.
Tsygan, A. I. (1980). The electric field near a rotating neutron star. *Soviet Astronomy*, **24**, 44–6.

Particle creation and radiation of γ-quanta was also analysed in the following.

Alber, Ya. I., Krotova, Z. N. & Eidman, V. Ya. (1975). On the cascade process in the strong magnetic and electric fields in astrophysical conditions. *Astrofizika*, **11**, 283–92.
Hinata, S. (1979). Inhomogeneous pulsar polar cap structure and radiation processes. *Astrophysical Journal*, **227**, 275–84.
Cheng, A. F. & Ruderman, M. A. (1980). Particle acceleration and radio emission above pulsar polar caps. *Astrophysical Journal*, **235**, 576–86.

3.2.2 The Arons scheme

The Arons scheme suggests a free particle outflow from the star surface. The theory of such a double layer was developed in the following.

Michel, F. C. (1974). Rotating magnetosphere: acceleration of plasma from the surface. *Astrophysical Journal*, **192**, 713–18.
Fawley, W. M., Arons, J. & Scharlemann, E. T. (1977). Potential drops above polar caps: acceleration of nonneutral beams. *Astrophysical Journal*, **217**, 227–43.
Scharlemann, E. T., Arons, J. & Fawley, W. M. (1978). Potential drops above polar caps: ultrarelativistic particle acceleration along the curved magnetic field. *Astrophysical Journal*, **222**, 297–316.
Arons, J. & Scharlemann, E. T. (1979). Pair formation above pulsar polar caps: structure of the low altitude acceleration zone. *Astrophysical Journal*, **231**, 854–79.
Arons, J. (1981). Pair creation above pulsar polar caps: steady flow in the surface acceleration zone and polar cap X-ray emission. *Astrophysical Journal*, **248**, 1099–116.

Barnard, J. J. & Arons, J. (1982). Pair production and pulsar cutoff in magnetized neutron stars with nondipolar magnetic geometry. *Astrophysical Journal*, **254**, 713–34.
Arons, J. (1983). Pair creation above pulsar polar caps. *Astrophysical Journal*, **266**, 215–41.

3.2.3 The Jones scheme

This scheme takes into account nuclear reactions in the surface layer of a neutron star, which are capable of contributing to the generation of secondary particles. Various aspects of this theory have been discussed in the following.

Jones, P. B. (1978). Particles acceleration at the magnetic poles of a neutron star. *Monthly Notices of the Royal Astronomical Society*, **184**, 807–23.
Jones, P. B. (1979). Pair production in the pulsar magnetosphere. *Astrophysical Journal*, **228**, 536–40.

3.3 The 'external' gap theory

The possibility of the existence of an 'external' gap, i.e. the region of acceleration and generation of particles in the external regions of pulsar magnetosphere has been discussed in the following.

Holloway, N. J. (1973). P–N junctions in pulsar magnetosphere? *Nature Physical Science*, **246**, 6–9.
Cheng, A. F., Ruderman, M. A. & Sutherland, P. (1976). Current flow in pulsar magnetospheres. *Astrophysical Journal*, **203**, 209–12.
Cheng, A. F. & Ruderman, M. A. (1977). Crab pulsar: X-ray, optical and radio emission. *Astrophysical Journal*, **216**, 865–72.
Holloway, N. J. & Pryce, M. H. L. (1981). Properties of gaps in pulsar magnetospheres. *Monthly Notices of the Royal Astronomical Society*, **194**, 95–110.
Ray, A. & Benford, G. (1981). Electron–positron cascade in pulsar outer gaps. *Physical Review D*, **23**, 2142–50.
Cheng, K. S., Ho, C. & Ruderman, M. A. (1986a). Energetic radiation from rapidly spinning pulsars. I. Outer magnetosphere gaps. *Astrophysical Journal*, **300**, 500–21.
Cheng, K. S., Ho, C. & Ruderman, M. A. (1986b). Energetic radiation from rapidly spinning pulsars. II. Vela and Crab. *Astrophysical Journal*, **300**, 522–39.
Ho, C. (1989). Spectra of Crab-like pulsars. *Astrophysical Journal*, **342**, 396–405.

3.4 Pulsars as sources of cosmic rays

Kennel, C. F., Schmidt, G. & Wilcox, T. (1973). Cosmic rays generation by pulsars. *Physical Review Letters*, **31**, 1364–7.
Rengarijan, T. N. (1975). Pulsars and the origin of cosmic rays. *Astrophysics and Space Science*, **32**, 55–75.
Sturrock, P. A. & Backer, K. B. (1979). Positron production by pulsars. *Astrophysical Journal*, **234**, 612–14.

4. Generation of radio emission and high-frequency radiation

4.1 Mechanisms of coherent radio emission

Simple models in which coherent radio emission of pulsars was associated with the existence in the radiating region of bunches of charged particles (the 'antenna' mechanism) have been considered in the following papers.

Komesaroff, M. M. (1970). Possible mechanism for the pulsar radio emission. *Nature*, **225**, 612–14.
Goldreich, P. & Keeley, D. A. (1971). Coherent synchrotron radiation. *Astrophysical Journal*, **170**, 463–77.
Eidman, V. Ya. (1971). On the problem of electromagnetic radiation of relativistic bunch of charged particles. *Astrofizika*, **7**, 135–42.
Buschauer, R. & Benford, G. (1976). General theory of coherent curvature radiation. *Monthly Notices of the Royal Astronomical Society*, **177**, 109–36.
Benford, G. & Buschauer, R. (1977). Coherent pulsar radio radiation by antenna mechanisms: general theory. *Monthly Notices of the Royal Astronomical Society*, **179**, 189–207.
Buschauer, R. & Benford, G. (1980). Narrow-band versus broad-band emission processes in pulsars. *Monthly Notices of the Royal Astronomical Society*, **190**, 945–59.
Ochelkov, Yu. P. & Usov, V. V. (1980). Curvature radiation of relativistic particles in the magnetosphere of pulsar. I. Theory. *Astrophysics and Space Science*, **69**, 439–60.
Ochelkov, Yu. P. & Usov, V. V. (1984). The nature of low-frequency cutoffs in the radio emission spectra of pulsars. *Nature*, **309**, 332–3.

Those models in which coherent radio emission was associated with the inverse energy population (the 'maser' mechanism) have been considered in the following papers.

Chiu, H. Y. & Canuto, V. (1971). Theory of radiation mechanisms of pulsars. *Astrophysical Journal*, **163**, 577–94.
Kaplan, S. A. & Tsytovich, V. N. (1973). Relativistic plasma and pulsar emission mechanisms. *Nature Physical Science*, **241**, 122–4.
Buschauer, R. & Benford, G. (1978). Physical mechanism of the Goldreich–Keeley radiative instability. *Monthly Notices of the Royal Astronomical Society*, **185**, 493–506.
Melrose, D. B. (1978). Amplified linear acceleration emission applied to pulsars. *Astrophysical Journal*, **225**, 557–73.

In the Smith model, the high directivity of radio emission was associated with relativistic contraction of directivity patterns due to the motion of the source in the region of the light cylinder. The following papers are devoted to this model.

Smith, F. G. (1969). Relativistic beaming of radiation from pulsars. *Nature*, **223**, 934–6.
Smith, F. G. (1970). The beaming of radio waves from pulsars. *Monthly Notices of the Royal Astronomical Society*, **149**, 1–15.
Smith, F. G. (1973). Emission mechanism in pulsars. *Nature*, **243**, 207–10.

A detailed survey of this scheme is given in

Ginzburg, V. L. & Zheleznyakov, V. V. (1975). On the pulsar emission mechanisms. *Annual Review of Astronomy and Astrophysics*, **13**, 511–35.

4.2 Plasma effects

4.2.1 The linear theory in a homogeneous magnetic field

In addition to the papers cited in the text, we point out the following papers where this theory has been developed.

Virtamo, J. & Jauho, P. (1973). Pulsar radio emission mechanism. *Astrophysical Journal*, **182**, 935–49.

Elitsur, M. (1974). Index of refraction of plasma in motion. *Astrophysical Journal*, **190**, 673–4.

Hinata, S. (1976). Relativistic plasma turbulence and its applications to pulsar phenomena. *Astrophysical Journal*, **206**, 282–94.

Blandford, R. D. & Scharlemann, E. T. (1976). On the scattering and absorption of electromagnetic radiation within pulsar magnetospheres. *Monthly Notices of the Royal Astronomical Society*, **174**, 59–85.

Kawamura, K. & Suzuki, I. (1977). A model of the radio emission mechanism in pulsar. *Astrophysical Journal*, **217**, 832–42.

Buschauer, R. & Benford, G. (1977). Temperature stabilization of instabilities in force-free magnetospheres. *Monthly Notices of the Royal Astronomical Society*, **179**, 99–103.

Hardee, P. E. & Rose, W. K. (1978). Wave production in an ultrarelativistic electron–positron plasma. *Astrophysical Journal*, **219**, 274–87.

Hardee, P. E. & Morrison, P. (1979). Plasma collective effects and pulsar emission models. *Astrophysical Journal*, **227**, 252–65.

Arons, J. & Smith, D. F. (1979). Electrostatic shear flow instability of relativistic nonneutral beam in pulsars. *Astrophysical Journal*, **229**, 728–33.

Cox, J. L. (1979). Coherent radiation from pulsars. *Astrophysical Journal*, **229**, 734–41.

Harding, A. K. & Tademaru, E. (1979). Propagating effects in a shearing field-free plasma and pulsar microstructure. *Astrophysical Journal*, **233**, 317–26.

Harding, A. K. & Tademaru, E. (1981). Propagation in a shearing plasma. Magnetic field effects and pulsar microstructure periods. *Astrophysical Journal*, **243**, 597–611.

Smith, D. F. (1981). Absence of cyclotron instability in some pulsar polar-cap models and its implications. *Astrophysical Journal*, **247**, 279–81.

Onishchenko, O. G. (1981). Cyclotron instability of relativistic plasma. *Soviet Astronomy Letters*, **7**, 28–9.

Mikhaylovsky, A. B., Onishchenko, O. G., Suramlishvili, G. I. & Sharapov, S. E. (1982). The emergence of electromagnetic waves from pulsar magnetosphere. *Soviet Astronomy Letters*, **8**, 369–71.

Arons, J. & Barnard, J. (1986). Wave propagation in pulsar magnetospheres: dispersion relation and normal modes of plasmas in superstrong magnetic fields. *Astrophysical Journal*, **302**, 120–37.

Usov, V. V. (1987). On two-stream instability in pulsar magnetospheres. *Astrophysical Journal*, **320**, 333–5.

Kazbegi, A. Z., Machabeli, G. Z. & Melikidze, G. I. (1987). Radio emission model of a typical pulsar. *Australian Journal of Physics*, **40**, 755–66.

Ursov, V. N. & Usov, V. V. (1988). Plasma flow nonstationarity in pulsar magnetospheres and two-stream instability. *Astrophysics and Space Science*, **140**, 325–36.

A detailed review of these papers is given in

Volokitin, A. S., Krasnoselskikh, V. V. & Machabeli, G. Z. (1985). Waves in a relativistic electron–positron plasma in pulsar. *Soviet Journal of Plasma Physics*, **11**, 531–8.

Lominadze, D. G., Machabeli, G. Z., Melikidze, G. I. & Pataraya, A. D. (1986). Plasma of the pulsar magnetosphere. *Soviet Journal of Plasma Physics*, **12**, 1233–49.

4.2.2 Non-linear processes in a homogeneous magnetic field

Non-linear wave interaction has been discussed in the following.

Mamradze, P. G., Machabeli, G. Z. & Melikidze, G. I. (1980). Generation of the infrared radiation of 0532 pulsar by Langmuir solitons. *Soviet Journal of Plasma Physics*, **6**, 1293–6.

ter Haar, D. & Tsytovich, V. N. (1981). Modulation instabilities in astrophysics. *Physics Reports*, **73**, 177–230.

Yu, M. Y., Shukla, P. K. & Rao, N. N. (1984). Strong electromagnetic relativistic electron–positron plasma. *Astrophysics and Space Science*, **107**, 327–32.

Melikidze, G. I. & Pataraya, A. D. (1984). Langmuir solitons propagated in the relativistic plasma at an angle to the external magnetic field. *Astrofizika*, **20**, 157–64.

Verga, A. D. & Fontan, C. F. (1985). Soliton turbulence in strongly magnetized plasma. Applications to the coherent radioemission of pulsars. *Plasma Physics and Controlled Fusion*, **27**, 19–45.

Mikhaylovsky, A. B. & Onishchenko, O. G. (1987). Drift instabilities of a relativistic plasma. *Journal of Plasma Physics*, **37**, 15–28.

The theory of high-frequency (X, gamma) radiation from young pulsars (Crab, Vela) has been considered in the following papers on the basis of the theory of quasi-linear stabilization of cyclotron instability (also with allowance for non-linear processes).

Mikhaylovsky, A. B. (1979). The hierarchy of instabilities in pulsar plasma. *Soviet Astronomy Letters*, **5**, 323–5.

Lominadze, D. G., Machabeli, G. Z. & Mikhaylovsky, A. B. (1979). Influence of magnetobremsstrahlung on the quasilinear relativistic plasma relaxation in a strong magnetic field. *Soviet Journal of Plasma Physics*, **5**, 1137–44.

Machabeli, G. Z. & Usov, V. V. (1979). Cyclotron instability in the magnetosphere of the Crab Nebula pulsar, and the origin of its radiation. *Soviet Astronomy Letters*, **5**, 238–41.

Mikhaylovsky, A. B. (1980). The nonlinear generation of electromagnetic waves in the relativistic electron–positron plasma. *Soviet Journal of Plasma Physics*, **6**, 613–20.

Onishchenko, O. G. (1981). On longitudinal waves and the beam instability in the relativistic anisotropic plasma. *Soviet Journal of Plasma Physics*, **7**, 1310–18.

Lominadze, D. G., Machabeli, G. Z. & Usov, V. V. (1983). Theory of NP0532 pulsar radiation and the nature of the activity of the Crab nebula. *Astrophysics and Space Science*, **90**, 19–43.

4.2.3 Plasma effects in a curved magnetic field

The influence of an inhomogeneous magnetic field upon plasma oscillations has been investigated in the following.

Blandford, R. D. (1975). Amplification of radiation by relativistic particles in a strong magnetic field. *Monthly Notices of the Royal Astronomical Society*, **170**, 551–7.

Hinata, S. (1976). Stability of a beam-plasma system against the excitation of the longitudinal mode around pulsars. *Astrophysical Journal*, **203**, 223–5.

Asseo, E., Pellat, R. & Rosado, M. (1980). Pulsar radio emission from beam plasma instability. *Astrophysical Journal*, **239**, 661–70.

Asseo, E., Pellat, R. & Sol, H. (1983). Radiative or two-stream instability as a source for pulsar radio emission. *Astrophysical Journal*, **266**, 201–14.

Laroche, O. & Pellat, R. (1987). Curvature instability of relativistic particle beam. *Physical Review Letters*, **59**, 1104–7.

Chugunov, Yu. V. & Shaposhnikov, V. E. (1988). Curvature radiation radio maser in a pulsar magnetosphere. *Astrofizika*, **28**, 98–106.

4.3 Geometric ideas

The phenomenological model of a hollow cone based on the paper by Ruderman and Sutherland has been developed in the following.

Oster, L. (1975). Pulsar geometrics. I. Basic model. *Astrophysical Journal*, **196**, 571–7.

Backer, D. C. (1976). Pulsar average wave forms and hollow-cone beam models. *Astrophysical Journal*, **209**, 895–907.

Oster, L. & Sieber, W. (1976). Pulsar geometrics. III. The hollow-cone model. *Astrophysical Journal*, **210**, 220–9.

Izvekova, V. A. & Malov, I. F. (1979). Some implications of the polar-cap model of pulsar magnetospheres. *Soviet Astronomy Letters*, **5**, 211–13.

Vivekanand, M. & Radhakrishnan, V. (1980). The structure of integrated pulse profiles. *Journal of Astrophysics and Astronomy*, **1**, 119–28.

Narayan, R. & Vivekanand, M. (1982). Geometry of pulsar beams: relative orientations of rotating axis, magnetic axis and line of sight. *Astronomy and Astrophysics*, **113**, L3–6.

4.4 High-frequency radiation

4.4.1 Optical radiation

Sturrock, P. A., Petrosian, V. & Turk, J. S. (1975). Optical radiation from the Crab pulsar. *Astrophysical Journal*, **196**, 73–82.

Benford, G. (1975). Crab pulsar optical and X-ray radiation through pinching instabilities. *Astrophysical Journal*, **201**, 419–24.

Ochelkov, Yu. P. & Usov, V. V. (1979). The nature of the radiation from the Crab Nebula pulsar. *Soviet Astronomy Letters*, **5**, 96–8.

Machabeli, G. Z. & Sakhokiya, D. M. (1982). An optical emission mechanism for the Crab pulsar. *Soviet Astronomy Letters*, **8**, 39–41.

Machabeli, G. Z. & Melikidze, G. I. (1988). On the mechanism and certain peculiarities of the optical emission of PSR 0531+21. *Soviet Astronomy*, **32**, 386–9.

4.4.2 X-ray and gamma-ray emission

Zhelezniakov, V. V. & Shaposhnikov, V. E. (1972). On the optical, X and γ-radiation mechanism in the Crab pulsar. *Astrophysics and Space Science*, **18**, 141–65.

Hardee, P. E. (1977). Production and beaming of pulsar γ-ray emission. *Astrophysical Journal*, **216**, 873–80.

Hardee, P. E. (1979). Production of pulsed emission from Crab and Vela pulsars by the synchrotron mechanism. *Astrophysical Journal*, **227**, 958–73.

Ayasly, S. (1981). The gamma-ray spectra of radio pulsars. *Astrophysical Journal*, **249**, 698–703.

Author index

Subject index